BOLYAI SOCIETY
MATHEMATICAL STUDIES

26

For further volumes:
www.springer.com/series/4706

BOLYAI SOCIETY MATHEMATICAL STUDIES

1. **Combinatorics, Paul Erdős is Eighty, Vol. 1**
   D. Miklós, V.T. Sós, T. Szőnyi (Eds.)
2. **Combinatorics, Paul Erdős is Eighty, Vol. 2**
   D. Miklós, V.T. Sós, T. Szőnyi (Eds.)
3. **Extremal Problems for Finite Sets**
   P. Frankl, Z. Füredi, G. Katona, D. Miklós (Eds.)
4. **Topology with Applications**
   A. Császár (Ed.)
5. **Approximation Theory and Function Series**
   P. Vértesi, L. Leindler, Sz. Révész, J. Szabados, V. Totik (Eds.)
6. **Intuitive Geometry**
   I. Bárány, K. Böröczky (Eds.)
7. **Graph Theory and Combinatorial Biology**
   L. Lovász, A. Gyárfás, G. Katona, A. Recski (Eds.)
8. **Low Dimensional Topology**
   K. Böröczky, Jr., W. Neumann, A. Stipsicz (Eds.)
9. **Random Walks**
   P. Révész, B. Tóth (Eds.)
10. **Contemporary Combinatorics**
    B. Bollobás (Ed.)
11. **Paul Erdős and His Mathematics I+II**
    G. Halász, L. Lovász, M. Simonovits, V. T. Sós (Eds.)
12. **Higher Dimensional Varieties and Rational Points**
    K. Böröczky, Jr., J. Kollár, T. Szamuely (Eds.)
13. **Surgery on Contact 3-Manifolds and Stein Surfaces**
    B. Ozbagci, A.I. Stipsicz
14. **A Panorama of Hungarian Mathematics in the Twentieth Century, Vol. 1**
    J. Horváth (Ed.)
15. **More Sets, Graphs and Numbers**
    E. Győri, G. Katona, L. Lovász (Eds.)
16. **Entropy, Search, Complexity**
    I. Csiszár, G. Katona, G. Tardos (Eds.)
17. **Horizons of Combinatorics**
    E. Győri, G. Katona, L. Lovász (Eds.)
18. **Handbook of Large-Scale Random Networks**
    B. Bollobás, R. Kozma, D. Miklós (Eds.)
19. **Building Bridges**
    M. Grötschel, G. Katona (Eds.)
20. **Fete of Combinatorics and Computer Science**
    G. Katona, A. Schrijver, T. Szonyi (Eds.)
21. **An Irregular Mind**
    I. Bárány, J. Solymosi (Eds.)
22. **Cylindric-like Algebras and Algebraic Logic**
    H. Andréka, M. Ferenczi, I. Németi (Eds.)
23. **Deformations of Surface Singularities**
    A. Némethi, Á. Szilárd (Eds.)
24. **Geometry – Intuitive, Discrete, and Convex**
    I. Bárány, K. Böröczky, G. Fejes Tóth, J. Pach (Eds.)
25. **Erdős Centennial**
    L. Lovász, I. Ruzsa, V.T. Sós (Eds.)

Frédéric Bourgeois
Vincent Colin
András Stipsicz
(Eds.)

Contact and Symplectic Topology

Springer

JÁNOS BOLYAI MATHEMATICAL SOCIETY

Frédéric Bourgeois
Laboratoire de Mathématiques d'Orsay
Université Paris-Sud
Bâtiment 425
91405 Orsay
France
bourgeois@math.u-psud.fr

András Stipsicz
Alfréd Rényi Institute of Mathematics
Hungarian Academy of Sciences
Reátanoda utca 13-15
1053 Budapest
Hungary
stipsicz.andras@renyi.mta.hu

Vincent Colin
Laboratoire de mathématiques Jean Leray
Université de Nantes
2 rue de la Houssinière
44322 Nantes
France
vincent.colin@univ-nantes.fr

ISSN 1217-4696 Bolyai Society Mathematical Studies
ISBN 978-3-319-02035-8 ISBN 978-3-319-02036-5 (eBook)
DOI 10.1007/978-3-319-02036-5

Springer Cham Heidelberg New York Dordrecht London

Library of Congress Control Number: 2014932294

Cover picture: An artist's view of a contact structure supported by a symplectic open book. © Otto van Koert

Printed on acid-free paper

Springer is part of Springer Science+Business Media (www.springer.com)

Dedicated to the memory of Vladimir Igorevich Arnold (1937–2010), founder of symplectic topology

Contents

Preface

Proceedings of the Conferences Nantes, 2011 and Budapest, 2012

The CAST (*Contact and Symplectic Topology*) Research Networking Programme has been established in 2010 as one of the ESF (European Science Foundation) sponsored networks. The network is financed by the support of 13 contributing European countries, embracing researchers from all over the globe. The main profile of the network is to foster collaboration throughout institutions in Europe. This aim has been achieved by supporting conferences, workshops, Summer Schools focusing on various aspects of contact and symplectic topology and by supporting research collaborations and exchanges of doctoral students and postdoctoral researchers within the field of symplectic and contact topology.

In particular, the network partially sponsored (together with the Pays de la Loire region, the ANR agency and the Institut Universitaire de France) the *Trimester on Contact and Symplectic Topology* in Nantes (March-June 2011), and (together with the *Lendület program* of the Hungarian Academy of Sciences, through the *Lendület* group ADT of the Rényi Institute) supported the CAST Summer School and Conference in Budapest (July 2012). Nantes' program has gathered, during five focused weeks, a summer school and an international conference, a total of 160 mathematicians. The Budapest event attracted more than 130 graduate students, postdoctoral researchers and senior mathematicians from around the globe. Both events provided lecture series in various current topics in contact and symplectic topology. The present volume is the compilation of the notes of these lecture series written by the lecturers. These notes provide a gentle introduction to topics which have developed in an amazing speed in the recent past. The surveys target both graduate students with solid contact and symplectic backgrounds, as well as senior researchers interested in certain aspects of the field. The topics of the lecture series include:

- contact topological questions in dimensions three and in dimensions greater than three,
- open book decompositions and Lefschetz fibrations in contact topology through asymptotically holomorphic techniques,
- Fukaya categories,
- Heegaard Floer homologies and embedded contact homologies (ECH) of 3-dimensional manifolds,
- Stein structures on manifolds of dimension at least six, and
- knot contact homologies.

We dedicate this volume to the memory of V.I. Arnold, whose ideas and results shaped the developement of symplectic and contact topology. The opening paper (by Michèle Audin) is a tribute to the influence of Arnold on symplectic topology, providing an account of the early days of the subject. It is followed by the contributions of the speakers of the Nantes and Budapest Summer Schools. Below we provide short abstracts of each of the contributions.

Orsay, France — Frédéric Bourgeois
Nantes, France — Vincent Colin
Budapest, Hungary — András Stipsicz

- **Patrick Massot (Université Paris Sud, Orsay, France) Topological methods in 3-dimensional contact geometry**

  These notes provide an introduction to Giroux's theory of convex surfaces in contact 3-manifolds and its simplest applications. They put a special emphasis on pictures and discussions of explicit examples. The first goal is to explain why all the information about a contact structure in a neighborhood of a generic surface is encoded by finitely many curves on the surface. Then we describe the bifurcations that happen in generic families of surfaces. As applications, we explain how Giroux used this technology to reprove Bennequin's theorem saying that the standard contact structure on S^3 is tight and Eliashberg's theorem saying that all tight contact structures on S^3 are isotopic to the standard one.

- **Denis Auroux (University of California, Berkeley, USA): A beginner's introduction to Fukaya categories**

  In these notes, we give a short introduction to Fukaya categories and some of their applications. We first briefly review the definition of Lagrangian Floer homology and its algebraic structures. Then we introduce the Fukaya category (informally and without a lot of the necessary technical detail), and discuss algebraic concepts such as exact triangles and generators. Finally, we outline a few applications to symplectic topology, mirror symmetry and low-dimensional topology.

- **Francisco Presas (ICMAT, Madrid, Spain): Geometric decompositions of almost contact manifolds**

  These notes are intended to be an introduction to the use of approximately holomorphic techniques in almost contact and contact geometry. We develop the setup of the approximately holomorphic geometry. Once done, we sketch the existence of the two main geometric decompositions available for an almost contact or contact manifold: open books and Lefschetz pencils. The possible use of the two decompositions for the problem of existence of contact structures is briefly explained.

- **Klaus Niederkrüger (Université de Toulouse, France): Higher dimensional contact topology via holomorphic disks**

  We will focus on fillability questions of higher dimensional contact manifolds. We start with an overview of some basic examples and theorems known so far, comparing them with analogous results in dimension three. We will also describe an easy construction of non-fillable manifolds by Fran Presas. Then we will explain how to use holomorphic curves with boundary to prove the fillability results stated earlier. No *a priori* knowledge of holomorphic curves will be required, and many properties will only be quoted.

- **Gordana Matić (University of Georgia): Contact invariants in Floer Homology**

  In a pair of seminal papers Peter Ozsváth and Zoltan Szabó defined a collection of homology groups associated to a 3-manifold they named

Heegaard-Floer homologies. Soon after, they associated to a contact structure ξ on a 3-manifold, an element of its Heegaard-Floer homology, the contact invariant $c(\xi)$. This invariant has been used to prove a plethora of results in contact topology of 3-manifolds. In this series of lectures we introduce and review some basic facts about Heegaard Floer Homology and its generalization to manifolds with boundary due to Andras Juhász, the Sutured Floer Homology. We use the open book decompositions in the case of closed manifolds, and partial open book decompositions in the case of contact manifolds with convex boundary to define contact invariants in both settings, and show some applications to fillability questions.

- **Robert Lipshitz (Columbia University, USA), Peter Ozsváth (Princeton University, USA) and Dylan Thurston (University of California, Berkeley, USA): Notes on bordered Floer homology**

  Bordered Heegaard Floer homology is an extension of Ozsváth-Szabó's Heegaard Floer homology to 3-manifolds with boundary, enjoying good properties with respect to gluings. In these notes we will introduce the key features of bordered Heegaard Floer homology: its formal structure, a precise definition of the invariants of surfaces, a sketch of the definitions of the 3-manifold invariants, and some hints at the analysis underlying the theory. We also talk about bordered Heegaard Floer homology as a computational tool, both in theory and practice.

- **Kai Cieliebak (Augsburg University, Germany) and Yakov Eliashberg (Stanford, USA): Stein structures: existence and flexibility**

  This survey on the topology of Stein manifolds is an extract from our book "From Stein to Weinstein and Back". It is compiled from two short lecture series given by the first author in 2012 at the Institute for Advanced Study, Princeton, and the Alfréd Rényi Institute of Mathematics, Budapest.

  The first part of this survey is devoted to the topological characterization of those smooth manifolds of real dimension greater than four that admit the structure of a Stein complex manifold. The second part discusses more recent results on the topology of Stein structures such as a Stein version of the h-cobordism theorem, a uniqueness theorem for subcritical Stein structures, and a remarkable class of "flexible" Stein structures that also satisfy uniqueness.

- **Michael Hutchings (University of California, Berkeley, USA): Lecture notes on embedded contact homology**

  These notes give an introduction to embedded contact homology (ECH) of contact three-manifolds, gathering together many basic notions which are scattered across a number of papers. We also discuss the origins of ECH, including various remarks and examples which have not been previously published. Finally, we review the recent application to four-dimensional symplectic embedding problems. This article is based on lectures given in Budapest and Munich in the summer of 2012, a series of accompanying blog postings at `floerhomology.wordpress.com`, and related lectures at UC Berkeley in Fall 2012. There is already a brief introduction to ECH in the article[1], but the present notes give much more background and detail.

- **Lenhard Ng (Duke University, USA): A topological introduction to knot contact homology**

  Knot contact homology is a Floer-theoretic knot invariant derived from counting holomorphic curves in the cotangent bundle of $\mathbb{R}^3$ with Lagrangian boundary condition on the conormal bundle to the knot. Among other things, this can be used to produce a three-variable polynomial that detects the unknot and conjecturally contains many known knot invariants; a different part of the package yields an effective invariant of transverse knots in $\mathbb{R}^3$.

  In these notes we will describe knot contact homology and the topology and algebra behind it, as well as connections to other knot invariants, transverse knot theory, and physics. Topics to be treated along the way include: Legendrian contact homology; the conormal construction; a combinatorial formulation of knot contact homology in terms of braids; the cord algebra, a topological interpretation of part of the invariant; transverse homology, a filtered version associated to transverse knots; and relations to the A-polynomial, the HOMFLY polynomial, and recent work in string theory.

[1] M. Hutchings: *Embedded contact homology and its applications*, Proceedings of the International Congress of Mathematicians, Volume II, 1022–1041, Hindustan Book Agency, New Delhi, 2010.

BOLYAI SOCIETY
MATHEMATICAL STUDIES, 26

Contact and Symplectic Topology
pp. 1–25.

Vladimir Igorevich Arnold and the Invention of Symplectic Topology

MICHÈLE AUDIN

1. First Step: A Definition (1986)

First steps in symplectic topology, this was the (English) title of a 1986 paper [14] of Vladimir Igorevich Arnold. Like any good mathematical paper, this one started with a definition:

> By symplectic topology, I mean the discipline having the same relation to ordinary topology as the theory of Hamiltonian dynamical systems has to the general theory of dynamical systems.

And, to make things clearer, the author added:

> The correspondence here is similar to that between real and complex geometry.

Well... this was Arnold's style. A definition by analogy (an analogy I am not sure I understand clearly). Nobody could accuse him of formalism or, worse, of Bourbakism.

However, this paper was, is, "stimulating" (as the reviewer in *Math. Reviews* would write[1]). Its first part (after the provocative introduction), entitled "Is there such a thing as symplectic topology?", even contains a proof of the "existence of symplectic topology" (hence the answer to the question is yes), that the author attributed to Gromov in [50] (as he notes, Eliashberg also contributed to the statement, see below):

[1] This one was Jean-Claude Sikorav.

F. Bourgeois et al. (eds.), *Contact and Symplectic Topology*,
Bolyai Society Mathematical Studies 26, DOI 10.1007/978-3-319-02036-5_1,

Theorem. *If the limit of a uniformly ($\mathcal{C}^0$) converging sequence of symplectomorphisms is a diffeomorphism, then it is symplectic.*

No geometer would contest that such a statement is indeed a proof: this is a theorem about the behavior of symplectic diffeomorphisms with respect to the $\mathcal{C}^0$-topology; the terms of the sequence are defined via their first derivatives while the convergence is in the $\mathcal{C}^0$-topology. This indeed belongs to symplectic topology. Hence the latter is not empty.

But, whatever the credit Arnold decided to give to Gromov and Eliashberg in this article, symplectic topology existed twenty years before Gromov's seminal paper [50] appeared: symplectic topology has an official birthdate, and this is October 27th, 1965.

In this paper, I plan to sketch a picture of how symplectic topology grew, in the hands of Arnold, his students, and followers, between his two papers [3] of 1965 and [14] of 1986.

2. October 27TH 1965

This is the day when a short paper by Vladimir Arnold (so the author's name was spelled, see Figure 1), *Sur une propriété topologique des applications globalement canoniques de la mécanique classique*, was presented to the Paris Academy of sciences by Academician Jean Leray and became the *Comptes rendus* note [3].

TOPOLOGIE DIFFÉRENTIELLE. — ***Sur une propriété topologique des applications globalement canoniques de la mécanique classique.*** **Note (*) de M. Vladimir Arnold, présentée par M. Jean Leray.**

On utilise les inégalités de M. Morse, concernant le nombre de points critiques d'une fonction sur une variété, afin de trouver les solutions périodiques des problèmes de la mécanique.

Fig. 1. A birth announcement (title and abstract of [3])

The so-called "applications globalement canoniques" would become symplectomorphisms, the topology was already in the title. Here are the statements of this note (my translation):

Theorem 1. *The tori T and AT have at least 2^n intersection points (counted with multiplicities) assuming that AT is given by*

$$p=p(q) \quad \left|\frac{\partial p}{\partial q}\right|<\infty. \tag{7}$$

Here T is the zero section $p=0$ in the "toric annulus" $\Omega = T^n \times B^n$ (with coordinates (q,p)) and the mapping $A:\Omega \to \Omega$ is globally canonical, namely, it is homotopic to the identity and satisfies

$$\oint_\gamma p\,dq = \oint_{A\gamma} p\,dq, \quad (p\,dq = p_1\,dq_1 + \cdots + p_n\,dq_n)$$

for any closed curve (possibly not nullhomologous) γ.

Hence, Theorem 1 asserts that the image of the zero section in $T^n \times B^n$ under a certain type of transformations should intersect the zero section itself. We shall come back to this later. The second statement concerned fixed points. To this also we shall come back.

Theorem 2. *Let A be a globally canonical mapping, close enough to A_0. The mapping A^N has at least 2^n fixed points (counted with multiplicities) in a neighborhood of the torus $p=p_0$.*

Here, A_0 has the form $(q,p) \mapsto (q+\omega(p), p)$, for a map $\omega: B^n \to \mathbf{R}^n$ such that $\det\left|\frac{\partial \omega}{\partial p}\right| \not\equiv 0$, so that there exist $p_0 \in B^n$ and integers $m_1, \ldots, m_n, N$ with

$$\omega_1(p_0) = \frac{2\pi m_1}{N}, \ldots, \omega_n(p_0) = \frac{2\pi m_n}{N}$$

(this defining the p_0 and the N in the statement).

Remark A. *Replacing in the proofs the theory of M. Morse by that of L.A. Lusternik and L.G. Schnirelman, we obtain, in Theorem 1, $(n+1)$ geometrically different intersection points of T and AT. One could wonder whether there exist $(n+1)$ intersection points of T and AT for the globally canonical homeomorphisms A?*

Remark B. *The existence of infinitely many periodic orbits near a generic elliptic orbit follows from Theorem 2 (extension of Birkhoff's Theorem to $n>1$).*

Remark C. *It is plausible that Theorem 1 is still true without the assumption (7), if A is a diffeomorphism*[2]*. From the proof, several "recurrence theorems" would follow.*

Remark D. *It also seems plausible that Poincaré's last theorem can be extended as follows:*
Let $A:\Omega \to \Omega$ ($\Omega = B^r \times T^n$; $B^n = \{p, |p| \leq 1\}$; $T^n = \{q \bmod 2\pi\}$) be a canonical diffeomorphism such that, for any $q \in T^n$ the spheres $S^{n-1}(q) = \partial B^n \times q$ and $AS^{n-1}(q)$ are linked in $\partial B^n \times \mathbf{R}^n$ ($\mathbf{R}^n$ being the universal cover of T^n). Then A has at least 2^n fixed points in Ω (counted with multiplicities).

[2] If A is not a diffeomorphism, counter-examples can be constructed with $n=1$. Note of V.I. Arnold.

Remark C is the statement that will become "Arnold's conjecture". The question in Remark A will also be part of this conjecture. Note that, twenty years after, when he wrote [14], Arnold mentioned that the statement in Remark D had still not been proved.

Before I comment more on the statements and their descendants, let me go back to one of their ancestors, the so-called last geometric theorem of Poincaré.

3. A Theorem of Geometry, 1912

On March 7th 1912, Henri Poincaré finished writing a paper and sent it to the *Rendiconti di Circolo matematico di Palermo.* It was accepted at the meeting of the Mathematical circle which took place three days later (adunanza del 10 marzo 1912), together, *e.g.* with papers of Francesco Severi and Paul Lévy, and it was printed in May[3] as [58]. In this paper, Poincaré stated what he called "un théorème de géométrie". Before that, he apologized for publishing a result

- that he would have liked to be true, because he had applications (to celestial mechanics) for it,
- that he believed to be true, because he was able to prove some special cases of it

but that he could not prove. Here is this statement (my translation). Poincaré denotes by x and y (mod 2π for the latter) the polar coordinates of a point. He considers an annulus $a \leq x \leq b$ and a transformation T of this annulus $(x, y) \mapsto (X(x, y), Y(x, y))$.

> *First condition.* As T transforms the annulus into itself, it preserves the two boundary circles $x = a$ and $x = b$. [He then explains that T moves one of the circle in a direction and the other in the opposite one. I shall (anachronistically) call this the twist condition.]

[3] All of this was very fast, including the mail from Paris to Palermo (recall that there was no air-mail and that Palermo was already on an island). All the dates given here can be found on the printed journal. For some reason (which I was unable to understand), they were cut out in Poincaré's complete works, even the date he probably wrote himself at the end of his paper.

Second condition. The transformation preserves the area, or, more generally, it admits a positive integral invariant, that is, there exists a positive function $f(x, y)$, so that

$$\iint f(x,y)\,dx\,dy = \iint f(X,Y)\,dX\,dY,$$

the two integrals being relative to any area and its transform.

If these two conditions are satisfied, I say that there will always exist in the interior of the annulus two points that are not modified by the transformation.

Clearly, the two conditions are necessary: there exists

- maps preserving the area without fixed points, a rotation for instance, but it does not satisfy the twist condition,
- twist maps without fixed points, *e.g.*[4] $(x, y) \mapsto (x^2, x + y - \pi)$, but it does not preserve the area.

Notice also that there exist twist maps preserving the area with exactly two fixed points, like the one evoked by Figure 2. The picture shows a part of an infinite strip. The diffeomorphism is the flow of the vector field drawn. It descends to the quotient (by the integral horizontal translation) annulus where it has two fixed points.

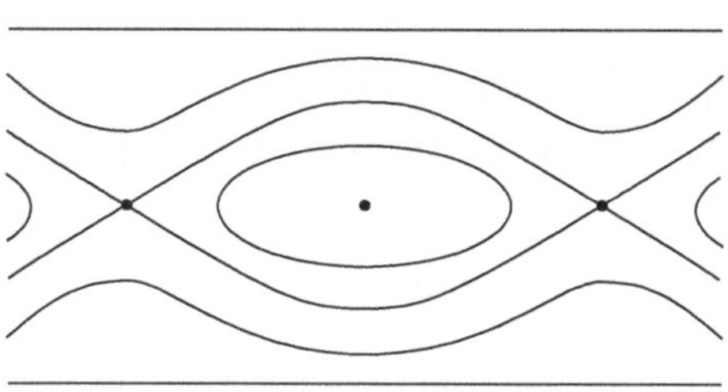

Fig. 2. A twist map with two fixed points

Such area preserving maps of the annulus arose as Poincaré sections for Hamiltonian systems with two degrees of freedom—namely, in dimension 4—and their fixed points would correspond to periodic orbits. Needless to say: celestial mechanicians love periodic orbits. Hence the Poincaré problem.

Let me add that, in the introduction of his paper, Poincaré wrote that he had thought of letting the problem mature for a few years and then of coming back to it more successfully, but that, at his age, he could not be

[4]I copied this example from [54].

sure. He was actually only 58, but he died, unexpectedly, four months later, on July 17th.

On October 26th, the same year, George David Birkhoff presented a proof of this theorem to the American mathematical society, and his paper *Proof of Poincaré's geometric theorem* was published in the Transactions of this society [30]. Birkhoff considered himself as a student (and even as the last student) of Poincaré. He and Jacques Hadamard were probably the two mathematicians who knew Poincaré's work best. Although this was not as easy as it is nowadays, Birkhoff would go very often to Paris and lecture at Hadamard's Seminar, on Poincaré's theorems, during the 1920's and 1930's. The main reference in his paper was a previous paper of him [29], published, in French, by the French mathematical society. No wonder that his proof of Poincaré's theorem was translated and republished, in French, as "Démonstration du dernier théorème de géométrie de Poincaré[5]" [31].

Note that, using a degree argument (that Poincaré attributed to Kronecker), the existence of one fixed point implies that there are two of them... except that they could coincide. It is not absolutely clear that the original proof of Birkhoff gave the existence of two geometrically distinct fixed points. This is why he himself came back to this theorem later. See his paper [32] and his book [33][6].

For modern symplectic readers: there is a proof of the existence of one fixed point in [54], which can be completed with [36].

Chapter VI of Birkhoff's book is devoted to the application of Poincaré's geometric theorem. It starts as follows:

> Poincaré's last geometric theorem and modifications thereof[7] yield an additional instrument for establishing the existence of periodic motions. Up to the present time no proper generalization of this theorem to higher dimensions has been found, so that its application remains limited to dynamical systems with two degrees of freedom.

At that time, the symplectic nature of Hamilton's equations still needed some clarification. Now we know that the good generalization of "preserv-

[5] "Dernier", which means last, was not in the American title. Also, the translation kept the original phrasing "théorème de géométrie" rather than "théorème géométrique", as in English.
[6] Note that, in the preface Marston Morse wrote for the 1966 edition of this 1927 book, he insisted on the relationship between Birkhoff's work on periodic orbits and "the work of Moser, Arnold and others on stability".
[7] See my paper, An extension of Poincaré's last geometric theorem, Acta Mathematica, vol. 47 (1926). Note of G.D. Birkhoff.

ing the area" is not "preserving the volume". And Arnold was (one of) the mathematicians who taught us that. A Hamiltonian flow, namely a solution $(q(t), p(t))$ of Hamilton's equations

$$\begin{cases} \dot{q} = \dfrac{\partial H}{\partial p} \\ \dot{p} = -\dfrac{\partial H}{\partial q} \end{cases}$$

preserves the symplectic form

$$\omega = dp_1 \wedge dq_1 + \cdots + dp_n \wedge dq_n$$

and not only the volume form

$$dp_1 \wedge dq_1 \wedge \cdots \wedge dp_n \wedge dq_n = \frac{\omega^{\wedge n}}{n!}.$$

This is written in $\mathbf{R}^n \times \mathbf{R}^n$, but could also be understood in $T^n \times \mathbf{R}^n$ (if H is periodic in q), which is the same as $T^\star T^n$, hence can be generalized to $T^\star V$ (which has a "$p\,dq$" and thus also a "$dp \wedge dq$" form), and to any symplectic manifold W. To a function $H : W \to \mathbf{R}$, the symplectic form ω associates a vector field (the Hamiltonian vector field) X_H by $dH = i_{X_H}\omega$ and thus a flow (the Hamiltonian flow) which preserves ω since

$$\mathcal{L}_{X_H}\omega = di_{X_H}\omega = d\,dH = 0.$$

4. Back to Arnold and His Golden Sixties

In 1965, although he was a young man of 28, Arnold was not a beginner. Ten years before, he had contributed (with his master Kolmogorov, as he would say) to Hilbert's thirteenth problem. Then he had worked on stability and had already proved the theorem on invariant tori that would soon be known, first as "Kolmogorov-Arnold-Moser", and later as "KAM". This was what he lectured on when he came to Paris at the Spring of 1965, as the book [16][8] shows (the "KAM" statement is Theorem 21.11 and there is a proof in Appendix 33). He had already published, for instance, the big paper [1][9], about which the reviewer of *Math. Reviews*[10] wrote:

[8]Soon translated in English as [17].
[9]This was also very fast: the translation in English in *Russian mathematical surveys* would arrive in the libraries less than one year after the publication of the Russian original.
[10]This one was Jürgen Moser.

It is to be hoped that this remarkable paper and exceptional work helps to arouse the interest of more mathematicians in this subject.

This might have been the first appearance of the famous cat of Arnold, and of a figure such as Figure 3[11].

Fig. 3. More fixed points... after [1]

Of course, KAM theorem was also the main topic of the half-an-hour talk Arnold gave at the ICM in Moscow in 1966, Проблема устойчивости и эргодические свойства классических динамических систем[12] [4]. However, there was a short section with the statements of (and reference to) the note [3].

5. Problems of Present Day Mathematics, 1974

In May 1974, the American mathematical society had a Symposium on developments arising from Hilbert problems. The organizers also intended to make another list of problems—for the present day. Arnold sent a problem (if I understand well, the problems were collected by Jean Dieudonné and edited by Felix Browder), which appeared in a list of "Problems for present day mathematics" in the book [35]. This is Problem XX, on page 66:

[11] Note that Figure 3 contains a 5-fold covering and a 3-fold covering of the map in Figure 2.
[12] A stability problem and ergodic properties of classical dynamical systems.

XX. Fixed points of symplectic diffeomorphisms (V. Arnold). The problem goes back to the "last geometric theorem" of Poincaré. The simplest case is the following problem: *Does every symplectic diffeomorphism of a 2-dimensional torus, which is homologous to the identity, have a fixed point?*
A symplectic diffeomorphism is a diffeomorphism which preserves a nondegenerate closed 2-form (the area in the 2-dimensional case). It is homologous to the identity iff it belongs to the commutator subgroup of the group of symplectic diffeomorphisms homotopical to the identity. With coordinates, such a diffeomorphism is given by $x \to x + f(x)$, where x is a point of the plane and f is periodic. It is symplectic iff the Jacobian $\det(D(x+f(x))/Dx)$ is identically 1, and it is homologous to the identity iff the mean value of f is 0.
The "last geometric theorem" of Poincaré (proved by G. D. Birkhoff) deals with a circular ring. The existence of 2 geometrically different fixed points for symplectic diffeomorphisms of the 2-sphere is also proved (A. Shnirelman, N. Nikishin). In the general case, one may conjecture that the number of fixed points is bounded from below by the number of critical points of a function (both algebraically and geometrically).

The AMS book appeared two years later, in 1976. Notice that the "simplest" question is asked in dimension 2, but that the general case, at the very end of the text, seems to refer to an arbitrary symplectic manifold. The complicated definition of "homologous to the identity" given shows that Arnold was indeed thinking of a general symplectic manifold. Note that, according to a theorem Augustin Banyaga [25] would prove in 1980, and that Arnold would quote in [14] and in 1986, these are the Hamiltonian diffeomorphisms.

Also note there was already a proof available, and this was for the S^2 case: Arnold was working... and his students were working too. The very first symplectic fixed point theorem (after [3]) was that of N.A. Nikishin [57]—note that, although published in 1974, the paper was submitted to the journal as soon as November 1972:

Theorem. *A diffeomorphism of S^2 which preserves the area has at least two geometrically distinct fixed points.*

Namely, at least as many as a function has critical points. The proof was not very hard: Nikishin proved that the index of a fixed point of such a diffeomorphism should be ≤ 1. But the Lefschetz number is 2.

Arnold was working. For instance on singularity (or catastrophe) theory. One of the people he met in Paris in Spring 1965 was René Thom (this we know at least from [16] and from [59]), whose seminar he attended. Arnold was working. Starting a seminar on singularity theory in Moscow[13]. Lecturing

[13]Let me mention here the beautiful little book [13] he wrote on this subject for a general audience in the eighties.

on classical mechanics in 1966–68. And writing up notes. Nikishin, in [57], quotes Arnold's *Lectures on classical mechanics*, dated 1968. They would become a famous book...

6. Mathematical Methods in Classical Mechanics, 1974 (Our Golden Seventies)

In 1974, the Soviet publishing house Наука published Arnold's Математические методы классической механики [8].

At that time, a wicked bureaucracy had decided not to allow Arnold to travel abroad anymore. However, his book was soon translated to French and published, in Moscow, by the foreign language Soviet publishing house Мир, Mir, and [9] was available in France, at a very low price, in 1976.

6.1. A Few Personal Remarks

In the seventies, the only math books we could afford, we Parisian students, were the Mir books. We would go quite often to their bookstore *la Librairie du Globe* rue de Buci to fetch the new books (whatever they were). The Soviet translation program was devoted helping French-speaking developing countries, not French students. So what?

The word "translation" was already used at least seven times in this text. A French mathematician publishing a paper in French in an Italian journal, an American mathematician writing papers in French and in English, a Russian one writing in Russian and in French. Before I leave the language question, let me comment on that. When I visited Arnold in Moscow in the Fall of 1986, he told me that he preferred to speak French than English, so we used to discuss in French. Of course, he asked me to lecture in English, because of his students. So I spoke English... but, he would interrupt quite often to ask a question or make a comment (well, this was Arnold's seminar, you know[14]), and, of course (?) he would do it in French, then I would answer (or not), and he would translate and comment in Russian, for his students[15]. And of course, I would try to understand the comment: I knew perfectly well that he was explaining things I was talking about but did not

[14]If you don't know, look at [59].
[15]Again, you should read [59].

quite understand[16]. Arnold's fast, intricate and subtle questions[17], plus two foreign languages at the same time—hard work!

⋆

There and then (I mean in [9] and in 1976), we discovered, after the Newtonian and the Lagrangian mechanics, the third part of the book, Hamiltonian mechanics[18], symplectic manifolds and action-angle variables, notably. So, mechanics was, after all, geometry! Good news! And you could put so much mathematics in a series of so-called "appendices".

7. The Symplectic Community

Two years later, Springer published a translation in English, by Karen Vogtmann and Alan Weinstein[19] [10]. In a letter to Alan Weinstein, Arnold complained:

> There is something wrong with the occidental scientific books editions: the prices are awful. e.g. my undergraduate ordinary differential equations textbook[20] costs here 0,67 rbls ($\sim 1/30$ the price of a pair of boots), and 40 000 exemplairs where sold in few months, so it is impossible to buy it at Moscow at present; the MIT Press translation by Silverman price was perhaps more than 20$ and the result – 650 sales the 1 year.
> Now the 17 000 exemplaires of the "mechanics" disappeared here at few days, the price being rbls 1,10. I think the right price for the translation must be less than 1$, then the students will buy it.

As Weinstein pointed out in his answer, books were unsubsidized in the U.S. economy. And, as it could be added, scientific publishers were not non-profitmaking organizations. And the price of the present Springer book is 100 euros (added in proof).

The English translation appeared. This time, this was no longer a short *Comptes rendus* note in French, a cheap translation made in the Soviet Union for developing countries or a paper in Russian. You (or your library) had to pay to read it. For instance, Helmut Hofer [52] would remember:

---

[16]In any case, you should read [59].
[17]Let me quote what I wrote at the very moment I learned his death in a short online paper [23]: he was charming, provocative, brilliant, cultured, funny, caustic sometimes even wicked, adorable, quick, lively, incisive, yes, all this together.
[18]Nothing is perfect. One thing I never understood and never dared to ask, is why there is a Lagrangian but no Hamiltonian treatment of the spinning top in this book.
[19]It seems that the idea was Jerry Marsden's. The translation was made by Karen Vogtmann and edited by Alan Weinstein, who knew the domain and its lexicon better.
[20]This one was [5–7], before becoming [15].

> As a student I read Arnold's wonderful book Mathematical Methods of Classical Mechanics.

After the AMS volume [35] and the Springer book [10], nobody in the West could ignore Arnold's question! It was more or less at the same time that Gromov emigrated[21], first to the States, then to Paris. Thirteen years after, things started to become serious[22].

In Appendix 9 of [10], one can read:

> Thus we come to the following generalization of Poincaré's theorem:
>
> **Theorem.** *Every symplectic diffeomorphism of a compact symplectic manifold, homologous to the identity, has at least as many fixed points as a smooth function on this manifold has critical points* (*at least if this diffeomorphism is not too far from the identity*).

Quoting Hofer again [52]:

> The symplectic community has been trying since 1965 to remove the parenthetical[23] part of the statement. After tough times from 1965 to 1982, an enormously fruitful period started with the Conley-Zehnder theorem in 1982–83.

It is not absolutely clear to me that there existed a symplectic community in the "tough times from 1965 to 1982". I may be wrong, so I will not insist on the precise date, but I would say that the "symplectic topology community" was born around 1982. So far, I have mainly mentioned Arnold[24] (and the Soviet Union). But there were indeed mathematicians working on celestial mechanics and stability questions elsewhere. The names of Marston Morse (who had been a student of Birkhoff) and Jürgen Moser have already been written in this paper. That of Michel Herman should be added. This would be connected to KAM rather than to actual symplectic geometry[25]. Working

[21] Mikhail Gromov's paper [49] (at ICM Nice 1970), where the h-principle for Lagrangian immersions was announced, should also be mentioned.

[22] *Math. Reviews* waited until May 1979 to publish a review of the 1974 Russian edition. The reviewer was very enthusiastic, so enthusiastic that he added a very elegant remark:

> The reader should be aware that the reviewer participated in the English translation of the work under review, and so has been prejudiced in favor of the book by the pleasure which that project provided.

This one was Alan Weinstein.

[23] The French translation has no parenthesis, only a comma.

[24] and Gromov.

[25] Not taking Moser's homotopy method [55] (see also [62]) into account.

on periodic orbits in the States and in the 1970's, Alan Weinstein not only solved problems [63, 64], but wrote a series of lectures [62], on symplectic geometry, which have also been quite useful. If I were to qualify all this activity in only two words, I would probably say "variational methods".

Well, another side of the story I have told so far, which also starts with the Poincaré-Birkhoff theorem and also ends with Weinstein's lecture notes, but is quite different—and complementary—is given in [54, p. 2].

There were some connections. Of course the name of Alan Weinstein must be repeated here. I should add that what we did not learn in [9], we learned it in [62].

However, it is around the Arnold conjecture (as it was named since then) that a community began to aggregate, and, if we needed a birthdate for this community, I would agree with Hofer and suggest March 1983, when Charles Conley and Eduard Zehnder sent their paper [40] to *Inventiones mathematicae.* This was soon reviewed by Marc Chaperon for the Bourbaki Seminar in Paris [37][26]. In this "report", Chaperon added a few personal (and new) ideas and results, in particular, he proved the non-displacement property for tori. At the same time, Daniel Bennequin [26] had succeeded in attacking the contact side of the story... and Gromov developed solutions of an elliptic operator, pseudo-holomorphic curves—the powerful new tool.

7.1. Symplectic Geometry/Topology

I am not sure I can date the locution "symplectic-topology".

I shall not take sides in the question "what is symplectic topology/what is symplectic geometry?". For instance, where should I put the symplectic reduction process [53]? And the glorious convexity theorem of Atiyah, Gullemin and Sternberg [19, 51], which appeared more or less at the same time as [40]? In geometry? But topologists use it a lot... And what about deformation quantization, which originated—in the Soviet Union and in the seventies—in Berezin's work [28]?

Let me just say that Arnold was a geometer in the widest possible sense of the word, and that he was very fast to make connections between different fields.

[26]Replacing Fourier series by a broken geodesics idea, Chaperon himself soon gave a more elementary proof in [39], which is the basis of the proof given in [54].

One of Arnold's important symplectic texts was published shortly before the "first steps" of [14]. Written in collaboration with the young Sasha Givental, it was still called "Symplectic geometry" [18]. This was in 1985. The Soviet Union was still publishing cheap books, in this case volumes of an "Encyclopedia"[27]. This is probably the best place to look at if you want to see the global idea Arnold had on the subject "symplectic geology-or-topometry". Note first that this is part of a series called "Dynamical systems". And then, let me make a list:

Well... integrable systems with the so-called Liouville Theorem (and the invariant tori some of which survive perturbations in KAM theory), Lagrangian and Legendrian submanifolds, caustics and wavefronts (and through generating functions, singularity theory, catastrophes and versal deformations), real algebraic geometry, the Maslov class (which he had defined in [2][28] and which is related to Fourier integral operators), Lagrange and Legendre cobordisms (this turned out to be symplectic algebraic topology[29]), generating functions, and, yes, fixed points of symplectic diffeomorphisms.

8. Lagrangian Submanifolds, Statements of Arnold's Conjecture

A Lagrangian in a symplectic manifold is a submanifold of the maximal possible dimension (which is half the dimension of the symplectic manifold) on which the symplectic form vanishes.

8.1. Sections of a Cotangent Bundle and Fixed Points

For instance, the zero section in a cotangent bundle $T^{\star}V$ is Lagrangian. Also the graph of a 1-form on V is Lagrangian if and only if this 1-form is closed. Notice, in connection with Theorem 1 in Arnold's note [3] (here page 2), that the graph of an exact 1-form df intersects the zero section precisely at the critical points of f.

[27] And this became one of the most expensive Springer series in the 1990's.

[28] The contents of [2] would deserve a whole paper... Note that the adjectives Lagrangian, Legendrian, in the sense used in symplectic geometry, were invented in [2].

[29] Allow me to mention that this was the way I entered symplectic geometry. See [44] and [20–22].

Let us now consider a Hamiltonian diffeomorphism φ of $T^\star V$, that is, a diffeomorphism generated by a Hamiltonian vector field X_H. A version of the Arnold conjecture would be:

Conjecture. *The Lagrangian submanifold $\varphi(V)$ intersects the zero section V of $T^\star V$ at least as many points as a function on V has critical points.*

Suppose that φ is $\mathcal{C}^1$-close to the identity. Then $\varphi(V)$ is a section of $T^\star V$. The fact that φ is symplectic implies that $\varphi(V)$ is Lagrangian and hence, the graph of a closed 1-form; the fact that φ is Hamiltonian implies that this is the graph of the differential of a function. Hence the result in this case. Note that the nondegenerate case, that is, when $\varphi(V)$ is transverse to V, is the case where the function is a Morse function. With the Morse inequalities, this leads to the weak (although nontrivial) form of the conjecture: the number intersection points is not less than the sum of the Betti numbers of V.

Now, according to a theorem of Weinstein [62], a tubular neighborhood of any Lagrangian submanifold L in any symplectic manifold is isomorphic (as a symplectic manifold) to a tubular neighborhood of the zero section in $T^\star L$. Generalizations of the statement above follow...

8.2. Graphs of Symplectic Diffeomorphisms

Another important class of examples is the following. Denote by W a manifold endowed with a symplectic form ω. Let $\varphi: W \to W$ be any map. Now, $W \times W$, endowed with $\omega \oplus -\omega$, is a symplectic manifold, and the graph of φ is a submanifold therein. Clearly, this is a Lagrangian submanifold if and only if $\varphi^\star\omega = \omega$, that is, if and only if φ is a symplectic diffeomorphism. And the intersection points of the graph with the diagonal are the fixed points of φ. Hence Lagrangian intersections are related to fixed points of symplectic diffeomorphisms.

Conjecture. *A Hamiltonian diffeomorphism of a compact symplectic manifold W has at least as many fixed points as a function on W has critical points.*

9. Generating Functions

A connection between symplectic geometry and catastrophe theory is via generating functions. Remember that, if S is a function, the graph of dS is a

Lagrangian submanifold of the cotangent bundle. Together with symplectic reduction, this has the following generalization (see [62]). Let $S : V \times \mathbf{R}^k \to \mathbf{R}$ be a function, so that the graph of dS is a Lagrangian submanifold in $T^\star(V \times \mathbf{R}^k)$. If this is transversal to the coisotropic submanifold $T^\star V \times \mathbf{R}^k$, the symplectic reduction process ensures that the projection

$$\operatorname{graph}(dS) \cap \left(T^\star V \times \mathbf{R}^k\right) \longrightarrow T^\star V$$

is a Lagrangian immersion. In coordinates $(q,a) \in V \times \mathbf{R}^k$, this is to say that, if

$$\Sigma_S = \left\{ (q,a) \in V \times \mathbf{R}^k \;\middle|\; \frac{\partial S}{\partial a} = 0 \right\}$$

is a submanifold, then

$$\begin{aligned} \Sigma_S &\longrightarrow T^\star V \\ (q,a) &\longmapsto \left(q, \frac{\partial S}{\partial q}\right) \end{aligned}$$

is a Lagrangian immersion. For instance (with $V = \mathbf{R}^n$ and $k = 1$), if we start from

$$\begin{aligned} S : \mathbf{R}^n \times \mathbf{R} &\longrightarrow \mathbf{R} \\ (q,a) &\longmapsto a\|q\|^2 + \frac{a^3}{3} - a \end{aligned}$$

then

$$\Sigma_S = \left\{ (q,a) \mid \|q\|^2 + a^2 = 1 \right\} = S^n$$

is an n-sphere and

$$\begin{aligned} S^n &\longrightarrow \mathbf{R}^n \times \mathbf{R}^n = T^\star \mathbf{R}^n \\ (q,a) &\longmapsto (q, 2aq) \end{aligned}$$

is a Lagrangian immersion. Note that it has a double point ($q = 0$, $a = \pm 1$): this is a Lagrangian version of the "Whitney immersion".

9.1. Caustics and Wave Fronts

The geometric version of a wave front is as follows. Start with $L \subset T^\star V$, a Lagrangian in a cotangent bundle (it may be only immersed) and look at the projection $L \to T^{\star}V \to V$. Using "canonical" coordinates (q,p), we are just forgetting the p. The caustic is the singular locus in the projection.

Now comes the contact structure. We rather look at the jet space $J^1(V;\mathbf{R})$, that is, $T^{\star}V \times \mathbf{R}$, with the 1-form $dz - p\,dq$. As the 2-form $dp \wedge dq$ vanishes on L, the 1-form $p\,dq$ is closed, hence (up to a covering) it is exact, $p\,dq = df$ and, well, now we can "draw" L in $V \times \mathbf{R}$, namely in codimension 1 rather than n.

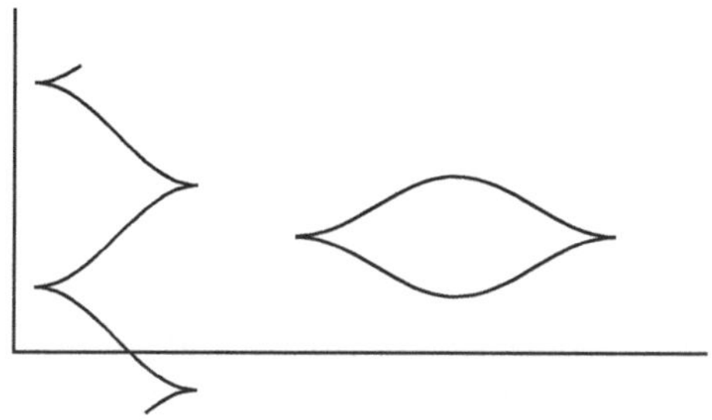

Fig. 4. Two wave fronts

For instance if S is a generating function

$$\Sigma_S \longrightarrow V \times \mathbf{R}$$

$$(q,a) \longmapsto \big(q, S(q,a)\big)$$

is the wave front of the Lagrangian immersion defined by S.

The pictures in Figure 4 represent (in coordinates (q,z)) a round circle and a figure eight (in coordinates (q,p)), the latter being the one-dimensional version of the Whitney immersion. Of course, only exact Lagrangians give closed wavefronts. Note also that any picture like the ones on Figures 4 or 5 would allow you to reconstruct a Lagrangian. Namely: knowing z and q, you get p by $dz = pdq$. For instance, to the two points with the same abscissa and horizontal tangents on the "smile" (right of Figure 4) correspond to the double point of the Whitney immersion.

Of course, this is related to the propagation, of light, say, this is related to evolvents, and to what Arnold calls "Singularities of ray systems" [12] and Daniel Bennequin the "Mystic caustic" [27].

So what? Well, this allowed Givental to construct examples of Lagrangian embeddings in $\mathbf{R}^4$ of all the surfaces which could have one, just by drawing them [48] in $\mathbf{R}^3$ ($4 = 2n \Rightarrow 3 = n+1$) (and leaving the Klein bottle case to posterity[30]).

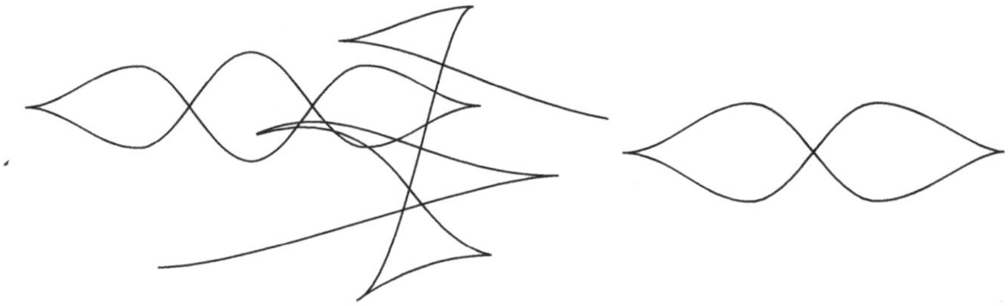

Fig. 5.

This also allowed Eliashberg to prove the Arnold conjecture for surfaces[31]—at the same time as Floer dit it. Eliashberg even had a proof [41] of the "existence theorem" of symplectic topology stated at the beginning of this article (see also [42]) using a decomposition of wave fronts.

9.2. Crossbows...

The last wave front drawn (right of Figure 5) represents an exact Lagrangian immersion of the circle with two double points, which is regularly homotopic to the standard embedding (exactness meaning that the total area enclosed by this curve is zero). It appeared in Arnold's papers on Lagrangian cobordisms [11]: this is the generator of the cobordism group in dimension 1. Arnold calls it "the crossbow". Which reminds me of something Stein is supposed to have told Remmert in 1953 when he learned the use Cartan and Serre made of sheaves and their cohomology to solve problems in complex analysis: "The French have tanks. We only have bows and arrows" [34].

[30] See [56].

[31] Note that Nikishin's article [57] quoted in [35] more or less disappeared from the literature. The statement and a (different) proof were given in [37] without any reference. A few years later the conjecture for $\mathbf{CP}^n$ was announced by Fortune and Weinstein [47] then published by Fortune [46] with no mention that the $\mathbf{CP}^1$-case was already known. Even in [14] the S^2-case is mentioned as an analogous of Poincaré's geometric theorem, but not in connection with the proof of the conjecture for surfaces (attributed both to Eliashberg [43] and Floer [45]).

9.3. ... and Tanks

This time the tank was Floer theory. Well, we were not anymore in 1953. And the war metaphor is not the best possible to speak of the Floer Power...

The starting point was the action functional, like

$$\mathcal{A}_H(x) = \int_0^1 (p\,dq - H_t\,dt)$$

where $x(t)$ is a path and H_t a (time-dependent) Hamiltonian...except that we are on a general symplectic manifold, where $p\,dq$ does not mean anything. Well this can be arranged and replaced by a (closed) action form α_H, defined on a path x and a vector field Y along this path by

$$(\alpha_H)_x(Y) = \int_0^1 \omega_{x(t)}\big(\dot{x}(t) - X_{H_t}(x(t)), Y(t)\big)\,dt.$$

The critical points are the solutions of the Hamilton equation. Once you have fixed a compatible almost complex structure, the gradient lines connecting the critical points are the solutions of the Floer equation:

$$\frac{\partial u}{\partial s} + J(u)\frac{\partial u}{\partial t} + \operatorname{grad} H_t(u) = 0.$$

Note that, when $H_t \equiv 0$, this is just the Cauchy-Riemann equation

$$\frac{\partial u}{\partial s} + J(u)\frac{\partial u}{\partial t} = 0$$

giving Gromov's pseudo-holomorphic curves.

Taking in his hands both the variational methods (Morse theory) used by Conley and Zehnder and the elliptic operators (pseudo-holomorphic curves) of Gromov, using the "characteristic class entering in quantization conditions" of [2], Andreas Floer built for us a Yellow-Brick-Road to prove the Arnold conjecture in greater and greater generality. (And this is what we[32] did.)

[32] By "we" here, I mean the community. I could also mention that some of us (and here, by "us", I mean the two authors of [24]) wrote a textbook to explain all this (a translation to English will be available soon).

10. Generating Functions (Continuation)

From the very description of wave fronts, it is clear that generating functions are a good tool for the study of contact geometry/topology. Note also that there are contact analogues of self-intersections of Lagrangians, namely chords of Legendrian knots.

Much progress has been done, but there is not enough space here to mention all this. The name of another former student of Arnold's, Yuri Chekanov, should be added here.

11. Twenty Years After... First Steps Again

Let us go back to the 1986 paper [14] we started with. Poincaré's geometric theorem was mentioned in the "Is there such a thing as symplectic topology?" section, but not its possible generalizations, which appeared only in Section 2, where, quoting [3, 4] for the statement and [40] for the proof, Arnold stated:

Theorem. *A symplectomorphism of the torus homologous to the identity has no fewer than four fixed points (taking multiplicities into account) and no fewer than three geometrically distinct fixed points.*

Four was for 2^n, three for $n+1$, hence the torus in the statement was 2-dimensional—this was the case, neither for the conjecture nor for the proof... The "multidimensional generalization" was more than just multidimensional, and for it Arnold quoted the problem in [35]... and his comments to the Russian edition of Poincaré's selected works[33], a book I never saw:

Conjecture. *A symplectomorphism of a compact manifold, homologous to the identity transformation*[34]*, has at least as many fixed points as a smooth function on the manifold has critical points.*

I think this was the first time the word "conjecture" (in reference to this problem) appeared in a paper by Arnold himself.

[33]See Review 52#5337 on *Math. Reviews.* Already in 1972, it was possible to publish double translations without checking the signification. The title of our favorite Poincaré paper [58] became there "A certain theorem of geometry".
[34]Joined by a one-parameter family of symplectomorphisms with single valued (but time-dependent) Hamiltonians. Note of V.I. Arnold.

And he listed the results obtained so far—a state of the art in 1986. That is, the torus ([37, 38, 40]), the surfaces ([43, 45]), the complex projective space ([47]), (many) Kähler manifolds of negative curvature ([45, 60]), diffeomorphisms that are $\mathcal{C}^0$-close to the identity ([65]).

12. Epilogue (2012)

And now, this is 2012. Twenty-six years after the "first steps". Three new appendices have been added to a second (1989) edition of [10]. Some, many versions of Arnold's conjecture have been proved. Others are still open. Many powerful techniques have been created, used, improved. Even the crossbows turned out to be very efficient. Helping to solve old problems, the new tools generated new ones.

Vladimir Igorevich died in Paris on June 3rd, 2010.

Symplectic topology is not standing still.

Acknowledgement. I thank Bob Stanton and Marcus Slupinski for their help with the translation of the adjectives in footnote 17.

Many thanks to Alan Weinstein and Karen Vogtmann, who were so kind to send me recollections and information and also to Alan, for allowing me to publish an excerpt of a letter Arnold had sent to him.

I am very grateful to Mihai Damian, Leonid Polterovich and Marc Chaperon, who kindly agreed to read preliminary versions of this paper, for their friendly comments and suggestions.

The last sentence in this paper was inspired by [61].

References

[1] V.I. Arnold, Small denominators and problems of stability of motion in classical and celestial mechanics. Usp. Mat. Nauk **18**(6(114)), 91–192 (1963)

[2] V.I. Arnold, A characteristic class entering in quantization conditions. Funct. Anal. Appl. **1** (1965)

[3] V.I. Arnold, Sur une propriété topologique des applications globalement canoniques de la mécanique classique. C. R. Acad. Sci. Paris **261**, 3719–3722 (1965)

[4] V.I. Arnold, Проблема устойчивости и эргодические свойства классических динамических систем (A stability problem and ergodic properties of classical dynam-

ical systems), in *Proc. Internat. Congr. Math.*, Moscow, 1966 (Izdat "Mir", Moscow, 1968), pp. 387–392

[5] V.I. Arnold, Обыкновенные дифференциальные уравнения (Izdat. "Nauka", Moscow, 1971)

[6] V.I. Arnold, *Ordinary Differential Equations* (MIT Press, Cambridge, 1973). Translated from the Russian and edited by Richard A. Silverman

[7] V.I. Arnold, *Équations différentielles ordinaires* (Mir, Moscow, 1974)

[8] V.I. Arnold, Математические методы классической механики (Izdat. "Nauka", Moscow, 1974)

[9] V.I. Arnold, *Méthodes mathématiques de la mécanique classique* (Mir, Moscow, 1976)

[10] V.I. Arnold, *Mathematical Methods in Classical Mechanics* (Springer, Berlin, 1978)

[11] V.I. Arnold, Lagrange and Legendre cobordisms I and II. Funct. Anal. Appl. **14**, 167–177, 252–260 (1980)

[12] V.I. Arnold, Singularities of ray systems. Usp. Mat. Nauk **38**, 77–147 (1983)

[13] V.I. Arnold, *Catastrophe Theory* (Springer, Berlin, 1984). Translated from the Russian by R.K. Thomas

[14] V.I. Arnold, First steps in symplectic topology. Russ. Math. Surv. **41**, 1–21 (1986)

[15] V.I. Arnold, *Ordinary Differential Equations.* Universitext (Springer, Berlin, 2006). Translated from the Russian by Roger Cooke, Second printing of the 1992 edition

[16] V.I. Arnold, A. Avez, *Problèmes ergodiques de la mécanique classique.* Monographies Internationales de Mathématiques Modernes, vol. 9 (Gauthier-Villars, Paris, 1967)

[17] V.I. Arnold, A. Avez, *Ergodic Problems of Classical Mechanics* (W. A. Benjamin, New York, 1968). Translated from the French by A. Avez

[18] V.I. Arnold, A.B. Givental, Symplectic geometry, in *Dynamical Systems IV.* Encyclopaedia of Math. Sci. (Springer, Berlin, 1985), pp. 1–138

[19] M.F. Atiyah, Convexity and commuting Hamiltonians. Bull. Lond. Math. Soc. **14**, 1–15 (1982)

[20] M. Audin, Quelques calculs en cobordisme lagrangien. Ann. Inst. Fourier **35**, 159–194 (1985)

[21] M. Audin, Cobordismes d'immersions lagrangiennes et legendriennes. Thèse d'état, Orsay, 1986, Travaux en cours, Hermann, Paris, 1987.

[22] M. Audin, Кобордизмы Лагранжевых иммерсий в пространство кокасательного расслоения многообразия. Funkc. Anal. Prilozh. **21**(3), 61–64 (1987)

[23] M. Audin, Vladimir Igorevich Arnold est mort, *Images des Mathématiques, CNRS* (2010), Available at http://images.math.cnrs.fr/Vladimir-Igorevich-Arnold-est-mort.html

[24] M. Audin, M. Damian, *Théorie de Morse et homologie de Floer.* Savoirs actuels (Edp-Sciences, Les Ulis, 2010)

[25] A. Banyaga, On fixed points of symplectic maps. Invent. Math. **56**, 215–229 (1980)

[26] D. Bennequin, Entrelacements et équations de Pfaff, in *Troisième conférence de géométrie du Schnepfenried*. Astérisque, vol. 107 (Soc. Math. France, Paris, 1983), pp. 87–161

[27] D. Bennequin, Caustique mystique, in *Séminaire Bourbaki*. Astérisque, vols. 133–134 (1986)

[28] F.A. Berezin, Quantization. Izv. Akad. Nauk SSSR, Ser. Mat. **38**, 1116–1175 (1974)

[29] G. Birkhoff, Quelques théorèmes sur le mouvement des systèmes dynamiques. Bull. Soc. Math. Fr. **40**, 305–323 (1912)

[30] G. Birkhoff, Proof of Poincaré's geometric theorem. Trans. Am. Math. Soc. **14**, 14–22 (1913)

[31] G. Birkhoff, Démonstration du dernier théorème de géométrie de Poincaré. Bull. Soc. Math. Fr. **42**, 1–12 (1914)

[32] G. Birkhoff, An extension of Poincaré's last geometric theorem. Acta Math. **47**, 297–311 (1926)

[33] G. Birkhoff, *Dynamical Systems*. American Mathematical Society Colloquium Publications, vol. 9 (American Mathematical Society, New York, 1927)

[34] J.-P. Bourguignon, R. Remmert, F. Hirzebruch, Henri Cartan 1904–2008. Newsl. - Eur. Math. Soc. **70**, 5–7 (2008)

[35] F.E. Browder (ed.), *Mathematical Developments Arising from Hilbert Problems*. Proceedings of Symposia in Pure Mathematics, vol. XXVIII (American Mathematical Society, Providence, 1976)

[36] M. Brown, W.D. Neumann, Proof of the Poincaré-Birkhoff fixed point theorem. Mich. Math. J. **24**, 21–31 (1977)

[37] M. Chaperon, Questions de géométrie symplectique, in *Séminaire Bourbaki*. Astérisque, vols. 105–106 (1983), pp. 231–249

[38] M. Chaperon, Une idée du type "géodésiques brisées" pour les systèmes hamiltoniens. C. R. Acad. Sci. Paris Sér. I Math. **298**(13), 293–296 (1984)

[39] M. Chaperon, An elementary proof of the Conley-Zehnder theorem in symplectic geometry, in *Dynamical Systems and Bifurcations*, Groningen, 1984. Lecture Notes in Math., vol. 1125 (Springer, Berlin, 1985), pp. 1–8

[40] C. Conley, E. Zehnder, The Birkhoff-Lewis fixed point theorem and a conjecture of V.I. Arnold. Invent. Math. **73**, 33–49 (1983)

[41] Y.M. Eliashberg, A theorem on the structure of wave fronts and its application in symplectic topology. Funkc. Anal. Prilozh. **21**, 65–72 (1987)

[42] Y.M. Eliashberg, The structure of 1-dimensional wave fronts, nonstandard Legendrian loops and Bennequin's theorem, in *Topology and Geometry—Rohlin Seminar*. Lecture Notes in Math., vol. 1346 (Springer, Berlin, 1988), pp. 7–12

[43] Y. Eliashberg, An estimate of the number of fixed points of transformations preserving area, Preprint, Syktyvkar (1978)

[44] Y. Eliashberg, Cobordisme des solutions de relations différentielles, in *Séminaire Sud-Rhodanien de géométrie, I*, Lyon, 1983. Travaux en Cours (Hermann, Paris, 1984), pp. 17–31, translated from the Russian by M. Audin

[45] A. Floer, Proof of the Arnold conjecture for surfaces and generalizations to certain Kähler manifolds. Duke Math. J. **53**, 1–32 (1986)

[46] B. Fortune, A symplectic fixed point theorem for $\mathbf{CP}^n$. Invent. Math. **81**, 29–46 (1985)

[47] B. Fortune, A. Weinstein, A symplectic fixed point theorem for complex projective spaces. Bull., New Ser., Am. Math. Soc. **12**, 128–130 (1985)

[48] A.B. Givental, Lagrangian imbeddings of surfaces and the open Whitney umbrella. Funkc. Anal. Prilozh. **20**(3), 35–41, 96 (1986)

[49] M. Gromov, A topological technique for the construction of solutions of differential equations and inequalities, in *Actes du Congrès International des Mathématiciens*, vol. 2, Nice, 1970 (Gauthier-Villars, Paris, 1971), pp. 221–225

[50] M. Gromov, Pseudo-holomorphic curves in symplectic manifolds. Invent. Math. **82**, 307–347 (1985)

[51] V. Guillemin, S. Sternberg, Convexity properties of the moment mapping. Invent. Math. **67**, 491–513 (1982)

[52] H. Hofer, Arnold and symplectic geometry. Not. Am. Math. Soc. **59**, 499–502 (2012)

[53] J. Marsden, A. Weinstein, Reduction of symplectic manifolds with symmetry. Rep. Math. Phys. **5**, 121–130 (1974)

[54] D. McDuff, D. Salamon, *Introduction to Symplectic Topology*. Oxford Science Publications (The Clarendon Press Oxford University Press, New York, 1995)

[55] J. Moser, On the volume elements on a manifold. Trans. Am. Math. Soc. **120**, 286–294 (1965)

[56] S. Nemirovski, Lagrangian Klein bottles in $\mathbb{R}^{2n}$. Geom. Funct. Anal. **19**, 902–909 (2009)

[57] N.A. Nikišin, Fixed points of the diffeomorphisms of the two-sphere that preserve oriented area. Funkc. Anal. Prilozh. **8**(1), 84–85 (1974)

[58] H. Poincaré, Sur un théorème de géométrie. Rend. Circ. Mat. Palermo **33**, 375–407 (1912)

[59] L. Polterovich, I. Scherbak, V.I. Arnold (1937–2010). Jahresber. Dtsch. Math.-Ver. **113**, 185–219 (2011)

[60] J.-C. Sikorav, Points fixes d'une application symplectique homologue à l'identité. J. Differ. Geom. **22**, 49–79 (1985)

[61] A. Weil, *Essais historiques sur la théorie des nombres*, L'Enseignement Mathématique (Université de Genève, Geneva, 1975), Extrait de l'Enseignement Math. **20** (1974), Monographie No. 22 de L'Enseignement Mathématique

[62] A. Weinstein, *Lectures on Symplectic Manifolds*. CBMS Regional Conference Series in Mathematics, vol. 29 (Amer. Math. Soc., Providence, 1977)

[63] A. Weinstein, Symplectic V-manifolds, periodic orbits of Hamiltonian systems, and the volume of certain Riemannian manifolds. Commun. Pure Appl. Math. **30**, 265–271 (1977)

[64] A. Weinstein, On the hypotheses of the Rabinowitz periodic orbit theorems. J. Differ. Equ. **33**, 353–358 (1979)

[65] A. Weinstein, C^0 perturbation theorems for symplectic fixed points and Lagrangian intersections, in *Séminaire Sud-Rhodanien de géométrie, III*, Lyon, 1983. Travaux en Cours (Hermann, Paris, 1984), pp. 140–144

M. Audin (✉)

Institut de Recherche Mathématique Avancée
Université de Strasbourg et CNRS
7 rue René Descartes
67084 Strasbourg cedex
France

e-mail: michele.audin@math.unistra.fr

BOLYAI SOCIETY
MATHEMATICAL STUDIES, 26

Contact and Symplectic Topology
pp. 27–83.

Topological Methods in 3-Dimensional Contact Geometry

An Illustrated Introduction to Giroux's Convex Surfaces Theory

PATRICK MASSOT

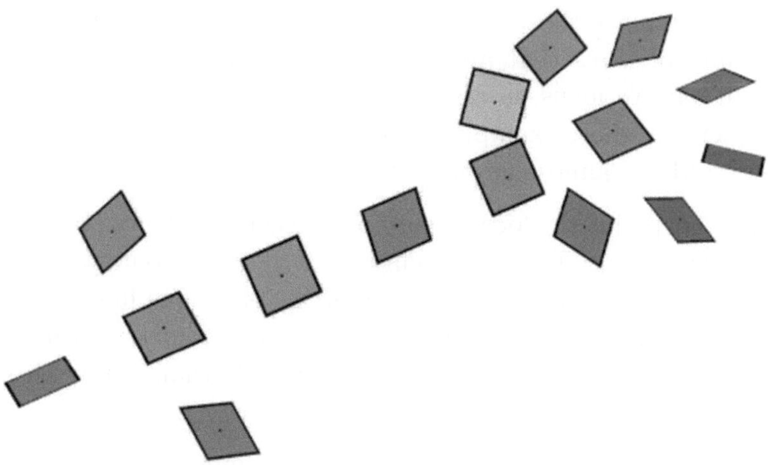

F. Bourgeois et al. (eds.), *Contact and Symplectic Topology*,
Bolyai Society Mathematical Studies 26, DOI 10.1007/978-3-319-02036-5_2,

1. Introduction

These lecture notes are an introduction to the study of global properties of contact structures on 3-manifolds using topological rather than analytical methods. From that perspective, the main tool to study a contact manifold (V, ξ) is the study of its ξ-convex surfaces. These surfaces embedded in V are useful because all the information about ξ near each of them is encoded into a surprisingly small combinatorial data. In order to illustrate the power of ξ-convex surfaces without long developments, we use them to reprove, following Giroux [9], two important theorems which were originally proved using different techniques by Bennequin [1] and Eliashberg [5].

Besides Giroux's original papers [8, 9], there are already two sets of lectures notes by Etnyre [6] and Honda [13] and a book by Geiges [7] which cover almost all topics we will discuss as well as more advanced topics. Our goal is not to replace those references but to complement them. Mostly, we include many pictures that are not easily found in print and can help to build intuition. We focus on a small set of contact manifolds and illustrate all phenomena on those examples by showing explicit embedded surfaces. On the other hand, we almost never give complete proofs.

Section 2 explains the local theory of contact structures starting with the most basic definitions. There are many ways to define contact structures and contact forms and we use unusual geometric definitions in order to complement existing sources. We also try to explain the geometric intuition behind the theorems of Darboux-Pfaff and Gray rather than using Moser's path method without explanation.

Once enough definitions are given, an interlude states the theorems of Bennequin and Eliashberg that are proved at the high point of these notes. It serves as motivation for the rather long developments of Section 4.

Section 4 begins the study of surfaces in contact manifolds. The starting point is the singular foliation printed by a contact structure on any surface. We then work towards ξ-convex surfaces theory by simplifying gradually the contact condition near a surface. Once the amazing realization lemma is proved, we investigate obstructions to ξ-convexity and prove these obstructions are generically not present. The last part of this section then gets the first fruits of this study by proving the Eliashberg-Bennequin inequalities.

Section 5 goes beyond the study of a single surface by studying some one-parameter families of surfaces. In particular we describe what happens exactly when one of the obstructions to ξ-convexity discussed in the preceding section arises. This allows us to prove the theorems of Bennequin and Eliashberg mentioned above. Until now, the proof of Bennequin's theorem using ξ-convex surfaces was explained only in [9].

Of course this is only the beginning of a story which continues both by itself and in combination with holomorphic curves techniques.

Conventions. A plane field ξ on a 3-manifold V is a (smooth) map associating to each point p of V a 2-dimensional subspace $\xi(p)$ of T_pV. All plane fields considered here will be coorientable, it means one can continuously choose one of the half spaces cut out by $\xi(p)$ in T_pV. In this situation, ξ can be defined as the kernel of some nowhere vanishing 1-form α: $\xi(p) = \ker \alpha(p)$. The coorientation is given by the sign of α. We will always assume that V is oriented. In this situation a coorientation of ξ combines with the ambient orientation to give an orientation on ξ. All contact structures in these notes will be cooriented.

Occasionally, we will include remarks or comments that are not part of the main flow of explanations. These remarks are typeset in small italic print.

2. Local Theory

2.1. Contact Structures as Rotating Plane Fields

The Canonical Contact Structure on the Space of Contact Elements. Let S be a surface and $\pi : ST^*S \to S$ the bundle of cooriented lines tangent to S (also called contact elements for S). It can be seen as the bundle of rays in T^*S, hence the notation. The canonical contact structure on ST^*S at a point d is defined as the inverse image under π_* of $d \subset T_{\pi(d)}S$, see Figure 1.

Suppose first that S is the torus $T^2 = \mathbb{R}^2/2\pi\mathbb{Z}^2$. Let x and y be the canonical $\mathbb{S}^1$-valued coordinates on T^2. A cooriented line tangent to T^2 at some point (x, y) can be seen as the kernel of a 1-form λ which has unit norm with respect to the canonical flat metric. So there is some angle z such that $\lambda = \cos(z)dx - \sin(z)dy$. Hence we have a natural identification of ST^*T^2 with T^3. In addition the canonical contact structure can be defined by $\cos(z)dx - \sin(z)dy$ now seen as a 1-form on T^3 called the canonical contact form on T^3, see Figure 2.

When S is the sphere $\mathbb{S}^2$, ST^*S is endowed with a free transitive action of $\mathrm{SO}_3(\mathbb{R})$ so it is diffeomorphic to $\mathrm{SO}_3(\mathbb{R})$. So there is a two-fold covering

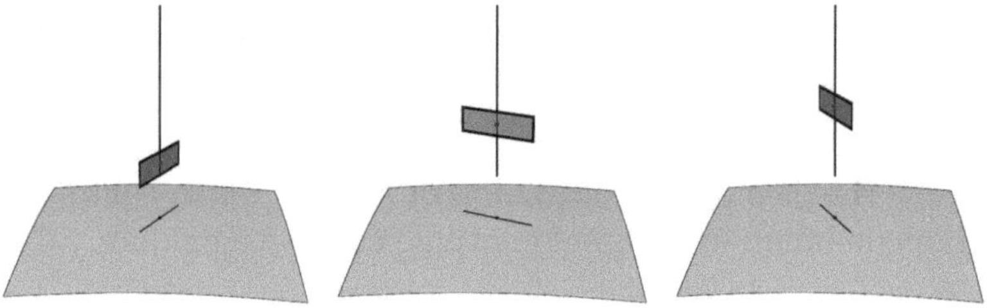

Fig. 1. Canonical contact structure on the bundle of cooriented lines. At bottom is a portion of S with a tangent line at some point. Above that point one gets the fiber by gluing top and bottom of the interval. The contact structure is shown at the point of the fiber corresponding to the line drawn below

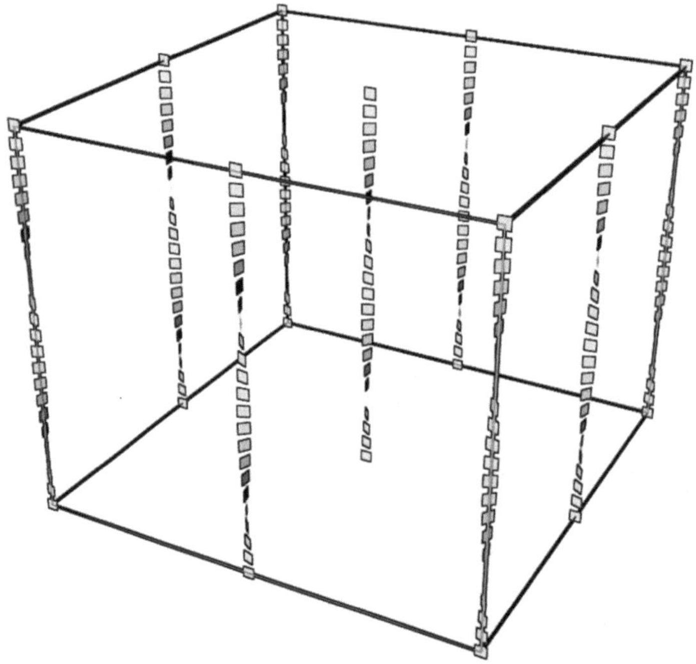

Fig. 2. Canonical contact structure on T^3. Opposite faces of the cube are glued to get T^3

map from $\mathbb{S}^3 \simeq \mathrm{SU}(2)$ to $ST^*\mathbb{S}^2$. The lifted plane field is called the canonical contact structure on $\mathbb{S}^3$. We will see different ways of describing this example later on.

Contact Structures and Contact Forms.

Definition 1. A *contact structure* on a 3-manifold is a plane field which is locally diffeomorphic to the canonical contact structure on ST^*T^2. A *contact* *form* is a 1-form whose kernel is a contact structure. A curve or a vector field is *Legendrian* if it is tangent to a given contact structure.

As noted above all our manifolds will be oriented and diffeomorphisms in the above definition shall preserve orientations.

Theorem 2 (Darboux–Pfaff theorem). *A 1-form α is a contact form if and only if $\alpha \wedge d\alpha$ is a positive volume form.*

Let ξ be the kernel of α. The condition $\alpha \wedge d\alpha > 0$ will henceforth be called the contact condition for α. It is equivalent to the requirement that $d\alpha_{|\xi}$ is non-degenerate and defines the orientation of ξ coming from the orientation of the ambient manifold and the coorientation of ξ.

Proof. If ξ is a contact structure then the image of α in the local model is $f\alpha_0$ where f is some nowhere vanishing function and $\alpha_0 = \cos(z)dx - \sin(z)dy$. So

$$\begin{aligned} \alpha \wedge d\alpha &= f\alpha_0 \wedge (f d\alpha_0 + df \wedge \alpha_0) = f^2 \alpha_0 \wedge d\alpha_0 \\ &= f^2\, dx \wedge dy \wedge dz \end{aligned}$$

which is a positive volume form. More generally the above computation proves that the contact condition for a nowhere vanishing one-form depends only on its kernel.

Conversely, suppose $\alpha \wedge d\alpha$ is positive. Let p be a point in M. We want to construct a coordinate chart around p such that $\xi = \ker(\cos(z)dx - \sin(z)dy)$. We first choose a small surface S containing p and transverse to ξ. Then we pick a non-singular vector field X tangent to S and ξ near p and a small curve c in S containing p and transverse to X, see Figure 3. Let y be a coordinate on c. The flow of X at time x starting from c gives coordinates (x, y) on S near p in which $X = \partial_x$.

We now consider a vector field V transverse to S and tangent to ξ. The flow of V at time t starting from S gives coordinates (x, y, t) near p such that $\alpha = f(x, y, t)dx + g(x, y, t)dy$ because $\alpha(\partial_t) = \alpha(V) = 0$. Up to rescaling, one can use instead $\alpha_1 = \cos z(x, y, t)dx - \sin z(x, y, t)dy$ for some function z such that $z(x, y, 0) = 0$. Now it is time to use the contact condition. We can compute

$$\alpha_1 \wedge d\alpha_1 = \frac{\partial z}{\partial t} dx \wedge dy \wedge dt.$$

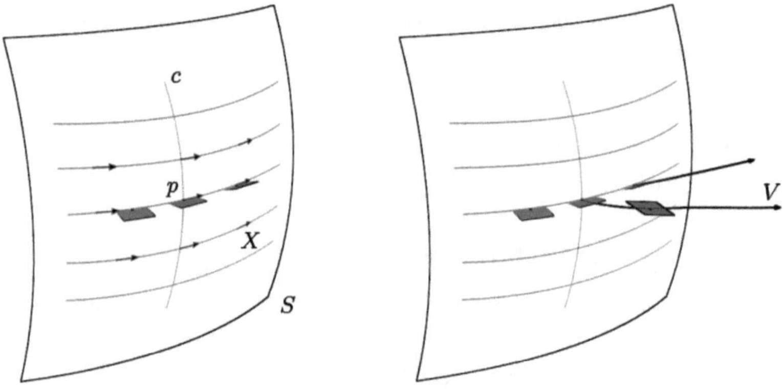

Fig. 3. Proof of the Darboux–Pfaff theorem

Remember the contact condition for α is equivalent to the contact condition for α_1. So $\frac{\partial z}{\partial t}$ is positive and the implicit function theorem then guaranties that we can use z as a coordinate instead of t. ■

In the above proof, $z(x, y, t)$ was the angle between ξ and the horizontal ∂_x is the plane normal to the Legendrian vector field ∂_t. We saw that the contact condition forces this angle to increase. This means that the contact structure rotates around ∂_t. The above proof essentially says that this rotation along Legendrian vector fields characterizes contact structures.

We now focus on the difference between contact structures and contact forms. The data of a contact form is equivalent to a contact structure and either a choice of a Reeb vector field or a section of its symplectization.

Definition 3. A Reeb vector field for a contact structure ξ is a vector field which is transverse to ξ and whose flow preserves ξ.

If one has a Riemannian metric on a surface S then the bundle of contact elements of S can be identified with the unit tangent bundle STS and the geodesic flow is then the flow of a Reeb vector field for the canonical contact structure.

One can easily prove that each contact form α comes with a canonical Reeb vector field R_α which is characterized by $d\alpha(R_\alpha, \cdot) = 0$ and $\alpha(R_\alpha) = 1$. All Reeb vector fields arise this way.

Next, for any co-oriented plane field ξ on a 3-manifold V, one can consider the annihilator of ξ in T^*V:

$$S_\xi := \{\lambda \in T^*V \mid \ker \lambda = \xi \text{ and } \lambda(v) > 0 \text{ if } v \text{ is positively transverse to } \xi\}.$$

It is a good exercise to check that a plane field ξ on V is a contact structure if and only if S_ξ is a symplectic submanifold of $(T^*V, \omega_{\text{can}})$. In this case S_ξ is called the *symplectization* of ξ. The manifold S_ξ is a principal $\mathbb{R}$-bundle where a real number t acts by $\lambda \mapsto e^t\lambda$. Any contact form α is a section of this $\mathbb{R}$-bundle, and thus determines a trivialization $\mathbb{R} \times V \to S_\xi$ given by $(t, v) \mapsto e^t\alpha_v$. In this trivialization, the restriction of the canonical symplectic form ω_{can} becomes $d(e^t\alpha)$.

2.2. Examples

The Canonical Contact Structure on $\mathbb{R}^3$. The universal cover of ST^*T^2 is of course $\mathbb{R}^3$ and the lifted contact structure is $\xi_0 = \ker(\cos(z)dx - \sin(z)dy)$ where x, y and z are now honest real-valued coordinates. The plane field ξ_0 is called the standard contact structure on $\mathbb{R}^3$.

Depending on context, it can be useful to have different ways of looking at ξ_0 using various diffeomorphisms of $\mathbb{R}^3$. The image of ξ_0 under the diffeomorphism

$$\begin{pmatrix} x \\ y \\ z \end{pmatrix} \mapsto \begin{pmatrix} t \\ p \\ q \end{pmatrix} = \begin{pmatrix} \cos(z) & -\sin(z) & 0 \\ \sin(z) & \cos(z) & 0 \\ 0 & 0 & 1 \end{pmatrix} \begin{pmatrix} x \\ y \\ z \end{pmatrix}$$

is drawn in Figure 5. It admits the contact form $dt + pdq$ and arises naturally on $\mathbb{R}^3$ seen as the space of 1-jets of functions from $\mathbb{R}$ to $\mathbb{R}$ (see e.g. [7, Example 2.5.11] for more information on this interpretation).

Figures 4 and 5 together are often confusing for beginners. First the thick black line $\{t = p = 0\}$ in Figure 5 is Legendrian yet the contact structure does not seem to rotate along it. Second, it seems the two pictures exhibit Legendrian foliations by lines with very different behavior. In the second picture the contact structure turns half a turn along each leave whereas it turns infinitely many turns in the first picture.

Both puzzles are solved by the same picture. The diffeomorphism we used above sends the foliation by Legendrian lines of Figure 4 to a foliation containing the mysterious line $\{t = p = 0\}$ in Figure 5 together with helices around that line, see Figure 6.

So we first see where is the foliation of Figure 4 inside Figure 5. And second we remember that it makes sense to say that a plane field rotates along a curve only compared to something else. Contact structures rotate along

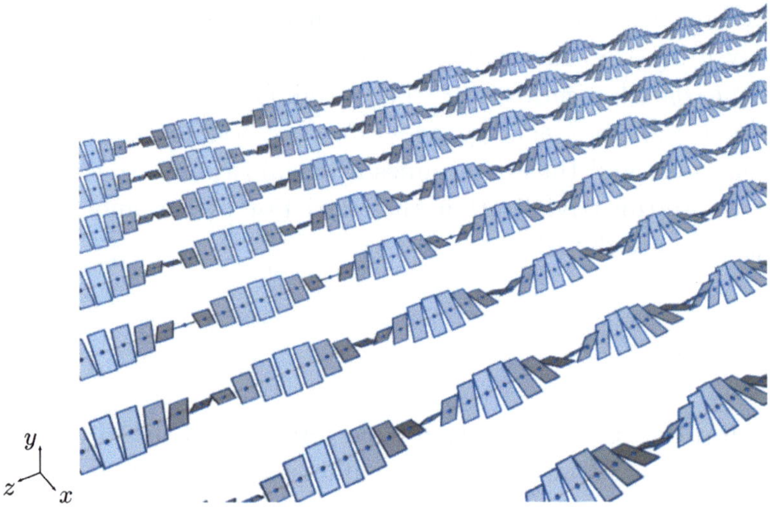

Fig. 4. Universal cover of the standard contact structure on $\mathbb{T}^3$ seen from the side. It is invariant under translation in the vertical direction

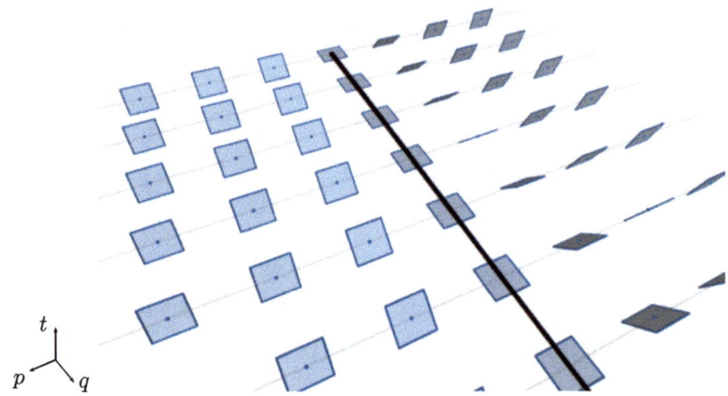

Fig. 5. $\ker(dt + pdq)$ on $\mathbb{R}^3$. It is invariant under translation in the vertical direction. It becomes vertical only if one goes all the way to $p = \pm\infty$

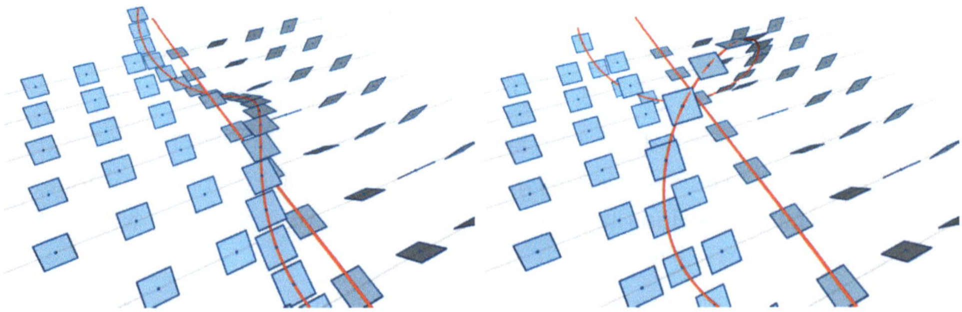

Fig. 6. The mysterious line in Figure 5 together with two helices coming from the lines of Figure 4

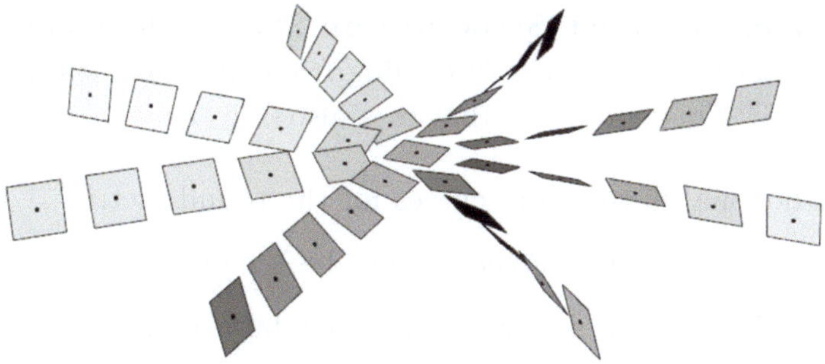

Fig. 7. Another view of the standard contact structure on $\mathbb{R}^3$

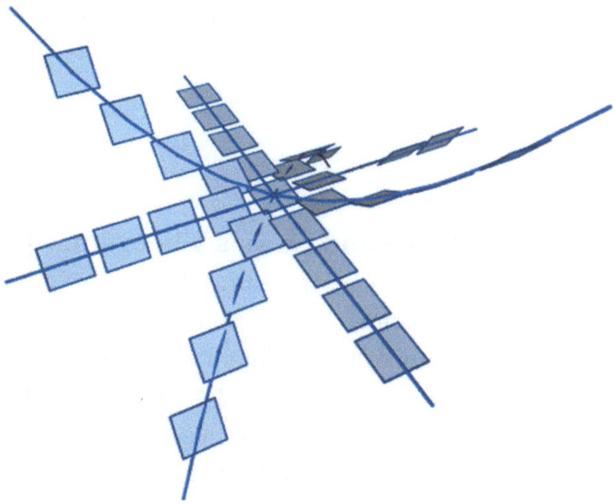

Fig. 8. Embedding of Figure 7 into Figure 5

Legendrian curves compared to neighborhood leaves of some Legendrian foliation. And indeed we see the contact structure turns infinitely many times along the mysterious line compared to the nearby Legendrian helices.

It is also sometimes convenient to consider the image of $\ker(dt + pdq)$ under the diffeomorphism $(t, p, q) \mapsto (q, -p, t + \frac{pq}{2})$. This image is the kernel of $dz + \frac{1}{2}r^2 d\theta$ in cylindrical coordinates, see Figure 7. In this model, one sees clearly that, at each point, there are Legendrian curves going in every possible direction.

Figure 8 shows how to deform Figure 7 to embed it inside Figure 5.

Either of these contact structures (which are diffeomorphic by construction) will be called the canonical contact structure on $\mathbb{R}^3$. Of course they can all be used as the local model in the definition of a contact structure.

The Canonical Contact Structure on $\mathbb{S}^3$. We have already met the canonical contact structure on $\mathbb{S}^3$ coming from the canonical contact structures on $ST^*\mathbb{S}^2$. One can prove that it is also

- the orthogonal of the Hopf circles for the round metric,
- a left-invariant contact structure on the Lie group SU(2),
- $T\mathbb{S}^3 \cap JT\mathbb{S}^3$ when $\mathbb{S}^3$ is seen as the boundary of the unit ball in $\mathbb{C}^2$ and J denotes the action of multiplication by i in $T\mathbb{C}^2$.

The complement of a point in the standard $\mathbb{S}^3$ is isomorphic to the standard $\mathbb{R}^3$, see [7, Proposition 2.1.8] for a computational proof valid in any dimension.

2.3. Isotopies

Isotopic Contact Structures and Gray's Theorem. Up to now we considered two contact structures to be the same if they are conjugated by some diffeomorphism. One can restrict this by considering only diffeomorphisms corresponding to deformations of the ambient manifold. An isotopy is a family of diffeomorphisms φ_t parametrized by $t \in [0,1]$ such that $(x,t) \mapsto \varphi_t(x)$ is smooth and $\varphi_0 = Id$. The time-dependent vector field generating φ_t is defined as $X_t = \frac{d}{dt}\varphi_t$. One says that two contact structures ξ_0 and ξ_1 are isotopic if there is an isotopy φ_t such that $\xi_1 = (\varphi_1)_*\xi_0$. In particular such contact structures can be connected by the path of contact structures $\xi_t := (\varphi_t)_*\xi_0$. It is then natural to consider the seemingly weaker equivalence relation of homotopy among contact structures. The next theorem says in particular that, on closed manifolds, this equivalence relation is actually the same as the isotopy relation.

Theorem 4 (Gray [12]). *For any path $(\xi_t)_{t\in[0,1]}$ of contact structures on a closed manifold, there is an isotopy φ_t such that $\varphi_t^*\xi_t = \xi_0$.*

The vector field X_t generating φ_t can be chosen in $\lim_{\varepsilon\to 0} \xi_t \cap \xi_{t+\varepsilon}$ at each time t.

Proof. The proof of this theorem can be found in many places but without much geometric explanations. So we now explain the picture behind it. The key is to be able to construct an isotopy pulling back $\xi_{t+\varepsilon}$ to ξ_t for infinitesimally small ε. It means we will construct the generating vector field

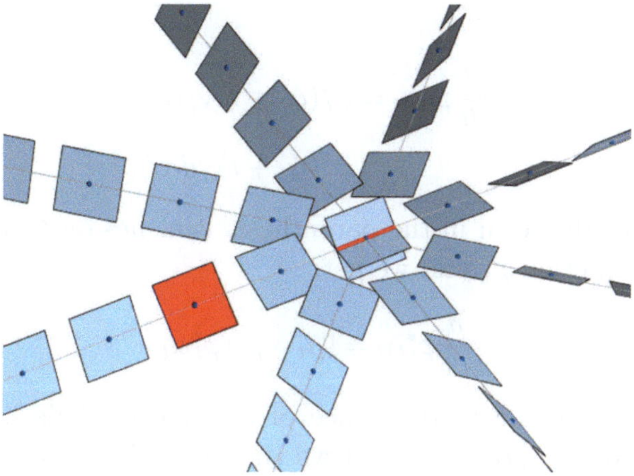

Fig. 9. Proof of Gray's theorem

X_t rather than φ_t directly. The compactness assumption will guaranty that the flow of X_t exists for all time.

At any point p, if the plane $\xi_{t+\varepsilon}$ coincides with ξ_t then we have nothing to do and set $X_t = 0$. Otherwise, these two planes intersect transversely along a line $d_{t,\varepsilon}$. The natural way to bring $\xi_{t+\varepsilon}$ back to ξ_t is to rotate it around $d_{t,\varepsilon}$. Since we know from the proof of Theorem 2 that the flow of Legendrian vector fields rotate the contact structure, we will choose X_t in the line $d_t := \lim_{\varepsilon\to 0} d_{t,\varepsilon}$, see Figure 9. Let us compute $d_{t,\varepsilon}$:

$$d_{t,\varepsilon} = \{v \mid \alpha_{t+\varepsilon}(v) = \alpha_t(v) = 0\} = \left\{ v \in \xi_t \;\middle|\; \frac{1}{\varepsilon}(\alpha_{t+\varepsilon} - \alpha_t)(v) = 0 \right\}$$

which gives, as ε goes to zero: $d_t = \xi_t \cap \ker(\dot\alpha_t)$.

The contact condition for α_t is equivalent to the fact that $(d\alpha_t)_{|\xi_t}$ is non-degenerate. So X_t belongs to $\xi_t \cap \ker(\dot\alpha_t)$ if and only if it belongs to ξ_t and $\iota_{X_t} d\alpha_t = f_t \dot\alpha_t$ on ξ_t for some function f_t.

Moreover, we want X_t to compensate the rotation expressed by $\dot\alpha_t$. A natural guess is then to pick the unique Legendrian vector field X_t such that $(\iota_{X_t} d\alpha_t)_{|\xi_t} = -(\dot\alpha_t)_{|\xi_t}$.

We now have a precise candidate for X_t and we can compute to prove that it does the job. Let φ_t be the flow of X_t. Using Cartan's formula, we get:

$$\begin{aligned}\frac{d}{dt}\varphi_t^*\alpha_t &= \varphi_t^*(\dot\alpha_t + \mathcal{L}_{X_t}\alpha_t)\\ &= \varphi_t^*(\dot\alpha_t + \iota_{X_t}d\alpha_t).\end{aligned}$$

By construction, the term in the parenthesis vanishes on ξ_t so it is α_t multiplied by some function μ_t and we get:

$$\frac{d}{dt}\varphi_t^*\alpha_t = (\mu_t \circ \varphi_t)\varphi_t^*\alpha_t.$$

So $\varphi_t^*\alpha_t$ stays on a line in the space of one forms. This line is obviously the line spanned by $\varphi_0^*\alpha_0 = \alpha_0$ and we then have $\ker\varphi_t^*\alpha_t = \ker\alpha_0 = \xi_0$ for all t. It is not hard to see that X_t is the only Legendrian vector field which works. ■

Some compactness assumption is indeed necessary in Gray's theorem. There are counter-examples on $\mathbb{R}^2 \times \mathbb{S}^1$ discovered in [4].
Contact structures form an open set in the space of all plane fields. Gray's theorem proves that isotopy classes of contact structures on a closed manifold are actually connected components of this open set. In particular there are only finitely many isotopy classes of contact structures on a closed manifold.
The example of linear foliations on T^3 proves that Gray's theorem wouldn't hold for foliations.

Libermann's Theorem on Contact Hamiltonians. Contact transformations of a contact manifold (V, ξ) are diffeomorphisms of V which preserve ξ. The infinitesimal version of these are vector fields whose flow consists of contact transformations. They are called contact vector fields and are exactly those X for which $(\mathcal{L}_X\alpha)_{|\xi} = 0$ for any contact form α defining ξ. Note that this condition is weaker than $\mathcal{L}_X\alpha = 0$ which would imply that the flow of X preserves α and not only its kernel ξ.

In the proof of Gray's theorem, we saw that one can rotate a contact structure at will using the flow of a Legendrian vector field uniquely determined by the rotation we want to achieve. The same idea allows to prove that any vector field on a contact manifold can be transformed into a contact vector field by adding a uniquely determined Legendrian vector field. This is the geometric fact underlying the existence of so-called contact Hamiltonians.

Theorem 5 (Libermann [14]). *On a contact manifold (V, ξ) the map which sends a contact vector field to its reduction modulo ξ is an isomorphism from the space of contact vector fields to the space of sections of the normal bundle TV/ξ.*

If we single out a contact form α then we get a trivialization $TV/\xi \to V \times \mathbb{R}$ given by $(x, [u]) \mapsto (x, \alpha(u))$. Sections of TV/ξ can then be seen as functions on V and the contact vector field X_f associated to a function f using the preceding theorem is called the Hamiltonian vector field coming from α and f. Libermann's theorem both implies existence of X_f and the fact that it is the unique contact vector field satisfying $\alpha(X) = f$. The situation is analogous to the case of Hamiltonian vector fields in symplectic geometry but in the symplectic case there are symplectic vector fields that are not Hamiltonian. Note that the above interpretation when a contact form is fixed is what Libermann originally discussed and also the most common use of the word contact Hamiltonian.

Proof of Theorem 5. Let X be any vector field on V. The theorem is equivalent to the assertion that there is a unique Legendrian vector field X_ξ such that $X + X_\xi$ is contact. Using any contact form α, we have equivalent reformulations:

$$\begin{aligned} X + X_\xi \text{ is contact} &\iff (\mathcal{L}_{X+X_\xi}\alpha)_{|\xi} = 0 \\ &\iff \big(\iota_{X+X_\xi} d\alpha + d(\iota_X \alpha)\big)_{|\xi} = 0 \\ &\iff (\iota_{X_\xi} d\alpha)_{|\xi} = -\big(\iota_X d\alpha + d(\iota_X \alpha)\big)_{|\xi} \end{aligned}$$

and the later condition defines uniquely X_ξ because $d\alpha_{|\xi}$ is non-degenerate. ■

Remark 6. A common use of contact Hamiltonians, and the only one we will need, is to cut-off or extend a contact vector field. For instance if X is a contact vector field defined on an open set $U \subset V$ and F is a closed subset of V contained in U then there is a contact vector field $\tilde{X}$ which vanishes outside U and equals X on F. If L denotes the isomorphism of Theorem 5 and ρ is a function with support in U such that $\rho_{|F} \equiv 1$ then we can use $\tilde{X} = L^{-1}(\rho L(X))$.

3. The Tight vs Overtwisted Dichotomy

After the local theory and before starting our study of convex surfaces, we need some motivation.

In Figure 7 showing $\ker(dz + r^2 d\theta)$, the contact planes rotate along rays perpendicular to the z-axis but are never horizontal away from the z-axis.

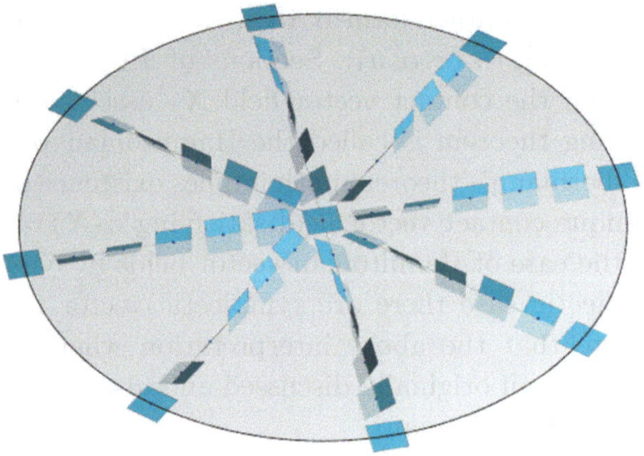

Fig. 10. An overtwisted contact structure

On the other extreme one can instead consider a contact structure which turns infinitely many times along these rays. A possible contact form for this is $\cos(r)\,dz + r\sin(r)\,d\theta$ which is horizontal for each r such that $\sin(r) = 0$, i.e. $r = k\pi$. Figure 10 shows what happens along $z = 0$ and $r \leq \pi$. One sees a disk whose tangent space agrees with ξ at the center and along the boundary.

Definition 7 (Eliashberg)**.** A contact manifold is *overtwisted* if it contains an embedded disk along which the contact structure is as in Figure 10: the contact structure ξ is tangent to the disk in the center and along the boundary and tangent to rays from the center to the boundary. A contact structure which is not overtwisted is called *tight.*

It may look like this is the beginning of an infinite series of definitions where ones looks at disks $z = 0$, $r \leq k\pi$ in the model above. But this would bring nothing new as can be seen from the following exercise.

Exercise. Prove that any neighborhood of an overtwisted disk in a contact manifold contains a whole copy of $(\mathbb{R}^3, \xi_{\mathrm{OT}})$ where $\xi_{\mathrm{OT}} = \ker(\cos(r)\,dz + r\sin(r)\,d\theta)$.

The above exercise is pretty challenging at this stage but it can serve as a motivation for the technology at the beginning of Section 4. And, most of all, it shows that not immediately seeing something in a contact manifold does not mean it is not there (recall also Figure 6). This begins to highlight the depth of the following two results whose proof is the main goal of these lecture notes.

Theorem 8 (Bennequin 1982 [1]). *The standard contact structures on $\mathbb{R}^3$ and $\mathbb{S}^3$ are tight.*

Theorem 9 (Eliashberg 1992 [5]). *All tight contact structures on $\mathbb{R}^3$ or $\mathbb{S}^3$ are isomorphic to the standard ones.*

Bennequin's theorem shows in particular that the standard contact structure on $\mathbb{R}^3$ is not isomorphic to the overtwisted structure of Figure 10. In order to put this in perspective, recall that Figures 4 and 5 show isomorphic contact structures. It may look like the difference between these is analogous to the difference between Figures 7 and 10. But Bennequin's theorem proves that the later two pictures are really different.

Eliashberg's theorem shows that tight contact structures on $\mathbb{S}^3$ are rare. By contrast, overtwisted contact structures abound. The Lutz–Martinet theorem, revisited by Eliashberg, says that, on a closed oriented manifold, any plane field is homotopic to an overtwisted contact structure [3]. Recall that, because the Euler characteristic of a 3-manifold always vanishes, all such manifolds have plane fields and even more, there are always infinitely many homotopy classes of plane fields (for the classification of homotopy classes of plane fields one can refer to [7, Section 4.2]).

In [2], Colin, Giroux and Honda proved that only finitely many homotopy classes of planes fields on each manifold can contain tight contact structures. This is far beyond the scope of these lectures but see Theorem 30 for a weaker version due to Eliashberg [5].

4. Convex Surfaces

The goal of this section is to explain the following crucial observation by Emmanuel Giroux in 1991:

> If S is a generic surface in a contact 3-manifold, all the information about the contact structure near S is contained in an isotopy class of curves on S.

All this section except the last subsection comes from Giroux's PhD thesis [8], see also the webpage of Daniel Mathews for his translation of that paper into English.

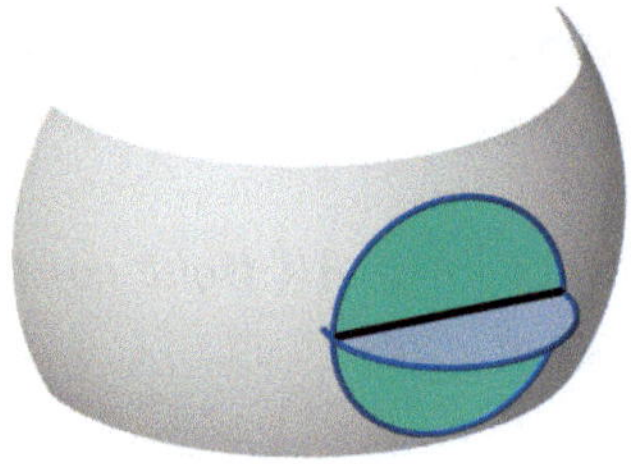

Fig. 11. Characteristic foliation of a surface as the intersection between the tangent space and the contact plane

4.1. Characteristic Foliations of Surfaces

After the local theory which explains what happens in neighborhoods of points in contact manifolds, we want to start the semi-local theory which deals with neighborhoods of surfaces.

The main tool will be characteristic foliations. The basic idea is to look at the singular foliation given on a surface S by the line field $TS \cap \xi$, see Figure 11.

In order to define precisely what is a line field with singularities, we see them as vector fields whose scale has been forgotten. It means they are equivalence classes of vector fields where $X \sim Y$ if there is a positive function f such that $X = fY$. A singularity is then a point where some, hence all, representative vanishes. Note that f should be positive everywhere, including singularities.

One can think of a line as the kernel of a linear form rather than a subspace spanned by a vector. This prompts an equivalent definition as an equivalence class of 1-forms where $\alpha \sim \beta$ if there is a positive function f such that $\alpha = f\beta$.

To go from one point of view to the dual one, we can use an area form ω on the surface. The correspondence between vector fields and 1-forms is then given by $X \mapsto \beta := \iota_X\omega$. The singular foliations $[X]$ defined by X and $[\beta]$ defined by β are indeed geometrically the same since X and β vanish at the same points and elsewhere X spans $\ker\beta$. In addition, one has the following commutative diagram which will be useful later.

$$\begin{array}{ccc} \text{vector fields} & \xrightarrow[\iota_\bullet\omega]{\sim} & \text{1-forms} \\ {\scriptstyle \mathrm{div}}\downarrow & & \downarrow{\scriptstyle d} \\ \text{functions} & \xrightarrow[\bullet\omega]{\sim} & \text{2-forms} \end{array} \tag{4.1}$$

The left-hand side vertical arrow is the divergence map defined by the equality $\mathcal{L}_X\,\omega = (\operatorname{div} X)\omega$. So positive divergence means the flow of X expands area while negative divergence means area contraction. Divergence is not well defined for a singular foliation because it depends on the representative vector field. However, at a singularity of a foliation, the sign of divergence is well defined because

$$\mathcal{L}_{fX}\,\omega = df \wedge \iota_X\omega + f(\operatorname{div} X)\omega$$

so, at points where X vanishes, $\operatorname{div} fX = f \operatorname{div} X$. The same kind of computation proves that this sign doesn't depend on the choice of the area form within a given orientation class.

Definition 10. Let S be an oriented surface in a contact manifold (M, ξ) with $\xi = \ker\alpha$, co-oriented by α. The characteristic foliation ξS of S is the equivalence class of the 1-form $\iota^*\alpha$ induced by α on S.

In particular, singularities of the characteristic foliation ξS are points where $\xi = TS$ (maybe with reversed orientation). At those points $d\iota^*\alpha = d\alpha_{|\xi}$ is non-degenerate so the above commutative diagram proves that singularities of characteristic foliations have non-zero divergence.

Examples. Figures 12, 13 and 14 show examples of characteristic foliations.

Leaves of Characteristics Foliations. The leaves (or orbits) of a singular foliation are the integral curves of any vector field representing it. The intuitive notion of a singular foliation is rather the data of leaves than an equivalence class of vector fields. In contact geometry, this discrepancy does not generate any confusion thanks to the following lemma. It is a rather technical point but we discuss it here anyway because it doesn't appear to be published anywhere else, although it is mentioned in [9, page 629].

Lemma 11 (Giroux). *If two singular foliations on a surface have the same leaves and if their singularities have non-zero divergence then they are equal.*

The following proof can be safely skipped on first reading.

Proof. The statement is clear away from singularities and a partition of unity argument brings it down to a purely local statement. So we focus on a neighborhood of a singularity (which may be non-isolated though).

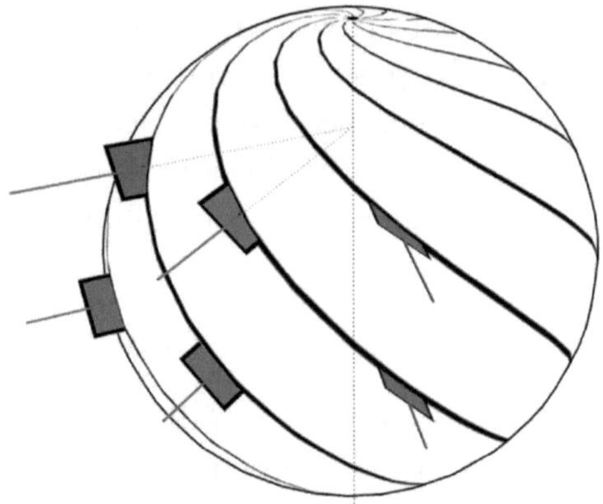

Fig. 12. Characteristic foliation of Euclidean spheres around the origin in $\mathbb{R}^3$ equipped with the canonical contact structure $\xi = \ker(dz + r^2 d\theta)$. There are singular points at the intersection with the z-axis and all regular leaves go from a singularity to the other one

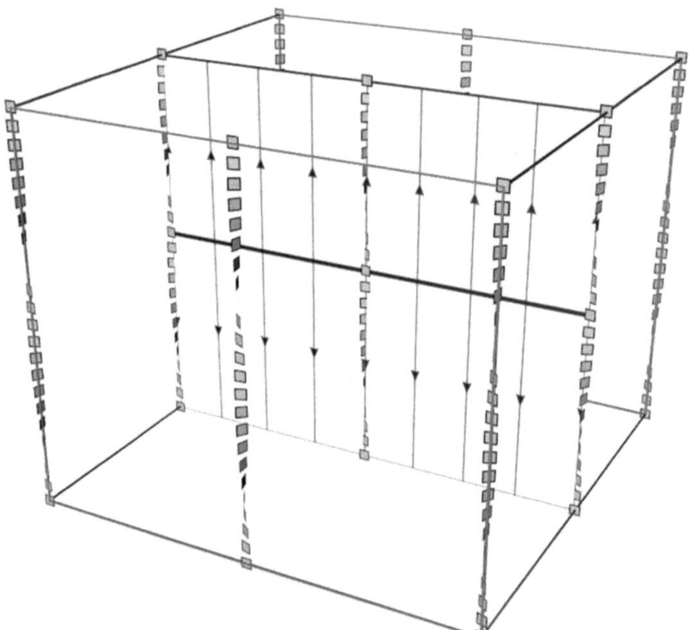

Fig. 13. Characteristic foliation of a torus $\{x = \text{constant}\}$ in T^3 equipped with its canonical contact structure $\xi = \ker(\cos(z)dx - \sin(z)dy)$. One can see two circles made entirely of singularities where $\sin(z) = 0$, one appear in the middle of the picture and the other one can be seen both at bottom and at top

Let Y and Y' be vector fields on $\mathbb{R}^2$ which vanish at the origin and have the same orbits.

$$Y = f\partial_x + g\partial_y \quad \text{and} \quad Y' = f'\partial_x + g'\partial_y.$$

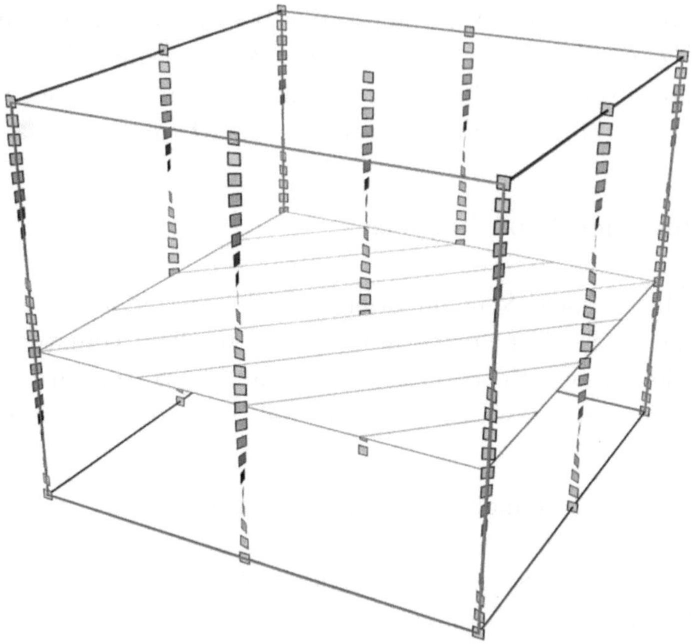

Fig. 14. Characteristic foliation of a torus $\{z = \text{constant}\}$ in T^3 equipped with its canonical contact structure $\xi = \ker(\cos(z)dx - \sin(z)dy)$

We will compute divergence using the Euclidean area form $\omega = dx \wedge dy$ (we know the sign of divergence of singular points does not depend on this choice). So $\operatorname{div} Y = \partial_x f + \partial_y g$. All the following assertions will be true in a neighborhood of the origin that will shrink only finitely many times. Since $\operatorname{div}(Y)$ is non-zero, we can use a linear coordinate change to ensure that $\partial_x f$ doesn't vanish. The implicit function theorem then gives new coordinates such that $f(x, y) = x$. Because

$$f'(x,y) = f'(0,y) + x\int_0^1 \partial_x f'(tx,y)dt$$

we can write $f' = xu(x,y) + v(y)$. Along the curve $\{x = 0\}$, the vector field Y is vertical (or zero) so the same is true for Y'. Hence f' also vanishes along this curve and v is identically zero. The condition that Y and Y' are either simultaneously zero or colinear is then:

$$\begin{vmatrix} x & xu \\ g & g' \end{vmatrix} = 0$$

which gives $g' = ug$ where x is non-zero hence everywhere by continuity. One then gets $Y' = uY$. In particular $\operatorname{div} Y' = u \operatorname{div} Y + du \wedge (\iota_Y dx \wedge dy)$.

Away from zeros of Y and Y', u is positive because Y and Y' have the same leaves. At a common zero, $\operatorname{div} Y' = u \operatorname{div} Y$ and, because singularities of Y' have non-zero divergence, the function u doesn't vanish. Hence it is positive everywhere (note that Y and Y' can't be everywhere zero). ■

4.2. Neighborhoods of Surfaces

Any orientable surface S in an orientable 3-manifold has a neighborhood diffeomorphic to $S \times \mathbb{R}$ (use the flow of a vector field transverse to S). We will always denote by t the coordinate on $\mathbb{R}$ and by S_t the surface $S \times \{t\}$ for a fixed t. From now on, we will assume that S is oriented and orient $S \times \mathbb{R}$ as a product.

Any plane field ξ defined near S has then an equation $\alpha = u_t dt + \beta_t$ where u_t is a family of functions on S and β_t is a family of 1-forms on S. Note that the characteristic foliation of S_t is the equivalence class of β_t since the latter is the 1-form induced by α on S_t.

The contact condition for ξ (with respect to the product orientation) is equivalent to

$$(\star) \qquad u_t d\beta_t + \beta_t \wedge (du_t - \dot{\beta}_t) > 0$$

where $\dot{\beta}_t$ denotes $\frac{\partial \beta_t}{\partial t}$. This condition is a non-linear partial differential relation which is not so simple. The main thrust of the following discussion will be to simplify it by fixing some of the terms.

4.3. Reconstruction Lemmas

The easiest case is to fix the whole family β_t. In this case the contact condition $(\star)$ is only about the family u_t and becomes convex. In particular the space of solutions u_t is connected and we get:

Lemma 12 (Global reconstruction). *If ξ and ξ' are positive contact structures on $S \times \mathbb{R}$ such that $\xi S_t = \xi' S_t$ for all t then ξ and ξ' are isotopic.*

We give a detailed proof since it is a model of several later proofs.

Proof. There are equations $u_t dt + \beta_t$ and $u'_t dt + \beta'_t$ of ξ and ξ'. The hypothesis of the lemma is that $\beta'_t = f_t \beta_t$ for some family of positive functions f_t on S. So another equation for ξ' is $u'_t / f_t dt + \beta_t$. We have two solutions u_t and

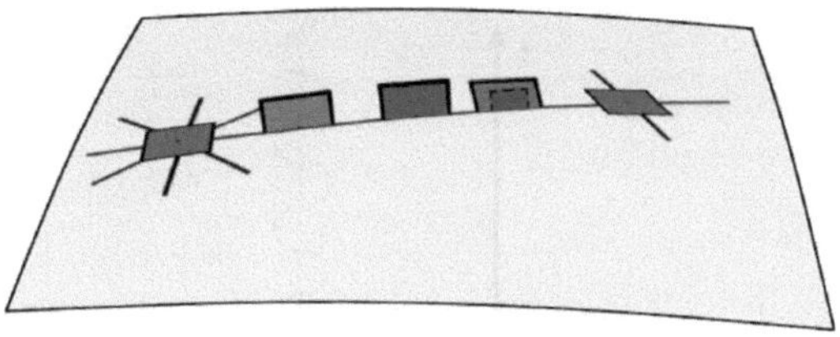

Fig. 15. Reconstruction lemmas. We have two contact structures printing the same characteristic foliation on a surface. One of them is drawn along an arc going from a singularity to another. The second one appears only at one point with dotted outline. At this point the isotopy constructed in the proof is tangent to the arc to make the contact structure rotate

u_t'/f_t of the contact condition, Equation $(\star)$, with β_t fixed. Since this condition is convex, the space of its solutions is connected so we can find a family of solution $(u_t^s)_{s\in[0,1]}$ relating them (a linear interpolation will do the job). This family corresponds to a family of contact structures $\xi_s = \ker(u_t^s dt + \beta_t)$ which Gray's theorem (Theorem 4) converts to an isotopy of contact structures[1].

Our discussion of Gray's theorem actually tells us more about what is going on. Recall the vector field generating the isotopy at time s can be chosen in the intersections of ξ_s and $\xi_{s+\varepsilon}$. So we see the isotopy is stationary at each singular point of the characteristic foliations $[\beta_t]$. At all other points it is tangent to the characteristic foliation and its flow makes the contact structures we want to relate to rotate toward each other, see Figure 15. ■

If instead of fixing the whole family β_t we fix only β_0 then we get the following lemma.

Lemma 13 (Local reconstruction). *If ξ and ξ' are positive contact structures which prints the same characteristic foliation on a closed embedded surface S then there is a neighborhood of S on which ξ and ξ' are isotopic (by an isotopy globally preserving S).*

Proof. The contact condition along S_0 becomes a convex condition on u_0 and $\dot\beta_0$. Again we can find a path of plane fields which, *along S*, are contact structures interpolating between ξ and ξ'. Because the contact condition is

[1]One may worry about the fact that $S \times \mathbb{R}$ is non-compact but here the vector field constructed during the proof of this theorem is tangent to S_t which is compact for all t hence its flow is well defined for all times.

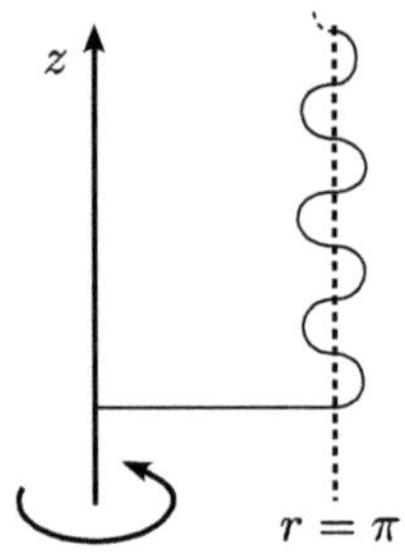

Fig. 16. Rotating the wavy curve around the z-axis in $(\mathbb{R}^3, \xi_{\mathrm{OT}})$ gives a plane having a characteristic foliation diffeomorphic to that of $\{z = 0\}$. Note that the curve is horizontal at each intersection with the $\{r = \pi\}$ axis

open, they will stay contact structures near S and we can use Gray's theorem again. ■

Exercise. Prove that the two preceding lemmas are false for foliations.

We can now return to the challenging exercise of Section 3 with much better chances of success. Recall that $\xi_{\mathrm{OT}} = \ker(\cos(r)dz + r\sin(r)d\theta)$.

Exercise. Use the local reconstruction lemma to prove that any neighborhood of an overtwisted disk in a contact manifold contains a copy of $(\mathbb{R}^3, \xi_{\mathrm{OT}})$. Hint: try to understand the characteristic foliation of the surface of Figure 16.

As illustrated by the previous exercise, the reconstruction lemmas are already quite useful by themselves. But the characteristic foliation is still a huge data and it is very sensitive to perturbations of the contact structure or the surface. This will be clear from the discussion of genericity of convex surfaces and of the realisation lemma below.

4.4. Convex Surfaces

Homogeneous Neighborhoods. The next step in our quest to simplify the contact condition $(\star)$ seems to be fixing u_t instead of β_t. But this still gives a non-linear equation on the family β_t if $\dot{\beta}_t$ is not zero. So we assume that β_t does not depend on t: $\beta_t = \beta$. In particular the families (u_0, β) and (u_t, β) both give contact structures with the same characteristic foliation $[\beta]$ on each S_t. Hence the global reconstruction Lemma tells us these contact structures are isotopic. So we now assume that u_t is also independent of t.

In this situation, the contact structure itself becomes invariant under $\mathbb{R}$ translations, one says that ∂_t is a contact vector field. Note that this vector field is transverse to all surfaces S_t. Conversely if a contact vector field is transverse to a surface then it can be cut-off away from the surface using Remark 6 and then its flow defines a tubular neighborhood $S \times \mathbb{R}$ with a t-invariant contact structure.

Definition 14 (Giroux [8]). A surface S in a contact 3-manifold (M, ξ) is ξ-convex if it is transverse to a contact vector field or, equivalently, if it has a so called homogeneous neighborhood: a tubular neighborhood $S \times \mathbb{R}$ where the restriction of ξ is $\mathbb{R}$-invariant.

Example 15. In T^3 with its canonical contact structure, all tori $\{x = \text{constant}\}$ as in Figure 13 are ξ-convex since they are transverse to the contact vector field ∂_x.

Example 16. In $(\mathbb{R}^3, \ker(dz + r^2 d\theta))$, any Euclidean sphere around the origin is ξ-convex since they are transverse to the contact vector field $x\partial_x + y\partial_y + 2z\partial_z$.

In the convex case, the contact condition becomes:

$$(\dagger) \qquad u d\beta + \beta \wedge du > 0$$

Using some area form ω and Equation (4.1), one can rephrase it in terms of the vector field Y ω-dual to β as:

$$(\dagger') \qquad u \operatorname{div}_\omega Y - du(Y) > 0$$

Analogously to the previous section we see that, u being fixed, the space of solutions β to $(\dagger)$ is contractible, this was our stated goal when we asked β_t to be independent of t. The miracle is that it essentially stays true if one fixes only the zero set Γ of u. Indeed, away from Γ, we can divide our contact form $udt + \beta$ by $|u|$ to replace it by $\ker(\pm dt + \beta')$ where $\beta' = \frac{1}{|u|}\beta$. The condition $(\dagger)$ for $(\pm 1, \beta')$ is simply $\pm d\beta' > 0$ which is not only convex, it does not depend on u! Of course this discussion needs some precise definitions which are provided below but the first miracle has already happened: near a ξ-convex surface S, all the information about ξ is contained in Γ. It remains to see that such surfaces are generic, the second miracle.

Dividing Sets. Let us take a look at $\Gamma = \{u = 0\}$. Along Γ, the contact condition $(\dagger')$ reads $-du(Y) > 0$. So Γ is a regular level set of u. Hence it is

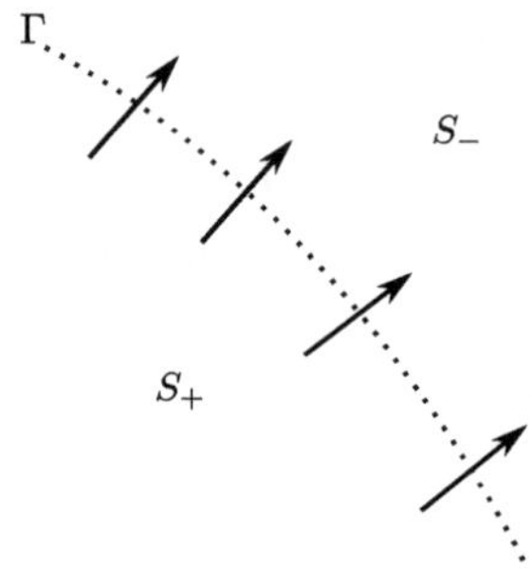

Fig. 17. Characteristic foliation near the dividing set Γ

a one-dimensional submanifold without boundary, i.e. a collection of disjoint simple closed curves in S. Such collections will be referred to as multi-curves.

The condition $-du(Y) > 0$ also implies that Γ is transverse to ξS. More precisely, Y goes from $S_+ = \{u > 0\}$ to $S_- = \{u < 0\}$ along Γ and the picture near Γ is always as in Figure 17. In the following discussion we will use several time the fact that this picture is very simple and controlled to be less precise about what happens near Γ.

The last remarkable property of the decomposition of S in S_+ and S_- is Y expands some area form in S_+ and contracts it in S_-. Indeed, if one sets $\Omega = \frac{1}{|u|}\omega$ on $S \setminus \Gamma$ then $\mathrm{div}_\Omega Y = \pm\frac{1}{u^2}$ on $S_\pm$. One can actually modify Ω near Γ so that $\mathrm{div}_\Omega Y$ is positive on S_+, negative on S_- and vanishes along Γ.

Definition 17. A singular foliation $\mathscr{F}$ of a surface S is divided by an (embedded) multi-curve Γ if there is some area form Ω on S and a vector field Y directing $\mathscr{F}$ such that:

- the divergence of Y does not vanish outside Γ—we set

$$S_\pm = \{p \in S;\ \pm\mathrm{div}_\Omega Y(p) > 0\}$$

- the vector field Y goes transversely out of S_+ and into S_- along Γ.

What we proved above is that the characteristic foliation of a ξ-convex surface is divided by some multi-curve. Using the local reconstruction lemma (Lemma 13), one can prove the converse to get:

Proposition 18. *A surface S is ξ-convex if and only if ξS is divided.*

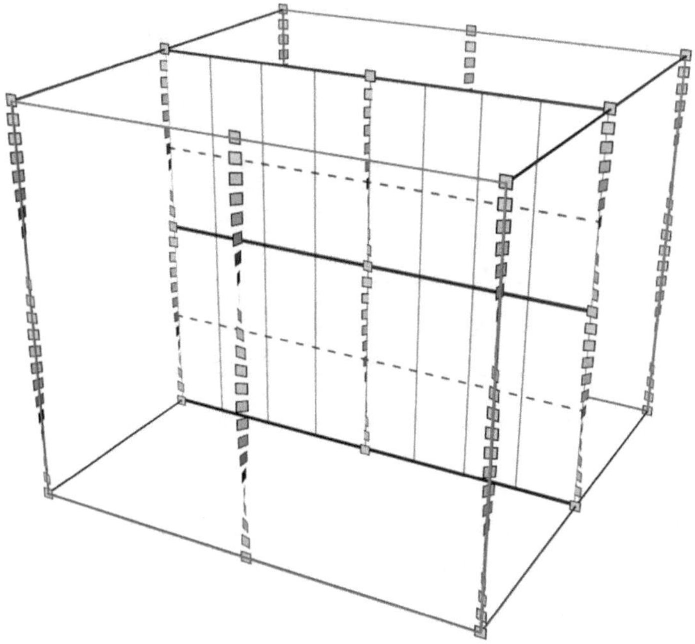

Fig. 18. A dividing set for the torus of Figure 13 (dashed on the picture)

Proof. We assume that ξS is divided by some multi-curve Γ. According to the local reconstruction lemma, we only need to prove that there is a contact structure ξ' defined near S such that S is ξ'-convex and $\xi' S = \xi S$. We set $\beta = \iota_Y \Omega$. In particular $\xi S = [\beta]$. On $S \setminus \Gamma$, $\xi' = \ker \pm dt + \beta$ is a contact structure which also prints $[\beta]$ on $S \setminus \Gamma$ and one can check that there is no problem to extend it along Γ. ■

Note that the dividing set is not unique for a given foliation. If X is a contact vector field transverse to the surface S then the considerations above prove that $\Gamma_X := \{s \in S;\ X(s) \in \xi\}$ is a dividing set for S.

However, if one fixes β in the contact condition (†), it becomes convex in u, hence the space of solutions u is connected. This implies that the space of multi-curves dividing a given foliation is connected (in fact contractible).

Examples. In the case of spheres of Example 16, the dividing set corresponding to the given vector field is the equator $\{z = 0\}$.

In the torus case of Figure 13, the dividing set coming from ∂_x is defined by $\cos(z) = 0$ so it is made of two circles sitting between the singularity circles defined by $\sin(z) = 0$, see Figure 18.

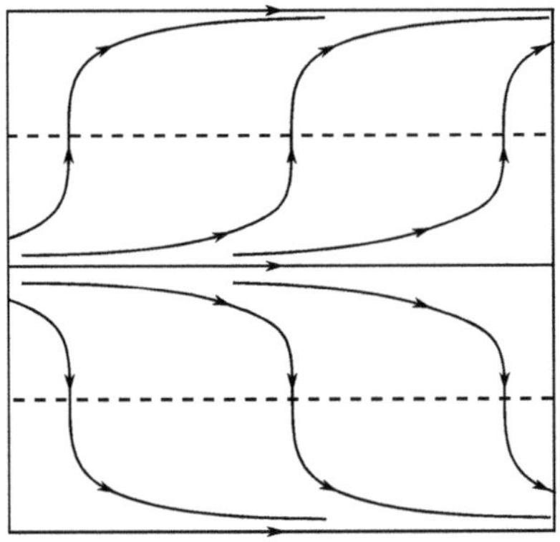

Fig. 19. A generic foliation of the torus divided by two curves

The Realization Lemma. We are now ready to make precise the fact that the dividing set contains all the information about the contact structure near a convex surface.

Lemma 19 (Realization Lemma). *Let S be a ξ-convex surface divided by some multi-curve Γ. For any singular foliation $\mathscr{F}$ divided by Γ, there is an isotopy δ_t with support in an arbitrarily small neighborhood of S and such that $\xi' = \delta_1^* \xi$ satisfies $\xi' S = \mathscr{F}$. Equivalently, one has $\xi \delta_1(S) = \delta_1(\mathscr{F})$.*

So any singular foliation divided by Γ is printed on S by some contact structure isotopic to ξ or, equivalently, it can be realized as the characteristic foliation of a surface isotopic to S.

The proof of this very important lemma has already been essentially explained right after stating condition (†). It follows from the fact that $\pm d\beta > 0$ is a convex condition and Gray's theorem as in the reconstruction lemmas.

This lemma is often called Giroux's flexibility theorem but one can argue that it is rather a rigidity result since all the information can be stored into a tiny combinatorial data: the isotopy class of the dividing set.

Example. Consider the convex torus of Figure 18. Its characteristic foliation is highly non generic since it has two circles of singularities. Yet it is divided by two circles parallel to the singularity circles. Figure 19 shows a generic foliation divided by the same curves but where singular circles have been replaced by regular closed leaves.

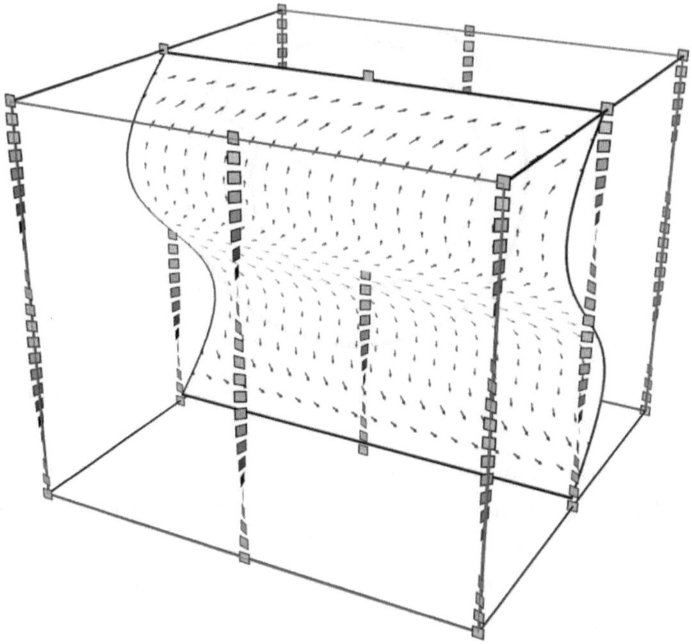

Fig. 20. A realization of Figure 19 as a deformation of the torus of Figure 13

The realization lemma implies that the surface of Figure 13 is isotopic to a surface which has Figure 19 as its characteristic foliation. Figure 20 shows this surface explicitly.

The transition between these foliations play an important role in the classification of tight contact structures on the product of a torus and an interval, see [9, Section 1.F].

In order to use the power of the realization lemma, we need to prove that ξ-convex surfaces exist in abundance. We will first discuss some obstructions to ξ-convexity then prove genericity of ξ-convex surfaces.

4.5. Obstructions to Convexity

Degenerate Closed Leaves. The most obvious obstruction to ξ-convexity for a closed surface S is when ξS is defined by some β with $d\beta = 0$, as in Figure 14, because then the contact condition (†) becomes $\beta \wedge du > 0$ which implies that u has no critical point.

Surfaces with such characteristic foliations are called pre-Lagrangian. They are either tori or Klein bottles and play an important role in some later part of the theory.

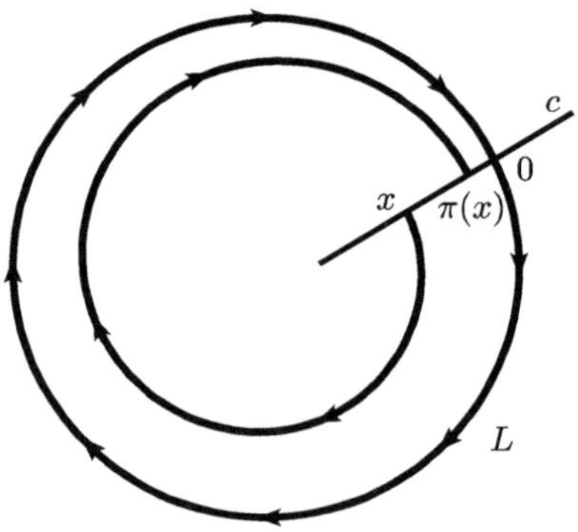

Fig. 21. Poincaré's first return map π on a transversal c to a closed leaf L

This obstruction idea can be extended remarking that it does not need the whole of S, it can be applied along a closed leaf L of ξS. This is easier to see in the dual picture of equation (†′). Indeed, if $\mathrm{div}_\omega(Y)$ vanishes along L, condition (†′) says that $-u'_{|L} > 0$ whereas the restriction $u_{|L}$ necessarily has some critical point.

Definition 20. A closed leaf L of a singular foliation is degenerate if there is a 1-form β defining the foliation near L and whose differential $d\beta$ vanishes along L. A non-degenerate leaf is called repelling (resp attracting) if there is some β such that $d\beta$ is positive (resp negative) along L.

The definition above is convenient for our purposes but one should keep in mind that it is equivalent to the more geometrical definition through Poincaré's first return map π on a transverse curve c, see Figure 21. A closed leaf is degenerate if $\pi'(0) = 1$. See Figures 22 and 23. A non-degenerate closed leaf is attracting if $\pi'(0) < 1$ and repelling if $\pi'(0) > 1$.

The discussion preceding the definition proves that if S is ξ-convex then ξS has no degenerate closed leaves.

Remark 21. Suppose now that S is indeed ξ-convex and L is a (non-degenerate) closed leaf of ξS. Let Γ be a dividing set for ξS. Because ξS is transverse to Γ and always goes out of S_+ and into S_-, L cannot meet Γ. Because L is compact, the restriction of u to L has at least one critical point. At this point, the contact condition gives $ud\beta > 0$. So repelling orbits are in S_+ and attracting orbits are in S_-.

Retrograde Connections. Recall from Section 4.1 that the contact condition ensures that all singularities of characteristic foliations have non-zero

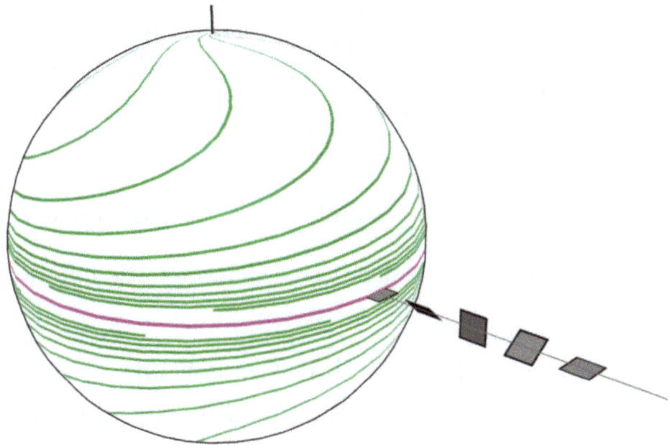

Fig. 22. A sphere or radius π in the overtwisted $\mathbb{R}^3$. The equator is a degenerate closed leaf. Note how leaves spiral a lot more around a degenerate leaf than around a non-degenerate

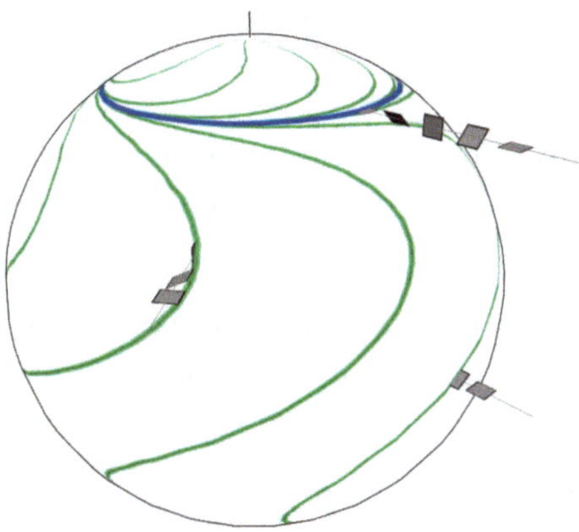

Fig. 23. A sphere or radius slightly less than 2π in the overtwisted $\mathbb{R}^3$. The intersection with the cylinder $\{r = \pi\}$ consists of two non-degenerate closed leaves (one of them is not visible in the picture)

divergence and hence have non-zero sign. Singularities of ξS correspond to points where S is tangent to ξ and they are positive or negative depending on whether the orientation of ξ and S match or not.

In generic characteristic foliations one sees only two topological types of singularities: nodes and saddles. If one considers generic families of characteristic foliations then saddle-nodes may appear, see Figure 24. Since the sign of singularities corresponds to their divergence, positive nodes are always sources while negative nodes are always sinks. The sign of saddles cannot be read from topological pictures only.

Fig. 24. Generic singularities of characteristic foliations

Let S be a ξ-convex surface so that $\xi = \ker(udt + \beta)$ near S. We begin by a remark analogous to Remark 21. At any singular point p of ξS, the contact condition (†) give $ud\beta(p) > 0$. So singularities are positive in S_+ and negative in S_-.

Suppose now that p and q are two singular points of ξS with opposite signs and there is a regular leaf L of ξS going from p to q. Because L has to be transverse to Γ and go from S_+ to S_-, the above discussion proves that p is positive and q is negative.

Definition 22. In the characteristic foliation of a surface, a retrograde connection is a leaf which goes from a negative singularity to a positive one.

The discussion above proves that ξ-convex surfaces have no retrograde connections. Note that retrograde connections cannot involve nodes since the sign of nodes determine the local orientations of the foliation.

Leaves of characteristic foliations between two singularities of opposite signs are always arcs tangent to the contact structure along which the contact structure rotates half a turn compared to the surface. What makes retrograde connections special is that the direction of rotation is opposite to the one around Legendrian foliations.

Example 23 ([9, Example 3.41]). In $\mathbb{R}^2 \times \mathbb{S}^1$ with contact structure $\xi = \ker(\cos(2\pi z)dx - \sin(2\pi z)dy)$, we consider the family of transformations

$$\varphi_t\big((x, y), z\big) = \big(R_{-4\pi t}(x, y), z + t\big)$$

where R_θ denotes the rotation of angle θ around the origin of $\mathbb{R}^2$. The orbit of a circle in $\mathbb{R}^2$ passing through the origin sweeps a torus S whose characteristic foliation has two retrograde saddle connections along the z-axis, see Figure 25. Indeed, along this axis, the tangent plane TS turns in the same direction as ξ but twice as fast. It means that, seen from TS, ξ rotates

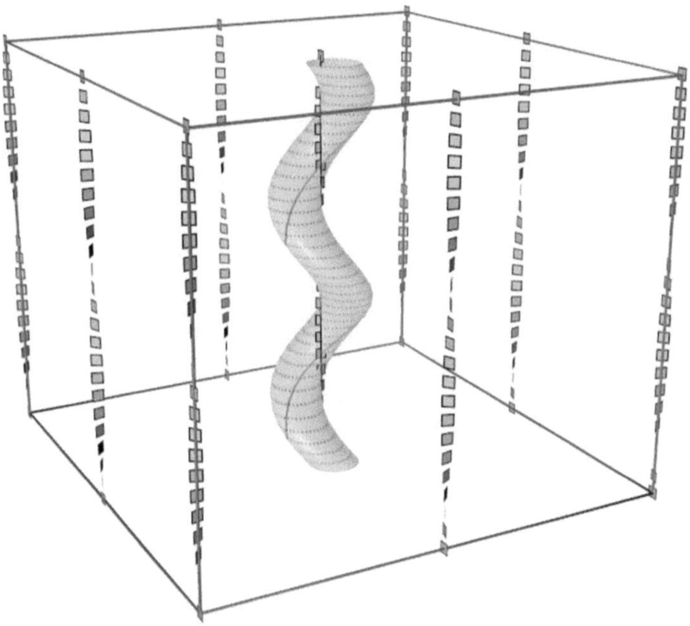

Fig. 25. A torus having a retrograde saddle connection

one turn in the opposite direction. See Figure 26 for a better view of the characteristic foliation.

4.6. Genericity of Convex Surfaces

We are now ready to use generic properties of vector fields on surfaces to prove that any surface in a contact manifold can be perturbed to a ξ-convex one. See Figures 27 and 28 for an example and [10, Proof of Proposition 2.10] for more examples of the same kind.

Proposition 24. *Any closed surface in a contact* 3*-manifold* (M, ξ) *is* C^∞*-close to a* ξ*-convex surface.*

Genericity of ξ-convex surfaces is a small dimensional phenomenon, it does not hold for hypersurfaces in higher dimensions [15]. In dimension 3, ξ-convexity is a degenerate notion, much like ordinary convexity in real dimension 1 and pseudo-convexity in complex dimension 1.

We first prove that any foliation sufficiently close to a characteristic foliation $\xi_0 S$ is the characteristic foliation ξS coming from some ξ isotopic to ξ_0.

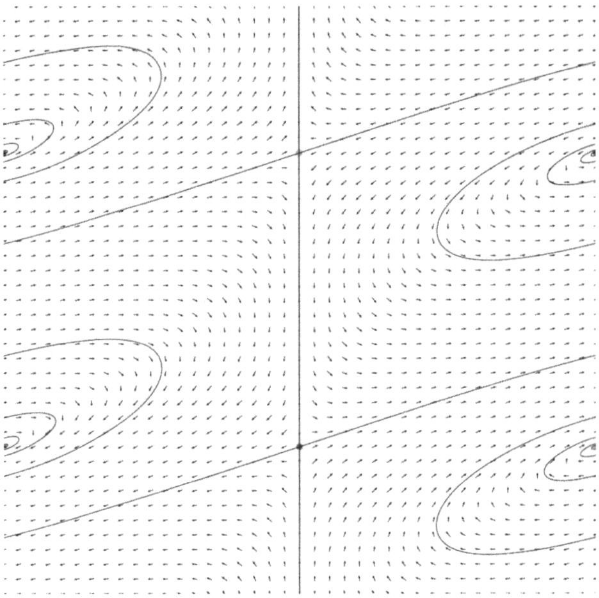

Fig. 26. A (double) saddle connection on the torus of Figure 25 after top/bottom and left/right are glued. The top saddle is negative, the bottom one positive. The top node is positive, the bottom one negative. The curves drawn are all the separatrices of the saddles

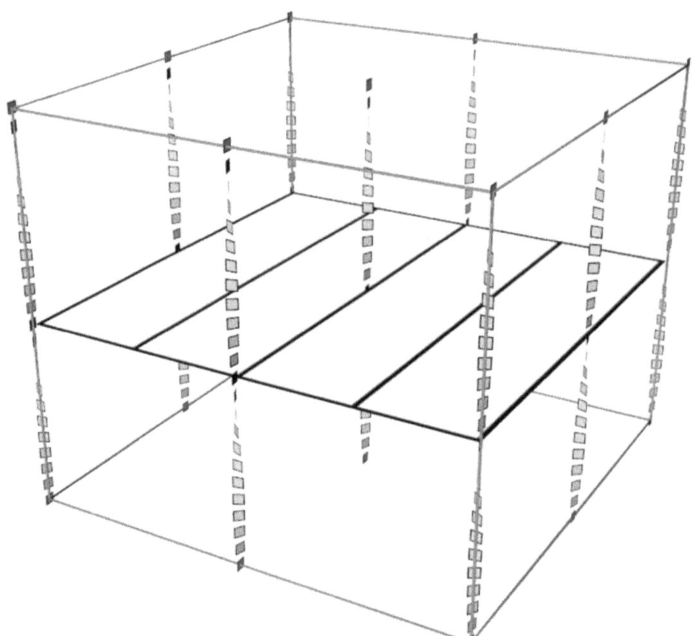

Fig. 27. A non-convex torus

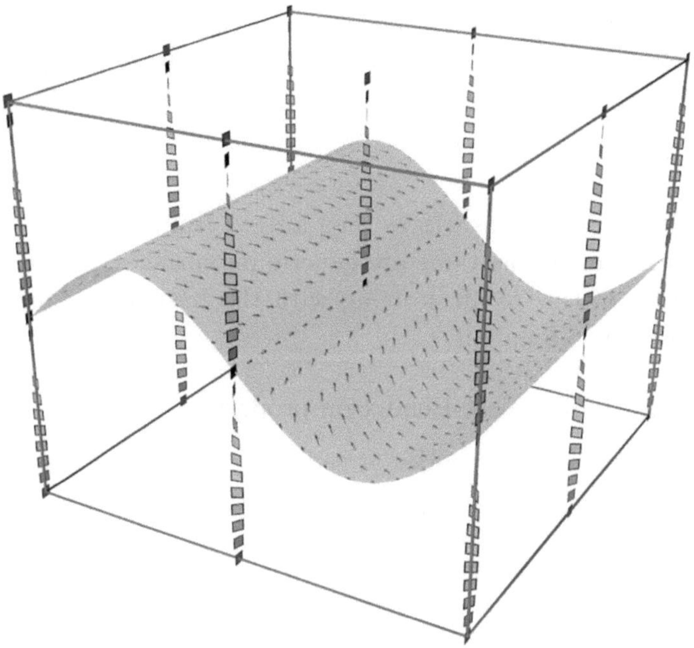

Fig. 28. Perturbation of the non-convex torus of Figure 27 into a convex torus

Equivalently it means it is the characteristic foliation printed by ξ_0 on some surface isotopic to S. Let C be the connected component of the space of contact structures which contains ξ_0. The first point is that the map which maps ξ in C to the characteristic foliation ξS is open. The second point is that Gray's theorem imply that all ξ in C are isotopic to ξ_0.

So the genericity of ξ-convex surfaces will follow from the one of divided foliations. Essentially we will see that the obstructions to the existence of a dividing set discussed above are the only ones provided that no non-trivial recurrence appear. The precise requirement is expressed in the following definition.

Definition 25. A singular foliation on a closed surface satisfies the *Poincaré–Bendixson property* if the limit set of any half orbit is either a singularity or a closed orbit or a union of singularities and orbits connecting them.

The Poincaré-Bendixson theorem thus says that a singular foliation on a sphere satisfies the Poincaré-Bendixson property as soon as its singularities are isolated, see e.g. [16].

Proposition 26. *Let S be a surface in a contact manifold (V, ξ). If the characteristic foliation ξS satisfies the Poincaré–Bendixson property then S is ξ-convex if and only if ξS has neither degenerate closed leaves nor retrograde connections.*

Genericity of ξ-convex surfaces then follows from Peixoto's theorem stating that Morse-Smale foliations are generic on surfaces, see [16] for a beautiful exposition of this result starting with the basic of dynamical systems. A foliation is Morse-Smale if

- it satisfies the Poincaré-Bendixson property,
- all its singularities are nodes or saddles,
- all its closed leaves are non-degenerate,
- it has no saddle connections.

Proof of Proposition 26. In the preceding sections, we have seen that the absence of degenerate closed leaves and retrograde connections is necessary for convexity.

We now prove that it is sufficient when the Poincaré-Bendixson property holds. In this proof we assume that all singularities are nodes, saddles or saddle-nodes. This is true for generic families of characteristic foliations with any number of parameters and is all we need in these lectures. In order to save some more words we will even pretend there are no saddle-nodes. The reader can replace any occurrence of the word "saddle" by "saddle or saddle-node" to get the more general proof.

During the discussion of obstructions to convexity, we have seen that singularities and closed leaves should be dispatched into S_+ or S_- according to their signs. Another constraint comes from separatrices of saddles: since we want the characteristic foliation to go transversely out of S_+ along Γ, stable separatrices of positive saddles and unstable separatrices of negative saddles cannot meet Γ.

So we build a subsurface S'_+ of S by putting a small disk around each positive singularity and narrow bands around positive closed leaves and stable separatrices of positive saddles. If all these elements are sufficiently small, the boundary of S'_+ can be smoothed to a curve transverse to the characteristic foliation, see Figure 29. In addition one can find an area form on S'_+ which is expanded by $\xi S'_+$. We can construct similarly a subsurface S'_- and

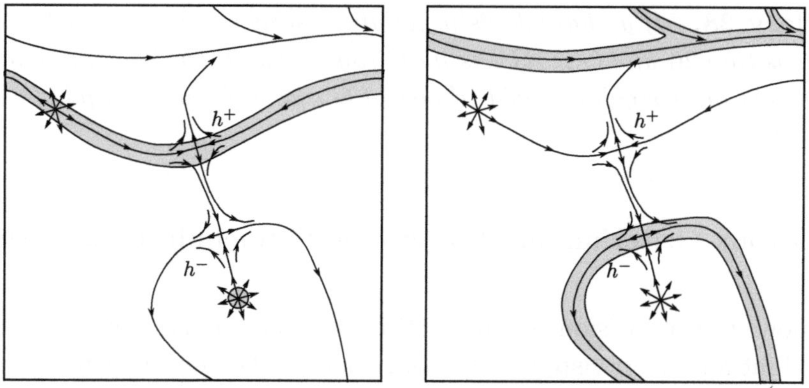

Fig. 29. Construction of a dividing set on a torus. One can check that $\partial S'_+$ and $\partial S'_-$ are indeed isotopic among dividing curves

a contracted area form on it. None of these subsurfaces is empty because of Stokes' theorem which guaranties that an area form on a closed surface is never exact.

Let A be a component of the complement of $S'_+ \cup S'_-$ in S. It has non-empty boundary and does not contain any singularity so A is an annulus. In addition it does not contain any closed leaf so Poincaré-Bendixson's theorem guaranties that all leaves of the characteristic foliation entering A along some boundary component leave it through the other boundary component. So we are indeed in the situation of Figure 17 and one can take the core of A as a dividing curve. The corresponding subsurfaces $S_\pm$ then retract onto $S'_\pm$. ■

The proof above contains some useful information about how a dividing set can be recovered from the important features of the characteristic foliation so we record this in a definition and a corollary.

Definition 27. Given a foliation $\mathscr{F}$ satisfying the Poincaré-Bendixson property, we denote by G_+ (resp G_-) the union of repelling (resp attracting) closed leaves, of positive (resp negative) singularities and of the stable (resp unstable) separatrices of these singularities. The union $G_+ \cup G_-$ is called the Giroux graph of $\mathscr{F}$.

Note that the terminology graph is a little stretched since one can have separatrices accumulating on closed orbits (like in Figure 29) or on connected singularities so the Giroux graph equipped with the induced topology is not necessarily homeomorphic to a CW-complex of dimension one.

Corollary 28. *If a characteristic foliation satisfies the convexity criterion of Proposition* 26 *and* $G_+ \cup G_-$ *is its Giroux graph then, for any dividing set,* S_+ *retracts on a regular neighborhood of* G_+ *and* S_- *on a regular neighborhood of* G_-.

4.7. Giroux Criterion and Eliashberg–Bennequin Inequalities

Until now, the discussion of this section does not make any distinction between tight and overtwisted contact structures. We now start to discuss how convex surfaces theory sees tightness.

Theorem 29 (Giroux criterion [10, Theorem 4.5a]). *In a contact manifold* (V, ξ), *a* ξ*-convex surface divided by some multi-curve* Γ *has a tight neighborhood if and only if one of the following conditions is satisfied:*

- *no component of* Γ *bounds a disk in* S
- S *is a sphere and* Γ *is connected.*

The only application of this theorem we will present in detail is in the classification of tight contact structures on $\mathbb{S}^3$ (existence by Bennequin and uniqueness by Eliashberg). There we will only need that, if S is a sphere, then it has a tight neighborhood only if its dividing set is connected. So we prove only this part of the theorem, we assume S is a sphere and Γ is not connected. Let S' be a component of $S \setminus \Gamma$ which is a disk and denote by γ its boundary. Let S'' be the other component containing γ in its boundary. Since Γ is not connected, S'' has more boundary components. Using this, one can construct a foliation $\mathscr{F}$ on S which is divided by Γ, has a circle of singularities L in S'', is radial inside a disk bounded by L and coincides with ξS outside $S' \cup S''$, see Figure 30. In any neighborhood U of S, the realization Lemma gives a surface $\delta_1(S)$ which has $\delta_1(\mathscr{F})$ as its characteristic foliation. Then $\delta_1(L)$ is the boundary of an overtwisted disk contained in $\delta_1(S)$ hence in U.

An important direct application of the Giroux criterion is Giroux's proof of the following constraint on the Euler class of a tight contact structure (originally due to Eliashberg). We will not use it in those notes but include it here since it now comes for free.

Theorem 30 (Eliashberg–Bennequin inequality [5]). *Let* (M, ξ) *be a 3-dimensional contact manifold. If* ξ *is tight and* S *is a closed surface embedded*

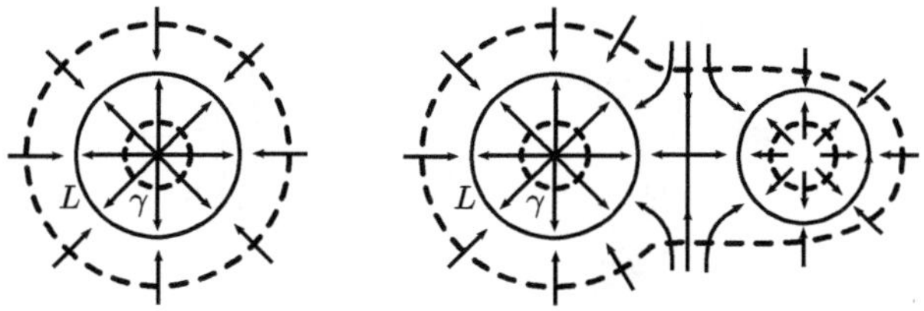

Fig. 30. Characteristic foliations for the Giroux criterion. The dividing set Γ is dashed. On the left-hand side one has the simplest case when S'' is an annulus. On the right hand-side one sees a possible foliation when S'' has one more boundary component (on the right). Note that the disk bounded by the small component of Γ on the right may contain more components of Γ. The extension to more boundary components uses the same idea

in M then the Euler class of ξ satisfies the following inequality:

$$\left|\langle e(\xi), S\rangle\right| \leq \max\big(0, -\chi(S)\big)$$

Proof. Using genericity of ξ-convex surfaces, one can homotop S until it is ξ-convex. This does not change the Euler class which can now be evaluated as $\chi(S_+)-\chi(S_-)$ since singularities are distributed among S_+ and S_- according to their signs. If S is a sphere then the Giroux criterion says that both S_+ and S_- are disks so $\langle e(\xi), S\rangle = 0$ and the inequality is proved. So suppose now that S has positive genus. The Giroux criterion says that no connected component of S_+ or S_- is a disk. This implies that both $\chi(S_+)$ and $\chi(S_-)$ are negative. Hence both $\chi(S_+) - \chi(S_-)$ and $-\chi(S_+) + \chi(S_-)$ are less than $-\chi(S_+) - \chi(S_-)$ which is $-\chi(S)$. ■

5. Bifurcations and First Classification Results

The goal of this section is to prove that any tight contact structure on $\mathbb{S}^3$ has to be isotopic to the standard contact structure and that the later is indeed tight. We will not give the original proofs due to Eliashberg [5] and Bennequin [1] respectively. We will rather use the technology of ξ-convex surfaces to prove them. These proofs were obtained by Giroux along its way towards more general classification results in [9]. The classification result is a comparatively easy special case of Giroux's preparation Lemma [9, Lemma 2.17] while the tightness result follows from the bifurcation lemmas [9, Lemmas 2.12 and 2.14].

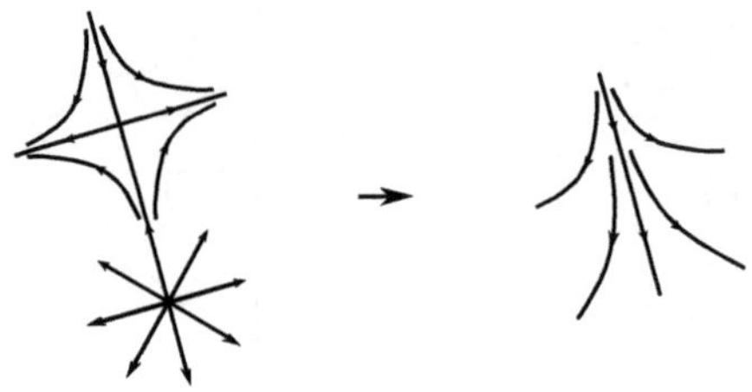

Fig. 31. Elimination of a pair of singular points

5.1. The Elimination Lemma

In the characteristic foliation of a surface, a saddle and a node are said to be in elimination position if they have the same sign and there is a leaf from one to the other. Such a leaf is called an elimination arc. Giroux's elimination lemma in its simplest form says one can perturb the surface to replace a neighborhood of the elimination arc by a region without singularity as in Figure 31.

For the classification of tight contact structures on $\mathbb{S}^3$ we will need a version of this process which keeps neighboring surfaces under control.

We do not need much control though and the following version is simpler than [9, Lemma 2.15] which is needed for the classification of tight contact structures on torus bundles.

Let ξ be a contact structure on $S\times[-1,1]$ and set $S_t := S\times\{t\}$. Suppose a node e_0 and a saddle h_0 are in elimination position on S_0. This configuration is stable so it persists for t in some interval $(-\varepsilon,\varepsilon)$. Let C_t denote a continuous family of elimination arcs between e_t and h_t on S_t.

Lemma 31 (Giroux elimination lemma). *Let δ be a positive number smaller than ε. Let U a neighborhood of $\bigcup_{|t|<\delta} C_t$ intersecting each S_t in a disk D_t whose characteristic foliation is as in the left hand side of Figure* 31. *One can deform ξ in U such that ξD_t has:*

- *no singular point when $|t| < \delta$,*
- *a saddle-node when $|t| = \delta$,*
- *a pair of singularities in elimination position when $|t| \in (\delta,\varepsilon)$.*

In addition, one can impose that separatrices facing the elimination arc are connected to the same points of ∂D_t as before the deformation, see Figure 33.

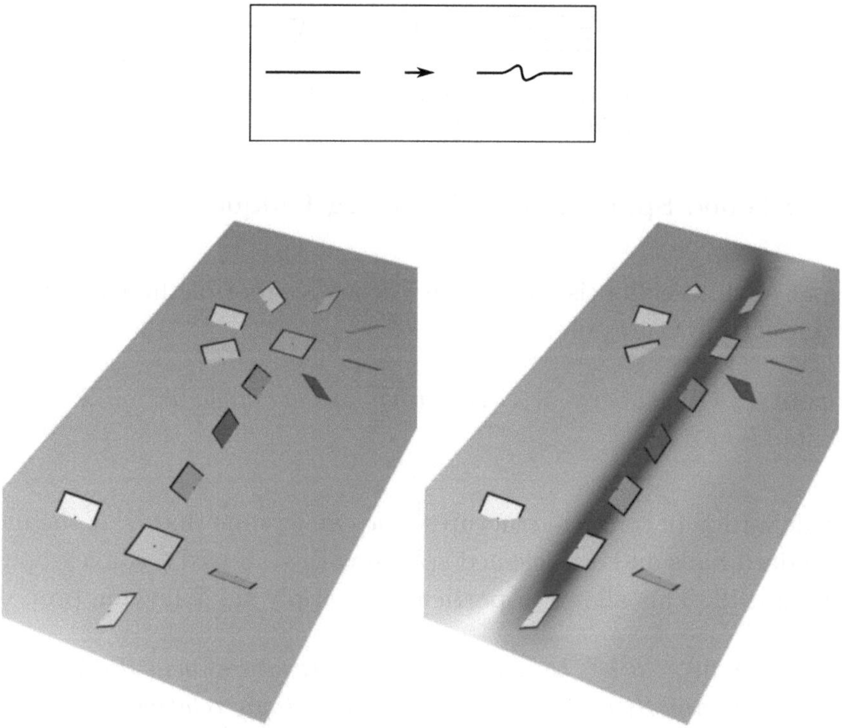

Fig. 32. The elimination move. The top box shows the move transverse to the elimination arc seen as the middle point of the segment. This move is cut off in the longitudinal direction

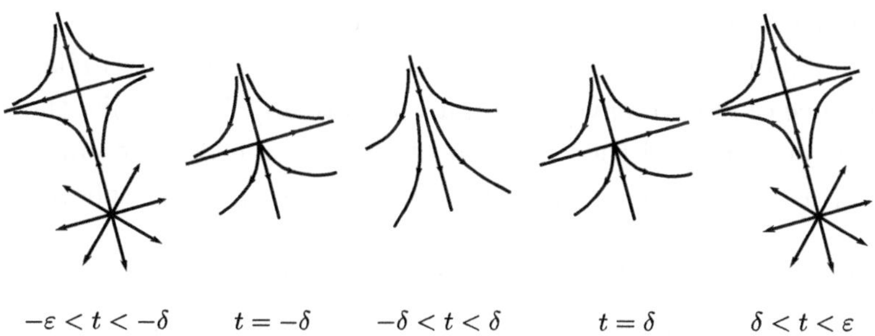

Fig. 33. Elimination in a family

The corresponding manipulation transverse to the elimination arc is explained in Figure 34.

Fig. 34. The elimination move in family. The left hand-side shows the original surfaces S_t stacked. The right hand-side performs the elimination, compare with top of Figure 32

5.2. Thickened Spheres and Eliashberg Uniqueness

The goal of this section is to explain Giroux's proof of the classification of tight contact structures on $\mathbb{S}^3$.

Theorem 32 (Eliashberg [5]). *Any tight contact structure on $\mathbb{S}^3$ is isotopic to the standard one.*

By definition of contact structures, one can assume that $\mathbb{S}^3$ is the union of two standard balls and a thickened sphere with standard ξ-convex boundary as in Figure 12. This allows in particular to apply the following proposition.

Proposition 33. *Let ξ be a tight contact structure on a thickened sphere $S \times [0, 1]$. If S_0 and S_1 are ξ-convex then ξ is isotopic relative to the boundary to a contact structure ξ' such that all spheres S_t are ξ'-convex.*

Proof. First note that tightness prevents the apparition of any closed leaf in any ξS_t since it would bound an overtwisted disk. Then we need some theory of one-parameter families of singular foliations on the sphere [17]. Specifically, one can assume that each ξS_t has finitely many singularities and at worse a saddle connection or a saddle-node (but not both at the same time). Note that finiteness of saddle connections can be achieved by perturbation thanks to the absence of closed leaves (compare Figure 40). Using this, the Poincaré-Bendixson theorem and the criterion of Proposition 26, one can see that all surfaces S_t are ξ-convex except for finitely many $t_1, \dots, t_k$ where:

- all singularities of ξS_{t_i} are saddles or nodes
- there is exactly one saddle connection on ξS_{t_i} and it is retrograde,

see Figure 35 for an example.

We will now modify ξ near each S_{t_i} in order to make all S_t ξ-convex. Since we know closed leaf or non-trivial recurrence cannot arise, it suffices to get rid of retrograde saddle connections. We concentrate on one t_i at a time. Let ε be a small positive number such that ξS_t does not change up to

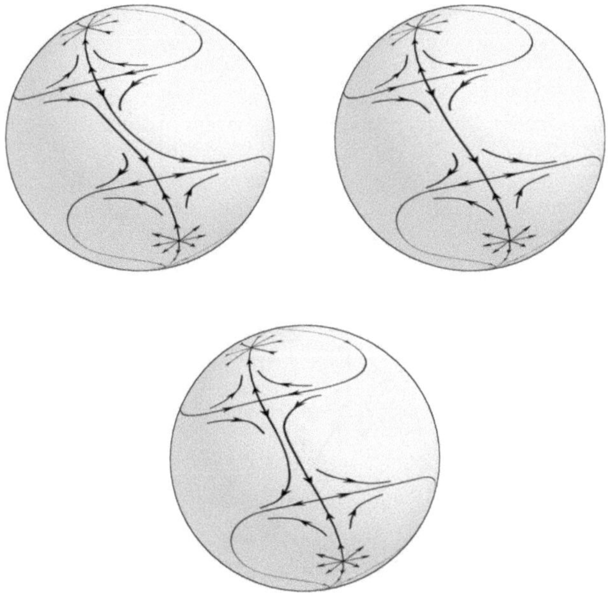

Fig. 35. Original movie

homeomorphism when t is either in $[t_i - \varepsilon, t_i)$ or $(t_i, t_i + \varepsilon]$. In particular the positive part G^+ of the Giroux graph deforms by isotopy in each of these intervals. Theorem 29, the Giroux criterion, and the link between the Giroux graph and the dividing set explained in Corollary 28 guarantee that G^+ is a tree in each interval. It implies that we can find elimination arcs between all positive saddles and all but one positive nodes without using the separatrix which enters the saddle connection at t_i (recall in particular that the number of vertices in a tree is exactly the number of edges plus one).

We now use Lemma 31, the elimination lemma, to get rid of all positive saddles for t in $[t - \delta, t + \delta]$ for some positive δ smaller than ε, see Figure 36. ■

Before continuing the proof of the theorem, we note two properties of the sphere which were somehow surreptitiously used in the above proof. After the elimination of the retrograde connections we needed the fact that no closed leaves could appear, this is due to Schönflies theorem which would have provided an overtwisted disk. We also needed the Poincaré-Bendixson theorem to prevent the apparition of non-trivial recurrence. Suppose one tries to use the elimination lemma to get rid of the bifurcation of Figure 39 (which is bound to fail since the isotopy class of the dividing set changes during this bifurcation). If one gets rid of both saddles then degenerate leaves arise. If

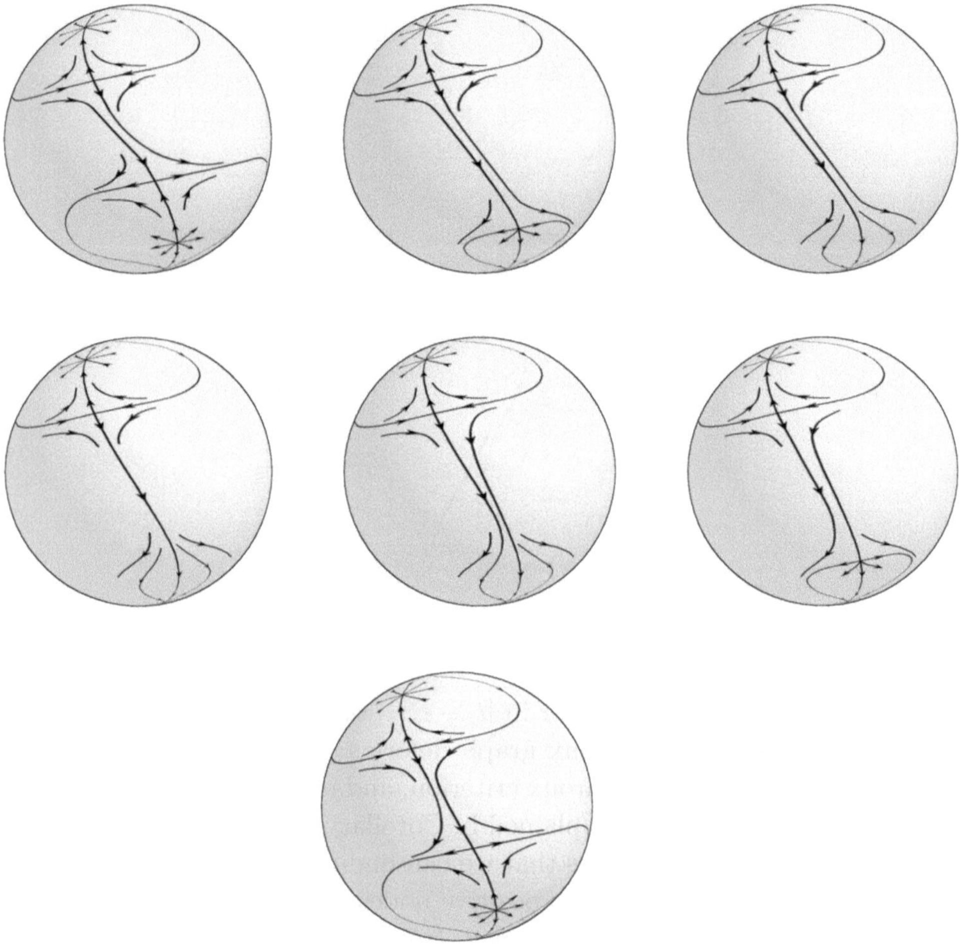

Fig. 36. Movie after elimination. The first picture is the same as in Figure 35 then a pair of singularity is replaced by a saddle-node then it disappears. The fourth picture corresponds to the central picture of Figure 35 but there is no more positive saddle so no saddle connection. The eliminated pair returns in the sixth picture as a saddle-node and the final picture is the same as in Figure 35

one gets rid of one saddle only (like we did for the sphere) then non-trivial recurrence appear: we get a Cherry flow on the torus, see [16].

The proof of Theorem 32 now follows from Giroux's uniqueness lemma which allows to replace the contact structure obtained on the thickened sphere of the previous proposition by the model.

Lemma 34 (Uniqueness lemma [9, Lemma 2.7])**.** *Let ξ_0 and ξ_1 be two contact structures printing the same characteristic foliations on the boundary of*

$S \times [0,1]$. *If there is a continuous family of multi-curves* Γ_t *dividing both* $\xi_0 S_t$ *and* $\xi_1 S_t$ *then* ξ_0 *and* ξ_1 *are isotopic relative to the boundary.*

The proof of this lemma is similar to the ones of the previous section but the path of contact structures is less obvious.

We now explain how to get the classification of tight contact structures on $\mathbb{S}^2 \times \mathbb{S}^1$ without extra effort. Let ξ be one of them and fix some $S = \mathbb{S}^2 \times \{\theta_0\}$. Using genericity of ξ-convex surfaces, we can perturb ξ to make S convex. Then the Giroux criterion tells us that its dividing set is connected. Using the realisation lemma, we change ξ by isotopy until ξS is standard, i.e. as in Figure 12. We can then remove a homogeneous neighborhood of S and we are back to a thickened sphere where we can apply Proposition 33 and the uniqueness lemma.

5.3. Bifurcation Lemmas

We now consider a general closed surface S and any contact structure ξ on $S \times I$ for some interval I. For each t in I, one has the surface $S_t := S \times \{t\}$ and its characteristic foliation ξS_t. If some S_{t_0} is not ξ-convex then the characteristic foliations for t close to t_0 are not all C^1-conjugate to ξS_{t_0}, otherwise the global reconstruction lemma (Lemma 12) would give a contradiction. We will now try to understand what really happens when this lack of ξ-convexity is explained by the obstructions we discussed in the previous section, i.e. it comes from a degenerate closed leaf or a retrograde connection. We will see in particular that the bifurcation is much sharper than expected: no foliation ξS_t is even C^0-conjuguate to ξS_{t_0} for t in a punctured neighborhood of t_0. Better, we will get a very precise description of what happens.

The Birth/Death Lemma. Let L be a degenerate closed leaf of the characteristic foliation ξS_t. This means that the Poincaré return map on any curve transverse to L is tangent to the identity. One says that L is positive (resp negative) if the second derivative of this map is positive (resp negative) at the intersection point between L and the transverse curve. If L is either positive or negative then one says that it is weakly degenerate.

Lemma 35 (Birth/Death Lemma [9, Lemma 2.12]). *A positive (resp negative) degenerate closed orbit indicates the birth (resp death) of a pair of non-degenerate closed leaves when* t *increases.*

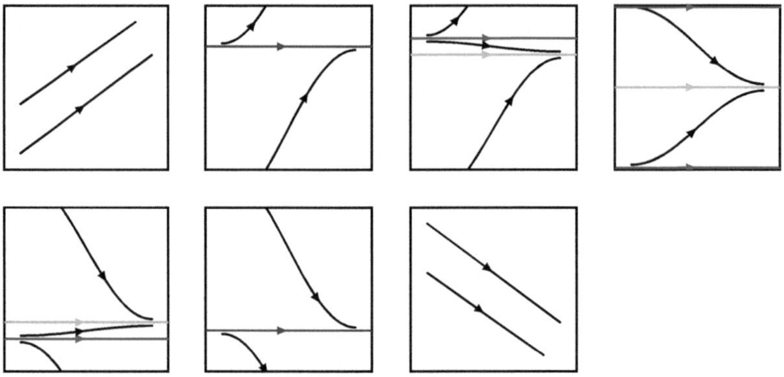

Fig. 37. Birth and death of closed leaves on a torus

See Figure 37 for examples of these situations on a thickened torus $T \times [0,1]$. Looking at these pictures it is easy to prove a weak form of the birth-death lemma which already shows how the contact condition enters. Since the contact structure is transverse to all tori T_t, $t \in [0,1]$, one can lift ∂_t to a vector field tangent to ξ. The flow of this lift defines a new product structure on $T \times [0,1]$ without changing the movie of singular foliations ξT_t up to diffeomorphism. So one can assume that all intervals $I_p = \{p\} \times [0,1]$ are Legendrian. If we think of foliations ξT_t as living all on T then the contact condition is equivalent to asking that, at each point p, $\xi T_t(p)$ rotates clockwise as t increases. Indeed, if x and y are coordinates on T, there is a function θ such that

$$\xi = \ker\bigl(\cos\theta(x,y,t)\,dx - \sin\theta(x,y,t)\,dy\bigr).$$

The contact condition is then equivalent to $\partial_t\theta > 0$, compare with the proof of the Darboux-Pfaff theorem (Theorem 2).

Now the second picture in Figure 37 shows a positive degenerate orbit L in some ξT_{t_0}. Let A be a small annulus around L. Along L, the slope of ξT_{t_0} is zero and it is positive in $A \setminus L$. So, for $t < t_0$ it was everywhere positive in A and there were no closed leaf at all in A. For $t > t_0$, the slope becomes negative along L and stays positive along the boundary of A. Then the complement of L in A is made of two (half-open) annuli whose boundary are transverse to ξT, see Figure 38. The Poincaré-Bendixson theorem guaranties that each of these two sub-annuli contain at least one closed leaf for $t > t_0$ sufficiently close to t_0.

So we proved the following weak version of the birth/death lemma which will be sufficient for our purposes: if there is a positive degenerate closed

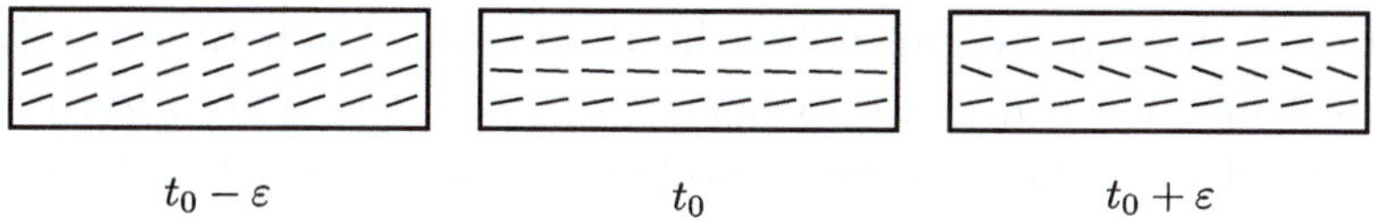

Fig. 38. Birth of at least a pair of periodic orbits. The annulus A is obtained by gluing left and right. The circle L is at mid-height of each annulus

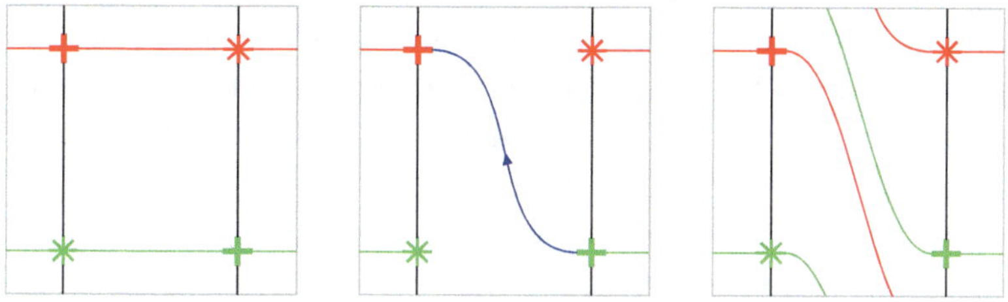

Fig. 39. Retrograde saddle connection on a torus

orbit L at time t_0 then there is an annulus A around L and some positive ε such that there is no closed leaves in A for t in $(t_0 - \varepsilon, t_0)$ and at least two for t in $(t_0, t_0 + \varepsilon)$. The death case on the bottom row of Figure 37 is explained similarly. Note that nothing required T to be a torus in this explanation, one only has to work near L.

The Crossing Lemma.

Lemma 36 (Crossing Lemma [9, Lemma 2.14]). *Assume that there is a retrograde connection at time t_0. For t close to t_0, there is a negative singularity b_t^-, a positive one b_t^+, an unstable separatrix c_t^- of b_t^- and a stable one c_t^+ of b_t^+ such that $c_{t_0}^- = c_{t_0}^+$.*

For t close to t_0, one can track separatrices using their intersection with an oriented curve positively transverse to ξS_t. Then, for $t < t_0$ (resp $t > t_0$), the separatrix c_t^- is below (resp above) c_t^+.

Figure 39 shows a retrograde saddle connection on a torus obtained by gluing top/bottom and left/right. Singularities in the lower part are negative while those in the upper part are positive. The saddle connection is marked by an arrow. The crossing Lemma tells us that the negative separatrix has to turn to its right after the connection.

The proof of the crossing lemma is rather delicate so we will only try to go as far as explaining how the contact condition and the fact that the connection is retrograde can enter the discussion. Each time we drop the t subscript it means $t = t_0$. Also we set $c = c^+ = c^-$. Compared to the situation of the birth/death lemma, there is no hope to have a neighborhood $S \times [0,1]$ with $[0,1]$ tangent to ξ near c since ξ is tangent to S at $b^\pm$. However we will find at least one point on c where the characteristic foliation has to turn clockwise. If Y_t is a vector field defining ξS_t, the contact condition ($\star$) can be expressed as: $u_t \operatorname{div} Y_t - du_t(Y_t) + \dot{\beta}_t(Y_t) > 0$. The sign of singularities is the sign of u_t so $u(b^-) < 0$ and $u(b^+) > 0$. Hence there is some point p on c such that $u(p) = 0$ and $du(Y) \geq 0$. Here we used that c, hence Y, is oriented from b^- to b^+. At p, the contact condition becomes $\dot{\beta}(Y) > du(Y)$ so $\dot{\beta}(Y) > 0$. This is the announced rotation. Since $\beta(Y) = 0$, we have that, at p, ξS_t is positively transverse to c for $t > t_0$ and negatively transverse for $t < t_0$. Of course this observation is very far from proving the crossing lemma, see [9, Lemma 2.14] for the full story.

5.4. Bennequin's Theorem

The goal of this section is to prove that the standard contact structure on $\mathbb{R}^3$ is tight. This was originally proved by Bennequin, without the word tight which was introduced by Eliashberg.

Suppose there is an overtwisted disk in the standard contact structure on $\mathbb{R}^3$. Since it is compact, it is contained in some finite radius ball. We can also assume it misses a small ball around the origin (for instance we can use the contact vector field ∂_z to push it upward until this is true). Recall we saw in Example 16 there is a contact vector field X on $\mathbb{R}^3$ which is transverse to all Euclidean spheres around the origin. So these spheres are all ξ-convex and divided by the equator $\{z = 0\}$ where X is tangent to ξ. The above discussion shows that Bennequin's theorem is a consequence of the following statement.

Theorem 37 (Bennequin seen by Giroux [9, Theorem 2.19]). *Let ξ be a contact structure on a thickened sphere $S \times [-1,1]$. If all spheres S_t are ξ-convex with connected dividing set then ξ is tight.*

Families of Movies. In order to prove Theorem 37, we first need some preparations from dynamical systems. Suppose that ξ_0 and ξ_1 are two contact structures which print generic movies on $S \times [-1,1]$. If they are isotopic, one gets a 2-parameters family $\xi_s S_t$ of characteristic foliations. Thom transversality and a little bit of normal form theory tells us that we can perturb the family until all these foliations have finitely many singularities which are either nodes, saddles or saddle-nodes. Further perturbations allow to make sure that all closed leaves have a Poincaré return map which is at worse tangent to the identity up to order 2, the worse case happening only for isolated values of (s, t).

Up to this point there was nothing specific to the sphere. The first special property of $\mathbb{S}^2$ which is crucial in the following is the Poincaré-Bendixson theorem which says that, since we have isolated singularities for all our foliations, the Poincaré-Bendixson property automatically holds. In particular we can apply the convexity criterion of Proposition 26. In the square $[0, 1] \times [-1, 1]$ the set Ω of points (s, t) such that S_t is ξ_s-convex is a dense open set. We denote by Σ the complement of Ω. It is a union of injectively immersed submanifolds of $[0, 1] \times [-1, 1]$. In codimension 1, one sees:

- Σ^1_{dl} where the characteristic foliation has a single weakly degenerate closed leaf and no retrograde saddle connection and no degenerate singularity, see Figure 22.
- Σ^1_{sc} where the characteristic foliation has a single retrograde saddle connection and no degenerate closed leaf or singularity, see Figure 39.

The bifurcation lemmas imply that these two subsets are injectively immersed submanifold of the square transverse to the t direction. In addition, the bifurcation lemmas imply that components of Σ^1_{dl} can accumulate only on Σ^1_{sc}, see Figure 40 for an example of accumulation. We set $\Sigma^1 = \Sigma^1_{\mathrm{dl}} \cup \Sigma^1_{\mathrm{sc}}$.

The accumulation of retrograde saddle connections in Figure 40 is not a phenomena which we can get rid of by perturbation: it is structurally stable in a 1-dimensional family, see [17]. However, Giroux's discretization lemma [11, Lemma 15] states that any contact structure on the product $F \times I$ of a closed surface and an interval with convex boundary is isotopic relative to the boundary to a contact structure such that only finitely many F_t are non-convex. This isotopy cannot be made arbitrarily small. It uses first the dynamics banalization lemma [9, Lemma 2.10] which gets rid of non-trivial recurrence and then replaces degenerate leaves with retrograde saddle connexions. Both moves are non-perturbative.

In codimension 2, one sees:

- Σ^{11} where two codimension one strata intersect transversely, see Figure 41 and also Figure 26 for a realistic view of the central picture in the case of Example 23.

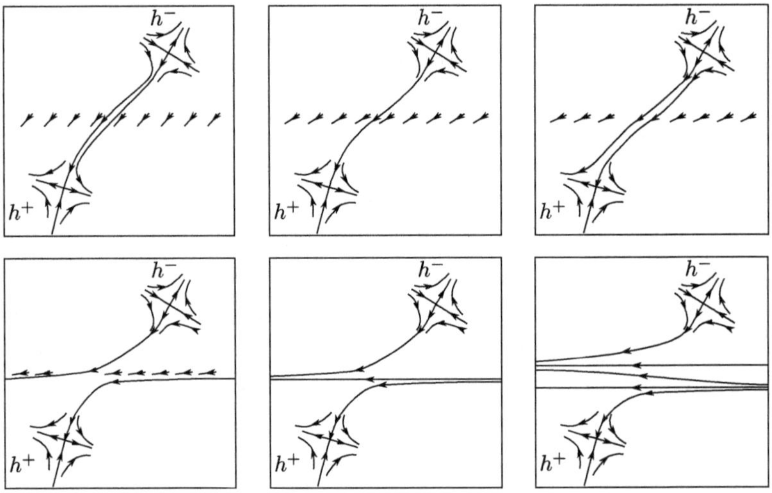

Fig. 40. Saddle connections accumulating a degenerate closed leaf. This is a movie of characteristic foliations on an annulus obtained by gluing the left and right sides of each square. A degenerate closed leaf is appearing in the middle. Leaves spiral more and more in this region, resulting in infinitely many retrograde saddle connections

- Σ^2_{sc} where there is a retrograde connection between a saddle and a saddle-node. These points adhere to exactly one stratum in Σ^1_{sc}, this typically happens in the proof of the classification on S^3 as an intermediate step between Figures 35 and 36.
- Σ^2_{dl} where there is a degenerate orbit corresponding to the fusion of two components of Σ^1_{dl}, see Figure 42 for the picture in the (s,t) square and Figure 43 for the corresponding foliations.

Proof Core. We now prove Theorem 37. Suppose there is some overtwisted disk in $(S \times [-1,1], \xi)$. Then there is some isotopy relative to the boundary bringing this disk onto the middle sphere S_0. So this isotopy sends $\xi_0 = \xi$ to a contact structure ξ_1 such that S_0 contains an overtwisted disk. Then it can be modified in the same way genericity of convex surfaces is proved until S_0 is ξ_1-convex and divided by a disconnected curve (use Corollary 28 to understand dividing sets here). We can perturb ξ_1 to make sure it also prints a generic movie of characteristic foliations and perturb the isotopy to be in the situation of the preceding discussion on families of movies.

The set Ω of (s,t) such that S_t is ξ_s-convex is the disjoint union of Ω_c corresponding to connected dividing sets and Ω_d corresponding to disconnected ones.

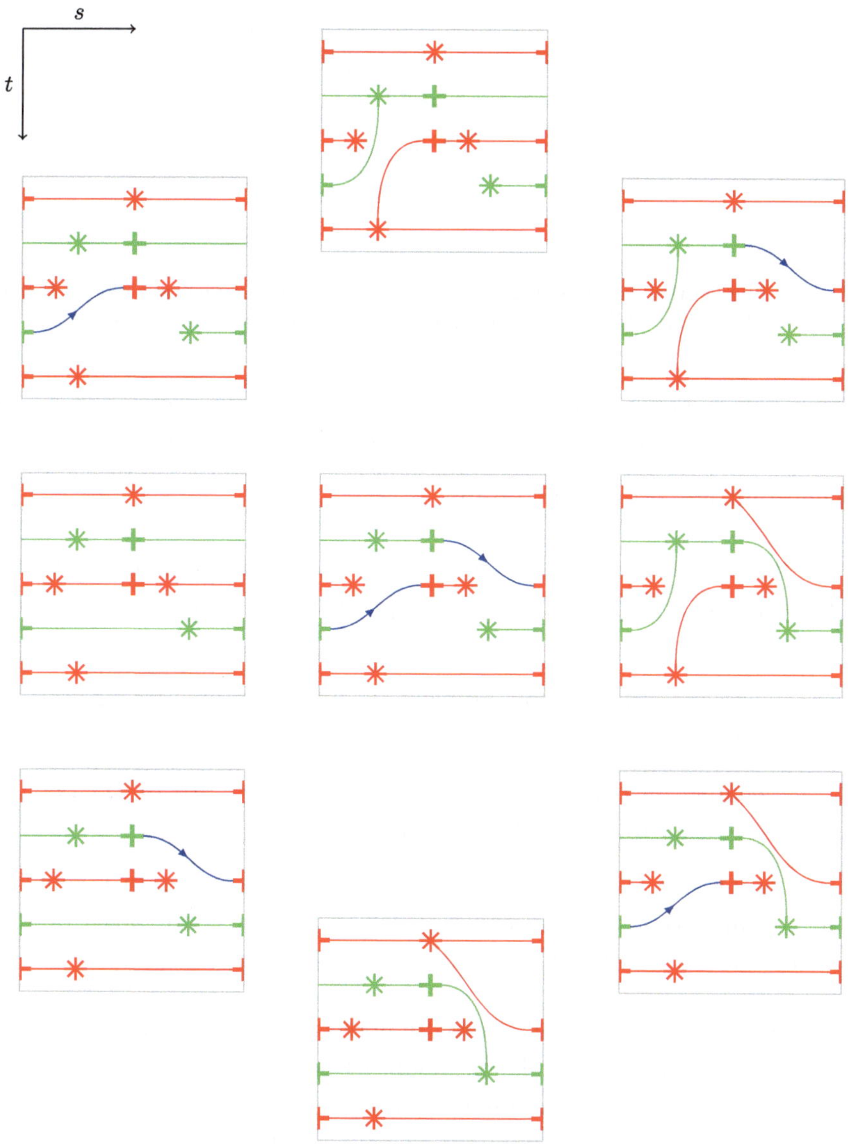

Fig. 41. Intersection of two strata of retrograde saddle connections on a torus. It is a good exercise to draw the Giroux graphs of all convex surfaces appearing to see the non-trivial effect of this codimension 2 phenomenon on the dividing sets, contrasting with the discussion below

In addition, we know by construction that Ω_d intersects the right vertical edge $\{s=1\}$ so it is not empty. But it does not intersect the left edge $\{s=0\}$ by hypothesis of the theorem. More precisely, we can assume the closure of

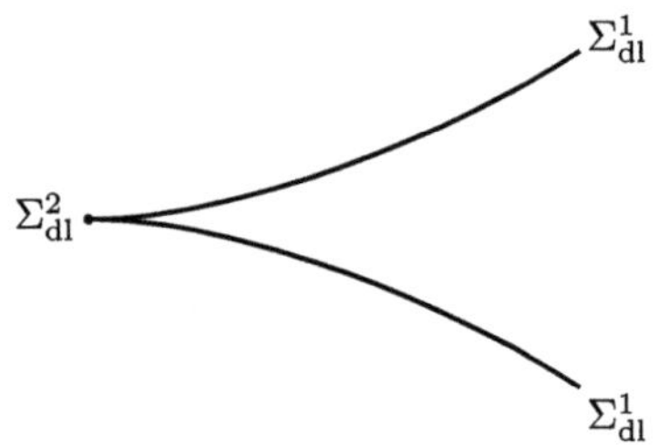

Fig. 42. The central point is in $\Sigma^2_{\rm dl}$. It corresponds to a degenerate closed leaf with $\pi''(0)=0$ but $\pi^{(3)}(0)<0$, see Figure 43 for the corresponding foliations

Ω_d does not meet $\{s=0\}$ so the minimum s_0 of its projection to $[0,1]$ is positive. Choose t_0 such that (s_0,t_0) is in the closure of Ω_d.

The point (s_0,t_0) cannot be in:

- Σ^1 because the later is transverse to the t direction so components of Ω adjacent to a point (s,t) in Σ^1 project to neighborhoods of s.
- $\Sigma^2_{\rm sc}$ because each point (s,t) in $\Sigma^2_{\rm sc}$ adheres to only one component of $\Sigma^1_{\rm sc}$ so the intersection between Ω and a small disc around (s,t) is connected and projects to a neighborhood of s.
- $\Sigma^2_{\rm dl}$ because all components of Ω touching $\Sigma^2_{\rm dl}$ are in Ω_d because the corresponding foliations have closed leaves.
- any point Σ^{11} involving degenerate closed leaves, again because strata in $\Sigma^1_{\rm dl}$ are transverse to the t-direction and indicate birth or death of stable closed leaves giving disconnected dividing sets.

The only configuration which really needs to be carefully ruled out is that of points in Σ^{11} involving only $\Sigma^1_{\rm sc}$ like in Figure 44. In this situation $\xi_{s_0}S_{t_0}$ has two retrograde saddle connections which happen on different surfaces S_t for s in a punctured neighborhood of s_0 and get swapped when s goes through s_0, as in Figure 41. Note that characteristic foliations around (s_0,t_0) have no closed leaf and we can also assume they do not have other saddle connections that the ones we explicitly study.

To ξ_sS_t we associate the oriented graph $\Gamma_+(s,t)$ (resp. $\Gamma^-(s,t)$) whose vertices are positive nodes and edges are the stable separatrices of positive saddles (resp. negative saddles). Since we do not have any closed leaf or degenerate singularities near (s_0,t_0), Γ_+ coincides as a set with G_+ from Definition 27 and Γ^- is somehow dual to G_-. So, according to Corollary 28, when S_t is ξ_s-convex, there is a regular neighborhood of $\Gamma_+(s,t)$ whose boundary

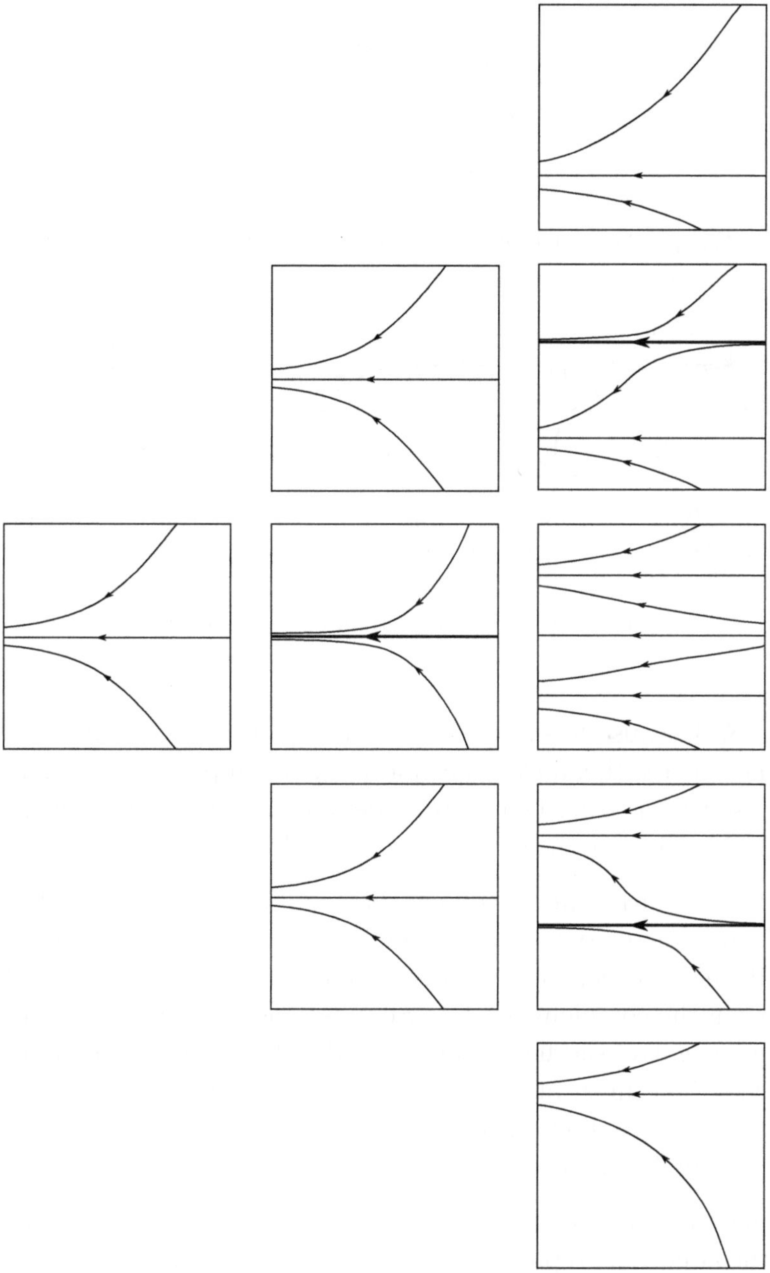

Fig. 43. Foliations corresponding to the strata of Figure 42. Left and right of each square are glued to get an annulus. Thick closed leaves are the degenerate ones. The central picture corresponds to the annihilation of a birth and a death of non-degenerate closed leaves

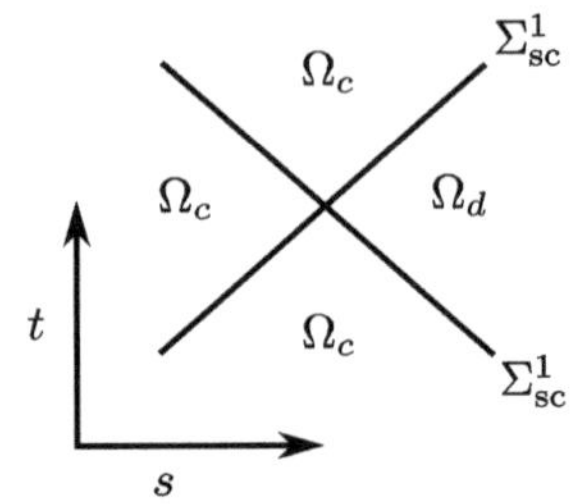

Fig. 44. The situation we must rule out for Bennequin's theorem

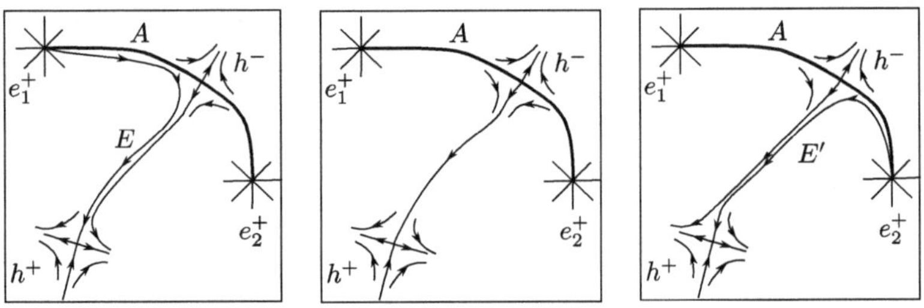

Fig. 45. Anatomy of a retrograde saddle connection

divides $\xi_s S_t$. Because S is a sphere, we then get that (s, t) is in Ω_c if and only if $\Gamma_+(s, t)$ is a tree (i.e. a closed connected and simply connected graph). We want to use the crossing lemma to understand how the graph changes when a retrograde saddle connection happens, see Figure 45.

First we remark that, if we focus on a sufficiently small neighborhood of (s_0, t_0) in parameter space, the graph $\Gamma^-(s, t)$ deforms by isotopy so we can assume it does not depend on s and t. The same is true for $\Gamma_+(s, t)$ as long as we stay in the complement of Σ. Suppose now there is a saddle connection involving a negative saddle h^-. Let A be the closure of the union of its stable separatrices. The unstable separatrix of h^- entering the saddle connection coorients A and, together with the orientation of S, this orients A. We denote by $o(A)$ and $d(A)$ the origin and destination of A.

During a bifurcation, exactly one edge E of Γ_+ changes. After the bifurcation, the edge E is replaced by an edge $A(E)$ which is obtained from the concatenation of E and A by a small push towards the right which makes it avoid $o(A)$, see Figure 46 which also explains how these things will be drawn schematically in the following. Note that the edge E is the edge which is immediately to the right of A at $o(A)$ with respect to the cyclic ordering of edges of $\Gamma_+ \cup \Gamma^-$ incident to $o(A)$. So the oriented arc A completely describes

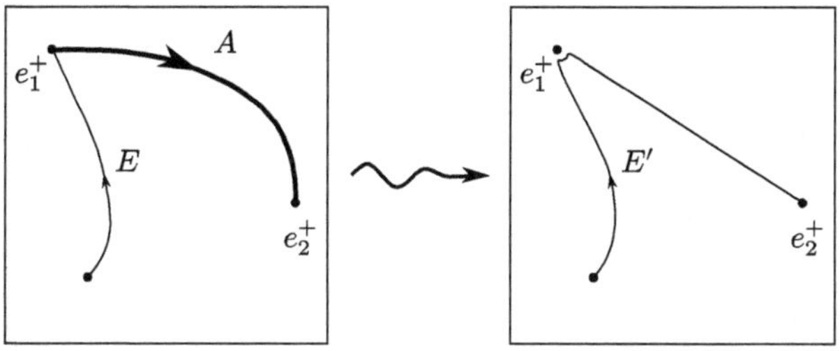

Fig. 46. A schematic view of the same retrograde saddle connection as in Figure 45

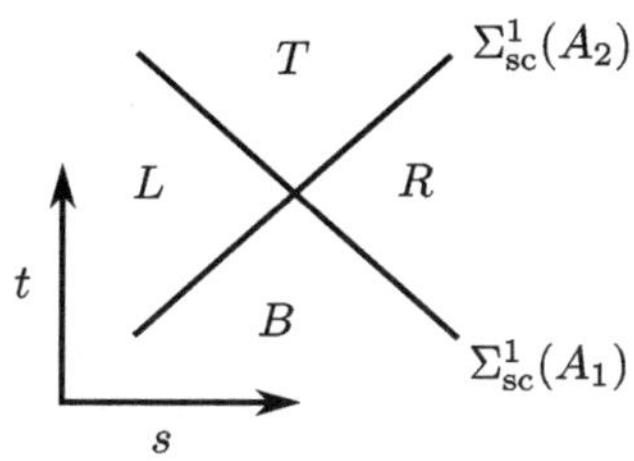

Fig. 47. Regions in the parameter space

the bifurcation. We will denote by $A(\Gamma_+)$ the graph obtained from Γ_+ after a bifurcation described by A (up to isotopy).

Returning to the codimension 2 bifurcation at (s_0, t_0) we have two distinct strata $\Sigma^1_{sc}(A_1)$ and $\Sigma^1_{sc}(A_2)$ corresponding to distinct (oriented) bifurcation arcs A_1 and A_2, see Figure 47. We take the graph Γ_+ of the Bottom region as a reference and apply to it the following proposition. Note that, on a tree, any ordered pair of vertices determines a unique oriented segment.

Proposition 38. *Suppose Γ is a tree and A_1 and A_2 are bifurcation arcs for Γ. The following properties are equivalent.*

1. *$A_1(\Gamma)$ is not a tree but $A_2(A_1(\Gamma))$ is a tree.*
2. *On Γ, the oriented segment S from $d(A_2)$ to $d(A_1)$ contains, in that order: $d(A_2) \le o(A_1) < o(A_2) \le d(A_1)$ and, furthermore, S is immediately to the right of A_1 at $o(A_1)$ and A_2 at $o(A_2)$.*

Note that condition 1 above holds if Γ is the tree Γ_+ coming from the Bottom region B since we assume T and B are in Ω_c while R is in Ω_d. This proposition concludes the proof of Theorem 37 because condition 2 above is

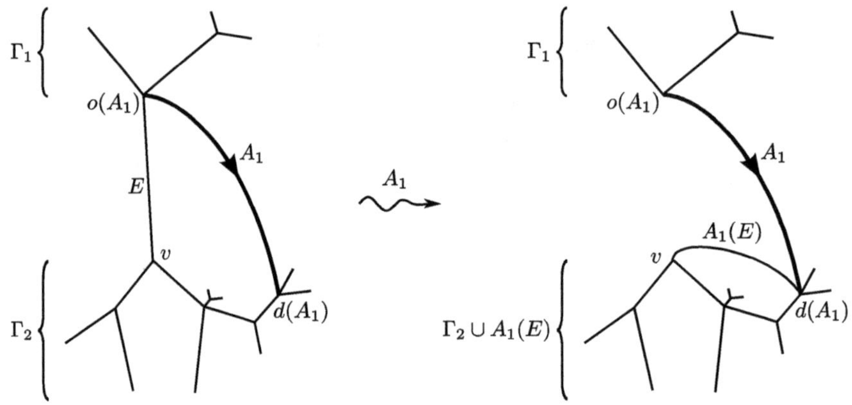

Fig. 48. Trees and graphs in the proof of Proposition 38

symmetric in A_1 and A_2 (here one should not forget that exchanging A_1 and A_2 will reverse the orientation on S). So the graph $A_2(\Gamma)$ corresponding to the left region L is not a tree and L is also in Ω_d.

Proof. We first prove that property 1 implies property 2. Let E be the edge of Γ modified by A_1. In particular E has vertices $o(A_1)$ and some other vertex v and E is immediately to the right of A_1 at $o(A_1)$. Because Γ is a tree, v can't be the same as $o(A_1)$ and (the closure of) $\Gamma \setminus E$ is the disjoint union of two trees Γ_1 containing $o(A_1)$ and Γ_2 containing v, see Figure 48.

Note that $d(A_1)$ cannot be in Γ_1 since otherwise $A_1(E)$ would go from Γ_1 to Γ_2 and $A_1(\Gamma)$ would be a tree.

So $d(A_1)$ is in Γ_2 and this implies that v in the segment $[o(A_1), d(A_1)] \subset \Gamma$. Also we learn that $A_1(\Gamma)$ is the disjoint union of the tree Γ_1 and the graph $\Gamma_2 \cup A_1(E)$ which contains exactly one cycle C. This cycle contains $A_1(E)$ and its vertices are all in $[v, d(A_1)] \subset \Gamma$, see Figure 48 again.

Since $A_2(A_1(\Gamma))$ is a tree, the edge E' modified by A_2 in $A_1(\Gamma)$ belongs to C otherwise C would persist in $A_2(A_1(\Gamma))$. So we get that $o(A_2)$ is in C (in particular it can't be the same as $o(A_1)$). In addition $d(A_2)$ is in Γ_1 otherwise $A_2(A_1(\Gamma))$ would stay disconnected. The last thing to check is that E' is part of the segment $[d(A_2), d(A_1)] \subset \Gamma$. The only edge of C which is not in this segment is $A_1(E)$. Remember E' is immediately to the right of A_2 at $o(A_2)$ so it cannot be $A_1(E)$ because that would force A_2 to go into the disk bounded by C which does not contain Γ_1 (surreptitiously using Schönflies theorem again).

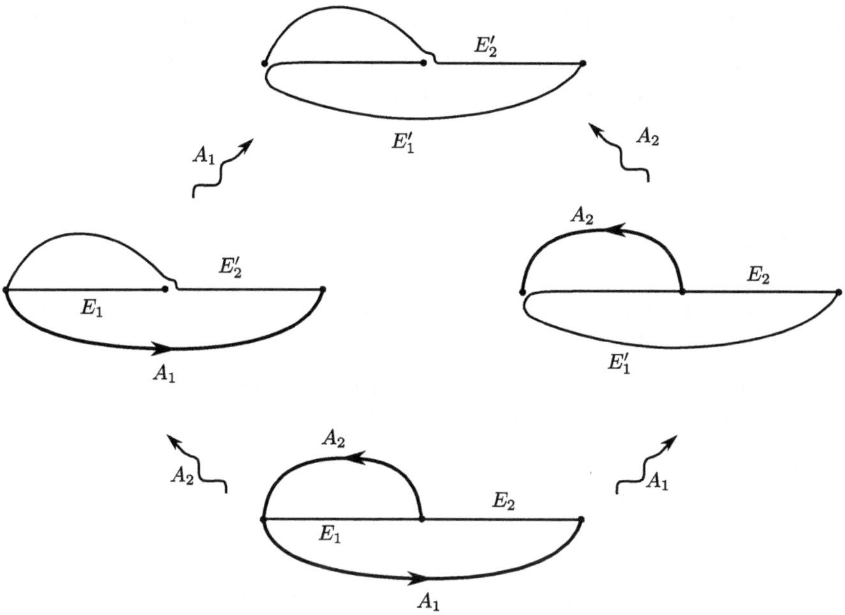

Fig. 49. How the discussion would fail if A_1 were reversed. In this example the reference graph has three vertices and two edges. Regions L, T and B are tight whereas R is overtwisted

We now prove the converse implication. Since S is immediately to the right of A_1 at $o(A_1)$, it contains the edge E of Γ moved by A_1. More precisely, E is in the segment $[o(A_1), d(A_1)] \subset \Gamma$. So $A_1(\Gamma)$ is the disjoint union of a tree Γ_1 and a graph Γ_2 containing a unique cycle C. Since S is immediately to the right of A_2 at $o(A_2)$ and $o(A_1) \neq o(A_2)$, the edge E' in $A_1(\Gamma)$ moved by A_2 is either an edge in S or $A_1(E)$. In both cases, it is contained in C. So the cycle C does not persist in $A_2(A_1(\Gamma))$ and $A_2(E_1)$ connects $\Gamma_2 \setminus E'$ to Γ_1. Hence $A_2(A_1(\Gamma))$ is a tree. ■

Now this proof is finished let us see where we used the contact condition and not only properties of generic families of foliations with two parameters. The first thing is that Σ^1 is transverse to the t direction because of the bifurcation lemmas. A second more subtle point is that the crossing lemma says more: it tells the direction of the bifurcations: separatrices turn to their right when t increases. Figure 49 show how the above proof would fail if A_1 and A_2 were allowed to act as switches in opposite direction. In that figure one sees an example of the bad situation of Figure 44. The explanation is that, if we assume that the bifurcation corresponding to A_1 acts in the wrong direction then, in Proposition 38, we must replace "to the right of

A_1" by "to the left of A_1" and we loose symmetry between A_1 of A_2. Of course if both A_1 and A_2 act in the wrong direction then we do not have any difference, this simply corresponds to considering negative tight contact structures on $\mathbb{S}^3$.

References

[1] D. Bennequin, Entrelacements et équations de Pfaff, in *Third Schnepfenried Geometry Conference*, vol. 1. Schnepfenried, 1982 (1983), pp. 87–161

[2] V. Colin, E. Giroux, K. Honda, Finitude homotopique et isotopique des structures de contact tendues. Publ. Math. Inst. Hautes Études Sci. **109**, 245–293 (2009). doi:10.1007/s10240-009-0022-y (French, with French summary)

[3] Y. Eliashberg, Classification of overtwisted contact structures on 3-manifolds. Invent. Math. **98**(3), 623–637 (1989)

[4] Y. Eliashberg, New invariants of open symplectic and contact manifolds. J. Am. Math. Soc. **4**(3), 513–520 (1991). doi:10.2307/2939267. MR1102580 (92c:58030)

[5] Y. Eliashberg, Contact 3-manifolds twenty years since J. Martinet's work. Ann. Inst. Fourier (Grenoble) **42**(1–2), 165–192 (1992)

[6] J. Etnyre, Convex surfaces in contact geometry (2004), available at http://people.math.gatech.edu/~etnyre/preprints/papers/surfaces.pdf

[7] H. Geiges, *An Introduction to Contact Topology*. Cambridge Studies in Advanced Mathematics, vol. 109 (Cambridge University Press, Cambridge, 2008)

[8] E. Giroux, Convexité en topologie de contact. Comment. Math. Helv. **66**(4), 637–677 (1991)

[9] E. Giroux, Structures de contact en dimension trois et bifurcations des feuilletages de surfaces. Invent. Math. **141**(3), 615–689 (2000)

[10] E. Giroux, Structures de contact sur les variétés fibrées en cercles au-dessus d'une surface. Comment. Math. Helv. **76**(2), 218–262 (2001)

[11] E. Giroux, Sur les transformations de contact au-dessus des surfaces, in *Essays on Geometry and Related Topics, Vol. 1, 2*. Monogr. Enseign. Math., vol. 38 (Enseignement Math, Geneva, 2001), pp. 329–350 (French)

[12] J.W. Gray, Some global properties of contact structures. Ann. Math. **69**, 421–450 (1959)

[13] K. Honda, Contact geometry, available at http://www-bcf.usc.edu/~khonda/math599/notes.pdf

[14] P. Libermann, Sur les automorphismes infinitésimaux des structures symplectiques et des structures de contact, in *Colloque Géom. Diff. Globale*, Bruxelles, 1958 (Centre Belge Rech. Math., Louvain, 1959), pp. 37–59 (French)

[15] A. Mori, On the violation of Thurston-Bennequin inequality for a certain non-convex hypersurface (2011), available at http://arxiv.org/abs/1111.0383

[16] J. Palis Jr., W. de Melo, *Geometric Theory of Dynamical Systems* (Springer, New York, 1982). An introduction; Translated from the Portuguese by A.K. Manning

[17] J. Sotomayor, Generic one-parameter families of vector fields on two-dimensional manifolds. Publ. Math. IHÉS **43**, 5–46 (1974)

P. Massot (✉)

Université Paris Sud
91405 Orsay
France

e-mail: patrick.massot@math.u-psud.fr

url:
http://www.math.u-psud.fr/~pmassot/

Current address:
Centre de Mathématiques Laurent Schwartz
École Polytechnique
91128 Palaiseau Cedex
France

BOLYAI SOCIETY
MATHEMATICAL STUDIES, 26

Contact and Symplectic Topology
pp. 85–136.

A Beginner's Introduction to Fukaya Categories

DENIS AUROUX

This text is based on a series of lectures given at a Summer School on Contact and Symplectic Topology at Université de Nantes in June 2011.

The goal of these notes is to give a short introduction to Fukaya categories and some of their applications. The first half of the text is devoted to a brief review of Lagrangian Floer (co)homology and product structures. Then we introduce the Fukaya category (informally and without a lot of the necessary technical detail), and briefly discuss algebraic concepts such as exact triangles and generators. Finally, we mention wrapped Fukaya categories and outline a few applications to symplectic topology, mirror symmetry and low-dimensional topology.

These notes are in no way a comprehensive text on the subject; however we hope that they will provide a useful introduction to Paul Seidel's book [42] and other texts on Floer homology, Fukaya categories, and their applications. We assume that the reader is generally familiar with the basics of symplectic geometry, and some prior exposure to pseudo-holomorphic curves is also helpful; the reader is referred to [28, 29] for background material.

The author was partially supported by NSF grant DMS-1007177.

F. Bourgeois et al. (eds.), *Contact and Symplectic Topology*,
Bolyai Society Mathematical Studies 26, DOI 10.1007/978-3-319-02036-5_3,

1. Lagrangian Floer (Co)homology

1.1. Motivation

Lagrangian Floer homology was introduced by Floer in the late 1980s in order to study the intersection properties of compact Lagrangian submanifolds in symplectic manifolds and prove an important case of Arnold's conjecture concerning intersections between Hamiltonian isotopic Lagrangian submanifolds [12].

Specifically, let (M, ω) be a symplectic manifold (compact, or satisfying a "bounded geometry" assumption), and let L be a compact Lagrangian submanifold of M. Let $\psi \in \mathrm{Ham}(M, \omega)$ be a Hamiltonian diffeomorphism. (Recall that a time-dependent Hamiltonian $H \in C^\infty(M \times [0,1], \mathbb{R})$ determines a family of Hamiltonian vector fields X_t via the equation $\omega(\cdot, X_t) = dH_t$, where $H_t = H(\cdot, t)$; integrating these vector fields over $t \in [0,1]$ yields the Hamiltonian diffeomorphism ψ *generated* by H.)

Theorem 1.1 (Floer [17]). *Assume that the symplectic area of any topological disc in M with boundary in L vanishes. Assume moreover that $\psi(L)$ and L intersect transversely. Then the number of intersection points of L and $\psi(L)$ satisfies the lower bound $|\psi(L) \cap L| \geq \sum_i \dim H^i(L; \mathbb{Z}_2)$.*

Note that, by Stokes' theorem, since $\omega_{|L} = 0$, the symplectic area of a disc with boundary on L only depends on its class in the relative homotopy group $\pi_2(M, L)$.

The bound given by Theorem 1.1 is stronger than what one could expect from purely topological considerations. The assumptions that the diffeomorphism ψ is Hamiltonian, and that L does not bound discs of positive symplectic area, are both essential (though the latter can be slightly relaxed in various ways).

Example 1.2. Consider the cylinder $M = \mathbb{R} \times S^1$, with the standard area form, and a simple closed curve L that goes around the cylinder once: then $\psi(L)$ is also a simple closed curve going around the cylinder once, and the assumption that $\psi \in \mathrm{Ham}(M)$ means that the total signed area of the 2-chain bounded by L and $\psi(L)$ is zero. It is then an elementary fact that $|\psi(L) \cap L| \geq 2$, as claimed by Theorem 1.1; see Figure 1 left. On the other hand, the result becomes false if we only assume that ψ is a symplectomorphism (a large vertical translation of the cylinder is area-preserving and eventually displaces

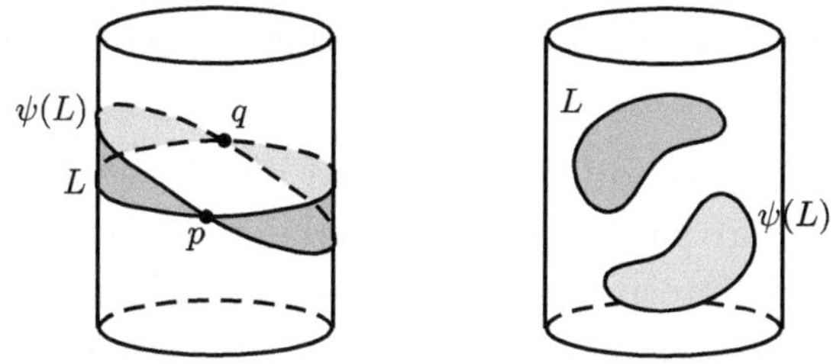

Fig. 1. Arnold's conjecture on the cylinder $\mathbb{R}\times S^1$: an example (left) and a non-example (right)

L away from itself); or if we take L to be a homotopically trivial simple closed curve, which bounds a disc of positive area (see Figure 1 right).

Floer's approach is to associate to the pair of Lagrangians $(L_0, L_1) = (L, \psi(L))$ a chain complex $CF(L_0, L_1)$, freely generated by the intersection points of L_0 and L_1, equipped with a differential $\partial : CF(L_0, L_1) \to CF(L_0, L_1)$, with the following properties:

(1) $\partial^2 = 0$, so the *Floer cohomology* $HF(L_0, L_1) = \operatorname{Ker}\partial/\operatorname{Im}\partial$ is well-defined;

(2) if L_1 and L_1' are Hamiltonian isotopic then $HF(L_0, L_1) \simeq HF(L_0, L_1')$;

(3) if L_1 is Hamiltonian isotopic to L_0, then $HF(L_0, L_1) \simeq H^*(L_0)$ (with suitable coefficients).

Theorem 1.1 then follows immediately, since the rank of $HF(L, \psi(L)) \simeq H^*(L)$ is bounded by that of the Floer complex $CF(L, \psi(L))$, which equals $|\psi(L)\cap L|$.

Formally, Lagrangian Floer (co)homology can be viewed as an infinite-dimensional analogue of Morse (co)homology for the *action functional* on (the universal cover of) the path space $\mathcal{P}(L_0, L_1) = \{\gamma : [0,1] \to M \mid \gamma(0) \in L_0, \gamma(1) \in L_1\}$,

$$\mathcal{A}(\gamma, [\Gamma]) = -\int_\Gamma \omega,$$

where $(\gamma, [\Gamma]) \in \tilde{\mathcal{P}}(L_0, L_1)$ consists of a path $\gamma \in \mathcal{P}(L_0, L_1)$ and an equivalence class $[\Gamma]$ of a homotopy $\Gamma : [0,1]\times[0,1] \to M$ between γ and a fixed base point in the connected component of $\mathcal{P}(L_0, L_1)$ containing γ. The critical points of $\mathcal{A}$ are (lifts of) constant paths at intersection points, and its gradient flow lines (with respect to the natural L^2-metric induced by ω and a compatible almost-complex structure) are pseudo-holomorphic strips bounded by L_0 and L_1.

However, the analytic difficulties posed by Morse theory in the infinite-dimensional setting are such that the actual definition of Floer (co)homology does not rely on this interpretation: instead, the Floer differential is defined in terms of moduli spaces of pseudo-holomorphic strips.

1.2. The Floer Differential

Let L_0, L_1 be compact Lagrangian submanifolds of a symplectic manifold (M,ω), and assume for now that L_0 and L_1 intersect transversely, hence at a finite set of points.

Before we introduce the Floer complex and the Floer differential, a brief discussion of coefficients is in order. In general, Floer cohomology is defined with *Novikov coefficients* (over some base field $\mathbb{K}$, for example $\mathbb{K}=\mathbb{Q}$, or $\mathbb{K}=\mathbb{Z}_2$).

Definition 1.3. The *Novikov ring* over a base field $\mathbb{K}$ is

$$\Lambda_0 = \left\{ \sum_{i=0}^{\infty} a_i T^{\lambda_i} \,\middle|\, a_i \in \mathbb{K},\ \lambda_i \in \mathbb{R}_{\geq 0},\ \lim_{i\to\infty} \lambda_i = +\infty \right\}.$$

The Novikov field Λ is the field of fractions of Λ_0, i.e.

$$\Lambda = \left\{ \sum_{i=0}^{\infty} a_i T^{\lambda_i} \,\middle|\, a_i \in \mathbb{K},\ \lambda_i \in \mathbb{R},\ \lim_{i\to\infty} \lambda_i = +\infty \right\}.$$

The Floer complex is then the free Λ-module generated by intersection points: we denote by $\mathcal{X}(L_0,L_1)=L_0\cap L_1$ the set of generators, and set

$$CF(L_0,L_1) = \bigoplus_{p\in\mathcal{X}(L_0,L_1)} \Lambda\cdot p.$$

Equip M with an ω-compatible almost-complex structure J. (By a classical result, the space of ω-compatible almost-complex structures $\mathcal{J}(M,\omega)=\{J\in \mathrm{End}(TM)\mid J^2=-1 \text{ and } g_J=\omega(\cdot,J\cdot) \text{ is a Riemannian metric}\}$ is non-empty and contractible [28].)

The Floer differential $\partial : CF(L_0,L_1)\to CF(L_0,L_1)$ is defined by counting pseudo-holomorphic strips in M with boundary in L_0 and L_1: namely, given intersection points $p, q \in \mathcal{X}(L_0, L_1)$, the coefficient of q in ∂p is obtained by considering the space of maps $u : \mathbb{R} \times [0,1] \to M$ which solve the Cauchy-Riemann equation $\bar{\partial}_J u = 0$, i.e.

$$\frac{\partial u}{\partial s} + J(u)\frac{\partial u}{\partial t} = 0, \tag{1.1}$$

subject to the boundary conditions

$$\begin{cases} u(s,0) \in L_0 \text{ and } u(s,1) \in L_1 \quad \forall s \in \mathbb{R}, \\ \lim\limits_{s \to +\infty} u(s,t) = p, \qquad \lim\limits_{s \to -\infty} u(s,t) = q, \end{cases} \tag{1.2}$$

and the *finite energy* condition

$$E(u) = \int u^*\omega = \iint \left|\frac{\partial u}{\partial s}\right|^2 ds\, dt < \infty. \tag{1.3}$$

(Note that, by the Riemann mapping theorem, the strip $\mathbb{R} \times [0,1]$ is biholomorphic to $D^2 \setminus \{\pm 1\}$, the closed unit disc minus two points on its boundary; the map u then extends to the closed disc, with the boundary marked points ± 1 mapping to p and q.)

Given a homotopy class $[u] \in \pi_2(M, L_0 \cup L_1)$, we denote by $\widehat{\mathcal{M}}(p,q;[u],J)$ the space of solutions of (1.1)–(1.3) representing the class $[u]$, and by $\mathcal{M}(p,q;[u],J)$ its quotient by the action of $\mathbb{R}$ by reparametrization (i.e., $a \in \mathbb{R}$ acts by $u \mapsto u_a(s,t) := u(s-a,t)$).

The boundary value problem (1.1)–(1.3) is a Fredholm problem, i.e. the linearization $D_{\bar{\partial}_J,u}$ of $\bar{\partial}_J$ at a given solution u is a Fredholm operator. Specifically, $D_{\bar{\partial}_J,u}$ is a $\bar{\partial}$-type first-order differential operator, whose domain is a suitable space of sections of the pullback bundle u^*TM (with Lagrangian boundary conditions), for example $W^{1,p}(\mathbb{R} \times [0,1], \mathbb{R} \times \{0,1\}; u^*TM, u^*_{|t=0}TL_0, u^*_{|t=1}TL_1)$. The Fredholm index $\mathrm{ind}([u]) := \mathrm{ind}_{\mathbb{R}}(D_{\bar{\partial}_J,u}) = \dim \mathrm{Ker}\, D_{\bar{\partial}_J,u} - \dim \mathrm{Coker}\, D_{\bar{\partial}_J,u}$ can be computed in terms of an invariant of the class $[u]$ called the *Maslov index*, which we discuss below.

The space of solutions $\widehat{\mathcal{M}}(p,q;[u],J)$ is then a smooth manifold of dimension $\mathrm{ind}([u])$, provided that all solutions to (1.1)–(1.3) are *regular*, i.e. the linearized operator $D_{\bar{\partial}_J,u}$ is surjective at each point of $\widehat{\mathcal{M}}(p,q;[u],J)$. This transversality property is one of three fundamental technical issues that need to be addressed for Floer (co)homology to be defined, the other two being the compactness of the moduli space $\mathcal{M}(p,q;[u],J)$, and its orientability (unless one is content to work over $\mathbb{K} = \mathbb{Z}_2$).

Transversality and compactness will be briefly discussed in Section 1.4 below. On the issue of orientations, we will only consider the case where L_0 and L_1 are oriented and spin. It is then known that the choice of spin structures on L_0 and L_1 determines a canonical orientation of the moduli spaces of J-holomorphic strips; the construction of this orientation is fairly technical, so we refer the reader to [19, 42] for details.

Assuming that all these issues have been taken care of, we observe that when $\mathrm{ind}([u])=1$ the moduli space $\mathcal{M}(p,q;[u],J)$ is a compact oriented 0-manifold, i.e. a finite set of points which can be counted with signs. We can then provisionally define:

Definition 1.4. The Floer differential $\partial : CF(L_0,L_1)\to CF(L_0,L_1)$ is the Λ-linear map defined by

$$\partial(p) = \sum_{\substack{q\in\mathcal{X}(L_0,L_1)\\ [u]:\,\mathrm{ind}([u])=1}} (\#\mathcal{M}(p,q;[u],J))\, T^{\omega([u])}\, q, \tag{1.4}$$

where $\#\mathcal{M}(p,q;[u],J)\in\mathbb{Z}$ (or $\mathbb{Z}_2$) is the signed (or unsigned) count of points in the moduli space of pseudo-holomorphic strips connecting p to q in the class $[u]$, and $\omega([u])=\int u^*\omega$ is the symplectic area of those strips.

In general, the definition needs to be modified by introducing a perturbation term into the Cauchy-Riemann equation in order to achieve transversality (see Section 1.4 below). Thus, the Floer differential actually counts *perturbed* pseudo-holomorphic strips connecting *perturbed* intersection points of L_0 and L_1.

The following result is due to Floer for $\mathbb{K}=\mathbb{Z}_2$:

Theorem 1.5. *Assume that* $[\omega]\cdot\pi_2(M,L_0)=0$ *and* $[\omega]\cdot\pi_2(M,L_1)=0$. *Moreover, when* $\mathrm{char}(\mathbb{K})\neq 2$ *assume that* L_0, L_1 *are oriented and equipped with spin structures. Then the Floer differential* ∂ *is well-defined, satisfies* $\partial^2=0$, *and the Floer cohomology* $HF(L_0,L_1)=H^*(CF(L_0,L_1),\partial)$ *is, up to isomorphism, independent of the chosen almost-complex structure* J *and invariant under Hamiltonian isotopies of* L_0 *or* L_1.

Remark 1.6. In this text we discuss the chain complex and differential for Floer *cohomology*, which is dual to Floer's original construction. Namely, in Floer homology, the strip shown on Figure 2 would be considered a trajectory from q to p rather than from p to q, and the grading conventions are reversed.

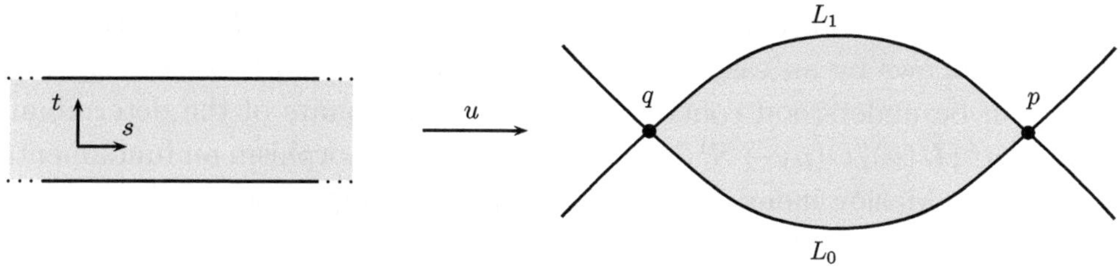

Fig. 2. A pseudo-holomorphic strip contributing to the Floer differential on $CF(L_0, L_1)$

Remark 1.7. In general, the sum in the right-hand side of (1.4) can be infinite. However, Gromov's compactness theorem ensures that, given any energy bound E_0, there are only finitely many homotopy classes $[u]$ with $\omega([u]) \leq E_0$ for which the moduli spaces $\mathcal{M}(p, q; [u], J)$ are non-empty. Thus, using Novikov coefficients and weighing counts of strips by area ensures that the sum in the right-hand side of (1.4) is well-defined.

However, it is sometimes possible to work over smaller coefficient fields. One such setting is that of *exact* Lagrangian submanifolds in an exact symplectic manifold. Namely, assume that $\omega = d\theta$ for some 1-form θ on M, and there exist functions $f_i \in C^\infty(L_i, \mathbb{R})$ such that $\theta_{|L_i} = df_i$ (for $i = 0, 1$). Then, by Stokes' theorem, any strip connecting intersection points p and q satisfies $\int u^*\omega = (f_1(q) - f_0(q)) - (f_1(p) - f_0(p))$. Thus, rescaling each generator by $p \mapsto T^{f_1(p) - f_0(p)} p$, we can eliminate the weights $T^{\omega([u])}$ from (1.4), and work directly over the coefficient field $\mathbb{K}$ instead of Λ.

Floer's construction [17] was subsequently extended to more general settings, beginning with Oh's result on monotone Lagrangians [32], and culminating with the sophisticated methods introduced by Fukaya, Oh, Ohta and Ono for the general case [19]; however as we will see below, Theorem 1.5 does not hold in full generality, as pseudo-holomorphic discs with boundary in L_0 or L_1 "obstruct" Floer cohomology.

1.3. Maslov Index and Grading

The Maslov index plays a similar role in the index formula for pseudo-holomorphic discs to that played by the first Chern class in that for closed pseudo-holomorphic curves; in fact it can be viewed as a relative version of the Chern class.

Denote by $LGr(n)$ the Grassmannian of Lagrangian n-planes in the symplectic vector space $(\mathbb{R}^{2n}, \omega_0)$. It is a classical fact that the unitary group $U(n)$ acts transitively on $LGr(n)$, so that $LGr(n) \simeq U(n)/O(n)$, from which it follows by an easy calculation that $\pi_1(LGr(n)) \simeq \mathbb{Z}$ (see e.g. [28]). This can be understood concretely by using the square of the determinant map, $\det^2 : U(n)/O(n) \to S^1$, which induces an isomorphism on fundamental groups; the Maslov index of a loop in $LGr(n)$ is simply the winding number of its image under this map.

In a similar vein, consider two paths $\ell_0, \ell_1 : [0,1] \to LGr(n)$ of Lagrangian subspaces in $\mathbb{R}^{2n}$, such that $\ell_0(0)$ is transverse to $\ell_1(0)$ and $\ell_0(1)$ is transverse to $\ell_1(1)$. The Maslov index of the path ℓ_1 relative to ℓ_0 is then the number of times (counting with signs and multiplicities) at which $\ell_0(t)$ and $\ell_1(t)$ are not transverse to each other. (More precisely, it is the intersection number of the path $(\ell_0(t), \ell_1(t))$ with the hypersurface in $LGr(n) \times LGr(n)$ consisting of non-transverse pairs of subspaces.)

We now return to our main discussion, and consider a map $u : \mathbb{R} \times [0,1] \to M$ satisfying the boundary conditions (1.2). Since $\mathbb{R} \times [0,1]$ is contractible, the pullback u^*TM is a trivial symplectic vector bundle; fixing a trivialization, we can view $\ell_0 = u^*_{|\mathbb{R}\times\{0\}} TL_0$ and $\ell_1 = u^*_{|\mathbb{R}\times\{1\}} TL_1$ as paths (oriented with s going from $+\infty$ to $-\infty$) in $LGr(n)$, one connecting T_pL_0 to T_qL_0 and the other connecting T_pL_1 to T_qL_1. The index of u can then be defined as the Maslov index of the path ℓ_1 relative to ℓ_0.

An equivalent definition, which generalizes more readily to the discs that appear in the definition of product operations, is as follows. Given a pair of transverse subspaces $\lambda_0, \lambda_1 \in LGr(n)$, and identifying $\mathbb{R}^{2n}$ with $\mathbb{C}^n$, there exists an element $A \in Sp(2n, \mathbb{R})$ which maps λ_0 to $\mathbb{R}^n \subset \mathbb{C}^n$ and λ_1 to $(i\mathbb{R})^n \subset \mathbb{C}^n$. The subspaces $\lambda_t = A^{-1}((e^{-i\pi t/2}\mathbb{R})^n)$, $t \in [0,1]$ then provide a distinguished homotopy class of path connecting λ_0 to λ_1 in $LGr(n)$, which we call the *canonical short path.*

Definition 1.8. Given $p, q \in L_0 \cap L_1$, denote by λ_p the canonical short path from T_pL_0 to T_pL_1 and by λ_q that from T_qL_0 to T_qL_1. Given a strip $u : \mathbb{R} \times [0,1] \to M$ connecting p to q, for $i \in \{0,1\}$, denote by ℓ_i the path $u^*_{|\mathbb{R}\times\{i\}} TL_i$ oriented with s going from $+\infty$ to $-\infty$, from T_pL_i to T_qL_i. View all these as paths in $LGr(n)$ by fixing a trivialization of u^*TM. The *index* of the strip u is then the Maslov index of the closed loop in $LGr(n)$ (based at T_qL_0) obtained by concatenating the paths $-\ell_0$ (i.e. ℓ_0 backwards), λ_p, ℓ_1, and finally $-\lambda_q$.

Example 1.9. Let $M = \mathbb{R}^2$, and consider the strip u depicted in Figure 2: then it is an easy exercise to check, using either definition, that $\operatorname{ind}(u) = 1$.

We now discuss the related issue of equipping Floer complexes with a grading. In order to obtain a $\mathbb{Z}$-grading on $CF(L_0,L_1)$, one needs to make sure that the index of a strip depends only on the difference between the degrees of the two generators it connects, rather than on its homotopy class. This is ensured by the following two requirements:

(1) The first Chern class of M must be 2-torsion: $2c_1(TM)=0$. This allows one to lift the Grassmannian $LGr(TM)$ of Lagrangian planes in TM (an $LGr(n)$-bundle over M) to a fiberwise universal cover $\widetilde{LGr}(TM)$, the Grassmannian of *graded Lagrangian planes* in TM (an $\widetilde{LGr}(n)$-bundle over M).

   Concretely, given a nowhere vanishing section Θ of $(\Lambda^n_{\mathbb{C}}T^*M)^{\otimes 2}$, the argument of Θ associates to any Lagrangian plane ℓ a *phase* $\varphi(\ell)=\arg(\Theta_{|\ell})\in S^1=\mathbb{R}/2\pi\mathbb{Z}$; a graded lift of ℓ is the choice of a real lift of $\tilde{\varphi}(\ell)\in\mathbb{R}$ of $\varphi(\ell)$.

(2) The *Maslov class* of L, $\mu_L\in\mathrm{Hom}(\pi_1(L),\mathbb{Z})=H^1(L,\mathbb{Z})$, vanishes. The Maslov class is by definition the obstruction to consistently choosing graded lifts of the tangent planes to L, i.e. lifting the section of $LGr(TM)$ over L given by $p\mapsto T_pL$ to a section of the infinite cyclic cover $\widetilde{LGr}(TM)$. The Lagrangian submanifold L together with the choice of such a lift is called a *graded Lagrangian submanifold* of M.

   Equivalently, given a nowhere vanishing section of $(\Lambda^n_{\mathbb{C}}T^*M)^{\otimes 2}$, we can associate to L its *phase function* $\varphi_L:L\to S^1$, which maps $p\in L$ to $\varphi(T_pL)\in S^1$; the Maslov class is then the homotopy class $[\varphi_L]\in[L,S^1]=H^1(L,\mathbb{Z})$, and a graded lift of L is the choice of a lift $\tilde{\varphi}_L: L\to\mathbb{R}$.

When these two assumptions are satisfied, fixing graded lifts $\tilde{L}_0,\tilde{L}_1$ of the Lagrangian submanifolds $L_0,L_1\subset M$ determines a natural $\mathbb{Z}$-grading on the Floer complex $CF(L_0,L_1)$ as follows. For all $p\in L_0\cap L_1$, we obtain a preferred homotopy class of path connecting T_pL_0 to T_pL_1 in $LGr(T_pM)$ by connecting the chosen graded lifts of the tangent spaces at p via a path in $\widetilde{LGr}(T_pM)$. Combining this path with $-\lambda_p$ (the canonical short path from T_pL_0 to T_pL_1, backwards), we obtain a closed loop in $LGr(T_pM)$; the degree of p is by definition the Maslov index of this loop. It is then easy to check that any strip connecting p to q satisfies

$$\mathrm{ind}(u)=\deg(q)-\deg(p). \tag{1.5}$$

In particular the Floer differential (1.4) has degree 1.

In general, if we do not restrict ourselves to symplectic manifolds with torsion $c_1(TM)$ and Lagrangian submanifolds with vanishing Maslov class, the natural grading on Floer cohomology is only by a finite cyclic group. As an important special case, if we simply assume that the Lagrangian submanifolds L_0, L_1 are oriented, then we have a $\mathbb{Z}/2$-grading, where the degree of a generator p of $CF(L_0, L_1)$ is determined by the sign of the intersection between L_0 and L_1 at p: namely $\deg(p) = 0$ if the canonical short path from T_pL_0 to T_pL_1 maps the given orientation of T_pL_0 to that of T_pL_1, and $\deg(p) = 1$ otherwise.

(Another approach, which we won't discuss further, is to enlarge the coefficient field by a formal variable of non-zero degree to keep track of the Maslov indices of different homotopy classes. In the monotone case, where index is proportional to symplectic area, it suffices to give a non-zero degree to the Novikov parameter T.)

1.4. Transversality and Compactness

We now discuss very briefly the fundamental technical issues of transversality and compactness.

Transversality of the moduli spaces of pseudo-holomorphic strips, i.e. the surjectivity of the linearized $\bar\partial$ operator at all solutions, is needed in order to ensure that the spaces $\widehat{\mathcal{M}}(p, q; [u], J)$ (and other moduli spaces we will introduce below) are smooth manifolds of the expected dimension. Still assuming that L_0 and L_1 intersect transversely, transversality for strips can be achieved by replacing the fixed almost-complex structure J in the Cauchy-Riemann equation (1.1) by a generic *family* of ω-compatible almost-complex structures which depend on the coordinate t in the strip $\mathbb{R} \times [0, 1]$.

A more basic issue is that of defining Floer cohomology for Lagrangian submanifolds which do not intersect transversely (in particular, one would like to be able to define the Floer cohomology of a Lagrangian with itself, i.e. the case $L_0 = L_1$). In view of the requirement of Hamiltonian isotopy invariance of the construction, the simplest approach is to introduce an inhomogeneous Hamiltonian perturbation term into the holomorphic curve equation: we fix a generic Hamiltonian $H \in C^\infty([0, 1] \times M, \mathbb{R})$, and consider the modified equation $(du - X_H \otimes dt)^{0,1} = 0$, i.e.

$$\frac{\partial u}{\partial s} + J(t, u)\left(\frac{\partial u}{\partial t} - X_H(t, u)\right) = 0, \tag{1.6}$$

still subject to the boundary conditions $u(s,0) \in L_0$ and $u(s,1) \in L_1$ and a finite energy condition. For $s \to \pm\infty$, the strip u converges no longer to intersection points but rather to trajectories of the flow of X_H which start on L_0 and end on L_1: thus the generators of the Floer complex $CF(L_0, L_1)$ are in fact defined to be flow lines $\gamma : [0,1] \to M$, $\dot\gamma(t) = X_H(t, \gamma(t))$, such that $\gamma(0) \in L_0$ and $\gamma(1) \in L_1$. Equivalently, by considering $\gamma(0)$, we set $\mathcal{X}(L_0, L_1) = L_0 \cap (\phi_H^1)^{-1}(L_1)$, where $\phi_H^1 \in \mathrm{Ham}(M, \omega)$ is the time 1 flow generated by H. In this sense, the generators are *perturbed* intersection points of L_0 with L_1, where the perturbation is given by the Hamiltonian diffeomorphism ϕ_H^1.

Remark 1.10. The perturbed equation (1.6) can be recast as a plain Cauchy-Riemann equation by the following trick: consider $\tilde u(s,t) = (\phi_H^t)^{-1}(u(s,t))$, where ϕ_H^t is the flow of X_H over the interval $[0,t]$. Then

$$\frac{\partial \tilde u}{\partial t} = (\phi_H^t)_*^{-1}\left(\frac{\partial u}{\partial t} - X_H\right),$$

so Floer's equation (1.6) becomes

$$\frac{\partial \tilde u}{\partial s} + \tilde J(t, \tilde u)\,\frac{\partial \tilde u}{\partial t} = 0,$$

where $\tilde J(t) = (\phi_H^t)_*^{-1}(J(t))$. Hence solutions to Floer's equation correspond to honest $\tilde J$-holomorphic strips with boundaries on L_0 and $(\phi_H^1)^{-1}(L_1)$.

Compactness of the moduli spaces is governed by Gromov's compactness theorem, according to which any sequence of J-holomorphic curves with uniformly bounded energy admits a subsequence which converges, up to reparametrization, to a nodal *tree* of J-holomorphic curves. The components of the limit curve are obtained as limits of different reparametrizations of the given sequence of curves, focusing on the different regions of the domain in which a non-zero amount of energy concentrates ("bubbling"). In the case of a sequence of J-holomorphic strips $u_n : \mathbb{R} \times [0,1] \to M$ with boundary on Lagrangian submanifolds L_0 and L_1, there are three types of phenomena to consider:

(1) *strip breaking*: energy concentrates at either end $s \to \pm\infty$, i.e. there is a sequence $a_n \to \pm\infty$ such that the translated strips $u_n(s - a_n, t)$ converge to a non-constant limit strip (Figure 3 left);

(2) *disc bubbling*: energy concentrates at a point on the boundary of the strip ($t \in \{0,1\}$), where suitable rescalings of u_n converge to a J-

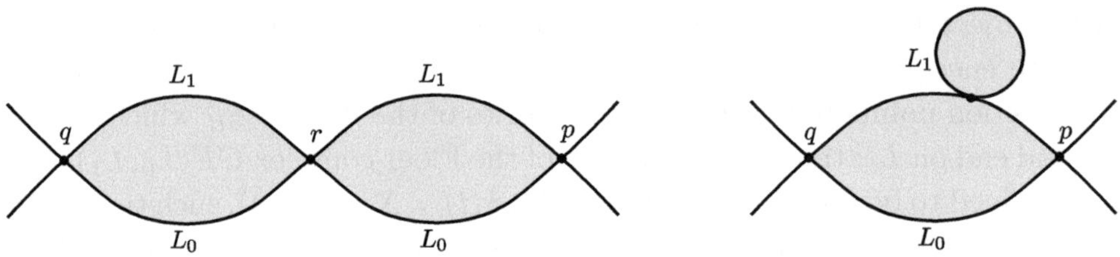

Fig. 3. Possible limits of pseudo-holomorphic strips: a broken strip (left) and a disc bubble (right)

holomorphic disc in M with boundary entirely contained in either L_0 or L_1 (Figure 3 right);

(3) *sphere bubbling*: energy concentrates at an interior point of the strip, where suitable rescalings of u_n converge to a J-holomorphic sphere in M.

As we will see below, strip breaking is the key geometric ingredient in the proof that the Floer differential squares to zero, *provided that disc bubbling can be excluded.* This is not simply a technical issue—in general the Floer differential does not square to zero, as illustrated by Example 1.11 below. Another issue posed by disc and sphere bubbling is that of transversality: the perturbation techniques we have outlined above are in general not sufficient to achieve transversality for limit curves that include disc or sphere bubble components. More sophisticated techniques, such as those proposed by Fukaya et al. [19][1], or the *polyfolds* developed by Hofer-Wysocki-Zehnder [23], are needed to extend Lagrangian Floer theory to the greatest possible level of generality.

In our case, the absence of disc and sphere bubbles is ensured by the assumption that $[\omega]\cdot\pi_2(M,L_i)=0$ in the statement of Theorem 1.5. A more general context in which the theory still works is when bubbling can be excluded for dimension reasons, for instance when all bubbles are guaranteed to have Maslov index greater than 2. (The important limit case where the minimal Maslov index is equal to 2 can also be handled by elementary methods; however, in that case disc bubbling can occur and the Floer differential does not automatically square to zero.) A common setting where an *a priori* lower bound on the Maslov index can be guaranteed is that of *monotone* Lagrangian submanifolds in monotone symplectic manifolds, i.e. when the

[1]The cautious reader should be aware that, as of this writing, the analytic foundations of this approach are still the subject of some controversy.

symplectic area of discs and their Maslov index are proportional to each other [32].

1.5. Sketch of Proof of Theorem 1.5

The proof that the Floer differential squares to zero (under the assumption that disc and sphere bubbling cannot occur) is conceptually similar to that for Morse (co)homology.

Fix Lagrangian submanifolds L_0 and L_1 as in Theorem 1.5, a generic almost-complex structure J and a Hamiltonian perturbation H so as to ensure transversality. Given two generators p, q of the Floer complex, and a homotopy class $[u]$ with $\mathrm{ind}([u]) = 2$, the moduli space $\mathcal{M}(p, q; [u], J)$ is a 1-dimensional manifold. Since our assumptions exclude the possibilities of disc or sphere bubbling, Gromov compactness implies that this moduli space can be compactified to a space $\overline{\mathcal{M}}(p, q; [u], J)$ whose elements are broken strips connecting p to q and representing the total class $[u]$.

Two-component broken strips of the sort depicted in Figure 3 (left) correspond to products of moduli spaces $\mathcal{M}(p, r; [u'], J) \times \mathcal{M}(r, q; [u''], J)$, where r is any generator of the Floer complex and $[u'] + [u''] = [u]$. Observe that the index is additive under such decompositions; moreover, transversality implies that any non-constant strip must have index at least 1. Thus, the only possibility is $\mathrm{ind}([u']) = \mathrm{ind}([u'']) = 1$, and broken configurations with more than two components cannot occur.

Conversely, a *gluing theorem* states that every broken strip is locally the limit of a unique family of index 2 strips, and $\overline{\mathcal{M}}(p, q; [u], J)$ is a 1-dimensional manifold with boundary, with

(1.7)
$$\partial\overline{\mathcal{M}}(p, q; [u], J) = \coprod_{\substack{r \in \mathcal{X}(L_0, L_1) \\ [u'] + [u''] = [u] \\ \mathrm{ind}([u']) = \mathrm{ind}([u'']) = 1}} \left(\mathcal{M}(p, r; [u'], J) \times \mathcal{M}(r, q; [u''], J)\right)$$

Moreover, the choice of orientations and spin structures on L_0 and L_1 equips all these moduli spaces with natural orientations, and (1.7) is compatible with these orientations (up to an overall sign). Since the total (signed) number of boundary points of a compact 1-manifold with boundary is always zero, we

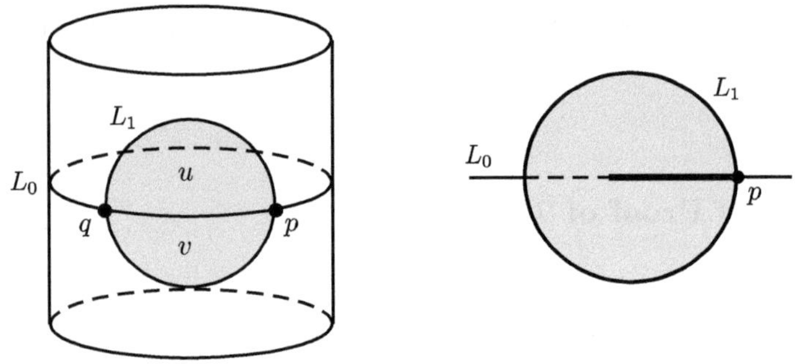

Fig. 4. A counterexample to $\partial^2 = 0$

conclude that

$$\sum_{\substack{r \in \mathcal{X}(L_0,L_1)\\ [u']+[u'']=[u]\\ \mathrm{ind}([u'])=\mathrm{ind}([u''])=1}} \big(\#\mathcal{M}(p,r;[u'],J)\big)\,\big(\#\mathcal{M}(r,q;[u''],J)\big)\,T^{\omega([u'])+\omega([u''])} = 0. \tag{1.8}$$

Summing over all possible $[u]$, the left-hand side is precisely the coefficient of q in $\partial^2(p)$; therefore $\partial^2 = 0$.

When L_0 and/or L_1 bound J-holomorphic discs, the sum (1.8) no longer cancels, because the boundary of the 1-dimensional moduli space $\mathcal{M}(p,q;[u],J)$ also contains configurations with disc bubbles. The following example shows that this is an issue even in the monotone case.

Example 1.11. Consider again the cylinder $M = \mathbb{R} \times S^1$, and let L_0 be a simple closed curve that goes around the cylinder once, and L_1 a homotopically trivial curve intersecting L_0 in two points p and q, as shown in Figure 4 left. Then L_0 and L_1 bound precisely two holomorphic strips of index 1, denoted by u and v in Figure 4. (There are other holomorphic discs with boundary on L_0 and L_1 but those have higher index.) Comparing with the convention depicted in Figure 2, u is a trajectory from p to q, while v is a trajectory from q to p: thus we have

$$\partial p = \pm T^{\omega(u)} q \quad \text{and} \quad \partial q = \pm T^{\omega(v)} p,$$

and $\partial^2 \neq 0$. To understand why $\partial^2(p) \neq 0$, consider the moduli space of index 2 holomorphic strips connecting p to itself. The images of these strips exactly cover the disc bounded by L_1, with a slit along L_0, as shown in Figure 4 right.

We can give an explicit description using local coordinates in which L_0 corresponds to the real axis and L_1 to the unit circle: using the upper half-disc minus the points ± 1 as domain of our maps instead of the usual $\mathbb{R} \times [0,1]$ (to which it is biholomorphic), one easily checks that any index 2 strip connecting p to itself can be parametrized as

$$u_\alpha(z) = \frac{z^2 + \alpha}{1 + \alpha z^2}$$

for some $\alpha \in (-1,1)$ (corresponding to the end point of the slit).

The two ends of this moduli space are different: when $\alpha \to -1$, energy concentrates at $z = \pm 1$, and the index 2 strips u_α converge to a broken strip whose nonconstant components are the index 1 strips u and v; whereas for $\alpha \to 1$ the maps u_α exhibit disc bubbling at $z = i$, the limit being a constant strip at p together with a disc bubble whose image is the disc bounded by L_1. Thus, broken strips do not cancel in pairs in the manner needed for $\partial^2 = 0$ to hold.

Once the Floer differential is shown to square to zero, it remains to prove that Floer cohomology does not depend on the choice of almost-complex structure and Hamiltonian perturbation. Recall that the spaces of such choices are contractible. Thus, given two choices (H,J) and (H',J') (for which we assume transversality holds), let $(H(\tau),J(\tau))$, $\tau \in [0,1]$ be a (generically chosen) smooth family which agrees with (H,J) for $\tau = 0$ and (H',J') for $\tau = 1$. One can then construct a *continuation map* $F : CF(L_0,L_1;H,J) \to CF(L_0,L_1;H',J')$ by counting solutions to the equation

$$\frac{\partial u}{\partial s} + J(\tau(s),t,u)\left(\frac{\partial u}{\partial t} - X_H(\tau(s),t,u)\right) = 0, \tag{1.9}$$

where $\tau(s)$ is a smooth function of s which equals 1 for $s \ll 0$ and 0 for $s \gg 0$. Unlike (1.6), the Equation (1.9) is not invariant under translations in the s direction. Given generators $p \in \mathcal{X}(L_0,L_1;H)$ and $p' \in \mathcal{X}(L_0,L_1;H')$ of the respective Floer complexes, the coefficient of p' in $F(p)$ is defined as a count of index 0 solutions to (1.9) which converge to p at $s \to +\infty$ and to p' at $s \to -\infty$ (weighted by energy as usual).

The proof that F is a chain map, i.e. satisfies $\partial' \circ F = F \circ \partial$ (again assuming the absence of bubbling), comes from studying spaces of index 1 solutions to (1.9). These spaces are 1-dimensional manifolds, whose end points correspond to broken trajectories where the main component is an index 0 solution to (1.9), either preceded by an index 1 J-holomorphic strip with perturbation data H (if energy concentrates at $s \to +\infty$), or followed by an index 1 J'-holomorphic strip with perturbation data H' (if energy concentrates at $s \to -\infty$). The composition $F \circ \partial$ counts the first type of limit configuration, while $\partial' \circ F$ counts the second type of limit configuration, and the equality between these two maps follows again from the statement that the total (signed) number of end points of a compact 1-manifold with boundary is zero.

Using the reverse homotopy, i.e., considering (1.9) with $\tau(s) = 0$ for $s \ll 0$ and 1 for $s \gg 0$, one similarly defines a chain map $F' : CF(L_0, L_1; H', J') \to CF(L_0, L_1; H, J)$. The chain maps F and F' are quasi-inverses, i.e. $F' \circ F$ is homotopic to identity (and similarly for $F \circ F'$). An explicit homotopy can be obtained by counting index -1 solutions to a one-parameter family of equations similar to (1.9) but where $\tau(s)$ is 0 near $\pm\infty$ and is nonzero over an interval of values of s of varying width.

1.6. The Floer Cohomology $HF(L, L)$

The Floer cohomology of a Lagrangian submanifold with itself is of particular interest in the context of Arnold's conjecture. By Weinstein's Lagrangian neighborhood theorem, a neighborhood of a Lagrangian submanifold L in (M, ω) is symplectomorphic to a neighborhood of the zero section of the cotangent bundle T^*L with its standard symplectic form. In light of this, we first consider the model case of the cotangent bundle.

Example 1.12. Let N be a compact real n-dimensional manifold, and consider the cotangent bundle T^*N, with its standard exact symplectic form (given locally by $\omega = \sum dq_i \wedge dp_i$, where q_i are local coordinates on N and p_i are the dual coordinates on the fibers of the cotangent bundle). Let L_0 be the zero section, and given a Morse function $f : N \to \mathbb{R}$ and a small $\epsilon > 0$, denote by L_1 the graph of the exact 1-form $\epsilon\, df$. Then L_0, L_1 are exact Lagrangian submanifolds of T^*N, Hamiltonian isotopic to each other (the Hamiltonian isotopy is generated by $H = \epsilon\, f \circ \pi$ where $\pi : T^*N \to N$ is the bundle map); L_0 and L_1 intersect transversely at the critical points of f.

Choosing a graded lift of L_0, and transporting it through the Hamiltonian isotopy to define a graded lift of L_1, we obtain a grading on the Floer complex $CF(L_0, L_1)$; by an explicit calculation, a critical point p of f of Morse index $i(p)$ defines a generator of the Floer complex of degree $\deg(p) = n - i(p)$. Thus, the grading on the Floer complex agrees with that on the complex $CM^*(f)$ which defines the Morse cohomology of f.

The Morse differential counts index 1 trajectories of the gradient flow between critical points of f, and depends on the choice of a Riemannian metric

g on N, which we assume to satisfy the Morse-Smale transversality condition. A result of Floer [18] is that, for a suitable choice of (time-dependent) almost-complex structure J on T^*N, solutions of Floer's equation

$$\frac{\partial u}{\partial s} + J(t,u)\frac{\partial u}{\partial t} = 0$$

with boundary on L_0 and L_1 are regular and in one-to-one correspondence with gradient flow trajectories

$$\dot{\gamma}(s) = \epsilon \nabla f\big(\gamma(s)\big)$$

on N, the correspondence being given by $\gamma(s) = u(s,0)$. (Note: an ascending gradient flow line with $\gamma(s)$ converging to p as $s \to +\infty$ and q as $s \to -\infty$ counts as a trajectory from p to q in the Morse differential.)

To understand this correspondence between moduli spaces, observe that, at any point x of the zero section, the natural almost-complex structure on T^*N induced by the metric g maps the horizontal vector $\epsilon\nabla f(x) \in T_xN \subset T_x(T^*N)$ to the vertical vector $X_H(x) = \epsilon\, df(x) \in T^*_xN \subset T_x(T^*N)$. This allows us to construct particularly simple solutions of (1.6) for this almost-complex structure and the Hamiltonian perturbation $-H$, with both boundaries of the strip mapping to L_0: for any gradient flow line γ of f, we obtain a solution of (1.6) by setting $u(s,t) = \gamma(s)$. Floer's construction of strips with boundary on L_0 and L_1 is equivalent to this via Remark 1.10.

Thus, for specific choices of perturbation data, after a rescaling of the generators by $p \mapsto T^{\epsilon f(p)}p$, the Floer complex of (L_0, L_1) is isomorphic to the Morse complex of f, and the Floer cohomology $HF^*(L_0, L_1)$ is isomorphic to the Morse cohomology of f (with coefficients in Λ). Using the independence of Floer cohomology under Hamiltonian isotopies and the isomorphism between Morse and ordinary cohomology, we conclude that $HF^*(L_0, L_0) \simeq HF^*(L_0, L_1) \simeq H^*(L_0; \Lambda)$.

(Since we are in the exact case, by Remark 1.7 one could actually work directly over $\mathbb{K}$ rather than over Novikov coefficients.)

Now we consider the general case of a compact Lagrangian submanifold L in a symplectic manifold (M,ω), under the assumption that $[\omega]\cdot\pi_2(M,L) = 0$. Energy estimates then imply that, for a sufficiently small Hamiltonian perturbation, the pseudo-holomorphic strips that determine the Floer cohomology $HF^*(L,L)$ must all be contained in a small tubular neighborhood of L, so that the calculation of Floer cohomology reduces to Example 1.12, and we get the following result (due to Floer in the exact case and for $\mathbb{K} = \mathbb{Z}_2$):

Proposition 1.13. *If* $[\omega]\cdot\pi_2(M,L)=0$, *then* $HF^*(L,L)\simeq H^*(L;\Lambda)$.

Together with Theorem 1.5, this implies Arnold's conjecture (Theorem 1.1).

Example 1.14. Let L be the zero section in $T^*S^1=\mathbb{R}\times S^1$ (see Figure 1 left), and consider the Hamiltonian perturbation depicted in the figure, which comes from a Morse function on $L=S^1$ with a maximum at p and a minimum at q. Then L and $\psi(L)$ bound two index 1 holomorphic strips (shaded on the figure), both connecting p to q, and with equal areas. However, the contributions of these two strips to the Floer differential cancel out (this is obvious over $\mathbb{K}=\mathbb{Z}_2$; when $\mathrm{char}(\mathbb{K})\neq 2$ a verification of signs is needed). Thus, $\partial p=0$, and $HF^*(L,\psi(L))\simeq H^*(S^1)$, as expected from Proposition 1.13.

Things are different when L bounds pseudo-holomorphic discs, and the Floer cohomology $HF^*(L,L)$ (when it is defined) is in general smaller than $H^*(L;\Lambda)$. For example, let L be a monotone Lagrangian submanifold in a monotone symplectic manifold, with minimal Maslov index at least 2; this is a setting where $HF^*(L,L)$ is well defined (though no longer $\mathbb{Z}$-graded), as disc bubbles either do not occur at all or occur in cancelling pairs. Using again a small multiple ϵf of a Morse function f on L as Hamiltonian perturbation, the Floer complex differs from the Morse complex $CM^*(f)$ by the presence of additional terms in the differential; namely there are index 1 Floer trajectories representing a class in $\pi_2(M,L)$ of Maslov index k and connecting a critical point p of Morse index $i(p)$ to a critical point q of index $i(q)=i(p)+k-1$. This situation was studied by Oh [32, 33], who showed that the Floer complex is filtered by index (or equivalently energy), and there is a *spectral sequence* starting with the Morse cohomology $HM^*(f)$ (or equivalently the ordinary cohomology of L), whose successive differentials account for classes of increasing Maslov index in $\pi_2(M,L)$, and converging to the Floer cohomology $HF^*(L,L)$.

It is often easier to study honest pseudo-holomorphic discs with boundary on L, rather than solutions of Floer's equation with a Hamiltonian perturbation, or strips with boundary on L and its image under a small isotopy. This has led to the development of alternative constructions of $HF^*(L,L)$. For instance, another model for the Floer cohomology of a monotone Lagrangian submanifold is the *pearl complex* first introduced in [34] (see also [15]). In this model, the generators of the Floer complex are again the critical points of a Morse function f on L, but the differential counts "pearly trajectories", which arise as limits of Floer trajectories of the sort considered above as $\epsilon \to 0$. Namely, a pearly trajectory between critical points p and q of f consists of $r \geq 0$ pseudo-holomorphic discs in M with boundary in L, connected to each other and to p and q by $r+1$ gradient flow lines of f in L. (When there are no discs, a pearly trajectory is simply a gradient flow line between p and q.) Yet another model, proposed by Fukaya-Oh-Ohta-Ono [19], uses a chain complex where $CF(L,L) = C_*(L)$ consists of chains in L, and the differential is the sum of the classical boundary map and a map defined in terms of moduli spaces of pseudo-holomorphic discs with boundary on L. This model is computationally convenient, but requires great care in its construction to address questions such as exactly what sort of chains are considered and, in the general (non-monotone) case, how to achieve transversality of the evaluation maps.

2. Product Operations

2.1. The Product

Let L_0, L_1, L_2 be three Lagrangian submanifolds of (M, ω), which we assume intersect each other transversely and do not bound any pseudo-holomorphic discs. We now define a product operation on their Floer complexes, i.e. a map

$$CF(L_1, L_2) \otimes CF(L_0, L_1) \longrightarrow CF(L_0, L_2).$$

Given intersection points $p_1 \in \mathcal{X}(L_0, L_1)$, $p_2 \in \mathcal{X}(L_1, L_2)$, and $q \in \mathcal{X}(L_0, L_2)$, the coefficient of q in $p_2 \cdot p_1$ is a weighted count of pseudo-holomorphic discs in M with boundary on $L_0 \cup L_1 \cup L_2$ and with corners at p_1, p_2, q. More precisely, let D be the closed unit disc minus three boundary points, say for instance $z_0 = -1, z_1 = e^{-i\pi/3}, z_2 = e^{i\pi/3}$, and observe that a neighborhood of each puncture in D is conformally equivalent to a strip (i.e., the product of an infinite interval with $[0,1]$).

Given an almost-complex structure J on M and a homotopy class $[u]$, we denote by $\mathcal{M}(p_1, p_2, q; [u], J)$ the space of finite energy J-holomorphic maps $u : D \to M$ which extend continuously to the closed disc, mapping the boundary arcs from z_0 to z_1, z_1 to z_2, z_2 to z_0 to L_0, L_1, L_2 respectively, and the boundary punctures z_1, z_2, z_0 to p_1, p_2, q respectively, in the given homotopy class $[u]$ (see Figure 5).

As in the case of strips, the expected dimension of $\mathcal{M}(p_1, p_2, q; [u], J)$ is given by the index of the linearized Cauchy-Riemann operator $D_{\bar{\partial}_J, u}$. This

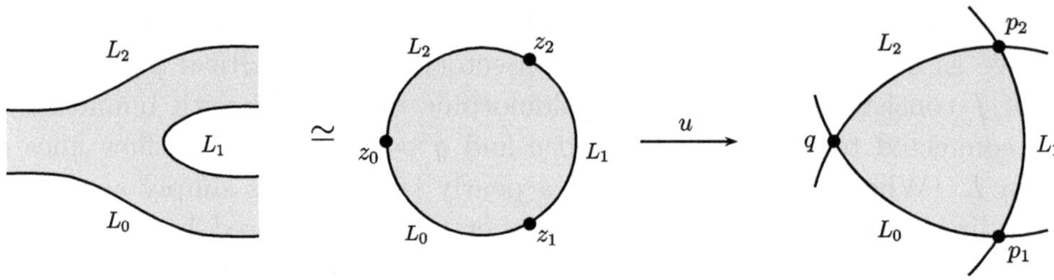

Fig. 5. A pseudo-holomorphic disc contributing to the product map

index can be expressed in terms of the Maslov index, exactly as in Definition 1.8: we now concatenate the paths given by the tangent spaces to L_0, L_1, L_2 going counterclockwise along the boundary of u, together with the appropriate canonical short paths at p_1, p_2, q, to obtain a closed loop in $LGr(n)$ whose Maslov index is equal to $\operatorname{ind}(u)$. If $c_1(TM)$ is 2-torsion and the Maslov classes of L_0, L_1, L_2 vanish, then after choosing graded lifts of the Lagrangians we have $\mathbb{Z}$-gradings on the Floer complexes, and one checks that

$$\operatorname{ind}(u) = \deg(q) - \deg(p_1) - \deg(p_2). \tag{2.1}$$

Remark 2.1. The apparent lack of symmetry in the index formula (2.1) is due to the difference between the gradings on $CF(L_0, L_2)$ and $CF(L_2, L_0)$. Namely, the given intersection point $q \in L_0 \cap L_2$ defines generators of both complexes, whose degrees sum to n (the dimension of L_i). In fact, the Floer complexes $CF(L_0, L_2)$ and $CF(L_2, L_0)$ and the differentials on them are dual to each other, provided that the almost-complex structures and perturbations are chosen suitably. For instance, the strip depicted in Figure 2 is a trajectory from p to q in the Floer complex $CF(L_0, L_1)$, and from q to p in $CF(L_1, L_0)$.

Assume that transversality holds, so that the moduli spaces $\mathcal{M}(p_1, p_2, q; [u], J)$ are smooth manifolds; if $\operatorname{char}(\mathbb{K}) \neq 2$, assume moreover that orientations and spin structures on L_0, L_1, L_2 have been chosen, so as to determine orientations of the moduli spaces. Then we define:

Definition 2.2. The Floer product is the Λ-linear map $CF(L_1, L_2) \otimes CF(L_0, L_1) \to CF(L_0, L_2)$ defined by

$$p_2 \cdot p_1 = \sum_{\substack{q \in \mathcal{X}(L_0, L_2) \\ [u]:\operatorname{ind}([u])=0}} \big(\#\mathcal{M}(p_1, p_2, q; [u], J)\big) T^{\omega([u])} q. \tag{2.2}$$

As in the previous section, in general this construction needs to be modified by introducing domain-dependent almost-complex structures and Hamiltonian perturbations to achieve transversality. We discuss this below, but for now we assume transversality holds without further perturbations and examine the properties of the Floer product.

Proposition 2.3. *If $[\omega] \cdot \pi_2(M, L_i) = 0$ for all i, then the Floer product satisfies the Leibniz rule (with suitable signs) with respect to the Floer differentials,*

$$\partial(p_2 \cdot p_1) = \pm(\partial p_2) \cdot p_1 \pm p_2 \cdot (\partial p_1), \tag{2.3}$$

and hence induces a well-defined product $HF(L_1, L_2) \otimes HF(L_0, L_1) \to HF(L_0, L_2)$. Moreover, this induced product on Floer cohomology groups is independent of the chosen almost-complex structure (and Hamiltonian perturbations) and associative.

(However, the chain-level product on Floer complexes is *not* associative, as we will see below.)

We now sketch the geometric argument behind the Leibniz rule, which relies on an examination of index 1 moduli spaces of J-holomorphic discs and their compactification. Namely, consider a triple of generators p_1, p_2, q as above, and let $[u]$ be a homotopy class with $\mathrm{ind}([u]) = 1$. Then (still assuming transversality) $\mathcal{M}(p_1, p_2, q; [u], J)$ is a smooth 1-dimensional manifold, and by Gromov compactness admits a compactification $\overline{\mathcal{M}}(p_1, p_2, q; [u], J)$ obtained by adding nodal trees of J-holomorphic curves.

Since our assumptions exclude bubbling of discs or spheres, the only phenomenon that can occur is strip-breaking (when energy concentrates at one of the three ends of the punctured disc D). Since transversality excludes the presence of discs of index less than 0 and nonconstant strips of index less than 1, and since the sum of the indices of the limit components must be 1, there are only three types of limit configurations to be considered, all consisting of an index 0 disc with boundary on L_0, L_1, L_2 and an index 1 strip with boundary on two of these three submanifolds; see Figure 6.

The three types of configurations contribute to the coefficient of $T^{\omega([u])}q$ in $\partial(p_2 \cdot p_1)$ (Figure 6 left), $(\partial p_2) \cdot p_1$ (middle), and $p_2 \cdot (\partial p_1)$ (right) respectively. On the other hand, a gluing theorem states that every such configuration arises as an end of $\mathcal{M}(p_1, p_2, q; [u], J)$, and that the compactified moduli space is a 1-dimensional compact manifold with boundary. Moreover,

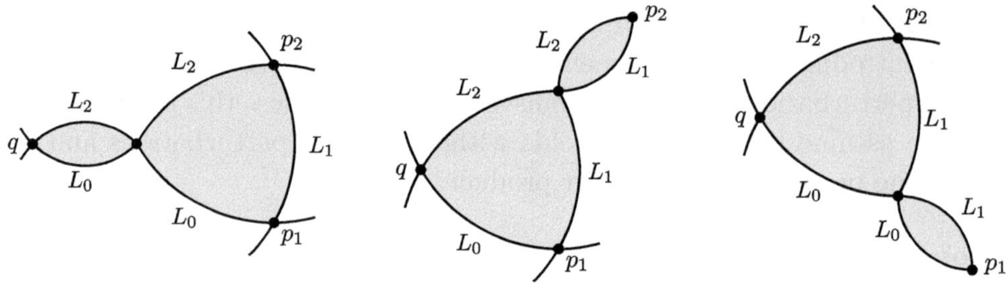

Fig. 6. The ends of a 1-dimensional moduli space $\mathcal{M}(p_1,p_2,q;[u],J)$

the orientations agree up to overall sign factors depending only on the degrees of p_1 and p_2. Since the (signed) total number of boundary points of $\overline{\mathcal{M}}(p_1,p_2,q;[u],J)$ is zero, the Leibniz rule (2.3) follows.

Before moving on to higher products, we briefly discuss the issue of transversality and compatibility in the choice of perturbations. As in the case of strips, even without assuming that L_0, L_1, L_2 intersect transversely, we can ensure transversality by introducing domain-dependent almost-complex structures and Hamiltonian perturbations; however, for the Leibniz rule to hold, these need to be chosen suitably near the punctures z_0, z_1, z_2. Fix once and for all "strip-like ends" near the punctures, i.e. biholomorphisms from $\mathbb{R}_+ \times [0,1]$ (resp. $\mathbb{R}_- \times [0,1]$) to neighborhoods of the punctures z_1 and z_2 (resp. z_0) in D; we denote by $s+it$ the natural complex coordinate in each strip-like end. Also fix a 1-form $\beta \in \Omega^1(D)$, such that $\beta_{|\partial D}=0$ and $\beta = dt$ in each strip-like end. Now, given L_0, L_1, L_2, we choose a family of ω-compatible almost-complex structures depending smoothly on $z \in D$, i.e. $J \in C^\infty(D, \mathcal{J}(M,\omega))$, and a family of Hamiltonians $H \in C^\infty(D \times M, \mathbb{R})$, with the property that in each strip-like end $J(z)$ and $H(z)$ depend only on the coordinate $t \in [0,1]$. We then perturb the Cauchy-Riemann equation to

$$(du - X_H \otimes \beta)^{0,1}_J = 0, \tag{2.4}$$

which in each strip-like end reduces to (1.6).

For $0 \le i < j \le 2$, denote by $H_{ij} \in C^\infty([0,1] \times M, \mathbb{R})$ and $J_{ij} \in C^\infty([0,1], \mathcal{J}(M,\omega))$ the time-dependent Hamiltonians and almost-complex structures on the strip-like end whose boundaries map to L_i and L_j. The solutions of (2.4) converge no longer to intersection points of $L_i \cap L_j$, but to trajectories of the time 1 flow generated by H_{ij} which begin on L_i and end on L_j, i.e. generators of the perturbed Floer complex of (L_i, L_j) with respect to the Hamiltonian perturbation H_{ij}. Moreover, when strip breaking occurs, the main component remains a solution of (2.4), while the strip component that breaks off is a solution of (1.6) with respect to H_{ij} and J_{ij}.

Thus, by considering the moduli spaces of solutions to the perturbed equation (2.4) and proceeding as in Definition 2.2, we obtain a product map

$$CF(L_1, L_2; H_{12}, J_{12}) \otimes CF(L_0, L_1; H_{01}, J_{01}) \longrightarrow CF(L_0, L_2; H_{02}, J_{02})$$

on the perturbed Floer complexes, and Proposition 2.3 still holds (with respect to the perturbed Floer differentials).

2.2. Higher Operations

Given $k+1$ Lagrangian submanifolds $L_0, \ldots, L_k$, a construction similar to those above allows us to define an operation

$$\mu^k : CF(L_{k-1}, L_k) \otimes \cdots \otimes CF(L_1, L_2) \otimes CF(L_0, L_1) \longrightarrow CF(L_0, L_k)$$

(of degree $2-k$ in the situation where the Floer complexes are graded), where μ^1 is the Floer differential and μ^2 is the product.

Given generators $p_i \in \mathcal{X}(L_{i-1}, L_i)$ $(i = 1, \ldots, k)$ and $q \in \mathcal{X}(L_0, L_k)$, the coefficient of q in $\mu^k(p_k, \ldots, p_1)$ is a count (weighted by area) of (perturbed) pseudo-holomorphic discs in M with boundary on $L_0 \cup \cdots \cup L_k$ and corners at $p_1, \ldots, p_k, q$.

Specifically, one considers maps $u : D \to M$ whose domain D is the closed unit disc minus $k+1$ boundary points $z_0, z_1, \ldots, z_k \in S^1$, lying in that order along the unit circle. The positions of these marked points are not fixed, and the moduli space $\mathcal{M}_{0,k+1}$ of conformal structures on the domain D, i.e., the quotient of the space of ordered $(k+1)$-tuples of points on S^1 by the action of $\mathrm{Aut}(D^2)$, is a contractible $(k-2)$-dimensional manifold.

Given an almost-complex structure J on M and a homotopy class $[u]$, we denote by $\mathcal{M}(p_1, \ldots, p_k, q; [u], J)$ the space of J-holomorphic maps $u : D \to M$ (where the positions of $z_0, \ldots, z_k$ are not fixed a priori) which extend continuously to the closed disc, mapping the boundary arcs from z_i to z_{i+1} (or z_0 for $i = k$) to L_i, and the boundary punctures $z_1, \ldots, z_k, z_0$ to $p_1, \ldots, p_k, q$ respectively, in the given homotopy class $[u]$, up to the action of $\mathrm{Aut}(D^2)$ by reparametrization. (Or, equivalently, one can avoid quotienting and instead take a slice for the reparametrization action by fixing the positions of three of the z_i.)

For a fixed conformal structure on D, the index of the linearized Cauchy-Riemann operator is again given by the Maslov index, as previously. Thus, accounting for deformations of the conformal structure on D, assuming transversality, the expected dimension of the moduli space is

$$\dim \mathcal{M}(p_1,\dots,p_k,q;[u],J) = k-2+\operatorname{ind}([u]) = k-2+\deg(q)-\sum_{i=1}^{k}\deg(p_i). \tag{2.5}$$

Thus, assuming transversality, and choosing orientations and spin structures on $L_0,\dots,L_k$ if $\operatorname{char}(\mathbb{K})\neq 2$, we define:

Definition 2.4. The operation $\mu^k : CF(L_{k-1},L_k)\otimes\cdots\otimes CF(L_0,L_1)\to CF(L_0,L_k)$ is the Λ-linear map defined by

$$\mu^k(p_k,\dots,p_1) = \sum_{\substack{q\in\mathcal{X}(L_0,L_k)\\ [u]:\operatorname{ind}([u])=2-k}} \big(\#\mathcal{M}(p_1,\dots,p_k,q;[u],J)\big)T^{\omega([u])}\,q. \tag{2.6}$$

Remark 2.5. As before, in general this construction needs to be modified by introducing domain-dependent almost-complex structures and Hamiltonian perturbations to achieve transversality. Thus, we actually count solutions of a perturbed Cauchy-Riemann equation similar to (2.4), involving a domain-dependent almost-complex structure $J\in C^\infty(D,\mathcal{J}(M,\omega))$ and Hamiltonian $H\in C^\infty(D\times M,\mathbb{R})$. As before, compatibility with strip-breaking requires that, in each of the $k+1$ strip-like ends near the punctures of D, the chosen J and H depend only on the coordinate $t\in[0,1]$ and agree with the almost-complex structures and Hamiltonians used to construct the Floer complexes $CF(L_i,L_{i+1})$ and $CF(L_0,L_k)$. An additional compatibility condition comes from the possible degenerations of the domain D to unions of discs with fewer punctures, as discussed below: we need to require that, when D degenerates in such a way, H and J are translation-invariant in the strip-like regions connecting the components and agree with the choices made in the construction of the Floer complexes $CF(L_i,L_j)$, while in each component H and J agree with the choices made for that moduli space of discs with fewer punctures. This forces the choices of H and J to further depend on the conformal structure of D. We refer the reader to [42] for a detailed construction (and proof of existence) of compatible and consistent choices of perturbation data (H,J).

The algebraic properties of μ^k follow from the study of the limit configurations that arise in compactifications of 1-dimensional moduli spaces

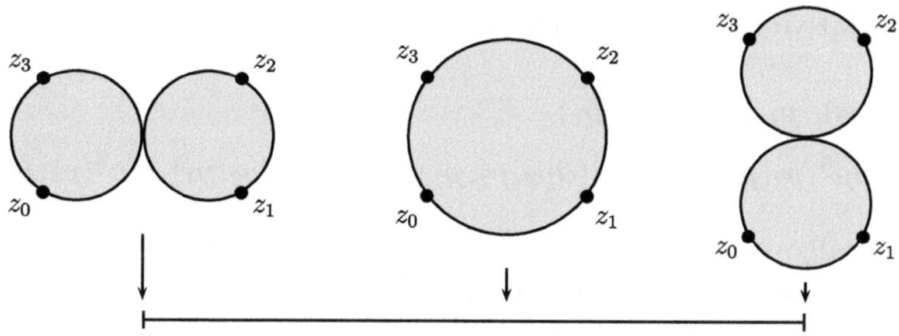

Fig. 7. The 1-dimensional associahedron $\overline{\mathcal{M}}_{0,4}$

of (perturbed) pseudo-holomorphic discs; besides strip breaking, there are now other possibilities, corresponding to cases where the domain D degenerates. The moduli space of conformal structures $\mathcal{M}_{0,k+1}$ admits a natural compactification to a $(k-2)$-dimensional polytope $\overline{\mathcal{M}}_{0,k+1}$, the Stasheff *associahedron*, whose top-dimensional facets correspond to nodal degenerations of D to a pair of discs $D_1 \cup D_2$, with each component carrying at least two of the marked points $z_0, \dots, z_k$; and the higher codimension faces correspond to nodal degenerations with more components.

Example 2.6. $\overline{\mathcal{M}}_{0,4}$ is homeomorphic to a closed interval, whose end points correspond to configurations where two adjacent marked points come together (Figure 7). For example, fixing the positions of z_0, z_1, z_2 on the unit circle and letting z_3 vary along the arc from z_2 to z_0, the right end point corresponds to the case where z_3 approaches z_2; the "main" component of the limit configuration carries the marked points z_0 and z_1, while the component carrying z_2 and z_3 arises from rescaling by suitable automorphisms of the disc. Equivalently up to automorphisms of the disc, one could instead fix the positions of z_1, z_2, z_3, and let z_0 vary along the arc from z_3 to z_1; the right end point then corresponds to the case where z_0 approaches z_1.

Proposition 2.7. *If* $[\omega] \cdot \pi_2(M, L_i) = 0$ *for all* i*, then the operations* μ^k *satisfy the* A_∞-relations

$$\sum_{\ell=1}^{k} \sum_{j=0}^{k-\ell} (-1)^* \mu^{k+1-\ell}(p_k, \dots, p_{j+\ell+1}, \mu^\ell(p_{j+\ell}, \dots, p_{j+1}), p_j, \dots, p_1) = 0, \tag{2.7}$$

where $* = j + \deg(p_1) + \dots + \deg(p_j)$.

The case $k=1$ of (2.7) is the identity $\partial^2 = 0$, while $k=2$ corresponds to the Leibniz rule (2.3). For $k=3$, it expresses the fact that the Floer product

μ^2 is associative up to an explicit homotopy given by μ^3:

(2.8)
$$\begin{aligned} &\pm (p_3 \cdot p_2) \cdot p_1 \pm p_3 \cdot (p_2 \cdot p_1) \\ &\quad = \pm \partial \mu^3(p_3, p_2, p_1) \pm \mu^3(\partial p_3, p_2, p_1) \pm \mu^3(p_3, \partial p_2, p_1) \pm \mu^3(p_3, p_2, \partial p_1). \end{aligned}$$

More generally, each operation μ^k gives an explicit homotopy for a certain compatibility property among the preceding ones.

The proof of Proposition 2.7 again relies on an analysis of 1-dimensional moduli spaces of (perturbed) J-holomorphic discs and their compactification. Fix generators $p_1, \ldots, p_k, q$ and a homotopy class $[u]$ with $\mathrm{ind}([u]) = 3-k$, and assume that J and H are chosen generically (so as to achieve transversality) and compatibly (see Remark 2.5). Then the moduli space $\mathcal{M}(p_1, \ldots, p_k, q; [u], J)$ compactifies to a 1-dimensional manifold with boundary, whose boundary points correspond either to an index 1 (perturbed) J-holomorphic strip breaking off at one of the $k+1$ marked points, or to a degeneration of the domain to the boundary of $\overline{\mathcal{M}}_{0,k+1}$, i.e. to a pair of discs with each component carrying at least two of the marked points. The first case corresponds to the terms involving μ^1 in (2.7), while the second case corresponds to the other terms.

Example 2.8. For $k=3$, limit configurations consisting of an index 1 strip together with an index -1 disc with 4 marked points account for the right-hand side in (2.8), while those consisting of a pair of index 0 discs with 3 marked points (when the domain degenerates to one of the two end points of $\overline{\mathcal{M}}_{0,4}$, see Figure 7) account for the two terms in the left-hand side.

2.3. The Fukaya Category

There are several variants of the Fukaya category of a symplectic manifold, depending on the desired level of generality and a number of implementation details. The common features are the following. The objects of the Fukaya category are suitable Lagrangian submanifolds, equipped with extra data, and morphism spaces are given by Floer complexes, endowed with the Floer differential. Composition of morphisms is given by the Floer product, which is only associative up to homotopy, and the Fukaya category is an A_∞-*category*, i.e. the differential and composition are the first two in a sequence of operations

$$\mu^k : \mathrm{hom}(L_{k-1}, L_k) \otimes \cdots \otimes \mathrm{hom}(L_0, L_1) \to \mathrm{hom}(L_0, L_k)$$

(of degree $2-k$ when a $\mathbb{Z}$-grading is available), satisfying the A_∞-relations (2.7).

Given the setting in which we have developed Floer theory in the preceding sections, the most natural definition is the following:

Definition 2.9. Let (M,ω) be a symplectic manifold with $2c_1(TM)=0$. The objects of the (compact) Fukaya category $\mathcal{F}(M,\omega)$ are compact closed, oriented, spin Lagrangian submanifolds $L \subset M$ such that $[\omega]\cdot\pi_2(M,L)=0$ and with vanishing Maslov class $\mu_L = 0 \in H^1(L,\mathbb{Z})$, together with extra data, namely the choice of a spin structure and a graded lift of L. (We will usually omit those from the notation and simply denote the object by L.)

For every pair of objects (L,L') (not necessarily distinct), we choose perturbation data $H_{L,L'} \in C^\infty([0,1]\times M,\mathbb{R})$ and $J_{L,L'} \in C^\infty([0,1],\mathcal{J}(M,\omega))$; and for all tuples of objects $(L_0,\dots,L_k)$ and all moduli spaces of discs, we choose consistent perturbation data (H,J) compatible with the choices made for the pairs of objects (L_i,L_j), so as to achieve transversality for all moduli spaces of perturbed J-holomorphic discs. (See [42, §9] for the existence of such perturbation data.)

Given this, we set $\hom(L,L') = CF(L,L';H_{L,L'},J_{L,L'})$; and the differential μ^1, composition μ^2, and higher operations μ^k are given by counts of perturbed pseudo-holomorphic discs as in Definition 2.4. By Proposition 2.7, this makes $\mathcal{F}(M,\omega)$ a Λ-linear, $\mathbb{Z}$-graded, non-unital (but cohomologically unital [42]) A_∞-category.

One can also consider other settings: for example, we can drop the requirement that $2c_1(TM)=0$ and the assumption of vanishing of the Maslov class if we are content with a $\mathbb{Z}/2$-grading; spin structures can be ignored if we work over a field of characteristic 2; and Novikov coefficients are unnecessary if we restrict ourselves to exact Lagrangian submanifolds in an exact symplectic manifold.

As is obvious from the definition, the actual chain-level details of the Fukaya category depend very much on the choice of perturbation data; however, the A_∞-categories obtained from various choices of perturbation data are *quasi-equivalent* (i.e., they are related by A_∞-functors which induce equivalences, in fact in this case isomorphisms, at the level of cohomology) [42].

We finish this section with a few remarks.

Remark 2.10. One can recover an honest category from an A_∞-category by taking the cohomology of morphism spaces with respect to the differential μ^1; the A_∞-relations imply that μ^2 descends to an associative composition operation on cohomology. The cohomology category of $\mathcal{F}(M,\omega)$, where $\hom(L,L') = HF(L,L')$ and composition is given by the cohomology-level Floer product, is sometimes called the Donaldson-Fukaya category. However, the higher operations contain important information that gets lost when passing to the cohomology category, and it is usually much better to work with the chain-level A_∞-category (see for instance the next section).

Remark 2.11. In the context of homological mirror symmetry, one is naturally led to consider a slightly richer version of the Fukaya category, whose objects are Lagrangian submanifolds equipped with *local systems*, i.e. flat vector bundles $\mathcal{E} \to L$ with unitary holonomy (over the Novikov field over $\mathbb{K} = \mathbb{C}$). In this situation, we define

$$CF\big((L_0,\mathcal{E}_0),(L_1,\mathcal{E}_1)\big) = \bigoplus_{p \in \mathcal{X}(L_0,L_1)} \hom(\mathcal{E}_{0|p},\mathcal{E}_{1|p}),$$

and modify the definition of μ^k as follows. Fix objects $(L_0,\mathcal{E}_0),\dots,(L_k,\mathcal{E}_k)$, intersections $p_1,\dots,p_k,q$, and a homotopy class $[u]$. Set $p_0 = p_{k+1} = q$ for simplicity. Parallel transport along the portion of the boundary of $[u]$ that lies on L_i yields an isomorphism $\gamma_i \in \hom(\mathcal{E}_{i|p_i},\mathcal{E}_{i|p_{i+1}})$ for each $i = 0,\dots,k$. Now, given elements $\rho_i \in \hom(\mathcal{E}_{i-1|p_i},\mathcal{E}_{i|p_i})$ $(i = 1,\dots,k)$, the composition of all these linear maps defines an element $\eta_{[u],\rho_k,\dots,\rho_1} = \gamma_k \cdot \rho_k \cdot \dots \cdot \gamma_1 \cdot \rho_1 \cdot \gamma_0 \in \hom(\mathcal{E}_{0|q},\mathcal{E}_{k|q})$. Then we set

$$\mu^k(\rho_k,\dots,\rho_1) = \sum_{\substack{q \in \mathcal{X}(L_0,L_k)\\ [u]:\mathrm{ind}([u])=2-k}} (\#\mathcal{M}(p_1,\dots,p_k,q;[u],J))\,T^{\omega([u])}\,\eta_{[u],\rho_k,\dots,\rho_1}.$$

Remark 2.12. It is in principle possible to lift the assumption $[\omega]\cdot\pi_2(M,L) = 0$ we have made throughout, at the expense of considerable analytic and algebraic difficulties in situations where disc bubbling occurs. Analytically, disc bubbles pose transversality problems that cannot be solved with the techniques we have described above. Algebraically, they lead to a *curved* A_∞-category, i.e. for each object L we have an element $\mu^0_L \in \hom(L,L)$ which encodes a weighted count of J-holomorphic discs bounded by L. The A_∞-relations (2.7) are then modified by allowing the case $\ell = 0$ in the sum. For example, the relation for $k = 1$ becomes

$$\mu^1(\mu^1(p)) + (-1)^{\deg p}\mu^2(\mu^0_{L_1},p) + \mu^2(p,\mu^0_{L_0}) = 0,$$

where the last two terms correspond to disc bubbling along either edge of an index 2 strip. To regain some sanity, one usually considers not arbitrary objects, but *weakly unobstructed* objects, i.e. those for which μ_L^0 is a scalar multiple of the (cohomological) unit of $\hom(L,L)$ (this multiple is sometimes called "central charge" or "superpotential" in the context of mirror symmetry); this happens for instance when the minimal Maslov index of a holomorphic disc with boundary on L is equal to two and Maslov index 2 discs are regular. Weakly unobstructed objects of fixed central charge then form an honest A_∞-category. The curious reader is referred to [19].

3. Exact Triangles and Generators

While it is usually impossible to classify all Lagrangian submanifolds of a given symplectic manifold, or even to directly compute Floer cohomology for all those we can find, it is often possible to understand the whole Fukaya category in terms of a small subset of *generating objects—provided that we understand not only differentials and products but also higher operations among those generators.* To understand how this comes about, a healthy dose of homological algebra is necessary; in this section we give a very brief and informal overview of exact triangles, twisted complexes and generators, in general and as they pertain to Fukaya categories in particular. The first part of [42] fills in the many details that we omit here, and more.

3.1. Exact Triangles and Mapping Cones

An exact triangle

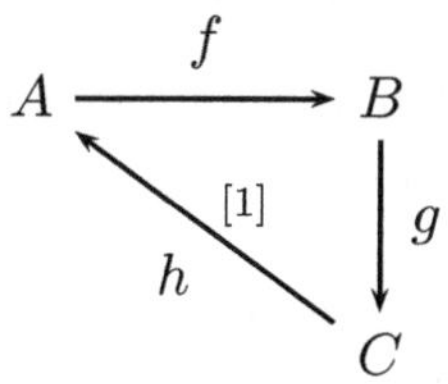

in an A_∞-category $\mathcal{A}$ consists of a triple of objects A, B, C and closed morphisms $f \in \hom^0(A,B)$, $g \in \hom^0(B,C)$, $h \in \hom^1(C,A)$ such that C is (up to quasi-isomorphism) a *mapping cone* of $f : A \to B$, with g and h the natural maps to and from it. We will clarify the meaning of this definition in the next section; for now, we simply mention some key features and motivate the concept.

Exactness means that the compositions $\mu^2(g,f)$, $\mu^2(h,g)$ and $\mu^2(f,h)$ are exact, i.e. in the cohomology category $H(\mathcal{A})$ the maps compose to zero. (However, their triple Massey product is typically nontrivial.) An exact triangle induces long exact sequences on morphism spaces in the cohomology category: for every test object T, we have a long exact sequence

$$(3.1) \qquad \cdots \to H^i \hom(T,A) \xrightarrow{f} H^i \hom(T,B) \xrightarrow{g} H^i \hom(T,C) \xrightarrow{h} H^{i+1} \hom(T,A) \xrightarrow{f} \cdots$$

where $H^i \hom(T,A)$ is the cohomology of $\hom(T,A)$ with respect to the differential μ^1, and the maps are given by composition (in the cohomology category) with f, g, and h; and similarly (in the contravariant direction) for morphisms from A,B,C to T. Moreover, as T varies these long exact sequences fit together naturally with respect to the multiplicative action of the groups $H^* \hom(T',T)$, i.e. (3.1) fits into an exact sequence of modules over $H(\mathcal{A})$.

Exact triangles can also be characterized as images under A_∞-functors of a "universal" abstract exact triangle living in an A_∞-category with three objects [42, §3g].

The A_∞-category $\mathcal{A}$ is said to be *triangulated* if every closed morphism $f: A \to B$ can be completed to an exact triangle (and the shift functor [1] acting on $\mathcal{A}$ by change of gradings is a quasi-equivalence); or, in other terms, if all morphisms in $\mathcal{A}$ have mapping cones. Here it is important to point out a key difference with the case of ordinary triangulated categories, where the triangles are an additional piece of structure on the category: the A_∞-structure is rich enough to "know" about triangles, and triangles automatically satisfy an analogue of the usual axioms. In the same vein, A_∞-functors are always exact, i.e. map exact triangles to exact triangles.

Before saying more about mapping cones in A_∞-categories, let us discuss some classical motivating examples.

Example 3.1. The mapping cone of a continuous map $f: X \to Y$ between topological spaces is, by definition, the space obtained from $X \times [0,1]$ by attaching Y to $X \times \{1\}$ via the map f and collapsing $X \times \{0\}$ to a point:

$$\mathrm{Cone}(f) = \big((X \times [0,1]) \sqcup Y\big)/(x,0) \sim (x',0),\ (x,1) \sim f(x) \quad \forall x, x' \in X.$$

We then have a sequence of maps

$$X \xrightarrow{f} Y \xrightarrow{i} \mathrm{Cone}(f) \xrightarrow{p} \Sigma X \xrightarrow{\Sigma f} \Sigma Y \to \cdots,$$

where i is the inclusion of Y into the mapping cone, and p is the projection to the suspension of X obtained by collapsing Y. The composition of any two of these maps is nullhomotopic, and the induced maps on (co)homology form a long exact sequence.

Example 3.2. The notion of mapping cone in the category of chain complexes is directly modelled on the previous example: let $A = (\bigoplus A^i, d_A)$ and $B = (\bigoplus B^i, d_B)$ be two chain complexes, and let $f : A \to B$ be a chain map (i.e., a collection of maps $f^i : A^i \to B^i$ satisfying $d_B f^i + f^{i+1} d_A = 0$). Then the mapping cone of f is, by definition, the chain complex $C = A[1] \oplus B$ (i.e., $C^i = A^{i+1} \oplus B^i$), equipped with the differential

$$d_C = \begin{pmatrix} d_A & 0 \\ f & d_B \end{pmatrix}.$$

The map f, the inclusion of B into C as a subcomplex, and the projection of C onto the quotient complex $A[1]$ then fit into an exact sequence.

Example 3.3. Let A be an algebra (resp. differential graded algebra or A_∞-algebra), and consider the category of differential graded modules (resp. A_∞-modules) over A. Recall that such a module M is a chain complex equipped with a degree 1 differential d_M and a multiplication map $A \otimes M \to M$, $(a, m) \mapsto a \cdot m$, satisfying the Leibniz rule and associative (up to homotopies given by higher structure maps $\mu_M^{k|1} : A^{\otimes k} \otimes M \to M[1-k]$, in the case of A_∞-modules). The mapping cone of a module homomorphism $f : M \to N$ can then be defined essentially as in the previous example. In the differential graded case, f is a chain map compatible with the multiplication, and the mapping cone of f as a chain complex inherits a natural module structure. For A_∞-modules, recalling that an A_∞-homomorphism is a collection of maps $f^{k|1} : A^{\otimes k} \otimes M \to N[-k]$ (where the linear term $f^{0|1}$ is a chain map compatible with the product $\mu^{1|1}$ up to a homotopy given by $f^{1|1}$, and so on), the structure maps $\mu_K^{k|1} : A^{\otimes k} \otimes K \to K[1-k]$ $(k \geq 0)$ of the mapping cone $K = M[1] \oplus N$ are given by

$$\begin{aligned} &\mu_K^{k|1}(a_1, \ldots, a_k, (m, n)) \\ &\quad = \big(\mu_M^{k|1}(a_1, \ldots, a_k, m), f^{k|1}(a_1, \ldots, a_k, m) + \mu_N^{k|1}(a_1, \ldots, a_k, n)\big). \end{aligned}$$

3.2. Twisted Complexes

When an A_∞-category $\mathcal{A}$ is not known to be triangulated, it is often advantageous to embed it into a larger category in which mapping cones are guaranteed to exist. For example, one can always do so by using the *Yoneda embedding* construction into the category of A_∞-modules over $\mathcal{A}$ (in which mapping cones always exist, cf. Example 3.3); see e.g. [42, §1]. A milder construction, which retains more features of the original category $\mathcal{A}$, involves *twisted complexes*. We give a brief outline, and refer the reader to [42, §3] for details.

Definition 3.4. A *twisted complex* (E, δ^E) consists of:

- a formal direct sum $E = \bigoplus_{i=1}^N E_i[k_i]$ of shifted objects of $\mathcal{A}$ (i.e., a finite collection of pairs (E_i, k_i) where $E_i \in \operatorname{ob}\mathcal{A}$ and $k_i \in \mathbb{Z}$);
- a strictly lower triangular differential $\delta^E \in \operatorname{End}^1(E)$, i.e. a collection of maps $\delta^E_{ij} \in \operatorname{Hom}^{k_j - k_i + 1}(E_i, E_j)$, $1 \le i < j \le N$, satisfying the equation

$$\sum_{k \ge 1} \mu^k(\delta^E, \ldots, \delta^E) = 0, \tag{3.2}$$

i.e., $\sum_{k\ge 1} \sum_{i=i_0<i_1<\cdots<i_k=j} \mu^k(\delta^E_{i_{k-1}i_k}, \ldots, \delta^E_{i_0 i_1}) = 0$ for all $1 \le i < j \le N$.

A degree d morphism of twisted complexes is simply a degree d map between the underlying formal direct sums, i.e. if $E = \bigoplus E_i[k_i]$ and $E' = \bigoplus E'_j[k'_j]$ then an element of $\operatorname{Hom}^d(E, E')$ is by definition a collection of morphisms $a_{ij} \in \operatorname{Hom}^{d + k'_j - k_i}(E_i, E'_j)$.

Finally, given twisted complexes $(E_0, \delta^0), \ldots, (E_k, \delta^k)$, $k \ge 1$, and morphisms $a_i \in \operatorname{Hom}(E_{i-1}, E_i)$, we set

$$\mu^k_{\mathrm{Tw}}(a_k, \ldots, a_1) = \sum_{j_0, \ldots, j_k \ge 0} \mu^{k + j_0 + \cdots + j_k}(\underbrace{\delta^k, \ldots, \delta^k}_{j_k}, a_k, \ldots, \underbrace{\delta^1, \ldots, \delta^1}_{j_1}, a_1, \underbrace{\delta^0, \ldots, \delta^0}_{j_0}).$$

(The sum is finite since each δ^i is strictly lower triangular.)

Proposition 3.5. *The above construction defines a triangulated A_∞-category which we denote by* $\operatorname{Tw}\mathcal{A}$, *and into which $\mathcal{A}$ embeds fully faithfully.*

It is instructive to see how twisted complexes relate to ordinary chain complexes:

Example 3.6. Given objects A, B, C of $\mathcal{A}$ and $f \in \hom^0(A, B)$, $g \in \hom^0(A, C)$, we can consider $(A[2] \oplus B[1] \oplus C, \delta = f + g)$, conventionally denoted by

$$\{A \xrightarrow{f} B \xrightarrow{g} C\}.$$

This forms a twisted complex if and only if $\mu^1(f) = \mu^1(g) = 0$ and $\mu^2(g,f) = 0$, i.e. f and g are closed morphisms and their composition is zero. However, we can also introduce an extra term $h \in \hom^{-1}(A,C)$ into the differential δ, in which case the last condition becomes $\mu^2(g,f) + \mu^1(h) = 0$: thus it is sufficient for the composition of f and g to be exact, with a homotopy given by h.

Definition 3.7. Given twisted complexes $(E,\delta), (E',\delta') \in \mathrm{Tw}\,\mathcal{A}$ and a closed morphism $f \in \hom^0(E,E')$ (i.e., such that $\mu^1_{\mathrm{Tw}}(f) = 0$), the *abstract mapping cone* of f is the twisted complex

$$\mathrm{Cone}(f) = \left(E[1] \oplus E', \begin{pmatrix} \delta & 0 \\ f & \delta' \end{pmatrix}\right).$$

Given objects A, B, C of $\mathcal{A}$ and a closed morphism $f \in \hom^0(A,B)$, we say that C is a *mapping cone* of f if, in the category of twisted complexes $\mathrm{Tw}\,\mathcal{A}$, the object C is quasi-isomorphic to the abstract mapping cone of f, $\{A \xrightarrow{f} B\} = (A[1] \oplus B, f)$.

When C is a mapping cone of $f : A \to B$, by composing the inclusion of B into the abstract mapping cone (resp. the projection to $A[1]$) with the given quasi-isomorphism from the abstract mapping cone to C (resp. its quasi-inverse) we obtain morphisms $i : B \to C$ and $p : C \to A[1]$, which sit with f in an exact triangle.

3.3. Exact Triangles in the Fukaya Category

The reader may legitimately wonder about the relevance of the above discussion to Fukaya categories. It turns out that at least some mapping cones in the Fukaya category of a symplectic manifold can be understood geometrically. There are two well-known sources of these: Dehn twists, and Lagrangian connected sums.

3.3.1. Dehn Twists. The symplectic geometry of Dehn twists was first considered by Arnold, and later studied extensively by Seidel [41, 42]. The local model is as follows. In the cotangent bundle T^*S^n equipped with

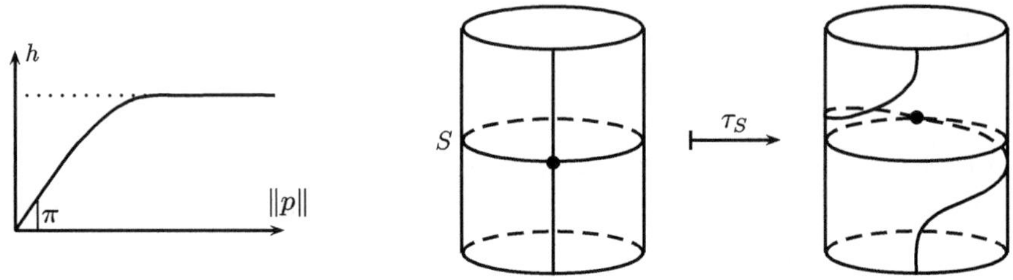

Fig. 8. The generating Hamiltonian on the complement of the zero section in T^*S^n, and the action of the Dehn twist on a cotangent fiber

its canonical symplectic form, a Hamiltonian of the form $H(p,q) = h(\|p\|)$ (where p is the fiber coordinate and $\|\cdot\|$ is the standard metric) generates a rescaled version of geodesic flow. Choosing $h : [0,\infty) \to \mathbb{R}$ so that $h'(0) = \pi$, $h'' \leq 0$, and h is constant outside of a neighborhood of zero, we obtain a Hamiltonian diffeomorphism of the complement of the zero section $T^*S^n \setminus S^n$, which can be extended across the zero section by the antipodal map on S^n to obtain a symplectomorphism of T^*S^n (see Figure 8).

Now, given a Lagrangian sphere S in a symplectic manifold (M,ω), by Weinstein's theorem a neighborhood of S in M is symplectomorphic to a neighborhood of the zero section in T^*S^n; thus, performing the above construction inside the standard neighborhood of S, we obtain a symplectomorphism τ_S, the *Dehn twist* about S, which is supported in a neighborhood of S and maps S to itself antipodally. (Note: τ_S depends on the choices made in the construction, but its isotopy class doesn't.)

Theorem 3.8 (Seidel [41, 42]). *Given a Lagrangian sphere S and any object L of $\mathcal{F}(M,\omega)$, there is an exact triangle in $\mathrm{Tw}\,\mathcal{F}(M,\omega)$,*

$$
\begin{array}{ccc}
HF^*(S,L)\otimes S & \xrightarrow{\ ev\ } & L \\
& {}_{[1]}\nwarrow & \downarrow \\
& & \tau_S(L)
\end{array}
\qquad (3.3)
$$

In other terms, the object $\tau_S(L)$ of $\mathcal{F}(M,\omega)$ is quasi-isomorphic in $\mathrm{Tw}\,\mathcal{F}(M,\omega)$ to the abstract mapping cone of ev.

In (3.3), $HF^*(S,L) \otimes S$ is a direct sum of shifted copies of S, with one summand for each generator of $HF^*(S,L)$, and *ev* is a tautological evaluation map, mapping each summand to L by a closed morphism representing the given generator of $HF^*(S,L) = H^*\mathrm{Hom}(S,L)$.

Given a test object T, the corresponding long exact sequence (3.1) is Seidel's long exact sequence in Floer cohomology [41] associated to the Dehn twist τ_S for all T, L:

$$\cdots \to HF^*(S,L) \otimes HF^*(T,S) \xrightarrow{\mu^2} HF^*(T,L) \longrightarrow HF^*\big(T, \tau_S(L)\big) \xrightarrow{[1]} \cdots$$

3.3.2. Lagrangian Connected Sums. Given two Lagrangian submanifolds L_1, L_2 which intersect transversely in a single point p, we can form the *Lagrangian connected sum* (or *surgery* in the terminology of [38] and [21]) $L_1 \# L_2$. One possible construction is as follows. For $\epsilon > 0$, the graph of the 1-form $\epsilon\, d\log\|x\|$ on $\mathbb{R}^n$, given by the equations $y_i = \epsilon\, x_i/\|x\|^2$, is a Lagrangian submanifold of $T^*\mathbb{R}^n \simeq \mathbb{C}^n$ which is asymptotic to the zero section (i.e., $\mathbb{R}^n \subset \mathbb{C}^n$) as $\|x\| \to \infty$ and to the cotangent fiber over zero (i.e., $(i\mathbb{R})^n \subset \mathbb{C}^n$) as $\|y\| \to \infty$; using suitable cut-off functions, we can modify this Lagrangian so that it agrees with $\mathbb{R}^n \cup (i\mathbb{R})^n$ outside of a small neighborhood of the origin. Pasting this local model into a suitable Darboux chart centered at the intersection point p and chosen so that $T_pL_1 = \mathbb{R}^n$ and $T_pL_2 = (i\mathbb{R})^n$ yields $L_1 \# L_2$. (Note that, for a single connected sum operation, the end result is independent of the size parameter ϵ and other choices up to Hamiltonian isotopy; not so when summing at multiple points. Also note that $L_2 \# L_1$ is *not* isotopic to $L_1 \# L_2$.)

Remark 3.9. When L_2 is a sphere, $L_1 \# L_2$ is Hamiltonian isotopic to $\tau_{L_2}(L_1)$; this provides the basis for an alternative description of the connected sum operation.

Given some other Lagrangian submanifold T (in generic position relatively to L_1 and L_2), choosing ϵ small enough in the above construction ensures that the intersections of T with $L_1 \# L_2$ are the same as with $L_1 \cup L_2$. Fukaya-Oh-Ohta-Ono [21] have studied the moduli spaces of J-holomorphic discs bounded by $L_1 \# L_2$ and T. Their main result is that, for suitable J and small enough ϵ, J-holomorphic strips with boundary on T and $L_1 \# L_2$ connecting an intersection in $T \cap L_2$ to one in $T \cap L_1$ are in bijection with J-holomorphic triangles bounded by T, L_2 and L_1 with a corner at p, whereas the counts of rigid strips in the other direction vanish. This is elementary in dimension 1, as illustrated by Figure 9, but much harder in higher dimensions.

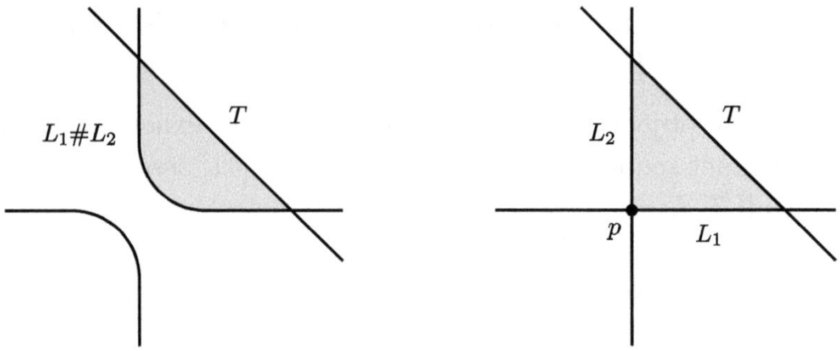

Fig. 9. The Lagrangian connected sum $L_1\#L_2$ vs. $L_1 \cup L_2$

The outcome is that, as a chain complex, $CF(T, L_1\#L_2)$ is the mapping cone of the map $\mu^2(p,\cdot): CF(T, L_2) \to CF(T, L_1)$ given by multiplication by the generator p of $CF(L_2, L_1)$. Hence, the short exact sequence

$$0 \to CF(T, L_1) \to CF(T, L_1\#L_2) \to CF(T, L_2) \to 0$$

induces a long exact sequence

$$\cdots \to HF(T, L_1) \longrightarrow HF(T, L_1\#L_2) \longrightarrow HF(T, L_2)$$
$$\xrightarrow{\mu^2([p],\cdot)} HF(T, L_1) \to \cdots$$

By an analogous argument for higher structure maps, one expects that this long exact sequence can be upgraded to an exact triangle in the Fukaya category,

$$\text{(3.4)} \qquad \begin{array}{ccc} L_2 & \xrightarrow{\;p\;} & L_1 \\ & {\scriptstyle [1]}\nwarrow & \downarrow \\ & & L_1\#L_2 \end{array}$$

i.e., $L_1\#L_2$ is quasi-isomorphic to the twisted complex $\mathrm{Cone}(p) = \{L_2 \xrightarrow{p} L_1\}$.

(If L_2 is a sphere, this is Seidel's exact triangle for the Dehn twist of L_1 about L_2.)

Remark 3.10. Recall that, by definition, the differential μ^1_{Tw} on $\hom(T, \mathrm{Cone}(p))$ involves not only the original Floer differential μ^1, but also multiplication by the differential of the twisted complex, i.e. $\mu^2(p,\cdot)$. This is exactly consistent with the above description of J-holomorphic strips with boundary on T and $L_1\#L_2$. Thus, replacing Lagrangian submanifolds by quasi-isomorphic twisted complexes built out of simpler Lagrangians, while computationally powerful, comes at the expense of having to consider higher operations on their Floer complexes (in this case, the expression for μ^1_{Tw} involves μ^2, and similarly that for μ^2_{Tw} involves μ^3).

3.4. Generation and Yoneda Embedding

3.4.1. Generators and Split-Generators.

Definition 3.11. The objects $G_1, \dots, G_r$ are said to *generate* the A_∞-category $\mathcal{A}$ if, in $\mathrm{Tw}\,\mathcal{A}$, every object of $\mathcal{A}$ is quasi-isomorphic to a twisted complex built from copies of $G_1, \dots, G_r$. (In other terms, every object of $\mathcal{A}$ can be obtained from $G_1, \dots, G_r$ by taking iterated mapping cones.)

The objects $G_1, \dots, G_r$ are said to *split-generate* $\mathcal{A}$ if every object of $\mathcal{A}$ is quasi-isomorphic to a direct summand in a twisted complex built from copies of $G_1, \dots, G_r$.

Example 3.12. Consider the Fukaya category of the torus T^2 with its standard area form. Starting from the standard curves α and β along the two factors of the torus, by taking iterated mapping cones we can obtain simple closed curves representing all nontrivial primitive elements in $\pi_1(T^2) = \mathbb{Z}^2$. For instance, the loop $\tau_\alpha(\beta) \simeq \beta \# \alpha$ (Figure 10 left) is quasi-isomorphic to the mapping cone of $p \in \mathrm{Hom}(\alpha, \beta)$; further applications of the Dehn twists τ_α and τ_β (which generate the mapping class group of T^2) eventually yield simple closed curves in all primitive homotopy classes. However, the objects obtained in this manner all satisfy a certain "balancing" condition: given a 1-form $\theta \in \Omega^1(T^2 \setminus \{pt\})$ with $d\theta = \omega$ and such that $\int_\alpha \theta = \int_\beta \theta = 0$, θ also integrates to zero on all iterated mapping cones built from α and β. For instance, all the simple closed curves that can be obtained in a given homotopy class are Hamiltonian isotopic to each other. Thus, α and β generate the subcategory of $\mathcal{F}(T^2)$ consisting of Lagrangians which are balanced with respect to θ, but not all of $\mathcal{F}(T^2)$.

On the other hand, given the two loops β and γ shown on Figure 10 right, the mapping cone of $T^{a_1} q_1 + T^{a_2} q_2 \in \mathrm{Hom}(\gamma, \beta)$ can be interpreted geometrically as the connected sum of β and γ at their two intersection points q_1 and q_2, with different gluing parameters. This mapping cone is therefore quasi-isomorphic to the direct sum of two simple closed curves in the homotopy class of α, but whose Hamiltonian isotopy classes depend on

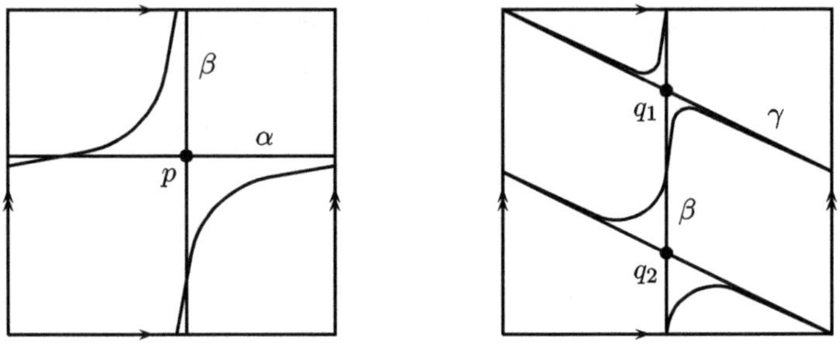

Fig. 10. Split-generating the Fukaya category of T^2

a_1 and a_2. Thus, by considering direct summands in mapping cones we can obtain all nontrivial simple closed curves up to Hamiltonian isotopy, rather than only those that are balanced: α and β split-generate $\mathcal{F}(T^2)$.

3.4.2. Yoneda Embedding. Let $G_1, \dots, G_r$ be split-generators of the A_∞-category $\mathcal{A}$. Then the endomorphism algebra of $G_1 \oplus \dots \oplus G_r$,

$$\mathcal{G} = \bigoplus_{i,j=1}^{r} \hom(G_i, G_j)$$

is an A_∞-algebra (with structure maps given by the operations μ^k of $\mathcal{A}$). Next, given any object L of $\mathcal{A}$,

$$\mathcal{Y}(L) = \bigoplus_{i=1}^{r} \hom(G_i, L)$$

is a (right) A_∞-module over $\mathcal{G}$, with differential given by μ^1, multiplication $\mu^{1|1}$ given by the operations

$$\hom(G_j, L) \otimes \hom(G_i, G_j) \xrightarrow{\mu^2} \hom(G_i, L),$$

and so on (the structure map $\mu^{1|k}$ of $\mathcal{Y}(L)$ is given by μ^{k+1}).

Moreover, to a morphism $a \in \hom(L, L')$ we can associate an A_∞-homomorphism $\mathcal{Y}(a) \in \hom_{\text{mod-}\mathcal{G}}(\mathcal{Y}(L), \mathcal{Y}(L'))$, whose linear term is given by composition with a.

The assignment $L \mapsto \mathcal{Y}(L)$, $a \mapsto \mathcal{Y}(a)$ is in turn the linear term of an A_∞-functor $\mathcal{Y}$, which is the restriction to the given set of objects $G_1, \dots, G_r$ of the A_∞ *Yoneda embedding* $\mathcal{A} \to \text{mod-}\mathcal{A}$ (see e.g. [42, §1]):

Proposition 3.13. *The above construction extends to an A_∞-functor $\mathcal{Y}$ from $\mathcal{A}$ to mod-$\mathcal{G}$. Moreover, if $G_1, \dots, G_r$ split-generate $\mathcal{A}$ then this A_∞-functor is a fully faithful quasi-embedding.*

4. The Wrapped Fukaya Category, Examples and Applications

In this section we assume that (M, ω) is a *Liouville manifold*, i.e. an exact symplectic manifold such that the Liouville vector field Z associated to the chosen primitive $\theta \in \Omega^1(M)$ of the symplectic form (i.e., the conformally symplectic vector field defined by $\iota_Z\omega = \theta$) is complete and outward pointing at infinity. More precisely, we require that M contains a compact domain M^{in} with boundary a smooth hypersurface ∂M on which $\alpha = \theta_{|\partial M}$ is a contact form, and Z is positively transverse to ∂M and has no zeroes outside of M^{in}. The flow of Z can then be used to identify $M \setminus M^{in}$ with the positive symplectization $(1, \infty) \times \partial M$ equipped with the exact symplectic form $\omega = d(r\alpha)$ and the Liouville field $Z = r\frac{\partial}{\partial r}$.

In this setting it is natural to consider not only compact exact Lagrangian submanifolds as we have done above, but also some noncompact ones with suitable behavior at infinity. There are two different types of such noncompact Fukaya categories, depending on the manner in which perturbations at infinity are used to define Floer complexes. One possibility is to perform "small" perturbations at infinity, restricting oneself to a smaller set of "admissible" objects which go to infinity along well-controlled directions. Two constructions that follow this philosophy are the "infinitesimal" Fukaya category first defined by Nadler and Zaslow for cotangent bundles [31] and later extended to Liouville manifolds equipped with a choice of Lagrangian skeleton; and Fukaya categories of Lefschetz fibrations as constructed by Seidel [42, 44], and their putative generalization to Landau-Ginzburg models, in which the behavior at infinity is controlled by a projection to the complex plane. Here we focus on the other approach, which is to consider *large* perturbations at infinity, leading to the *wrapped Fukaya category* of Abouzaid and Seidel [3, 9]. For completeness we mention the nascent subject of *partially wrapped* Fukaya categories, which attempt to interpolate between these two approaches (cf. e.g. [13]).

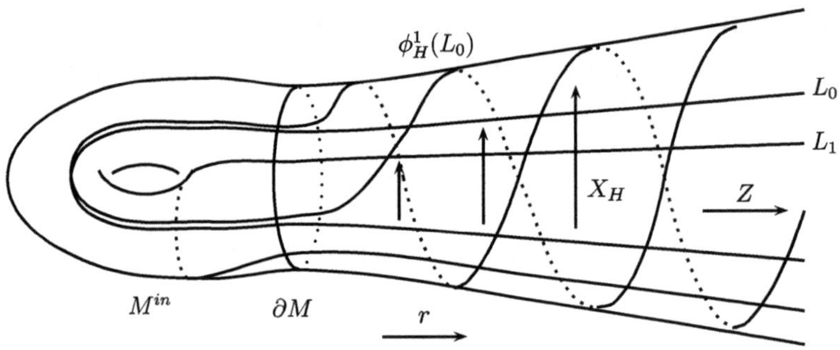

Fig. 11. Wrapping by a quadratic Hamiltonian

4.1. The Wrapped Fukaya Category

The objects of the wrapped Fukaya category $\mathcal{W}(M)$ of a Liouville manifold $(M, \omega = d\theta)$ are exact Lagrangian submanifolds $L \subset M$ which are conical at infinity, i.e. invariant under the flow of the Liouville vector field outside of a compact subset, and such that the exact 1-form $\theta_{|L}$ vanishes outside of a compact set. In other terms, if L is noncompact then at infinity it must coincide with the cone $(1, \infty) \times \partial L$ over some Legendrian submanifold ∂L of ∂M.

The Hamiltonian perturbations used to define Floer complexes in the wrapped setting are very specific: namely, we only consider Hamiltonians $H : M \to \mathbb{R}$ which, outside of a compact subset of M, satisfy $H = r^2$ where $r \in (1, \infty)$ is the radial coordinate of the symplectization $(1, \infty) \times \partial M$. Thus, outside of a compact set the Hamiltonian vector field X_H is equal to $2r$ times the Reeb vector field R_α of the contact form α on ∂M.

Given two objects L_0, L_1, the generating set $\mathcal{X}(L_0, L_1)$ of the *wrapped Floer complex* $CW(L_0, L_1) = CW(L_0, L_1; H)$ consists of time 1 trajectories of the flow of X_H which start on L_0 and end on L_1, i.e. points of $\phi^1_H(L_0) \cap L_1$. More concretely, these consist of (perturbed) intersections between L_0 and L_1 in the interior M^{in} on one hand, and Reeb chords (of arbitrary length) from ∂L_0 to ∂L_1 on the other hand (see Figure 11). Thus, wrapped Floer cohomology is closely related to Legendrian contact homology. (Of course, we need to assume that $\phi^1_H(L_0)$ intersects L_1 transversely, and in particular that the Reeb chords from ∂L_0 to ∂L_1 are non-degenerate; otherwise a small modification of H is required.)

The differential on the wrapped Floer complex counts solutions to Floer's equation (1.6), i.e. perturbed J-holomorphic strips with boundary on L_0 and

L_1, as in Section 1. (Note: due to exactness we can work directly over the field $\mathbb{K}$, without resorting to Novikov coefficients.) As in Remark 1.10, these can equivalently be viewed as $(\phi_H^{1-t})_*J$-holomorphic strips with boundary on $\phi_H^1(L_0)$ and L_1. The assumptions made on the objects of $\mathcal{W}(M)$ and on the Hamiltonian H ensure that, for suitably chosen J, perturbed J-holomorphic strips are well-behaved: an *a priori* energy estimate ensures that all solutions of (1.6) which converge to a given generator $p \in \mathcal{X}(L_0, L_1)$ as $s \to +\infty$ remain within a bounded subset of M (see e.g. [3]). Thus, ∂p is a finite linear combination of generators of the wrapped Floer complex.

A subtlety comes up when we attempt to define the product operation on wrapped Floer complexes,

$$CW(L_1, L_2; H) \otimes CW(L_0, L_1; H) \to CW(L_0, L_2; H). \tag{4.1}$$

For the perturbed Cauchy-Riemann equation (2.4) to be well-behaved and satisfy *a priori* energy estimates in spite of the non-compactness of M, one needs the 1-form β that appears in the perturbation term $X_H \otimes \beta$ to satisfy $d\beta \leq 0$ (cf. [3, 9]). In other terms, the naturally defined product map would take values in $CW(L_0, L_2; 2H)$, and the usual continuation map from this complex to $CW(L_0, L_2; H)$ fails to be well-defined. This can be remedied using the following rescaling trick alluded to in [20] and systematically developed in [3].

Recall that the flow of the Liouville vector field is conformally symplectic and, in the symplectization $(1, \infty) \times \partial M$ where $Z = r\frac{\partial}{\partial r}$, simply amounts to rescaling in the r direction. For $\rho > 1$, denote by ψ^ρ the time $\log \rho$ flow of Z, which rescales r by a factor of ρ. Then there is a natural isomorphism

$$CW(L_0, L_1; H, J) \cong CW\big(\psi^\rho(L_0), \psi^\rho(L_1); \rho^{-1} H \circ \psi^\rho, \psi^\rho_* J\big). \tag{4.2}$$

Moreover, our assumptions imply that $\psi^\rho(L_i)$ is exact Lagrangian isotopic to L_i by a compactly supported isotopy, and $\rho^{-1} H \circ \psi^\rho$ coincides with ρH at infinity. Abouzaid shows that these properties ensure the existence of a well-defined product map

$$\begin{aligned} CW(L_1, L_2; H, J) &\otimes CW(L_0, L_1; H, J) \\ &\to CW\left(\psi^2(L_0), \psi^2(L_2); \frac{1}{2} H \circ \psi^2, \psi^2_* J\right), \end{aligned} \tag{4.3}$$

determined by counts of index 0 finite energy maps $u : D \to M$ from a disc with three strip-like ends to M, mapping the three components of ∂D to the images of the respective Lagrangians under suitable Liouville rescalings, and solving the perturbed Cauchy-Riemann equation

$$(du - X_{\tilde{H}} \otimes \beta)^{0,1}_{\tilde{J}} = 0,$$

where β is a closed 1-form on D with $\beta_{|\partial D} = 0$ which is standard in the strip-like ends (modelled on dt for the input ends, $2\,dt$ for the output end), and $\tilde{H}$ and $\tilde{J}$ are obtained from H and J by suitable rescalings ($\tilde{H} = H$ and $\tilde{J} = J$ near the input punctures; $\tilde{H} = \frac{1}{4} H \circ \psi^2$ and $\tilde{J} = \psi^2_* J$ near the output puncture; see [3]). The map (4.3), composed with the isomorphism (4.2), yields the desired product map (4.1). The higher products

$$\mu^k : CW(L_{k-1}, L_k; H) \otimes \cdots \otimes CW(L_0, L_1; H) \to CW(L_0, L_k; H)$$

are constructed in the same manner [3]. These structure maps make $\mathcal{W}(M)$ an A_∞-category, the wrapped Fukaya category of the Liouville manifold M.

Remark 4.1. The rescaling trick can be informally understood as follows. As mentioned above, the naturally defined product map on wrapped Floer complexes takes values in $CW(L_0, L_2; 2H)$; while the usual construction of a continuation map cannot be used to map this complex to $CW(L_0, L_2; H)$, the fact that $\frac{1}{2} H \circ \psi^2 = 2H$ at infinity and the assumptions made on L_0 and L_2 imply that there is a well-defined continuation map to $CW(\psi^2(L_0), \psi^2(L_2); \frac{1}{2} H \circ \psi^2)$, which by (4.2) is isomorphic to $CW(L_0, L_2; H)$. (Note: while this is a slightly simpler way to describe the cohomology-level product, it lacks the compatibility and consistency features needed to construct the chain-level A_∞-structure, hence the slightly more complicated construction in [3]).

Remark 4.2. Since compact exact Lagrangian submanifolds of M^{in} are not affected by the wrapping at infinity, $\mathcal{W}(M)$ contains the ordinary Fukaya category (of compact exact Lagrangian submanifolds) as a full A_∞-subcategory.

4.2. An Example

Let $M = T^*S^1 = \mathbb{R} \times S^1$, equipped with the standard Liouville form $r\,d\theta$ and the wrapping Hamiltonian $H = r^2$, and consider the exact Lagrangian submanifold $L = \mathbb{R} \times \{pt\}$. We can label the intersection points of $\phi^1_H(L)$ with L by integers, $\mathcal{X}(L, L) = \{x_i,\ i \in \mathbb{Z}\}$, in increasing order along the real axis, where x_0 is the intersection occurring at the minimum of H; in other terms, x_0 is an interior intersection of L with a small pushoff of it, while the

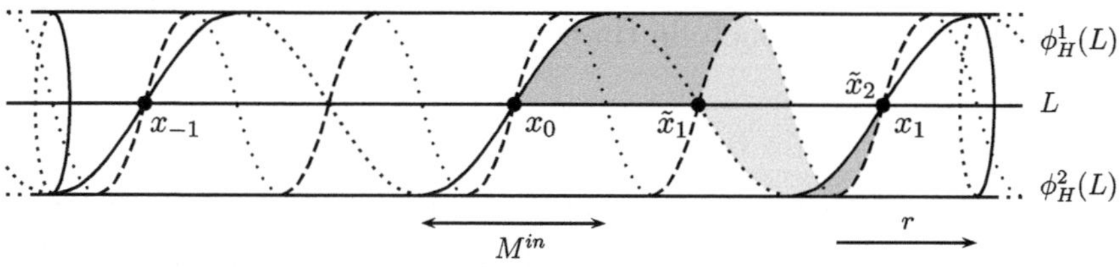

Fig. 12. The wrapped Floer cohomology of $L = \mathbb{R} \times \{pt\}$ in $\mathbb{R} \times S^1$

other generators correspond to Reeb chords from $\partial L = \{pt\} \sqcup \{pt\}$ to itself in the contact manifold $\partial M = S^1 \sqcup S^1$ (see Figure 12).

Recall that the differential on $CW(L, L)$ counts rigid pseudo-holomorphic strips (for a t-dependent almost-complex structure) with boundary on L and $\phi_H^1(L)$. Since there are no such strips (see Figure 12), the Floer differential on $CW(L, L)$ vanishes identically, and $HW(L, L) \simeq CW(L, L) = \operatorname{span}\{x_i, i \in \mathbb{Z}\}$. (This can also be seen by observing that all generators of $CW(L, L)$ have degree 0 for the natural $\mathbb{Z}$-grading.)

The product structure on $CW(L, L)$ counts perturbed pseudo-holomorphic discs with three strip-like ends, as explained above; in the present case, L is invariant under the Liouville flow $\psi^\rho : (r, \theta) \mapsto (\rho r, \theta)$, while $H \circ \psi^\rho = \rho^2 H$. Thus, the rescaling trick only affects the almost-complex structure (i.e., ψ^2 intertwines $CW(L, L; H, J)$ and $CW(L, L; 2H, \psi^2_* J)$), and otherwise simply amounts to identifying $\mathcal{X}(L, L; 2H) = \phi_H^2(L) \cap L$ with $\mathcal{X}(L, L; H) = \phi_H^1(L) \cap L$ via the radial rescaling $r \mapsto 2r$.

Proceeding as in Remark 1.10, the perturbed pseudo-holomorphic discs with boundary on L which determine the product on $CW(L, L)$ can then be reinterpreted as genuine pseudo-holomorphic discs (with respect to a modified family of almost-complex structures) with boundaries on $\phi_H^2(L)$, $\phi_H^1(L)$ and L. Specifically, the coefficient of a generator $q \in \mathcal{X}(L, L)$ in the product $p_2 \cdot p_1$ of two generators $p_1, p_2 \in \mathcal{X}(L, L)$ is given by a count of index 0 pseudo-holomorphic discs with boundaries on $\phi_H^2(L)$, $\phi_H^1(L)$ and L, and with strip-like ends converging to the intersection points $\phi_H^1(p_1) \in \phi_H^2(L) \cap \phi_H^1(L)$, $p_2 \in \phi_H^1(L) \cap L$, and $\tilde{q} \in \phi_H^2(L) \cap L$, where $\tilde{q}$ corresponds to $q \in \phi_H^1(L) \cap L$ under the Liouville rescaling.

With this understood, the product structure can be determined directly by looking at Figure 12. Observe that any two input intersections $\phi_H^1(x_i) \in \phi_H^2(L) \cap \phi_H^1(L)$ and $x_j \in \phi_H^1(L) \cap L$ are the vertices of a unique immersed triangle, whose third vertex is $\tilde{x}_{i+j} \in \phi_H^2(L) \cap L$. (This is easiest to see by lifting the diagram of Figure 12 to the universal cover of M.) These triangles are all regular, and we conclude that

$$x_j \cdot x_i = x_{i+j}.$$

(Recall that thanks to exactness we are working over $\mathbb{K}$ and not keeping track of symplectic areas.) For example, the triangle shaded in Figure 12 illustrates the identity $x_0 \cdot x_1 = x_1$. In other terms, renaming the generator x_i to x^i, we have a ring isomorphism

$$CW(L, L) \simeq \mathbb{K}[x, x^{-1}]. \tag{4.4}$$

Furthermore, the higher products on $CW(L, L)$ are all identically zero, as can be checked either by drawing the successive images of L under the wrapping flow and looking for rigid holomorphic polygons (there are none), or more directly by recalling that $\deg(x^i) = 0$ for all $i \in \mathbb{Z}$ whereas $\deg(\mu^k) = 2 - k$. Thus (4.4) is in fact an isomorphism of A_∞-algebras.

4.3. Cotangent Bundles

The previous example is the simplest case of a general result about cotangent bundles. Let N be a compact spin manifold, and let $M = T^*N$ equipped with its standard Liouville form $p\,dq$ and the wrapping Hamiltonian $H = \|p\|^2$ (for some choice of Riemannian metric on N). Then we have:

Theorem 4.3 (Abouzaid [6]). *Let $L = T_q^*N$, the cotangent fiber at some point $q \in N$. Then there is a quasi-isomorphism of A_∞-algebras*

$$CW^*(L, L) \simeq C_{-*}(\Omega_q N) \tag{4.5}$$

between the wrapped Floer complex of $L = T_q^*N$ and chains on the based loop space $\Omega_q N$ equipped with (an A_∞-refinement of) the usual Pontryagin product.

(The corresponding statement for cohomology is an earlier result of Abbondandolo and Schwarz [2].)

For instance, in the case of $N = S^1$, the based loop space $\Omega_q S^1$ has countably many components, each of which is contractible, thus $\Omega_q S^1 \sim \mathbb{Z}$, and (4.5) reduces to (4.4). In fact, the assumption that N is spin can be removed; in that case, $CW^*(L, L)$ is related to chains on $\Omega_q N$ twisted by the $\mathbb{Z}$-local system determined by $w_2(N)$ [6].

Furthermore, Abouzaid has shown that the fiber $L = T_q^*N$ generates the wrapped Fukaya category $\mathcal{W}(T^*N)$ [4]. Using Yoneda embedding (cf. Section 3.4.2), we conclude:

Corollary 4.4 (Abouzaid). *The wrapped Fukaya category $\mathcal{W}(T^*N)$ quasi-embeds fully faithfully into the category of A_∞-modules over $C_{-*}(\Omega_q N)$.*

(Here again, when N is not spin a twist by a suitable local system is required.)

This and other related results can be viewed as the culmination of over a decade of investigations of the deep connections between the symplectic topology of T^*N and the algebraic topology of the loop space of N, as previously studied by Viterbo [48], Salamon-Weber [39], Abbondandolo-Schwarz [1, 2], Cieliebak-Latschev [16], etc.

At the same time, studying Fukaya categories of cotangent bundles has led to much progress on Arnold's conjecture on exact Lagrangian submanifolds:

Conjecture 4.5 (Arnold). *Let N be a compact closed manifold: then any compact closed exact Lagrangian submanifold of T^*N (with its standard Liouville form) is Hamiltonian isotopic to the zero section.*

Theorem 4.6 (Fukaya-Seidel-Smith [20], Nadler-Zaslow [31], Abouzaid [7], Kragh [26]). *Let L be a compact connected exact Lagrangian submanifold of T^*N. Then as an object of $\mathcal{W}(T^*N)$, L is quasi-isomorphic to the zero section, and the restriction of the bundle projection $\pi_{|L} : L \to N$ is a homotopy equivalence.*

Abouzaid has further shown that Floer theory detects more than purely topological information about exact Lagrangians in cotangent bundles: certain exotic spheres (in dimensions ≥ 9) do not admit Lagrangian embeddings into T^*S^{4k+1} [8].

However, in spite of all the recent progress, Conjecture 4.5 appears to remain out of reach of current technology.

4.4. Homological Mirror Symmetry

Kontsevich's homological mirror symmetry conjecture [24] asserts that the main manifestation of the phenomenon of mirror symmetry is as a derived equivalence between the Fukaya category of a symplectic manifold and the category of coherent sheaves of its mirror. While this conjecture was initially stated for compact Calabi-Yau manifolds (and recently proved for the quintic 3-fold by Sheridan [46]), it also holds (and is often easier to prove) for non-compact manifolds (in which case one should consider the *wrapped* Fukaya category), and outside of the Calabi-Yau case (in which case the mirror is a *Landau-Ginzburg model*, for which one should consider Orlov's derived category of singularities [35, 36] rather than the ordinary derived category of coherent sheaves).

The calculation we have performed in Section 4.2, together with Abouzaid's generation statement, essentially proves homological mirror symmetry for the cylinder $\mathbb{C}^* = T^*S^1$, and its mirror $\mathbb{C}^* = \operatorname{Spec}\mathbb{C}[x^{\pm 1}]$. Namely, coherent sheaves over $\mathbb{C}^*$ are the same thing as finite rank $\mathbb{C}[x^{\pm 1}]$-modules. However, since the object L considered in Section 4.2 generates the wrapped Fukaya category, $\mathcal{W}(T^*S^1)$ quasi-embeds into the category of modules over $CW(L,L) \simeq \mathbb{C}[x^{\pm 1}]$, and the image can be characterized explicitly enough to prove the desired equivalence between $\mathcal{W}(T^*S^1)$ and $D^b\mathrm{Coh}(\mathbb{C}^*)$.

This general approach extends to other examples, with the caveat that in general there are infinitely many non-trivial higher A_∞-operations; one then needs to rely on an algebraic classification result in order to determine which structure coefficients need to be computed in order to fully determine the A_∞-structure up to homotopy. Symplectic manifolds whose Fukaya categories have been determined in this manner include (but are not limited to) pairs of pants [10], genus 2 curves [43], and Calabi-Yau hypersurfaces in projective space [46].

4.5. An Application to Heegaard-Floer Homology

Heegaard-Floer homology associates to a closed 3-manifold Y a graded abelian group $\widehat{HF}(Y)$. This invariant is constructed by considering a Heegaard splitting $Y = Y_1 \cup_{\bar\Sigma} Y_2$ of Y into two genus g handlebodies Y_i, each of which determines a product torus T_i in the g-fold symmetric product of the Heegaard surface $\bar\Sigma = \partial Y_1 = -\partial Y_2$. Deleting a marked point z from $\bar\Sigma$ to obtain an open surface Σ, $\widehat{HF}(Y)$ is then defined as the Floer cohomology of the Lagrangian tori T_1, T_2 in the symplectic manifold $\mathrm{Sym}^g(\Sigma)$, see [37].

In this context it is natural to study the Fukaya category (ordinary or wrapped) of $\mathrm{Sym}^g(\Sigma)$ (equipped with a Kähler form which agrees with the product one away from the diagonal). It turns out that the wrapped category

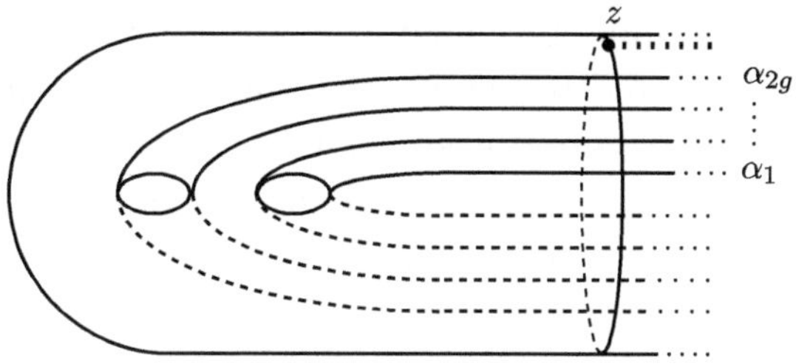

Fig. 13. Generating $\mathcal{W}(\mathrm{Sym}^g(\Sigma))$

has a particularly nice set of generators. Namely, consider a collection of $2g$ disjoint properly embedded arcs $\alpha_1, \dots, \alpha_{2g}$ in Σ such that $\Sigma \setminus (\alpha_1 \cup \dots \cup \alpha_{2g})$ is homeomorphic to a disc, see e.g. Figure 13. Given a g-element subset $s \subseteq \{1, \dots, 2g\}$, the product $D_s = \prod_{i \in s} \alpha_i$ is an exact Lagrangian submanifold of $\mathrm{Sym}^g(\Sigma)$, and we have:

Theorem 4.7 [13, 14]. *The Lagrangian submanifolds $D_s = \prod_{i \in s} \alpha_i$, $s \subseteq \{1, \dots, 2g\}$, $|s| = g$ generate $\mathcal{W}(\mathrm{Sym}^g(\Sigma))$.*

Thus, by Yoneda embedding, Lagrangian submanifolds of $\mathrm{Sym}^g(\Sigma)$ can be viewed as modules over the A_∞-algebra $\bigoplus_{s,s'} \hom(D_s, D_{s'})$.

Determining this A_∞-algebra is not completely hopeless, as the wrapping Hamiltonian H on $\mathrm{Sym}^g(\Sigma)$ can be chosen in a manner compatible with the product structure so that $\phi_H^1(D_s) = \prod_{i \in s} \phi_h^1(\alpha_i)$, where h is a Hamiltonian on Σ that grows quadratically in the cylindrical end, and pseudo-holomorphic discs in the symmetric product can be viewed by projecting them to Σ as is customary in Heegaard-Floer theory; nonetheless, things are complicated by the presence of many nontrivial A_∞-products.

It is easier to study a *partially wrapped* version of the Fukaya category, in which the wrapping "stops" along a ray $\{z\} \times (1, \infty)$ in the cylindrical end of Σ; i.e., the Hamiltonian is again chosen to be compatible with the product structure away from the diagonal, but the effect on each component is to push the ends of the arc α_i in the positive direction towards the ray $\{z\} \times (1, \infty)$, without ever crossing it: see [13]. Theorem 4.7 continues to hold in this setting: the product Lagrangians D_s also generate the partially wrapped Fukaya category. Furthermore, in the partially wrapped case the A_∞-algebra $\mathcal{A} = \bigoplus_{s,s'} \hom(D_s, D_{s'})$ turns out to be a finite-dimensional dg-algebra (i.e., $\mu^k = 0$ for $k \geq 3$) which admits a simple explicit combinatorial description [13]; in fact, $\mathcal{A}$ is precisely the *strands algebra* first introduced by Lipshitz, Ozsváth and Thurston [27].

By Yoneda embedding, Lagrangian submanifolds of $\mathrm{Sym}^g(\Sigma)$, such as the product tori associated to genus g handlebodies in Heegaard-Floer theory, can be viewed as A_∞-modules over the strands algebra. Moreover, the same holds true for *generalized Lagrangian submanifolds* of $\mathrm{Sym}^g(\Sigma)$ (i.e., formal images of Lagrangian submanifolds under sequences of Lagrangian correspondences, cf. [49]), such as those associated to arbitrary 3-manifolds with boundary $\bar{\Sigma}$ (not just handlebodies) according to ongoing work of Lekili and Perutz. This provides a symplectic geometry interpretation of Lipshitz-Ozsváth-Thurston's *bordered Heegaard-Floer homology* [27], which associates to a 3-manifold Y with boundary $\partial Y = \bar{\Sigma}$ an A_∞-module $\widehat{CFA}(Y)$ over the strands algebra. Namely, Lekili and Perutz's construction associates to such a 3-manifold a generalized Lagrangian submanifold of $\mathrm{Sym}^g(\Sigma)$, whose image under Yoneda embedding (as in Section 3.4.2, but using *quilted Floer cohomology* of Lagrangian correspondences) is the A_∞-module $\widehat{CFA}(Y)$; see [13, 14].

4.6. A Closing Remark

The methods available to calculate Floer cohomology and Fukaya categories are still evolving rapidly. Besides the use of algebraic generation statements such as those in [3] and [42] to reduce to a simpler set of Lagrangian submanifolds, there are at least two key ideas that have made calculations possible.

On one hand, it is often possible to find holomorphic projection maps (to the complex plane or to other Riemann surfaces) under which the given Lagrangians project to arcs or curves, in which case holomorphic discs can be studied by looking at their projections to the base and by reducing to the symplectic geometry of the fiber; this is e.g. the guiding principle of Seidel's work on Lefschetz fibrations [42, 44] and the various calculations done using that framework.

At the same time, since such holomorphic projections are easier to come by on open manifolds, another idea that nicely complements this one is to carry out calculations for an exact open subdomain M^0 of the given symplectic manifold M obtained by deleting some complex hypersurface, and then use abstract deformation theory to view the Fukaya category of M as an A_∞-deformation of that of M^0 (cf. [40]). The Hochschild cohomology class that determines the deformation is then often determined by symmetry considerations and/or by studying specific A_∞-structure maps (i.e., certain counts of holomorphic discs in M). See e.g. [43, 46] for an illustration of this approach. (One guiding principle which might explain why this approach is so successful is that algebraic deformations of Fukaya categories are often *geometric*: natural "closed-open" maps from the quantum or symplectic cohomology of M to the Hochschild cohomology of its ordinary or wrapped Fukaya category often turn out to be isomorphisms [11, 22].)

Going forward, there is hope that sheaf-theoretic methods will lead to completely new methods of computation of Fukaya categories (at least for Liouville manifolds) in terms of the topology of a Lagrangian "skeleton". This is an idea that to our knowledge originated with Kontsevich [25], and was subsequently developed by various other authors (see e.g. [5, 30, 45, 47]); the ultimate goal being to bypass the analysis of pseudo-holomorphic curves in favor of algebraic and topological methods. It is too early to tell how successful these approaches will be, but it is entirely possible that they will ultimately supplant the techniques we have described in this text.

Acknowledgements. The author wishes to thank the organizers of the Nantes Trimester on Contact and Symplectic Topology for the pleasant atmosphere at the Summer School, and Ailsa Keating for providing a copy of the excellent notes she took during the lectures. Much of the material presented here I initially learned from Paul Seidel and Mohammed Abouzaid, whom I thank for their superbly written papers and their patient explanations. Finally, the author was partially supported by an NSF grant (DMS-1007177).

References

[1] A. Abbondandolo, M. Schwarz, On the Floer homology of cotangent bundles. Commun. Pure Appl. Math. **59**, 254–316 (2006)

[2] A. Abbondandolo, M. Schwarz, Floer homology of cotangent bundles and the loop product. Geom. Topol. **14**, 1569–1722 (2010)

[3] M. Abouzaid, A geometric criterion for generating the Fukaya category. Publ. Math. IHÉS **112**, 191–240 (2010)

[4] M. Abouzaid, A cotangent fibre generates the Fukaya category. Adv. Math. **228**, 894–939 (2011)

[5] M. Abouzaid, A topological model for the Fukaya categories of plumbings. J. Differ. Geom. **87**, 1–80 (2011)

[6] M. Abouzaid, On the wrapped Fukaya category and based loops. J. Symplectic Geom. **10**, 27–79 (2012)

[7] M. Abouzaid, Nearby Lagrangians with vanishing Maslov class are homotopy equivalent. Invent. Math. **189**, 251–313 (2012)

[8] M. Abouzaid, Framed bordism and Lagrangian embeddings of exotic spheres. Ann. Math. **175**, 71–185 (2012)

[9] M. Abouzaid, P. Seidel, An open string analogue of Viterbo functoriality. Geom. Topol. **14**, 627–718 (2010)

[10] M. Abouzaid, D. Auroux, A.I. Efimov, L. Katzarkov, D. Orlov, Homological mirror symmetry for punctured spheres. arXiv:1103.4322

[11] M. Abouzaid, K. Fukaya, Y.-G. Oh, H. Ohta, K. Ono, J. Am. Math. Soc. **26**, 1051–1083 (2013)

[12] V. Arnol'd, Sur une propriété topologique des applications globalement canoniques de la mécanique classique. C. R. Acad. Sci. Paris **261**, 3719–3722 (1965)

[13] D. Auroux, Fukaya categories of symmetric products and bordered Heegaard-Floer homology. J. Gökova Geom. Topol. **4**, 1–54 (2010)

[14] D. Auroux, Fukaya categories and bordered Heegaard-Floer homology, in *Proc. International Congress of Mathematicians*, vol. II, Hyderabad, 2010 (Hindustan Book Agency, Gurgaon, 2010), pp. 917–941

[15] P. Biran, O. Cornea, A Lagrangian quantum homology, in *New Perspectives and Challenges in Symplectic Field Theory*. CRM Proc. Lecture Notes, vol. 49 (Amer. Math. Soc., Providence, 2009), pp. 1–44

[16] K. Cieliebak, J. Latschev, The role of string topology in symplectic field theory, in *New Perspectives and Challenges in Symplectic Field Theory*. CRM Proc. Lecture Notes, vol. 49 (Amer. Math. Soc., Providence, 2009), pp. 113–146

[17] A. Floer, Morse theory for Lagrangian intersections. J. Differ. Geom. **28**, 513–547 (1988)

[18] A. Floer, Witten's complex and infinite-dimensional Morse theory. J. Differ. Geom. **30**, 207–221 (1989)

[19] K. Fukaya, Y.-G. Oh, H. Ohta, K. Ono, *Lagrangian Intersection Floer Theory: Anomaly and Obstruction I and II*. AMS/IP Studies in Advanced Math., vol. 46 (Amer. Math. Soc. and International Press, Somerville, 2009)

[20] K. Fukaya, P. Seidel, I. Smith, The symplectic geometry of cotangent bundles from a categorical viewpoint, in *Homological Mirror Symmetry*. Lecture Notes in Phys., vol. 757 (Springer, Berlin, 2009), pp. 1–26

[21] K. Fukaya, Y.-G. Oh, H. Ohta, K. Ono, Lagrangian surgery and holomorphic discs. Chapter 10 of preprint version of [19]

[22] S. Ganatra, Symplectic cohomology and duality for the wrapped Fukaya category. Ph.D. thesis, MIT, 2012

[23] H. Hofer, Polyfolds and a general Fredholm theory. arXiv:0809.3753

[24] M. Kontsevich, Homological algebra of mirror symmetry, in *Proc. International Congress of Mathematicians*, Zürich, 1994 (Birkhäuser, Basel, 1995), pp. 120–139

[25] M. Kontsevich, Symplectic geometry of homological algebra. Preprint (2009)

[26] T. Kragh, Parametrized ring-spectra and the nearby Lagrangian conjecture. Geom. Topol. **17**, 639–731 (2013)

[27] R. Lipshitz, P. Ozsváth, D. Thurston, Bordered Heegaard Floer homology: invariance and pairing. arXiv:0810.0687

[28] D. McDuff, D. Salamon, *Introduction to Symplectic Topology* (Oxford University Press, New York, 1998)

[29] D. McDuff, D. Salamon, *J-Holomorphic Curves and Symplectic Topology.* Amer. Math. Soc. Colloq. Publ., vol. 52 (Amer Math Soc., Providence, 2004)

[30] D. Nadler, Fukaya categories as categorical Morse homology. arXiv:1109.4848

[31] D. Nadler, E. Zaslow, Constructible sheaves and the Fukaya category. J. Am. Math. Soc. **22**, 233–286 (2009)

[32] Y.-G. Oh, Floer cohomology of Lagrangian intersections and pseudo-holomorphic disks I and II. Commun. Pure Appl. Math. **46**, 949–993, 995–1012 (1993)

[33] Y.-G. Oh, Floer cohomology, spectral sequences, and the Maslov class of Lagrangian embeddings. Int. Math. Res. Not., 305–346 (1996)

[34] Y.-G. Oh, Relative Floer and quantum cohomology and the symplectic topology of Lagrangian submanifolds, in *Contact and Symplectic Geometry*, Cambridge, 1994. Publ. Newton Inst., vol. 8 (Cambridge Univ. Press, Cambridge, 1996), pp. 201–267

[35] D. Orlov, Triangulated categories of singularities and D-branes in Landau-Ginzburg models. Proc. Steklov Inst. Math. **246**, 227–249 (2004)

[36] D. Orlov, Matrix factorizations for nonaffine LG-models. Math. Ann. **353**, 95–108 (2012)

[37] P. Ozsváth, Z. Szabó, Holomorphic disks and topological invariants for closed three-manifolds. Ann. Math. **159**, 1027–1158 (2004)

[38] L. Polterovich, The surgery of Lagrange submanifolds. Geom. Funct. Anal. **1**, 198–210 (1991)

[39] D. Salamon, J. Weber, Floer homology and the heat flow. Geom. Funct. Anal. **16**, 1050–1138 (2006)

[40] P. Seidel, Fukaya categories and deformations, in *Proc. International Congress of Mathematicians*, vol. II, Beijing, 2002 (Higher Ed. Press, Beijing, 2002), pp. 351–360

[41] P. Seidel, A long exact sequence for symplectic Floer cohomology. Topology **42**, 1003–1063 (2003)

[42] P. Seidel, *Fukaya Categories and Picard-Lefschetz Theory.* Zurich Lect. in Adv. Math. (European Math. Soc., Zürich, 2008)

[43] P. Seidel, Homological mirror symmetry for the genus two curve. J. Algebr. Geom. **20**, 727–769 (2011)

[44] P. Seidel, Fukaya A_∞ structures associated to Lefschetz fibrations I. J. Symplectic Geom. **10**, 325–388 (2012)

[45] P. Seidel, Some speculations on pairs of pants decompositions and Fukaya categories. arXiv:1004.0906

[46] N. Sheridan, Homological mirror symmetry for Calabi-Yau hypersurfaces in projective space. arXiv:1111.0632

[47] N. Sibilla, D. Treumann, E. Zaslow, Ribbon graphs and mirror symmetry I. arXiv: 1103.2462

[48] C. Viterbo, Functors and computations in Floer homology with applications II. Preprint (1996)

[49] K. Wehrheim, C. Woodward, Functoriality for Lagrangian correspondences in Floer theory. Quantum Topol. **1**, 129–170 (2010)

D. Auroux (✉)

Department of Mathematics
UC Berkeley
Berkeley CA 94720-3840
USA

e-mail: auroux@math.berkeley.edu

BOLYAI SOCIETY
MATHEMATICAL STUDIES, 26

Contact and Symplectic Topology
pp. 137–172.

Geometric Decompositions of Almost Contact Manifolds

FRANCISCO PRESAS

Mathematics Subject Classification Primary 53D10 · Secondary 53D15 · 57R17

1. Introduction

Projective algebraic geometry is a field in which meaningful classification and existence questions of manifolds have been answered. Complete theories have been developed in the last two centuries: from the classification of algebraic curves already completed by Riemann, the study and classification of surfaces by the Italian school at the beginning of the 20th century, to the more recent high–dimensional analogues studied by means of Mori theory. There are two central ingredients in these theories:

- The existence of algebraic curves in abundance in a projective variety.
- The theory of divisors: the algebraic understanding of the codimension one subvarieties of a projective variety.

Symplectic manifolds can be thought as topological generalizations of the projective varieties. In the projective setting the essential geometric object

---

The author was supported by the Spanish National Research Project MTM2010-17389 and the ESF Network Contact and Symplectic Topology.

F. Bourgeois et al. (eds.), *Contact and Symplectic Topology*,
Bolyai Society Mathematical Studies 26, DOI 10.1007/978-3-319-02036-5_4,

from which the theory is developed is the hyperplane divisor. In symplectic geometry the symplectic form is the topological analogue of this divisor: in a projective variety the Poincaré dual of the hyperplane section is the induced Fubini–Study symplectic form. Therefore, for a while, it was thought that the classification of symplectic manifolds could be achieved by the same methods as in the projective case. For that to work a correct generalization of the concept of algebraic curve and divisor had to be provided. The algebraic curves concept was generalized by the notion of pseudo-holomorphic curves introduced by Gromov [18] and has been shown to be central in the development of the symplectic topology. In real dimension 4, a divisor coincides with an algebraic curve and this has been enough to push a meaningful theory in such a situation. There was left to procure an analogue of the Riemann–Roch theorem providing the existence of algebraic curves when topologically expected. In the nineties, C. Taubes showed how to handle this by introducing a relation with the Seiberg–Witten invariants [31]. From that point onwards a partial classification of 4-dimensional symplectic manifolds has been achieved, e.g. see [22].

In higher dimensions a correct theory of divisors is lacking and probably it is not reasonable to expect it, since the symplectic geography problem in high dimensions is considerably wild, cf. [17, 26]. However, the particular case of very ample divisors was worked out by S. Donaldson in a series of foundational articles [7, 9]. The claim is that a theory of asymptotically very ample divisors can be developed in symplectic geometry, in other words very ample linear systems are available. The notion of ampleness is related to positivity, which holds due to the non-degeneracy of the symplectic form. The implications of these results are the same as in projective geometry:

- Bertini's theorem on the existence and genericity of smooth very ample divisors [7].
- Existence of symplectic Lefschetz pencils [9] and associated symplectic invariants [3, 8].
- Connectedness of the space of very ample divisors [1].
- High-dimensional linear systems in the symplectic setting [2].

Maybe, the main conclusion is the existence of nice decompositions of a symplectic manifold in the same fashion as in the projective setting. This is not enough to classify though, but it provides a better understanding of the symplectic topology. In other words, a Lefschetz pencil is a clever way of trivializing a symplectic manifold. Therefore, the implications of the existence result are the usual implications of a statement providing a combinatorial description of a geometric object: construction of solutions of equations [10] and building blocks for the definition of new theories [30].

Contact geometry can be understood as a conformal analogue of the symplectic geometry and from this understanding the Donaldson techniques have been adapted to the contact setting. However, the general picture was initially far less clear since there is no classical analogue of the projective setting for contact manifolds. It turned out that there is one: the goal of these notes is to show its behaviour and the results it produces. The history developed as follows. The first attempt was to study the existence of codimension 2 contact submanifolds on a general contact manifold [21], this was non-achievable by the h-principle and it is the expected analogue of the Bertini's theorem in contact geometry. Not surprisingly, the contact picture was much more flexible than the symplectic one and it was shown that any codimension 2 integer homology class on a closed contact manifold admits a smooth contact representative. The next two constructions to be worked out are the analogues of the:

- Lefschetz pencil decomposition of a symplectic manifold [9].
- Decomposition of a symplectic manifold in terms of a very ample divisor and its Stein complementary [4].

The equivalent of the Stein-divisor decomposition for a contact manifold is the open book decomposition constructed by Giroux and Mohsen [15, 16]. There, the Stein manifold in which the symplectic manifold is trivialized is substituted by a 1-parametric family of Stein manifolds, the so-called leaves of the open book, and the divisor becomes a codimension 2 contact submanifold. The equivalent of the Donaldson's construction is straightforward and introduces the concept of a contact Lefschetz pencil [28]. The idea in that case is to produce a codimension 2 fibration over the sphere whose fibers are contact manifolds, special singular fibers are also allowed corresponding to parametric holomorphic singularities.

The central question concerns the possible uses of these constructions, the essential feature being that these constructions are almost topological. In other words, they are h-principle achievable: there is no need for a contact structure in order to produce them, an almost contact structure is enough for them to exist. In case the contact structure could be recovered, they would produce an existence result in contact topology: any almost contact structure could be deformed to a contact one. This was highly unexpected a decade ago but nowadays it seems to be a reasonable statement in contact topology. The reason for the old perception is based on the fact that the almost symplectic condition does not imply the existence of a symplectic structure, which was proved in the late nineties. Hence it was believed to be a matter of time to find equivalent examples in the contact category. Nevertheless, the appearance of the previously mentioned decompositions gave support to the idea of almost contact implying contact being conceivable. The reason is that both the Stein-divisor decomposition and the Lefschetz pencil decomposition are not doable in the almost symplectic setting and they constitute an actual geometric obstruction to the existence of a symplectic structure on a general almost symplectic manifold.

This approach has been successful in dimension 5, where any almost contact manifold has been proved to be contact [5] through the appropriate use of an almost contact pencil decomposition. The hope is that any of the two decompositions will eventually succeed to prove the existence of a contact structure in higher dimensions. Based on that, we detail in these notes the construction of the two decompositions.

The structure of the article reads as follows. In Section 2 we establish the foundations of the approximately holomorphic theory for almost contact geometry. In Section 3 we provide the argument of E. Giroux for the existence of open book decompositions adapted to a contact structure. In Section 4 we detail the construction of almost contact Lefschetz pencils after [28] and [24].

2. Approximately Holomorphic Techniques

Let M be a $(2n+1)$-dimensional smooth manifold. A global distribution $\xi \subset TM$ is said to be a contact structure if it admits a global 1-form $\alpha \in \Omega^1(M)$ such that $\xi = \ker \alpha$ and $\alpha \wedge (d\alpha)^n > 0$ everywhere. A contact manifold is a manifold with a contact structure. The 1-form α defining the contact structure is said to be a contact form for the distribution.

In the literature this definition corresponds to the notion of a cooriented contact distribution, we will restrict ourselves to this case[1]. Note that the

[1] Just once and for all it is important to mention that all the results in these notes can be easily adapted to the non-coorientable case. The essential point being that any non-coorientable contact manifold admits a coorientable double-cover. Therefore to study non-coorientable manifolds is reduced to study coorientable ones with free $\mathbb{Z}/2\mathbb{Z}$-actions. See [21] for details.

contact condition does not strictly depend on the choice of the 1-form α, any other 1-form $\alpha' = f\alpha$, with $f : M \longrightarrow \mathbb{R}^+$ a smooth function, also satisfies

$$\alpha' \wedge \left(d\alpha'\right)^n = f^{n+1}\alpha \wedge (d\alpha)^n > 0.$$

Let us emphasize the topological features of a contact distribution. There are two topological objects appearing in the definition

(i) The distribution ξ, a real codimension 1 subbundle of TM.

(ii) The symplectic structure induced in the bundle ξ by $d\alpha$. To be precise, only the conformal symplectic class is determined: a change of form $\alpha' = f\alpha$ as above does change the representatives from the symplectic bundle $(\xi, d\alpha)$ to $(\xi, fd\alpha)$.

We therefore define an almost contact manifold as a $(2n+1)$-dimensional manifold M with a codimension-1 cooriented[2] distribution ξ and a conformally symplectic class on ξ, understood as an abstract bundle. A distribution admitting a conformally symplectic class is called an almost contact structure. The almost contact condition might be seen as the formal necessary condition for the existence of a contact structure. The long standing conjecture in contact topology is

Conjecture 2.1. *Any almost contact structure on a manifold M admits a deformation in its homotopy class of almost contact structures to a contact structure.*

In the case of open manifolds M, the result is true and it is one of the first applications of Gromov's h-principle, see [19]. The situation is not as established for closed manifolds. The conjecture was proven by R. Lutz [23] for 3-dimensional closed manifolds. The classification of simply connected 5-dimensional contact closed manifolds allowed H. Geiges [12] to also answer positively in these cases. The general 5-dimensional case was recently proved by Casals et al., see [5]. The conjecture remains open in general, some recent progress has been obtained by E. Giroux using the techniques described in Section 3.

[2]The normal bundle TM/ξ of ξ as a subbundle of TM is trivial.

2.1. The Quasi-contact Category

An initial strategy in a geometric setting consists in understanding the implications of the h-principle; in our case we start with an almost-contact manifold. The further structure that the h-principle offers is provided in the following

Definition 2.2. A quasi-contact structure on a $(2n+1)$-dimensional manifold M is a pair (ξ, β) satisfying:

- ξ is a cooriented distribution.
- β is a 1-form on M such that $(\xi, d\beta)$ is a symplectic bundle.

Observe that the condition is stronger than the almost-contact one for $d\beta$ is necessarily closed, and not just a non-degenerate 2-form. However, it is still weaker than the contact condition since (ξ, β) inducing a contact structure would imply $\alpha = \beta$, with the previous notations. As previously mentioned, the quasi-contact condition can be reached through the h-principle, indeed one may show:

Lemma 2.3. *Any almost-contact structure admits a quasi-contact structure in its homotopy class of symplectic hyperplane fields.*

For the proof see Lemma 2.2 in [5] (see also [11]). In the article [5] the definition of quasi-contact structure is given in a slightly more general setting, however no further applications are obtained and so we may concentrate in the more adapted definition above.

The fundamental property of quasi-contact manifolds is the closedness condition $d(d\beta) = 0$. This is the precise piece of data we require to develop the theory of approximately holomorphic bundles: to begin with, a closed 2-form topologically induces a complex line bundle. Let us start with the definitions: the pre-quantizable line bundle associated to the quasi-contact structure (ξ, β) is the hermitian line bundle $L := M \times \mathbb{C}$ with the choice of connection $\nabla_L = d - i\beta$.

A compatible almost complex structure for the quasi-contact structure (ξ, β) is a compatible complex structure J for the symplectic bundle $(\xi, d\beta)$. Also, a compatible metric for (ξ, β, J) is any Riemannian metric g such that

$$g(u, v) = d\beta(u, Jv),$$

for all $u, v \in \xi$ and such that $\ker d\beta$ is orthogonal to ξ. This amounts to a choice of a unitary vector field in $\ker d\beta$. Let us fix the unitary vector field orthogonal to ξ, it will be referred as the Reeb vector field R. We define the 1-form

$$\alpha(v) := g(R, v),$$

that clearly satisfies $\ker \alpha = \xi$. We suppose that set of objects (ξ, β, g) and the induced R and α are given. For convenience, we also fix the following sequence of Riemannian metrics $g_k = k \cdot g$.

Let E be a hermitian complex bundle with connection ∇, we can split the connection along ξ in its holomorphic and antiholomorphic parts since ∇ restricted to ξ is an operator between complex linear spaces and therefore admits a decomposition

$$\nabla_{|\xi} = \partial + \bar{\partial}.$$

As explained in Section 1, we should be able to produce symplectic and contact divisors. In analogy with the projective setting, the procedure Donaldson developed provides such divisors as vanishing loci of sections of a vector bundle. Instead of complex submanifolds from holomorphic sections we procure to obtain symplectic and contact submanifolds from asymptotically holomorphic sections. For a symplectic or contact structure to be induced in the vanishing locus the intersection of the asymptotically holomorphic section with the base manifold has to satisfy certain transversality condition. Let us recall the following ideas from linear algebra:

Definition 2.4. A linear map $f : \mathbb{R}^n \longrightarrow \mathbb{R}^r$ is said to be ε-transverse to zero if it admits a right inverse of norm smaller than ε^{-1}.

There is a more geometric way of understanding the previous property

Lemma 2.5. *A linear map $f : \mathbb{R}^n \longrightarrow \mathbb{R}^r$ is ε-transverse to zero if and only if there exists an r-dimensional subspace $W \subset \mathbb{R}^n$ such that for any $w \in W$, we have*

$$|f(w)| \geq \varepsilon |w|.$$

Proof. If there exists a right inverse $g : \mathbb{R}^r \longrightarrow \mathbb{R}^n$, set $W = g(\mathbb{R}^r)$. This satisfies the required property. Conversely, suppose that such subspace $W \subset \mathbb{R}^n$ exists. Define g as the right inverse map of the restriction $f : W \longrightarrow \mathbb{R}^r$, which exists since $\dim_{\mathbb{R}}(W) = r$. ■

The linear condition required for the tangent bundle of the submanifold to be a symplectic subbundle can be stated as

Lemma 2.6 (Proposition 3 in [7]). *Let $f : \mathbb{C}^n \to \mathbb{C}^r$ be an $\mathbb{R}$-linear map ε-transverse to zero. Suppose there exists a constant $\delta > 0$, depending only on ε, such that the antiholomorphic part of f satisfies*

$$|f^{0,1}| \leq \delta,$$

then $\ker f$ *is a symplectic subspace of* $\mathbb{C}^n$.

The proof is based on the fact that the condition $f^{0,1} = 0$ implies that the subspace is complex and therefore symplectic for the compatible symplectic structure and the symplectic condition is open. We are in position to describe the suitable transversality condition:

Definition 2.7. Let $E \longrightarrow M$ be a hermitian complex bundle with connection ∇. A section $s : M \longrightarrow E$ is said to be ε-transverse to zero along ξ if $\forall x \in M$ any of the following conditions hold:

- $|s(x)| > \varepsilon$,
- $\nabla_\xi(s)(x) : \xi_x \to E_x$ is ε-transverse to zero.

A submanifold $S \overset{e}{\hookrightarrow} (M, \xi, \beta)$ is said to be quasi-contact if ξ is everywhere transverse to S and $(e^*(\xi), e^*\beta)$ is a quasi-contact structure on S. Let us provide a simple way to decide whether the zero locus of a section is a quasi-contact submanifold:

Lemma 2.8. *For any $\varepsilon > 0$, there exists a $\delta > 0$ such that if $s : M \longrightarrow E$ is ε-transverse to zero along ξ and $|\bar{\partial} s| \leq \delta$, then the zero set $Z(s)$ is a smooth quasi-contact submanifold of M.*

Proof. Denote by r the rank of the complex bundle E. The ε-transversality along ξ, in particular, implies that the section is transverse to zero in the usual sense, i.e. for any $x \in Z(s)$, the linear map $\nabla s(x)$ is surjective. Therefore, the set $Z(s)$ is a smooth submanifold of dimension $2(n-r)+1$. The transversality along ξ further implies that the submanifold $Z(s)$ is transverse to ξ, since for any point $x \in Z(s)$ we have that the induced distribution $e^*(\xi) = \ker \nabla_\xi s(x)$ on x has real dimension $2(n-r)$.

It is left to verify that $e^*(\xi)=\ker \nabla_\xi s$ is symplectic everywhere, but for a suitable choice of δ we are in the hypothesis of the Lemma 2.6. ∎

Note that ε-transversality is a C^1-stable notion. Indeed, let $s: M \longrightarrow E$ be a section ε-transverse along ξ and $s_\delta : M \longrightarrow E$ a perturbative section satisfying that $|s_\delta|_{C^1} \leq \delta$. Then the perturbed section $s+s_\delta$ still is $(\varepsilon - c\delta)$-transverse to zero along ξ for some universal constant $c>0$. Thus, C^1-perturbations do not destroy the estimated transversality along ξ.

Let us define a central notion in the approximately holomorphic theory, these are also referred as asymptotically holomorphic techniques since the produced objects acquire a holomorphic behaviour at the limit. Given a line bundle L on M constructed as above, we can associate to the hermitian bundle E the following sequence of bundles $E_k := E \otimes L^{\otimes k}$ for $k \in \mathbb{N}$. This is related to the twisting sheaf in projective geometry, allowing to shift a coherent sheaf to an affine behaviour. In quasi-contact geometry, our aim is to produce the following objects:

Definition 2.9. A sequence of sections $s_k : M \longrightarrow E_k$ is C^r-asymptotically holomorphic if the following estimates hold

$$|s_k| = O(1), \qquad \left|\nabla^l s_k\right| = O(1), \qquad \left|\nabla^{l-1}\bar{\partial} s_k\right| = O\left(k^{-1/2}\right), \quad \forall l \leq r,$$

the norms being measured with respect to the g_k-metric.

The index l will be omitted if it is clear from the context. As a consequence of the previous discussion we conclude:

Corollary 2.10. *Let $s_k : M \longrightarrow E_k$ be an asymptotically holomorphic sequence of sections ε-transverse to zero along ξ. For k large enough, the set $Z(s_k)$ is a smooth quasi-contact submanifold.*

The existence of ε-transverse asymptotically holomorphic sections is partially guaranteed due to the following:

Theorem 2.11. *Let E be a vector bundle, $\delta > 0$ and $s_k : M \longrightarrow E_k$ be an asymptotically holomorphic sequence of sections. There exists a constant $\varepsilon > 0$ and an asymptotically holomorphic sequence of sections $\sigma_k : M \longrightarrow E_k$ such that they are ε-transverse to zero along ξ and $|\sigma_k - s_k|_{C^2} \leq \delta$.*

We will give an overview of the proof of this Theorem in the rest of this Section. However, it is just the generalization to the quasi-contact category of the main result in [21]. The result implies the existence of quasi-contact divisors with prescribed topology:

Corollary 2.12. *Fix an integer homology class $A \in H_{2n-1}(M,\mathbb{Z})$, there exists a smooth quasi-contact submanifold S representing it.*

Proof. Let $\gamma \in H^2(M,\mathbb{Z})$ be the Poincaré dual of A. Construct a hermitian line bundle V with $c_1(V) = \gamma$. Apply the Theorem 2.11 to the sequence $V_k = V \otimes L^{\otimes k}$ with starting sequence $s_k = 0$. Since $c_1(L_k) = c_1(L) = \gamma = PD(A)$, any submanifold $Z(\sigma_k)$, for k large enough, fulfills the requirements. ■

The previous construction can also be made relative to a complex distribution. Let $s_k : M \longrightarrow E \otimes L^{\otimes k}$ be a C^r-asymptotically holomorphic sequence of sections and $D \subset \xi$ any fixed complex distribution, it is simple to verify that $\partial_D s_k : M \longrightarrow D^* \otimes E \otimes L^{\otimes k}$ is a C^{r-1}-asymptotically holomorphic sequence of sections. There is also an existence result for this case:

Theorem 2.13. *Let V be a line bundle, $D \subset \xi$ complex distribution and $\delta > 0$. Consider $s_k : M \longrightarrow V \otimes L^{\otimes k} \otimes \mathbb{C}^2$ an asymptotically holomorphic sequence of sections. There exists a constant ε and an asymptotically holomorphic sequence of sections $\sigma_k = (\sigma_{k,0}, \sigma_{k,1}) : M \longrightarrow V \otimes L^{\otimes k} \otimes \mathbb{C}^2$ such that $|s_k - \sigma_k|_{C^2} \leq \delta$ and $\partial_D \sigma_{k,0} \otimes \sigma_{k,1} - \sigma_{k,0} \otimes \partial_D \sigma_{k,1}$ are ε-transverse to zero along ξ.*

The previous results also hold for the particular setting in which $\alpha = \beta$, systematically replacing the word quasi-contact by contact. This is the formulation found in [21, 28]. However, the proofs remain practically unchanged in this more general setting. D. Martínez-Torres has provided a general theory for the quasi-contact case in different articles [20, 24]. In his notation the quasi-contact structures are called 2-calibrated structures. This Section is intended to provide a short version of [24] centered just in 0 and 1-dimensional linear systems; in that article the general theory for r-dimensional linear systems is worked out. The essential problem to overcome is to obtain transversality for sequences of sections, there are two available techniques to do it:

- The one developed in [21], strictly working in the quasi-contact manifold itself.

– The relative version developed in [27] in which we embed the quasi-contact structure in a symplectic manifold and the transversality is achieved there.

We shall use the second alternative to deduce Theorem 2.11. The definitions required for the symplectization setting are now provided.

2.2. The Symplectization of a Quasi-contact Structure

We define the symplectization of a quasi-contact structure $(M, \xi = \ker \alpha, d\beta)$ as the symplectic manifold $S(M) := M \times [-\tau, \tau]$ with the form $\omega_S := d(\beta + t\alpha)$, where $\tau > 0$ is chosen small enough so that the form ω_S symplectic. We can construct the required data as restrictions of objects in the symplectization. In particular, we will need:

(i) Any choice of compatible J can be extended to an almost-complex structure $\hat{J}$ on $TS(M) = \xi \oplus R \oplus \partial_t$ by declaring $J\partial_t = R$. This almost-complex structure is compatible with ω_S.

(ii) The associated Riemannian metric $\hat{g} = \omega_S(\cdot, \hat{J}\cdot)$ extends g setting ∂_t to be orthonormal to $TM \subset TS(M)$. Also, we define $\hat{g}_k = k\hat{g}$.

(iii) The prequantizable bundle on the quasi-contact manifold is the restriction of the prequantizable bundle over $S(M)$ defined as the trivial bundle $L = S(M) \times \mathbb{C}$ with connection $\nabla = d - i\hat{\beta}$, for the primitive form $\hat{\beta} = \beta + t\alpha$.

These choices are to be assumed in the following discussion. Thus, we may use the definitions of transversality and asymptotically holomorphic sequences in the symplectic case, see [1]. We briefly recall them; given a hermitian bundle E over $S(M)$, denote $E_k := E \otimes L^{\otimes k}$. The fundamental notion in the symplectic case is contained in the following:

Definition 2.14. A sequence of sections $s_k : S(M) \longrightarrow E_k$ is C^r-asymptotically holomorphic if the following estimates hold,

$$|s_k| = O(1), \qquad \left|\nabla^l s_k\right| = O(1), \qquad \left|\nabla^{l-1}\bar{\partial} s_k\right| = O\big(k^{-1/2}\big), \quad \forall l \leq r,$$

the norms being measured with respect to the $\hat{g}_k$-metric.

The splitting of the connection in holomorphic and antiholomorphic parts is a consequence of the usual decomposition $\nabla = \partial + \bar{\partial}$. The transversality condition is analogously stated as

Definition 2.15. Let $E \longrightarrow S(M)$ be a hermitian complex bundle with connection ∇. A section $s : S(M) \longrightarrow E$ is said to be ε-transverse to zero over M, if $\forall x \in M \times \{0\}$ any of the following conditions hold:

- $|s(x)| > \varepsilon$,
- $\nabla_M(s)(x) : (T(M \times \{0\}))_x \to E_x$ is ε-transverse to zero.

Observe that an asymptotically holomorphic sequence of sections in the symplectization $s_k : S(M) \longrightarrow E_k$ restricts to $M \times \{0\}$ as an asymptotically holomorphic sequence of sections $e^* s_k$. It is less clear that transversality along the quasi-contact distribution can be also achieved. Let us prove the following

Proposition 2.16. *Let $s : S(M) \longrightarrow E$ be an ε-transverse section over M. Assume that $|\bar{\partial} s| \leq \delta$ for some $\delta > 0$ small enough and depending only on ε. Then the restriction of the section*

$$e^* s : M \longrightarrow e^* E$$

is an $\frac{\varepsilon}{2}$-transverse section along ξ.

Proof. This is essentially a linear algebra question. Let $2r = rk(E)$ and fix a point $x \in M$ for which $|s(x)| \leq \varepsilon$, then the linear map

$$\nabla_M s(x) : \xi_x \oplus \langle R \rangle \to E_x$$

is ε-transverse to zero. Define $\varepsilon' = \frac{3}{4}\varepsilon$. Let $\delta > 0$ be small enough such that $f = \partial s(x)$ is ε'-transverse to zero. Thus, there exists a right inverse of norm smaller than $(\varepsilon')^{-1}$. By Lemma 2.5, this implies that there exists a $2r$-dimensional subspace $W \subset \xi_x \oplus \langle R \rangle$ such that for any $w \in W$,

$$|f(w)| \geq \varepsilon' |w|.$$

If $W \subset \xi_x$ we are done. Otherwise, define $V = W \cap \xi_x$ and let $U \subset V$ be an isotropic r-dimensional subspace. Consider $\hat{U} = f(U)$, then we obtain the splitting $E_x = \hat{U} \oplus i\hat{U}$. The map f restricted to the subspace $U_{\mathbb{C}} = U \oplus JU \subset \xi$ satisfies

$$|f(u)|^2 = |f(u_1) + if(u_2)|^2 = |f(u_1)|^2 + |f(u_2)|^2 \geq (\varepsilon')^2 |u|^2,$$

where $u = u_1 + Ju_2 \in U_{\mathbb{C}}$. Therefore $f_{|\xi}$ is ε'-transverse to zero. Again, for fixed $\delta > 0$ small enough, the linear map $\nabla_\xi s(x)$ is $\frac{1}{2}\varepsilon$-transverse to zero. ■

Since we may achieve transversality with respect to the distribution, to conclude the proof of Theorem 2.11 we must ensure the existence of uniformly transverse asymptotically holomorphic sections. For that purpose, we refer to the main result in [27]:

Theorem 2.17 (Mohsen). *Let (W,ω) be a symplectic manifold of integer class. Fix a compatible almost complex structure J, a closed submanifold S and a hermitian vector bundle E. Then for any asymptotically holomorphic sequence of sections $s_k : W \longrightarrow E \otimes L^{\otimes k}$ and for any $\delta > 0$, there exists a C^3-asymptotically holomorphic sequence of sections $\sigma_k : W \to E \otimes L^{\otimes k}$ satisfying*

- $|\sigma_k - s_k|_{C^2} \leq \delta$,
- *The sequence $(\sigma_k)_{|S}$ is ε-transverse to zero in S, for some uniform constant $\varepsilon > 0$ not depending on k.*

It is now immediate to conclude the existence of transverse asymptotically holomorphic sections:

Proof of Theorem 2.11. Let us describe the elements appearing in the hypothesis of Theorem 2.17. The symplectization $(S(M), d(\beta + t\alpha))$ will be the symplectic manifold. The submanifold will be the quasi-contact manifold, hence $S = M \times \{0\}$. Finally, pull-back the vector bundle $E \longrightarrow M$ to a bundle in $S(M)$, still denoted E. As a consequence of the theorem applied to the constant sequence $s_k = 0$ we obtain a C^2-small sequence $\sigma_k : S(M) \longrightarrow E \otimes L^{\otimes k}$ which is ε-transverse to zero in M. After Proposition 2.16 the sequence $e^*\sigma_k : M \to E \otimes L^{\otimes k}$ is $\frac{1}{2}\varepsilon$-transverse to ξ. ■

To conclude Theorem 2.13 we have to slightly generalize Theorem 2.17 to allow certain control for the derivative of the quotient along the complex distribution. The precise statement we require is the following

Theorem 2.18. *Let (W,ω) be a symplectic manifold of integer class. Fix a compatible almost complex structure J, a closed submanifold S, a hermitian line bundle V and a complex distribution $D \subset TS$ over the submanifold. Then, for any asymptotically holomorphic sequence of sections $s_k : W \longrightarrow V \otimes L^{\otimes k} \otimes \mathbb{C}^2$ and for any $\delta > 0$, there exists an asymptotically holomorphic sequence of sections $\sigma_k = (\sigma_{k,0}, \sigma_{k,1}) : W \longrightarrow V \otimes L^{\otimes k} \otimes \mathbb{C}^2$ satisfying*

- $|\sigma_k - s_k|_{C^2} \leq \delta$,

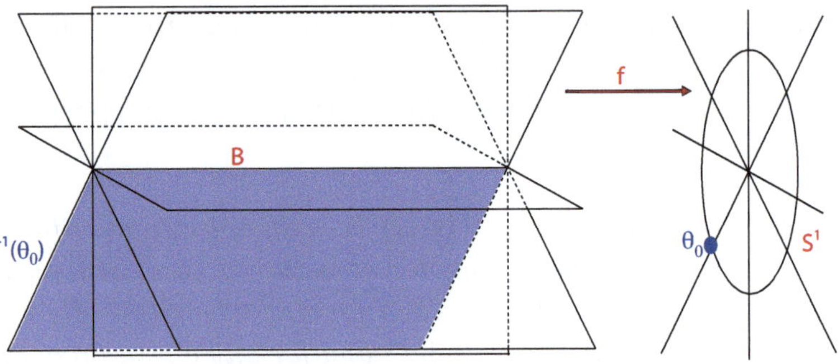

Fig. 1. Open book close to the binding

- $\partial_D \sigma_{k,0} \otimes \sigma_{k,1} - \sigma_{k,0} \otimes \partial_D \sigma_{k,1}$ *is* ε*-transverse to zero in* S, *for some uniform constant* $\varepsilon > 0$ *not depending on* k.

The proof of this Theorem, in a more general case, can be found in [24]. Finally, Theorem 2.13 is an immediate consequence of this result.

3. Open Books in Contact Geometry

This section develops the open book decomposition mentioned in Section 1. We begin with the main definition:

Definition 3.1. Let M be a smooth closed manifold. A pair of objects (B, π) is called an open book decomposition if they satisfy:

- B is a codimension-2 closed submanifold.
- $\pi : M \setminus B \to S^1$ is a submersion.
- The normal bundle of B is trivial and there exists a tubular neighborhood U with a trivializing diffeomorphism $\phi : B \times B^2(\delta) \to U$ such that

$$(\pi \circ \phi)(p, r, \theta) = \theta,$$

where $p \in B$ and (r, θ) are polar coordinates in $B^2(\delta)$ (see Figure 1).

The divisor B is referred to as the binding. The closure of the fibers of π in M are called the pages of the open book (B, π).

Let us describe an equivalent construction. Consider a smooth manifold P with boundary $B = \partial P$, and $\Psi : P \to P$ a diffeomorphism restricting to the identity close to the boundary. We then construct a closed manifold M from the pair (P, Ψ): as a topological space $M = (P \times [0,1])/\sim$ where the equivalence relation is defined as

$$(p,t) \sim (q,s) \Longleftrightarrow \begin{cases} p = q \in B, \\ \text{or} \\ p = \Psi(q) \text{ and } t = 0, s = 1. \end{cases}$$

This produces a manifold. Indeed, before quotienting it is certainly a manifold and then the quotient can be understood as a two-step process. In the first step $P \times \{0\}$ and $P \times \{1\}$ are identified by means of the diffeomorphism Ψ to produce a manifold P_Ψ that fibers over S^1, its boundary is diffeomorphic to $B \times S^1$. Secondly, in order to obtain the collapse from the first condition in the equivalence relation, we fill the boundary of P_Ψ with $B \times D^2$. This produces a smooth manifold M without boundary. Define the map

$$\begin{aligned} \pi : \big((P \times [0,1])/\sim\big) \setminus B &\longrightarrow S^1 \\ (p,t) &\longmapsto t. \end{aligned}$$

Then (B, π) is an open book decomposition of M with pages diffeomorphic to P. Conversely, given an open book decomposition (B, π) of a manifold M, we may recover $P = \overline{\pi^{-1}(0)}$ and Ψ. For the diffeomorphism, consider a connection for the fibration $\pi : M \setminus B \to S^1$, thus providing a notion of parallel transport, and then $\Psi \in \mathrm{Diff}(P)$ is obtained as the time-1 flow of the lifting of ∂_t with respect to the chosen connection. Hence, we can define an open book decomposition either by providing the pair (B, π) or the pair (P, Ψ).

The notion of an open book decomposition is essentially topological. We now follow E. Giroux [15] to relate it with contact geometry (see also [25]). Given a contact form α for a contact structure, let $R = R_\alpha$ be the unique vector field such that $d\alpha(R, \cdot) = 0$ and $\alpha(R) = 1$. This is called the Reeb vector field of α. The interaction between contact geometry and open book decompositions is based on the following

Definition 3.2. Let (M, ξ) be a contact manifold. A contact form α supports an open book decomposition (B, π) if:

- $(B, \alpha_B = \alpha_{|B})$ is a contact submanifold.
- The Reeb vector field R is positively transverse to the projection π, i.e. $d\pi(R) > 0$ everywhere, and tangent to the submanifold B.

Given a fixed contact structure and a supporting contact form, the open book is said to be adapted to the contact structure through the contact form. The open book is called adapted to a contact structure if it is adapted through a contact form inducing the given contact structure. The requirements in the definition have the following implications:

- The pages $P_t = \overline{\pi^{-1}(t)}$ inherit an exact symplectic structure provided by the restriction of $d\alpha$.
- The associated flow Ψ is a symplectomorphism since the generating vector field is of the form $X = f \cdot R$ and $\mathcal{L}_R d\alpha = 0$.
- The boundary of any page is $\partial P_t = B$, and it is of convex type with respect to the symplectic structure $(d\alpha)|_{P_t}$.

Recall that an exact symplectic manifold $(M, \omega = d\alpha)$ has a boundary of convex type with respect to the Liouville form α if the associated Liouville vector field X, defined by $\alpha = i_X d\alpha$, is transverse to the boundary of M and points outwards. Convexity is a relevant property in procedures such as gluing or filling constructions and has a fundamental role in the understanding of Conjecture 2.1. The first two assertions are readily seen to hold, let us detail the third statement:

Lemma 3.3. *Let (B, π) be an open book decomposition adapted to $(M, \ker\alpha)$. There exists a neighborhood U of the binding B and a trivializing diffeomorphism $\psi : B \times B^2(\delta) \to M$ such that*

$$\psi^*\alpha = g \cdot \left(\alpha_{|B} + r^2 d\theta\right),$$

where $g : B \times B^2(\delta) \longrightarrow \mathbb{R}^+$ satisfies $\partial_r g < 0$ for $r > 0$.

Proof. Consider the trivializing map $\phi : B \times B^2(\delta) \to U$ provided by the definition of an open book decomposition. Note that for $\delta > 0$ small enough the fibers are contact submanifolds. Let $\pi_2 : B \times B^2(\delta) \longrightarrow B^2(\delta)$ be the projection onto the second factor, then the projection

$$\pi_U = \pi_2 \circ \phi^{-1} : U \to B^2(\delta)$$

is a contact fibration in the sense of [29]. As such, there is an associated contact connection. Certainly, at a point $p \in U$ the vertical subspace is $V_p = \ker d\pi_U(p)$. Since the fiber is a contact submanifold, $(\xi_B)_p = V_p \cap \xi_p$ is a symplectic subspace of $(\xi_p, d\alpha_p)$ and therefore we may define the horizontal subspace as the symplectic orthogonal $H_p = (\xi_B)_p^{\perp d\alpha_p}$. This defines a contact connection for the contact fibration. In particular, the induced parallel transport is by contactomorphisms. We use this connection to suitably trivialize the fibration π_U. Lifting the radial vector field $r\partial_r$ on the disk provides a flow on U: the associated contactomorphism from the central fiber $\pi_U^{-1}(0)$ to the general fiber $\pi_U^{-1}(r,\theta)$ will be denoted by $\Phi_{(r,\theta)}$. The appropriate trivialization is provided by the contactomorphism

$$\begin{aligned} \Phi: B\times B^2(\delta) &\longrightarrow B\times B^2(\delta) \\ (p,r,\theta) &\longmapsto \big(\Phi_{(r,\theta)}(p),r,\theta\big). \end{aligned}$$

The composition $\widetilde{\Phi}=\Phi\circ\phi$ satisfies

$$\widetilde{\Phi}^*\alpha=\widetilde{g}\cdot\big(\alpha_B+rH(p,r,\theta)d\theta\big),$$

where $\widetilde{g}$ is a strictly positive smooth function and H is a function with the following properties:

- The identity being induced in the central fiber, $H(p,0,0)=0$.
- After the contact condition, $\partial_r(rH)>0$ in $r>0$.
- It achieves a radial minimum in the central fiber, $\partial_r H(p,0,0)>0$.

In order to suppress the H factor we further compose with

$$\begin{aligned} f: B\times B^2(\delta) &\longrightarrow B\times B^2(\delta) \\ (p,r,\theta) &\longmapsto (p,\sqrt{rH},\theta), \end{aligned}$$

which is injective for r small enough. We then obtain the diffeomorphism

$$\psi=\widetilde{\Phi}\circ f: B\times B^2(\delta)\to U$$

satisfying $\psi^*\alpha=g\cdot(\alpha_{|B}+r^2d\theta)$, for some positive function g. Denote this form by α_g, it remains to verify that the radial derivative of g is negative. Let us express this in terms of the Reeb vector fields. Note that the open book map restricts as $(\pi\circ\psi)(p,r,\theta)=\theta$ and thus R_α satisfies

$$\text{(1)} \qquad \partial_\theta\big(\psi^*R_\alpha\big)>0, \quad \text{for } r>0,$$

since the set $\{r=0\}$ is the binding B in these coordinates. This condition implies $\partial_r g<0$. Indeed, decompose the Reeb vector field R_g of α_g as

$$\psi^*R=R_g=V+b\partial_r+c\partial_\theta, \quad \text{for some } V\in\Gamma(TB) \text{ and } b,c\in\mathbb{R}.$$

Condition (1) translates into $c > 0$. The symplectic form is written as

$$d\alpha_g = dg \wedge \left(\alpha_B + r^2 d\theta\right) + g(d\alpha_B + 2r dr \wedge d\theta)$$

and from the defining equations of the Reeb vector field we obtain

$$0 = d\alpha_g(R_g, \partial_r) = -\partial_r g \cdot \frac{1}{g} - crg.$$

Consequently, the condition on g is verified as

$$c = -\partial_r g \cdot \frac{1}{g^2 r} > 0 \Longleftrightarrow \partial_r g < 0. \quad \blacksquare$$

The third assertion regarding the convexity of the boundary can be deduced as follows:

Corollary 3.4. *Let (B, π) be an open book decomposition supported by (M, α). The binding B is a convex boundary of any page $P_t = \overline{\pi^{-1}(t)}$ with respect to the Liouville vector field associated to the Liouville form $\alpha|_{P_t}$.*

Proof. This is a computation close to the boundary, as such we may use the trivialization model provided in Lemma 3.3. In this chart a page is defined as the set

$$\widetilde{P}_t = \psi^{-1}(P_t) = \left\{(p, r, \theta) \in B \times B^2(\delta) : \ r > 0, \ \theta = t\right\}.$$

In these coordinates the Liouville vector field X is given by the equation

$$(\alpha_g)_{|P_t} = i_X (d\alpha_g)_{|P_t},$$

and the solution can be explicitly written as

$$X = (\partial_r g)^{-1} g \cdot \partial_r,$$

which is certainly outwards-transverse to the boundary. Thus, the boundary is convex with respect to the stated Liouville structure. $\blacksquare$

Given a contact structure and a choice of contact form, we have described the geometric properties of an open book decomposition supported by them. An open book decomposition supported by a contact structure will be shown to exist at the end of this Section. Part of the relevance of the open book decompositions in contact geometry also resides on the converse construction: we will able to obtain contact structures from the symplectic data associated to an open book allegedly supported by a contact form. To be precise, an open book decomposition (P, Ψ) is said to be symplectic if $(P, d\beta)$ is an exact symplectic manifold with convex boundary and $\Psi \in \mathrm{Symp}(P, \partial P; d\beta)$ is a symplectomorphism supported away from the boundary. Then we obtain the following

Proposition 3.5. *Let $M = (P, \Psi)$ be a symplectic open book decomposition. Then, there exists a contact structure with contact forms supporting the open book decomposition. Further, any two such adapted contact forms that induce symplectomorphic (relative to the boundary) pages are isotopic through contact structures.*

Proof. Let us first show existence. The contact structure will be constructed from a deformation of the constant distribution $\ker(\beta)$. Note that the case in which Ψ is an exact symplectomorphism is particularly simple. Consider a smooth decreasing cut-off function $c : [0,1] \longrightarrow [0,1]$ such that $c(t)|_{[0,0.1]} = 1$ and $c(t)|_{[0.9,1]} = 0$. Define on $P \times [0,1]$ the interpolating 1-form

$$\beta_t = c(t)\Psi^*(\beta) + \big(1 - c(t)\big)\beta.$$

Then the form

$$\alpha_m = \beta_t + mdt \tag{2}$$

is a contact form for m large enough. Indeed, since

$$d\alpha_m = dt \wedge \big(\dot{c}(t)\Psi^*\beta + \big(1 - \dot{c}(t)\big)\beta\big) + d\beta$$

the contact condition reads

$$\alpha_m \wedge (d\alpha_m)^n = mdt \wedge (d\beta)^n + \eta$$

where η is a $(2n+1)$-form independent of m. It remains to extend the form α_m to the relative suspension, that is to say, to fill the mapping torus P_Ψ. This will be done explicitly.

We use the characterization of the convex boundary of a symplectic manifold in terms of the symplectization, cf. [13]. Let $(M, d\alpha)$ be an exact symplectic manifold with convex boundary $B = \partial M$, then there exists a neighborhood U of the boundary symplectomorphic to

$$\big(B \times (-\varepsilon, 0], d(e^s \cdot \alpha_{|B})\big), \quad s \in (-\epsilon, 0].$$

In other words, a neighborhood is symplectomorphic to the symplectization of the contact manifold $(B,\alpha_{|B})$. In particular, the Liouville vector field reads $X=\partial_s$ in these coordinates.

Let us fill the mapping torus P_Ψ. A neighborhood V of its boundary is of the form $V \overset{\varphi}{\cong} B\times(-\varepsilon,0]\times S^1$. Fix coordinates $(p,s,t)\in B\times(-\varepsilon,0]\times S^1$, then the contact form α_m in (2) is written as

$$\varphi^*\alpha_m = e^s(\alpha_{|B}) + mdt.$$

Defining the form in the filling is tantamount to an extension in a neighborhood of the boundary. Geometrically, we invert the model away from the section $B\times\{0\}\times S^1$ and glue it from the other side. In explicit terms, we consider the change of coordinates

$$\begin{aligned}\rho : B\times(-\varepsilon,0)\times S^1 &\longrightarrow B\times(0,\varepsilon)\times S^1\\ (p,s,t) &\longmapsto (p,-s,t).\end{aligned}$$

In these coordinates our aim is to extend the form

$$\eta = e^{-s}\alpha_{|B} + mdt = e^{-s}\cdot\big(\alpha_{|B} + me^s dt\big)$$

to the gluing area $s=0$ preserving the contact condition. The contact structure will be defined on the whole open book since we may understand the (s,t)-coordinates as polar coordinates in the disk $B^2(\varepsilon)$. In the spirit of the proof of Lemma 3.3, we define two smooth functions

$$H:[0,\varepsilon)\longrightarrow[0,1], \qquad g:[0,\varepsilon)\longrightarrow\big[0,me^{\varepsilon}\big],$$

that contact interpolate between η and the contact form in the boundary. Being precise, the functions must satisfy the following conditions:

- $H|_{[\varepsilon/2,\varepsilon)} = me^s$ and $H|_{[0,\tau]} = s^2$ for an arbitrarily small $\tau<\epsilon/2$. For the contact condition, we require $\partial_s H>0$, for $s>0$.
- $g|_{[\varepsilon/2,\varepsilon)} = e^{-s}$ and $g|_{[0,\tau]} = 1-s^2$ for an arbitrarily small $\tau<\epsilon/2$. After the lemma, since s is the radial coordinate, g should also satisfy $\partial_s g<0$, for $s>0$.

Finally we may construct the form

$$\widetilde{\eta} = g(s)\big(\alpha_{|B} + H(s)dt\big),$$

coinciding with η on the domain $B \times [\varepsilon/2, \varepsilon] \times S^1$ and extending the contact form α_m to a neighborhood of the boundary. The contact structure is adapted to the open book by construction.

Let us focus on the uniqueness statement. Consider two contact forms α_0 and α_1 adapted to the same open book (B, π) and inducing the same symplectic structure on the leaves. Endow the manifold with a Riemmanian metric and define the function $d: M_B \to \mathbb{R}^+$ measuring the square of the distance from a point to the binding. This function is smooth close to the binding; smoothly deform d so it becomes constant away from a small neighborhood of B, denote this deformation by $\widetilde{d}$. Consider the 1-form $\nu = \widetilde{d} \cdot \pi^*(d\theta)$ defined over $M \setminus B$. We construct the following deformation of any compatible contact form α:

$$\alpha^t = \alpha + t\nu,$$

for $t \in [0, K]$ where the constant $K > 0$ is arbitrarily large. Observe that the family α^t is a family of compatible contact forms. In order to connect the forms α_0 and α_1 we use the following linear family of compatible contact structures: $\widetilde{\alpha}_t = (1-t)\alpha_0^K + t\alpha_1^K$. By Gray's stability we conclude the uniqueness of the contact structure. ■

As previously mentioned, we will explain a converse of this result. There are two different cases depending on the dimension of the manifold being 3 or higher. In dimension 3 there is a strong statement that ensures a complete equivalence:

Theorem 3.6 (Giroux). *Let M be a smooth manifold. There exists a one-to-one correspondence between contact structures over M up to isotopy and symplectic open book pairs (P, ϕ) associated to M up to positive stabilization.*

For the notion of stabilization and an account of the proof of this result, see [6]. The higher-dimensional analogue is weaker. The statement is:

Theorem 3.7 (Giroux). *Let (M, ξ) be a contact manifold. There exists a contact form α for the contact structure ξ supporting an open book.*

Proof. The proof constructs the open book decomposition using the theory of asymptotically holomorphic sections. Let us divide the argument in 3 parts: construction of the binding as a contact divisor, obtaining the topological fibration over the circle and description of the contact form following Lemma 3.3.

Step 1: The binding. We first describe the input data require to use the methods of Section 2: we select a contact form α for the contact distribution ξ, fix a compatible almost complex structure J, construct the prequantizable bundle $L = M \times \mathbb{C}$ with associated connection $\nabla = d - i\alpha$ and fix the sequence of metrics g_k that we will compute the norms with. Theorem 2.11 provides us with an asymptotically holomorphic sequence of sections $s_k : M \to L^{\otimes k}$ which are ε-transverse to zero along ξ. For k large enough, the zero locus $B_k = Z(s_k)$ is a contact submanifold with trivial normal bundle. This is the contact divisor that will be used as binding.

Step 2: The topological open book. Consider the following sequence of maps

$$\begin{aligned} \pi_k : M \setminus B_k &\longrightarrow S^1, \\ p &\longmapsto \frac{s_k(p)}{|s_k(p)|}. \end{aligned}$$

The section $s_k : M \to L^{\otimes k} = M \times \mathbb{C}$ is being understood as a $\mathbb{C}$-valued smooth function since L is topologically trivial. Consider the following sequence of open covers $M = U_k \cup V_k$, where

$$U_k = \left\{ p \in M : |s_k| < \frac{\varepsilon}{2} \right\} \quad \text{and} \quad V_k = \left\{ p \in M : |s_k| > \frac{\varepsilon}{4} \right\}.$$

The covariant derivative reads as

$$\nabla s_k(p) = ds_k(p) - ik\alpha s_k(p),$$

and thus differentiating in the Reeb vector field direction we obtain

$$\nabla_R s_k(p) = d_R s_k(p) - iks_k(p).$$

Since the sections satisfy the asymptotically holomorphic bounds $|\nabla s_k| = O(1)$ and $|R|_k = k^{1/2}$, the g_k-norm of the derivative in the Reeb direction can be estimated as

$$\left| d_R s_k(p) + iks_k(p) \right| = O\left(k^{1/2}\right).$$

Hence $d_R s_k(p) \approx -iks_k(p)$ in the open set V_k. A brief computation shows that π_k is a submersion in this situation and that the Reeb vector field R is transverse to the fibers.

To conclude analogously for U_k we use the directions in the distribution. For any vector $e_p \in \xi_p$ the covariant derivative reads $\nabla_{e_p} s_k = d_{e_p} s_k$, therefore the ε-transversality ensures that for any point p in the region U_k, there are two vectors $u_k, v_k \in \xi_p$ such that the map $\nabla s_k : \xi_p \longrightarrow \mathbb{C}$ is surjective restricted to them. Consequently so is the map $ds_k : \xi_p \longrightarrow \mathbb{C}$ when restricted to them, consequently the map π_k is a submersion on U_k. The implicit function theorem provides the topological local model for the function close to the binding.

Step 3: Contact form close to the binding. The contact form in a neighborhood over the binding will be constructed as a contact fibration over the normal disk. We should ensure that the fibers close to the binding are also contact submanifolds, i.e. the sets $B_k(t) = Z(s_k - t)$ for $t \in B^2(\varepsilon/2) \subset \mathbb{C}$ are contact submanifolds. It is important to notice that the sequence $s_k - t$ is no longer asymptotically holomorphic, since the sequence of sections $\sigma_k = t$ is not asymptotically holomorphic because the derivatives in the Reeb direction do not satisfy the asymptotically holomorphic bounds. Since $s_k - t$ are $\frac{\varepsilon}{2}$-transverse to zero along ξ, the sets $B_k(t)$ are at least smooth submanifolds. However, the antiholomorphic part is bounded as $|\bar{\partial}(s_k - t)| = O(k^{-1/2})$ and therefore the sets $B_k(t)$ are also contact submanifolds. In particular, the projection map

$$\begin{aligned} \Pi_k : V_k &\longrightarrow D^2(\varepsilon) \\ p &\longmapsto s_k(p) \end{aligned}$$

is a contact fibration. We now use the contact fibration methods from the proof of Lemma 3.3 to obtain a positive function $\delta : B \longrightarrow \mathbb{R}^+$ such that the domain $\widetilde{V}_k = \{(p, v) \in B \times \mathbb{R}^2 : |v| \leq \delta(p)\}$ admits a diffeomorphism

$$\phi_k : B_k \times D^2(\delta) \longrightarrow \widetilde{V}_k,$$

such that $\phi_k^*(\alpha) = \widetilde{g} \cdot (\alpha_B + r^2 d\theta)$, for a positive function $\widetilde{g}$. This diffeomorphism is also compatible with the circle projection, i.e. if

$$\pi_\theta : \widetilde{V}_k \setminus B \times \{0\} \longrightarrow S^1$$

denotes the projection into the angular coordinates, then $\pi_\theta = \pi_k \circ \phi_k$. Note that there is no a priori guarantee that $\partial_r \widetilde{g} < 0$ for $r > 0$, which is the condition for the contact form to be adapted. However, we have already verified that α supports the open book in the neighborhood of the boundary $\phi_k^{-1}(U_k \cap \widetilde{V}_k)$, in particular $\partial_r \widetilde{g} < 0$ in this region. Thus, it remains to find a function $g : \widetilde{V}_k \longrightarrow \mathbb{R}^+$ such that:

- g extends $\widetilde{g}$, i.e. $g = \widetilde{g}$ over $\phi_k^{-1}(U_k \cap \widetilde{V}_k)$.

- It satisfies the contact condition $\partial_r g < 0$ for $r > 0$ and for small radii $0 \leq r < r_0$ it coincides with the local model $g(p,r) = k - r^2$ for some large $k > 0$.

The function g always exists. Finally, we define the contact form

$$\widetilde{\alpha} = g \cdot \left(\alpha_B + r^2 d\theta\right),$$

extending $\phi_k^*(\alpha)$ beyond $\phi_k^{-1}(U_k \cap \widetilde{V}_k)$. This form can be extended through ϕ_k to an adapted global form on the open book decomposition $M = (B_k, \pi_k)$. ■

Remark 3.8. We believe that is possible to construct the adapted contact form to be C^0-close to the arbitrary initial contact form. This would require the use of the asymptotically holomorphic theory as developed in [21], since we need a better control for the derivatives in the Reeb direction to ensure the bound

$$\left|\nabla_R s_k(p)\right| = O\left(k^s\right),$$

where $0 \leq s < 1/2$. This would imply that the Reeb vector field would be asymptotically tangent to the submanifold B_k and therefore, for k large enough, we could obtain $|\partial_r \widetilde{g}| < \gamma$, for $\gamma > 0$ arbitrarily small.

As mentioned in the introduction this existence result was essentially proved by E. Giroux almost 10 years ago. E. Giroux and J.P. Mohsen are writing a monography [16] containing this result and a more complete dictionary relating open books and contact structures. We briefly cite some of the main results in the area:

- Uniqueness up to stabilization of the open books constructed in the preceding Theorem 3.7.
- Equivalence between convex decompositions, as defined in [14], and adapted open book decompositions.
- In the case of a Stein fillable contact manifold, the open book can always be understood as the boundary of a Lefschetz pencil over the disk. As a corollary, they obtain that a contact manifold is Stein fillable if and only if it admits an open book whose monodromy map is generated by positive Dehn–Seidel twists.
- Relations with the existence of contact structures in higher dimensions. In particular, existence of contact fibrations.

4. Pencils in Quasi-contact and Contact Geometry

In this section we explain the second type of decomposition mentioned in Section 1: the analogue of the Lefschetz pencil in a symplectic manifold. They were initially introduced in [28] for the contact case and in [24] for the quasi-contact one. The geometric construction still consists in projecting the manifold to reduce the dimension. In this case we will produce a projection onto $\mathbb{CP}^1$, thus the fibers become real codimension-2 submanifolds.

4.1. Definitions

Let $(M, \xi, d\beta)$ be a quasi-contact structure and fix a compatible almost-complex structure for the symplectic bundle $(\xi, d\beta)$. A chart $\phi : (U, p) \longrightarrow V \subset (\mathbb{C}^n \times \mathbb{R}, 0)$ is said to be compatible with the quasi-contact structure at a point $p \in U \subset M$ if the push-forward at p of ξ_p is the hyperplane $\mathbb{C}^n \times \{0\}$ and $\phi_* d\beta_p$ is a positive $(1,1)$-form. The central notion in this section is the content of the following:

Definition 4.1. A quasi-contact pencil on a closed quasi-contact manifold $(M^{2n+1}, \xi, d\beta)$ is a triple (f, B, C) consisting of a codimension-4 quasi-contact submanifold B, called the base locus, a finite set C of smooth transverse curves and a map $f : M \backslash B \longrightarrow \mathbb{CP}^1$ conforming the following conditions:

(1) The set $f(C)$ contains locally smooth curves with transverse self-intersections and the map f is a submersion on the complement of C.

(2) Each $p \in B$ has a compatible local coordinate map to $(\mathbb{C}^n \times \mathbb{R}, 0)$ under which B is locally cut out by $\{z_1 = z_2 = 0\}$ and f corresponds to the projectivization of the first two coordinates, i.e. locally $f(z_1, \ldots, z_n, t) = \frac{z_2}{z_1}$.

(3) At a critical point $p \in \gamma \subset M$ there exists a compatible local coordinate chart ϕ_P such that

$$(f \circ \phi_P^{-1})(z_1, \ldots, z_n, s) = f(p) + z_1^2 + \cdots + z_n^2 + g(s)$$

where $g : (\mathbb{R}, 0) \longrightarrow (\mathbb{C}, 0)$ is an embedding at the origin (see Figure 2).

(4) The fibers $f^{-1}(P)$, for any $P \in \mathbb{CP}^1$, are quasi-contact submanifolds at the regular points.

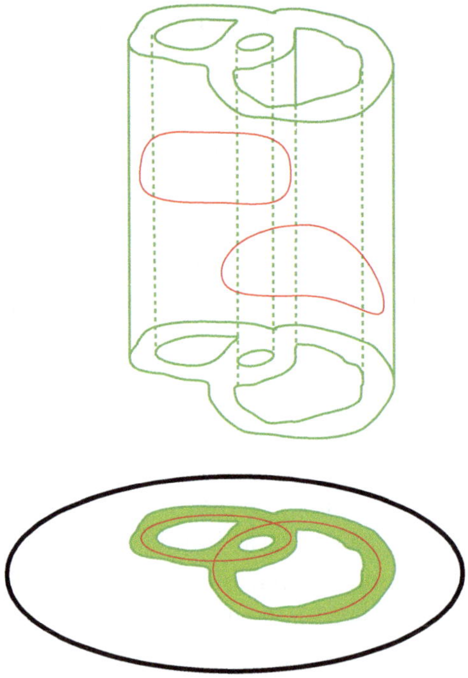

Fig. 2. Counter-image of a neighborhood of two curves of critical values

These objects always exist on a quasi-contact manifold. Actually, it is even possible to partially prescribe the topology of the fibres:

Theorem 4.2. *Let $(M, \xi, d\beta)$ be a quasi-contact manifold. Given an integral class $a \in H^2(M, \mathbb{Z})$, there exists a quasi-contact pencil (f, B, C) such that the fibers are Poincaré dual to the class a.*

This existence result can be readily extended to a more general notion of quasi-contact structures, i.e. triples of objects (M, ξ, ω), with ξ a codimension-1 cooriented distribution and ω a closed 2-form of integral class such that (ξ, ω) is a symplectic bundle. In [24] the theory is developed for these objects, though no further applications have been found in that more general setting.

The remaining part of this section is dedicated to the proof of Theorem 4.2. The strategy mimics the construction of Lefschetz pencils in projective geometry: we will produce a pair (s_0, s_1) of suitable sections of a complex line bundle, thought of as a basis for a 1-dimensional linear system, and use them to map the quasi-contact manifold onto $\mathbb{CP}^1$. As in Section 3, the asymptotically holomorphic theory from Section 2 will provide the sections. Observe that in this occasion we will produce a pair of sections and there should be further control for the behaviour of their quotient.

The initial data to obtain the sections is a quasi-contact form β, a compatible almost complex structure J for the symplectic bundle $(\xi, d\beta)$, a sequence of metrics g_k and the prequantizable bundle $L = M \times \mathbb{C}$ associated to the connection $\nabla = d - i\beta$. In order to prescribe the Poincaré dual of the fibers, let V be a fixed hermitian line bundle with a connection such that the associated curvature Θ_V satisfies $[\Theta_V] = a$. Then the existence theorems from Section 2 allow us to prove the following

Proposition 4.3. *With the data as described above, there exists an asymptotically holomorphic sequence of sections*

$$s_k = (s_{k,0} \oplus s_{k,1}) : M \longrightarrow V \otimes L^{\otimes k} \otimes \mathbb{C}^2$$

and two fixed constants $\varepsilon, \varepsilon' > 0$ satisfying that

- *s_k is ε-transverse to zero along ξ over M,*
- *$s_{k,0}$ is ε-transverse to zero along ξ over M,*
- *Consider the set $W_\infty^{s_k} = \{p \in M : s_{k,0}(p) = 0\}$. Then the holomorphic part $\partial(\frac{s_{k,1}}{s_{k,0}})$ of the covariant derivative is ε'-transverse to zero along ξ over $M \setminus W_\infty^{s_k}$.*

Proof. The ε-transversality is a C^1-stable property, thus we may systematically perform C^1-perturbations and it will be preserved. Consider the asymptotically holomorphic null-constant sequence of sections $(0) : M \longrightarrow L^{\otimes k} \otimes \mathbb{C}^2$, then Theorem 2.11 provides an asymptotically holomorphic sequence of sections s_k which are ε_1-transverse to zero.

Let $\delta = \varepsilon_1/2$ and apply Theorem 2.11 to the sequence $s_{k,0} : M \longrightarrow L^{\otimes k}$ in order to obtain an asymptotically holomorphic sequence of sections $\sigma_{k,0} : M \longrightarrow L^{\otimes k}$ such that $|s_{k,0} - \sigma_{k,0}|_{C^2} \leq \delta$ and ε_2-transverse to zero along ξ. The pair $\tilde{s}_k = (\tilde{s}_{k,0}, \tilde{s}_{k,1}) = (\sigma_{k,0}, s_{k,1}) : M \longrightarrow L^{\otimes k} \otimes \mathbb{C}^2$ satisfies the first two transversality properties for $\varepsilon_3 = \min(\delta, \varepsilon_2)$.

Observe that $\partial(\frac{\tilde{s}_{k,1}}{\tilde{s}_{k,0}})$ is an asymptotically holomorphic sequence on the open set $U_\varepsilon^{s_k} = \{p \in M : |s_{k,0}| > \varepsilon\}$ since it is

$$\partial \tilde{s}_{k,1} \otimes \tilde{s}_{k,0} - \tilde{s}_{k,1} \otimes \partial \tilde{s}_{k,0} : M \longrightarrow \xi^{1,0} \otimes L^{\otimes k} \otimes V \otimes \mathbb{C}^2,$$

divided by the bounded below asymptotically holomorphic sequence $\tilde{s}_{k,0}$. Finally, apply Theorem 2.13 to the section $\tilde{s}_k$ and the constant $\varepsilon_3/2$: there exists an asymptotically holomorphic sequence $\tilde{\sigma}_k$ such that $\partial\tilde{\sigma}_{k,1} \otimes \tilde{\sigma}_{k,0} - \tilde{\sigma}_{k,1} \otimes \partial\tilde{\sigma}_{k,0}$ is ε_4-transverse to zero along ξ over M. This sequence satisfies that $\partial(\frac{\tilde{\sigma}_{k,1}}{\tilde{\sigma}_{k,0}})$ is ε'-transverse to zero along $U^{\sigma_k}_{\varepsilon_3/2}$. Thus, the sequence $\tilde{\sigma}_k$ satisfies the required properties for the chosen ε' and $\varepsilon = \varepsilon_3/2$. ■

4.2. Existence of Quasi-contact Pencils

Let us briefly describe the argument. Apply Proposition 4.3 to the data induced by the quasi-contact manifold $(M, \xi, d\beta)$ and $a \in H^2(M, \mathbb{Z})$ as previously explained. This provides a pair of suitably transverse sections inducing the potential quasi-contact pencil. The first step is the structure of the fibers, which should satisfy (4) in Definition 4.1. Secondly we focus on the base locus and obtain the required local model. Finally, it is ensured that the Morse model around the singularities can be achieved.

Step 1: Analysis. Since the sections s_k provided by Proposition 4.3 are ε-transverse, the zero set $B_k = Z(s_k)$ is a codimension 4 quasi-contact manifold. Also, the set $W^{s_k}_\infty$ is a codimension 2 quasi-contact submanifold. Let us define the sequence of maps

$$F_k : M \setminus W^{s_k}_\infty \longrightarrow \mathbb{C}$$
$$p \longmapsto \frac{s_{k,1}}{s_{k,0}}.$$

These are our candidates for quasi-contact pencil structures. Define the set

$$\Gamma = \{p \in M : |\partial F_k| \le |\bar{\partial} F_k|\}.$$

If we are able to show that $\Gamma = \{p \in M : d_\xi F_k = 0\}$ then the fibers of F_k will be quasi-contact submanifolds at the regular points. We will actually justify that the set Γ lies arbitrarily close to the critical curves. First, a bound from below for the norm of $s_{k,0}$ on Γ:

Lemma 4.4. *There is a constant $\eta > 0$, depending only on ε' and ε, such that if k is large enough, then $|s_{k,0}| \ge \eta$ on Γ.*

Proof. The section s_k is ε-transverse to zero along ξ and thus at any point $p \in M$ with $|s_k(p)| < \varepsilon$ the map $\partial s_k(p)$ is ε-transverse. Without loss of generality suppose that $|s_{k,0}| \le |s_{k,1}|$. By surjectivity there exists a unitary vector

$v \in \xi_p$ such that $|\partial_v s_{k,0}(p)| \geq \varepsilon$ and $|\partial_v s_{k,1}(p)| = 0$ and thus

$$\big|\partial s_{k,1}(p) \otimes s_{k,0}(p) - s_{k,1}(p) \otimes \partial s_{k,0}(p)\big| \geq \frac{|s_k(p)|}{2}\varepsilon. \tag{3}$$

The asymptotically holomorphic bounds impose

$$\big|\bar{\partial} s_{k,1}(p) \otimes s_{k,0}(p) - s_{k,1}(p) \otimes \bar{\partial} s_{k,0}(p)\big| \leq ck^{-1/2}|s_k(p)| \tag{4}$$

and so combining the inequalities (3) and (4), for k large enough, we obtain

$$\big|\partial s_{k,1}(p) \otimes s_{k,0}(p) - s_{k,1}(p) \otimes \partial s_{k,0}(p)\big| > \big|\bar{\partial} s_{k,1}(p) \otimes s_{k,0}(p) - s_{k,1}(p) \otimes \bar{\partial} s_{k,0}(p)\big|.$$

This implies $|\partial F_k(p)| > |\bar{\partial} F_k(p)|$ at any point $p \in M \setminus W_\infty^{s_k}$ with $|s_k(p)| < \varepsilon$.

Consider $p \in M$ with $|s_k(p)| > \varepsilon$, say $|s_{k,1}(p)| \geq \varepsilon/2$. Let $\eta \leq \varepsilon$ and suppose further that $|s_{k,0}(p)| < \eta$, by ε-transversality of $s_{k,0}$ the inequality $|\partial s_{k,0}(p)| > \varepsilon$ holds. At the same time the asymptotically holomorphic bounds require $|\partial s_{k,1}(p)| \leq c$, for some fixed constant $c > 0$. Fix $\eta = \varepsilon^2/4c$, then the reverse triangle inequality yields

$$\big|\partial s_{k,1}(p) \otimes s_{k,0}(p) - s_{k,1}(p) \otimes \partial s_{k,0}(p)\big| \geq \frac{\varepsilon^2}{4}.$$

Again (4), for k large enough, gives $|\partial F_k(p)| > |\bar{\partial} F_k(p)|$. ■

Let $\Delta \subset \Gamma$ be the set of points where $\partial F_k = 0$. The connected components of Δ form a discrete set of smooth transverse curves since ∂F_k satisfies the adequate transversality condition. Observe that $\pi_0(\Delta)$ is finite because $\Delta \subset \Gamma$ and the set Γ is contained in the complementary of a τ-neighborhood of the compact manifold $W_\infty = Z(s_k^0)$, for $\tau > 0$ a constant small enough, after Lemma 4.4. In order to understand the behaviour of the set Γ, consider the set $\Omega_\eta = \{p \in C, |s_0(p)| > \eta/2\}$. The following statement describes the neighborhoods of the elements in Δ and in particular the geometry of Γ:

Proposition 4.5. *With the above notation, there exists a uniform constant $\rho_0 > 0$ such that the ρ_0-neighborhoods of each connected component $\gamma_i \in \Delta$ are disjoint and contained in Ω_η. Further, given any $\rho < \rho_0$, for $k = k(\rho)$ large enough, the set Γ is contained in a ρ-neighborhood of Δ.*

This is essentially Proposition 9 in [9], instead of the distance to a finite number of points we use the distance from a point to a curve. Geometrically, in a point of Γ the norm $|\partial F_k|$ is bounded by $|\overline{\partial} F_k| = O(k^{-1/2})$ and thus can be arbitrarily small, transversality then provides a solution for the equation $\partial F_k = 0$ and the norm being arbitrarily small ensures its existence nearby.

The detailed argument requires an estimated form of the Inverse Function Theorem, cf. [9].

Step 2: Perturbation at the Base Point Set. The sequence ∂F_k will be perturbed in arbitrarily small neighborhoods of the alleged base locus B_k in order to achieve the local model in Definition 4.1. To ease notation we write $B = B_k$, as k is thought as fixed if large enough. We describe the local model in terms of the equation for the tangent space at a point $p \in B$

$$\nabla s_k(p) = \nabla s_{k,0}(p) \oplus \nabla s_{k,1}(p) : T_pM \longrightarrow L_p^{\otimes k} \oplus L_p^{\otimes k}.$$

If $T_pB \subset \xi_p$ were a symplectic subspace, the $\mathbb{R}$-linear map $\nabla s_k(p)$ provides an isomorphism as $\mathbb{R}$-vector spaces of the symplectic orthogonal with the $\mathbb{C}$-vector space $L_p^{\otimes k} \oplus L_p^{\otimes k}$. In particular the symplectic orthogonal is endowed with a complex structure. Let us prove a linear characterization of the model with respect to the base locus:

Lemma 4.6. *Let $p \in B$. Then F_k can be represented around p in the standard local model of Definition 4.1 if and only if $T_pB \cap (\xi_p, d\beta_p)$ is a symplectic subspace and the restriction of $d\beta$ to the symplectic orthogonal $N_p = (T_pB \cap \xi)_x^{\perp d\beta}$ is a positive form of type $(1,1)$ with respect to the complex structure of N_p induced by $\nabla s_k(p)$.*

Proof. It is readily seen that the condition is necessary since the existence of a local model provided in the Definition 4.1 implies the properties in the statement. Conversely, suppose that B is a quasi-contact submanifold near p and let $(z_3, \dots, z_n, t)$ be local coordinates at p such that

$$(d\beta)_{|B} = \frac{i}{2} \sum_{j=3}^{n} dz_j \wedge d\bar{z}_j \quad \text{and} \quad \xi_p = \ker dt.$$

Extend the coordinate functions $(z_3, \dots, z_n)$ ensuring that the their derivatives vanish in the normal directions of N_p. Locally trivialize the bundle $L^{\otimes k}$ via a non-vanishing section σ and define functions $z_0 = s_{k,0} \cdot \sigma$, $z_1 = s_{k,1} \cdot \sigma$. These provide a complete set of coordinates $(z_1, \dots, z_n, t)$ around p in which the symplectic form is expressed as

$$(d\beta)_{|B} = (d\beta)_{|N_p} + \sum_{j=3}^{n} dz_j \wedge d\bar{z}_j.$$

Thus we obtain the required local model. ■

To achieve the local model at B, it remains to perturb F_k such that it satisfies the hypothesis of the linear characterization. More precisely, there exists a perturbation $D_p : T_pM \longrightarrow L_p^{\otimes k} \oplus L_p^{\otimes k}$ of the map $\nabla s_k(p)$ conforming

$$\left|(\nabla s_k(p) - D_p)v\right| \leq ck^{-1/2}\left|\nabla s_k(p)(v)\right|, \quad \forall v \in T_pM,$$

and the requirements of the Lemma 4.6. Indeed, it is simple to perturb the pair $s_k = s_{k,0} \oplus s_{k,1}$ to $\tilde{s}_k = \tilde{s}_{k,0} \otimes \tilde{s}_{k,1}$ at distance $O(k^{-1/2})$ in C^3-norm with D_p as the linearization at p of the associated pencil–map $\tilde{F}_k$. This perturbation fulfills the property (2) of the Definition 4.1. The perturbation will still be referred to as F_k.

Step 3: Local Model for the Singularities. It remains to study the map F_k near the singular set Δ. Let $\gamma \in \Delta$ be a smooth connected curve. The perturbation of F_k will occur in a δ-neighborhood of Δ, Proposition 4.5 implies that the perturbations can be independently performed in each connected component of Δ. We will describe the associated quasi-contact data around the curve γ, define a general perturbation well-behaved with respect to a simple integral distribution and then prove that it can be chosen to induce a Morse model with respect to the actual quasi-contact distribution.

Let us specify the information contained in a trivialization. For k large enough, the curve γ is a transverse contact loop, equivalently $TM|_\gamma = T\gamma \oplus \xi$, and the angle between $T\gamma$ and ξ is bounded below by a uniform constant because of the transversality of the sequence. To trivialize we use the geodesic flow of the metric g_J associated to the fixed almost complex structure J and obtain a diffeomorphism

$$\phi : U_\rho \longrightarrow V_\rho \subset S^1 \times \mathbb{C}^n,$$

where U_ρ is a ρ-neighborhood of γ, measured with the g_k metric, and V_ρ its image by the flow, which is an open neighborhood of $S^1 \times \{0\}$. In the neighborhood $S^1 \times \mathbb{C}^n$ we consider the product metric with first component the image of g_k through ϕ and second component the standard hermitian metric in $\mathbb{C}^n$. Suppose also that $\phi_* J_{|\gamma} = J_0$, J_0 being the standard complex structure of $\mathbb{C}^n$. In particular, the model being isometric allows us to explicitly measure in V_ρ.

Regarding the distributions, denote $\xi_k = \phi_*\xi$ and let ξ_h be the integrable distribution given as $\{p\} \times \mathbb{C}^n \subset S^1 \times \mathbb{C}^n$. Possibly after a uniform shrinking of ρ, the angle[3] between ξ_k and ξ_h tends to zero; more precisely, there exists a uniform constant $C>0$ such that

$$\angle_M(\xi_k(s,z),\xi_h(s,z)) < C|z|k^{-1/2}, \quad \forall (s,z)\in V_\rho \subset S^1\times\mathbb{C}^n. \tag{5}$$

Hence, we are able to project orthogonally the almost complex structure $\phi_* J$ on ξ_k to an almost complex structure J_h in the distribution ξ_h. Let $\mu:\bigwedge^{(1,0)}_{J_0}\longrightarrow\bigwedge^{(0,1)}_{J_0}$ be the defining function of J_h with respect to J_0 as almost complex structures in ξ_h, cf. [7]. Denote by ∂ and ∂_0 the associated holomorphic parts of the covariant derivative with respect to J_h and J_0. The holomorphic and antiholomorphic parts of the operator corresponding to $\phi_* J$ in ξ_k will be denoted ∂_k and $\bar\partial_k$. In particular

$$\partial=\partial_0+\bar\mu\bar\partial_0, \qquad \bar\partial=\bar\partial_0+\mu\partial_0. \tag{6}$$

It follows that $|\mu(z)|\le C|z|k^{-1/2}$, C being a uniform constant.

Let us define the perturbation of F_k, as mentioned above we will construct the perturbation in the isometric neighborhood $S^1\times\mathbb{C}^n$. The context being clear, we will still denote by F_k the pull-back $(\phi^{-1})^* F_k$. Since the local model required in Definition 4.1 is quadratic, we will deform F_k to approximately its complex Hessian $H=\frac{1}{2}\nabla_\partial(\partial F_k)$. Locally, it is expressed as

$$H(s,z)=\sum H_{\alpha\beta}(s)z_\alpha z_\beta.$$

Consider a cut-off function $\beta_\rho : S^1\times\mathbb{C}^n\longrightarrow[0,1]$ satisfying

- $\beta_\rho(\phi(p))=1$ if $d_k(p,\gamma)\le\rho/2$, and $\beta_\rho(\phi(p))=0$ if $d_k(p,\gamma)\ge\rho$.
- $|\nabla\beta_\rho|=O(\rho^{-1})$.

The second condition can be ensured due to the choice of metrics. The constant $\rho<\rho_0$ will be shrunk in a uniform way, and so we consider it fixed assuring that the conditions are satisfied. The perturbation of F_k will be of the form

$$\widetilde{F}_k(s,z)=\beta_\rho\big(w(s)+H(s,z)\big)+(1-\beta_\rho)F_k(s,z),$$

where $w:S^1\longrightarrow\mathbb{C}$ is any smooth function. The only remaining issue is the verification of $\Gamma=\Delta$ for the perturbed $\widetilde{F}_k(s,z)$, this is the content of the following

[3]The maximum angle between two subspaces $U,V\subset\mathbb{R}^m$ of the Euclidean space is by definition $\angle_M(U,V)=\max_{u\in U}\{\angle(u,V)\}$.

Lemma 4.7. *With the above notations, let $\rho > 0$ and $|w(s) - F_k(s,0)|$ be small enough and $|\dot{w}(s)| = O(1)$. Then for sufficiently large k the inequality $|\partial_k \widetilde{F}_k| \leq |\bar{\partial}_k \widetilde{F}_k|$ is only satisfied in γ.*

Proof. There are two different scenarios, close to the curve where $\beta_\rho \equiv 1$ and the transition area where $\nabla\beta_p$ does not vanish. Let us first consider the former. Then the perturbation reads $\widetilde{F}_k = w + H$ and so $\partial\widetilde{F}_k = \partial H$, $\bar{\partial}\widetilde{F}_k = \bar{\partial}H$. The η-transversality of ∂F_k yields the following bound

$$\left|\partial H(s,z)\right| \geq \eta|z| - \left|\bar{\partial}(\partial F)\right|_{(z=0)}|z|.$$

Since $\bar{\partial}\partial + \partial\bar{\partial} \equiv 0$ on functions and the norm $\partial\bar{\partial}F$ is controlled, as F has uniform C^3-bounds, we obtain

$$\left|\partial H(s,z)\right| \geq \eta|z| - Ck^{-1/2}|z|.$$

We need to relate this to the distribution ξ_k. By hypothesis $|\dot{w}(s)| = O(1)$ and we also know $|\partial s H_{\alpha\beta}(s)| = O(1)$ due to the C^3-bounds of F_k. Then the angle inequality (5) implies

$$\left|\partial_k H(s,z)\right| \geq \eta|z| - Ck^{-1/2}|z|, \tag{7}$$

where $C > 0$ is another suitable uniform constant; the uniform constants appearing on the bounds will deliberately be referred to as C. The asymptotically holomorphic bounds also imply that $|\bar{\partial}H| \leq C|z|^2k^{-1/2}$ and we analogously deduce

$$|\bar{\partial}_k H| \leq C\big(|z|^2k^{-1/2} + |z|k^{-1/2}\big). \tag{8}$$

The condition $|\partial_k H| \leq |\bar{\partial}_k H|$ along with (7) and (8) implies $z = 0$ for k large enough, concluding the statement in this case.

We focus on the latter situation, i.e. the behaviour of Γ around points in the annulus containing the support of $\nabla\beta_\rho$. The antiholomorphic derivative of the perturbation reads

$$\bar{\partial}\widetilde{F}_k = \bar{\partial}\beta_\rho(w + H - F_k) + \beta_\rho\bar{\partial}H + (1 - \beta_\rho)\bar{\partial}F_k.$$

As before this concerns ξ_h and we may bound the norm $|\bar{\partial}f_0|$ as in [9]. Again the hypothesis $|\dot{w}(s)| = O(1)$ and the asymptotically holomorphic estimate $|\bar{\partial}_k F_k| = O(k^{-1/2})$ combine with the angle inequality (5) to conclude

$$|\bar{\partial}_k\widetilde{F}_k| \leq C\big(\rho^2 + k^{-1/2} + \big|\widetilde{F}_k(s,0) - w\big|\rho^{-1}\big).$$

The direct computation $\partial_k \widetilde{F}_k = \partial_k \beta_\rho (w + H - F_k) + \beta_\rho \partial_k H + (1 - \beta_\rho)\partial_k F_k$ and the transversality of F_k yield a lower bound for $|\partial_k \widetilde{F}_k|$. The argument follows as in [9] until

$$|\partial_k \widetilde{F}_k| - |\bar{\partial}_k \widetilde{F}_k| \geq \frac{\eta\rho}{2} - C\big(\rho^2 + k^{-1/2} + \big|w - F_k(s,0)\big|\rho^{-1}\big).$$

By the hypothesis $|w - F_k(s,0)|$ is small enough, and once fixed a sufficiently small ρ, for k large enough the right hand side of the inequality is strictly positive over the annulus. This concludes the statement in the second case. ■

Note that w can be chosen generic enough to ensure that the projection of the critical points is a family of immersed curves. Also, the perturbation satisfies the local model around the curves because a real generic S^1-family of non-degenerate quadratic forms can be diagonalised. This proves the existence of quasi-contact pencils. The statement concerning the Poincaré dual of the fibers follows from the fact that first Chern class of the normal bundle to the section is the Poincaré dual of its vanishing locus.

Acknowledgements. I want to thank Roger Casals by his proof-reading that has greatly improved the final version of this article. Special thanks also to Emmanuel Giroux who patiently explained to me the results contained in Section 3. Most of the ideas in these notes come from him. J.P. Mohsen kindly offered a copy of his thesis [27] to me. Discussions with Vincent Colin, Dishant Pancholi, Klaus Niederkrüeger, Thomas Vogel and Eva Miranda have been really helpful to write down these notes.

References

[1] D. Auroux, Asymptotically holomorphic families of symplectic submanifolds. Geom. Funct. Anal. **7**(6), 971–995 (1997)

[2] D. Auroux, Estimated transversality in symplectic geometry and projective maps, in *Symplectic Geometry and Mirror Symmetry*, Seoul (2000), pp. 1–30

[3] D. Auroux, S. Donaldson, L. Katzarkov, Luttinger surgery along Lagrangian tori and non-isotopy for singular symplectic plane curves. Math. Ann. **326**(1), 185–203 (2003)

[4] P. Biran, Lagrangian barriers and symplectic embeddings. Geom. Funct. Anal. **11**(3), 407–464 (2001)

[5] R. Casals, D. Pancholi, F. Presas, Almost contact 5-folds are contact. arXiv preprint

[6] V. Colin, Livres ouverts en géométrie de contact. Sémin. Bourbaki **59**(969), 91–118 (2006)

[7] S. Donaldson, Symplectic submanifolds and almost-complex geometry. J. Differ. Geom. **44**, 666–705 (1996)

[8] S. Donaldson, Lefschetz fibrations in symplectic geometry, in *Proceedings of the International Congress of Mathematicians*, vol. II, Berlin (1998)

[9] S. Donaldson, Lefschetz pencils on symplectic manifolds. J. Differ. Geom. **53**, 205–236 (1999)

[10] S. Donaldson, I. Smith, Lefschetz pencils and the canonical class for symplectic four-manifolds. Topology **42**(4), 743–785 (2003)

[11] Y. Eliashberg, N. Mishachev, *Introduction to the h-Principle*. Graduate Studies in Mathematics, vol. 48 (AMS, Providence, 2002)

[12] H. Geiges, Contact structures on 1-connected 5-manifolds. Mathematica **38**, 303–311 (1991)

[13] H. Geiges, *An Introduction to Contact Topology*. Cambridge Studies in Advanced Mathematics, vol. 109 (Cambridge University Press, Cambridge, 2008)

[14] E. Giroux, Convexité en topologie de contact. Comment. Math. Helv. **66**, 637–677 (1991)

[15] E. Giroux, Géométrie de contact: de la dimension trois vers les dimensions supérieures, in *Proceedings of the International Congress of Mathematicians*, vol. II, Beijing, 2002 (Higher Ed. Press, Beijing, 2002), pp. 405–414

[16] E. Giroux, J.P. Mohsen, preprint

[17] R. Gompf, A new construction of symplectic manifolds. Ann. Math. **142**(3), 527–595 (1995)

[18] M. Gromov, Pseudoholomorphic curves in symplectic manifolds. Invent. Math. **82**(2), 307–347 (1985)

[19] M. Gromov, *Partial Differential Relations*. Ergeb. Math. Grenzgeb., vol. 9 (Springer, Berlin, 1986)

[20] A. Ibort, D. Martínez-Torres, Lefschetz pencil structures for 2-calibrated manifolds. C. R. Acad. Sci. Paris, Ser. I. **339**, 215–218 (2004)

[21] A. Ibort, D. Martínez-Torres, F. Presas, On the construction of contact submanifolds with prescribed topology. J. Differ. Geom. **56**(2), 235–283 (2000)

[22] F. Lalonde, D. McDuff, The classification of ruled symplectic 4-manifolds. Math. Res. Lett. **3**(6), 769–778 (1996)

[23] R. Lutz, Structures de contact sur les fibré principaux en cercles de dimension trois. Ann. Inst. Fourier (Grenoble) **27**(3), 1–15 (1977)

[24] D. Martínez Torres, The geometry of 2-calibrated manifolds. Port. Math. **66**(4), 427–512 (2009)

[25] D. Martínez-Torres, V. Muñoz, F. Presas, Open book decompositions for almost contact manifolds, in *Proceedings of the XI Fall Workshop on Geometry and Physics*, Oviedo, 2002. Publicaciones de la RSME

[26] D. McDuff, Examples of simply-connected symplectic non-Kählerian manifolds. J. Differ. Geom. **20**(1), 267–277 (1984)

[27] J.-P. Mohsen, Transversalitè quantitative et sous–variétés isotropes. Ph.D. Thesis, École Normale Supérieure de Lyon, 2001

[28] F. Presas, Lefschetz type pencils on contact manifolds. Asian J. Math. **6**(2), 277–301 (2002)

[29] F. Presas, On a class of non-fillable contact manifolds. Geom. Topol. **11**, 2203–2225 (2007)

[30] P. Seidel, *Fukaya Categories and Picard-Lefschetz Theory*. Zurich Lectures in Advanced Mathematics (European Mathematical Society (EMS), Zürich, 2008)

[31] C.H. Taubes, GR=SW: counting curves and connections. J. Differ. Geom. **52**(3), 453–609 (1999)

F. Presas (✉)

Instituto de Ciencias Matemáticas
CSIC–UAM–UC3M–UCM
C. Nicolás Cabrera, 13–15
28049, Madrid
Spain

e-mail: fpresas@icmat.es

BOLYAI SOCIETY
MATHEMATICAL STUDIES, 26

Contact and Symplectic Topology
pp. 173–244.

Higher Dimensional Contact Topology via Holomorphic Disks

KLAUS NIEDERKRÜGER

1. Introduction

In '85 Gromov published his article on pseudo-holomorphic curves [17] that made symplectic topology as we know it today only possible. Using these techniques, Gromov presented in his initial paper many spectacular results, and soon many other people started using these methods to settle questions that before had been out of reach [1, 9, 10, 18, 22, 23] and many others; for more recent results in this vein we refer to [31, 36].

While the references above rely on studying the topology of the moduli space itself, Gromov's J-holomorphic methods have also been used to develop powerful algebraic theories like Floer Homology, Gromov-Witten Theory, Symplectic Field Theory, Fukaya Theory etc. that basically rely on counting rigid holomorphic curves (that means holomorphic curves that are isolated). Note though that we will completely ignore such algebraic techniques in these notes.

Gromov's approach for studying a symplectic manifold (W,ω) consists in choosing an *auxiliary* almost complex structure J on W that is compatible with ω in a certain way. This auxiliary structure allows us to study so called J-holomorphic curves, that means, equivalence classes of maps

$$u\colon (\Sigma, j) \to (W, J)$$

F. Bourgeois et al. (eds.), *Contact and Symplectic Topology*,
Bolyai Society Mathematical Studies 26, DOI 10.1007/978-3-319-02036-5_5,

from a Riemann surface (Σ, j) to W whose differential at every point $x \in \Sigma$ is a (j, J)-complex map

$$Du_x \colon T_x\Sigma \to T_{u(x)}W.$$

Conceivable generalizations of such a theory based on studying J-holomorphic surfaces or even higher dimensional J-complex manifolds only work for *integrable* complex structures; otherwise generically such submanifolds do not exist. A different approach has been developed by Donaldson [7, 8], and consists in studying approximately holomorphic sections in a line bundle over W. This theory yields many important results, but has a very different flavor than the one discussed here by Gromov.

The J-holomorphic curves are relatively rare and usually come in finite dimensional families. Technical problems aside, one tries to understand the symplectic manifold (W, ω) by studying how these curves move through W.

Let us illustrate this strategy with the well-known example of $\mathbb{C}P^n$. We know that there is exactly one complex line through any two points of $\mathbb{C}P^n$. We fix a point $z_0 \in \mathbb{C}P^n$, and study the space of all holomorphic lines going through z_0. It follows directly that $\mathbb{C}P^n \setminus \{z_0\}$ is foliated by these holomorphic lines, and every line with z_0 removed is a disk. Using that the lines are parametrized by the corresponding complex line in $T_{z_0}\mathbb{C}P^n$ that is tangent to them, we see that the space of holomorphic lines is diffeomorphic to $\mathbb{C}P^{n-1}$, and that $\mathbb{C}P^n \setminus \{z_0\}$ will be a disk bundle over $\mathbb{C}P^{n-1}$.

In this example, we have used an ambient manifold that we understand rather well, $\mathbb{C}P^n$, to compute the topology of the space of complex lines. So far, it might seem unclear how one could obtain information about the topology of the space of complex lines in an ambient space that we do not understand equally well, to then extract in a second step missing information about the ambient manifold.

The common strategy is to assume that the almost complex manifold we want to study already contains a family of holomorphic curves. We then observe how this family evolves, hoping that it will eventually "fill up" the entire symplectic manifold (or produce other interesting effects).

To briefly sketch the type of arguments used in general, consider now a symplectic manifold W with a compatible almost complex structure, and suppose that it contains an open subset U diffeomorphic to a neighborhood of $\mathbb{C}P^1 \times \{0\}$ in $\mathbb{C}P^1 \times \mathbb{C}$ (see [22]). In this neighborhood we find a family of holomorphic spheres $\mathbb{C}P^1 \times \{z\}$ parametrized by the points z. We can explicitly write down the holomorphic spheres that lie completely inside U, but Gromov compactness tells us that as the holomorphic curves approach the boundary of U, they cannot just cease to exist but instead there is a well understood way in which they can degenerate, which is called *bubbling*. Bubbling means that a family of holomorphic curves decomposes in the limit into several smaller ones. Sometimes bubbling can be controlled or even excluded by imposing technical conditions, and in this case, the limit curve will just be a regular holomorphic curve.

In the example we were sketching above, this means that if no bubbling can happen, there will be regular holomorphic spheres (partially) outside U that are obtained by pushing the given ones towards the boundary of U. This limit curve is also part of the 2-parameter space of spheres, and thus it will be surrounded by other holomorphic spheres of the same family. As long as we do not have any bubbling, we can thus extend the family by pushing the spheres to the limit and then obtain a new regular sphere, which again is surrounded by other holomorphic spheres. This way, we can eventually show that the whole symplectic manifold is filled up by a 2-dimensional family of holomorphic spheres. Furthermore the holomorphic spheres do not intersect each other (in dimension 4), and this way we obtain a 2-sphere fibration of the symplectic manifold.

In conclusion, we obtain in this example just from the existence of the chart U, and the conditions that exclude bubbling that the symplectic manifold needs to be a 2-sphere bundle over a compact surface (the space of spheres).

Note that many arguments in the example above (in particular the idea that the moduli spaces foliate the ambient manifold) do not hold in general, that means for generic almost complex structures in manifolds of dimension more than 4. Either one needs to weaken the desired statements or find suitable work arounds. The principle that is universal is the use of a well understood local model in which we can detect a family of holomorphic curves. If bubbling can be excluded, this family extends into the unknown parts of the symplectic manifold, and can be used to understand certain topological properties of this manifold.

These notes are based on a course that took place at the Université de Nantes in June 2011 during the *Trimester on Contact and Symplectic Topology*. We will explain how holomorphic curves can be used to study symplectic fillings of a given contact manifold. Our main goal consists in showing that certain contact manifolds do not admit any symplectic filling at all. Since closed symplectic manifolds are usually studied using closed holomorphic curves, it is natural to study symplectic fillings by using holomorphic curves with boundary. We will explain how the existence of so called *Legendrian open books* (Lobs) and *bordered Legendrian open books* (bLobs) controls the behavior of holomorphic disks, and what properties we can deduce from families of such disks. The notions are direct generalizations of the overtwisted disk [9, 17] and standardly embedded 2-spheres in a contact 3-manifold [4, 17, 18].

For completeness, we would like to mention that symplectic fillings have also been studied successfully via punctured holomorphic curves whose behavior is linked to Reeb orbit dynamics, and via closed holomorphic curves by first capping off the symplectic filling to create a closed symplectic manifold.

1.1. Outline of the Notes

In the first part of these notes we will talk about Legendrian foliations, and in particular about Lobs and bLobs. We will not consider any holomorphic curves here, but the main aim will be instead to illustrate examples where these objects can be localized. In Section 3, we study the properties of holomorphic disks imposed by Legendrian foliations and convex boundaries. In the last section, we use this information to understand moduli spaces of holomorphic disks obtained from a Lob or a bLob, and we prove some basic results about symplectic fillings.

The content of these notes are based on an unfinished manuscript of [28].

1.2. Notation

We assume throughout a certain working knowledge on contact topology (for a reference see for example [24, Chapter 3.4] and [12]) and on holomorphic curves [3, 25]. The contact structures we consider in this text are always *cooriented.* Remember that by choice of a coorientation, (M, ξ) always obtains a natural orientation and its contact structure ξ carries a natural *conformal* symplectic structure. For both, it suffices to choose a **positive** contact form α, that means, a 1-form with $\xi = \ker \alpha$ that evaluates positively on vectors that are positively transverse to the contact structure. The orientation on M is then given by the volume form

$$\alpha \wedge d\alpha^n,$$

where $\dim M = 2n+1$, while the conformal symplectic structure is represented by $d\alpha|_\xi$.

One can easily check that these notions are well-defined by choosing any other positive contact form α' so that there exists a smooth function $f\colon M \to \mathbb{R}$ such that $\alpha' = e^f \alpha$.

Further Conventions. Note that $\mathbb{D}^2$ denotes in this text the *closed* unit disk.

I owe it to Patrick Massot to have been converted to the following *jargon*.

Definition. The term **regular equation** can refer in this text to any of the following objects:

(1) When Σ is a cooriented hypersurface in a manifold M, then we call a smooth function $h\colon M \to \mathbb{R}$ a **regular equation for** Σ, if 0 is a regular value of h and $h^{-1}(0) = \Sigma$.

(2) When $\mathcal{D} \leq TM$ is a singular codimension 1 distribution, then we say that a 1-form β is a **regular equation for** $\mathcal{D}$, if $\mathcal{D} = \ker \beta$ and if $d\beta \neq 0$ at singular points of $\mathcal{D}$.

According to this definition, an equation of a contact structure is just a contact form.

2. Lobs & bLobs: Legendrian Open Books and Bordered Legendrian Open Books

2.1. Legendrian Foliations

2.1.1. General Facts about Legendrian Foliations. Let (M,ξ) be a contact manifold that contains a submanifold N. Generically, if we look at any point $p \in N$ the intersection between ξ_p and the tangent space T_pN will be a codimension 1 hyperplane. Globally though, the distribution $\mathcal{D} = \xi \cap TN$ may be singular, because there can be points $p \in N$ where $T_pN \subset \xi_p$, and equally important the distribution $\mathcal{D}$ will only be in very rare cases a foliation. In fact, if we choose a contact form α for ξ, then we obtain by the Frobenius theorem that $\mathcal{D}$ will only be a (singular) foliation if

$$(\alpha \wedge d\alpha)|_{TN} \equiv 0.$$

Another way to state this condition is to say that we have $d\alpha|_{\mathcal{D}_p} = 0$ at every regular point $p \in N$ of $\mathcal{D}$, so that $\mathcal{D}_p$ has to be an isotropic subspace of $(\xi_p, d\alpha_p)$. In particular, this shows that the induced distribution $\mathcal{D}$ can never be integrable if $\dim \mathcal{D} > \frac{1}{2} \dim \xi$.

We will usually denote the distribution $\xi \cap TN$ by $\mathcal{F}$ whenever it is a singular foliation. Furthermore, we will call such an $\mathcal{F}$ a **Legendrian foliation** if $\dim \mathcal{F} = \frac{1}{2} \dim \xi$, which implies that N has to be a submanifold of dimension $n+1$ if the dimension of the ambient contact manifold is $2n+1$. For reasons that we will briefly sketch below, but that will be treated extensively from Section 3 on, we will be mostly interested in submanifolds carrying such a Legendrian foliation. Note in particular that in a contact 3-manifold every hypersurface N carries automatically a Legendrian foliation.

Denote the set of points $p \in N$ where $\mathcal{F}$ is singular by $\operatorname{Sing}(\mathcal{F})$. One of the basic properties of a Legendrian foliation is that for any contact form α, the restriction $d\alpha|_{TN}$ does not vanish on $\operatorname{Sing}(\mathcal{F})$, because otherwise $T_pN \subset \xi_p$ would be an isotropic subspace of $(\xi_p, d\alpha_p)$ which is impossible for dimensional reasons. Since $d\alpha|_{TN}$ does not vanish on $\operatorname{Sing}(\mathcal{F})$, we deduce in particular that $N \setminus \operatorname{Sing}(\mathcal{F})$ is a dense and open subset of N.

The main reason, why we are interested in submanifolds that have a Legendrian foliation is that they often allow us to successfully use J-holomorphic curve techniques. On one side, such submanifolds will be automatically totally real for any suitable almost complex structure on a symplectic filling, thus posing a good boundary condition for the Cauchy-Riemann equation: The solution space of a Cauchy-Riemann equation with totally real boundary condition is often a finite dimensional smooth manifold, so that it follows that the moduli spaces of J-holomorphic curves whose boundaries lie in a submanifold with a Legendrian foliation will have a nice local structure. A second important property is that the topology of the Legendrian foliation controls the behavior of J-holomorphic curves, and will allow us to obtain many results in contact and symplectic topology. Elliptic codimension 2 singularities of the Legendrian foliation "emit" families of holomorphic disks; suitable codimension 1 singularities form "walls" that cannot be crossed by holomorphic disks.

In the rest of this section, we will state some general properties of Legendrian foliations. Theorem 2.2 shows that a manifold with a Legendrian foliation determines the germ of the contact structure on its neighborhood. This allows us to describe small deformations of the Legendrian foliation, and study almost complex structures more explicitly (see Section 3.2). Theo-

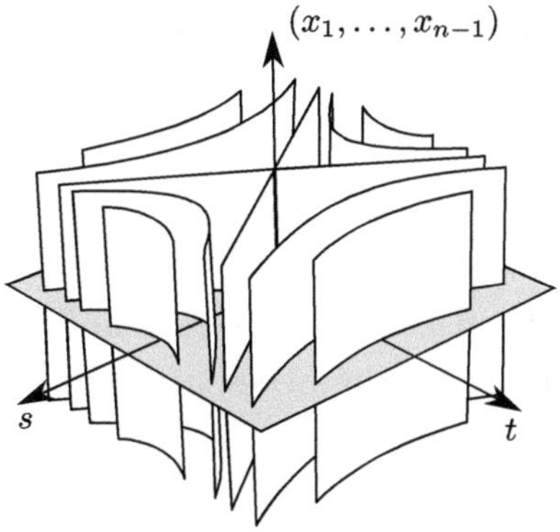

Fig. 1. The singularities of a Legendrian foliation look locally like the product of $\mathbb{R}^{n-1}$ with a foliation in the plane

rem 2.3 gives a precise characterization of the foliations that can be realized as Legendrian ones.

2.1.2. Singular Codimension 1 Foliations. The principal aim of this section will be to explain the following result due to Kupka [20] that tells us that the behavior of a Legendrian foliation close to a singular point can always be reduced to the 2-dimensional situation (see Figure 1).

Theorem 2.1. *Let N be a manifold with a singular foliation $\mathcal{F}$ that admits a regular equation β. Then we find around any $p \in \operatorname{Sing}(\mathcal{F})$ a chart with coordinates $(s, t, x_1, \dots, x_{n-1})$, such that β is represented by the 1-form*

$$a(s,t)\,ds + b(s,t)\,dt$$

for smooth functions a and b.

We will call any chart of N of the form described in the theorem a **Kupka chart**. Note that the foliation in a Kupka chart restricts on every 2-dimensional slice $\{(x_1, \dots, x_{n-1}) = \text{const}\}$ to one that does not have any isochore singularities (a term introduced in [15]).

Proof. From the Frobenius condition $\beta \wedge d\beta \equiv 0$, it follows that $d\beta^2 = 0$, so that if $\dim N > 2$, there is a non-vanishing vector field X on a neighborhood of p with $d\beta(X, \cdot) = 0$. We can also easily see that $X \in \ker \beta$ and $\mathcal{L}_X \beta = 0$, because

$$0 = \iota_X(\beta \wedge d\beta) = \beta(X)\,d\beta - \beta \wedge (\iota_X d\beta) = \beta(X)\,d\beta,$$

and $d\beta$ does not vanish on a neighborhood of p.

Let Φ_t^X be the flow of X, and choose a small hypersurface Σ transverse to X. Using the diffeomorphism

$$\Psi\colon \Sigma \times (-\varepsilon,\varepsilon) \to N, \quad (p,t) \mapsto \Phi_t^X(p)$$

we can pull back the 1-form β to $\Sigma \times (-\varepsilon,\varepsilon)$ and we see it reduces to $\beta|_{T\Sigma}$. By repeating this construction the necessary number of times we obtain the desired statement. ■

2.1.3. Local Behavior of Legendrian Foliations. We state the following two theorems without proof, and point the interested reader to [28] for more details. The situation in Section 2.2.2 is treated in these notes in full completeness to illustrate the flavor of the necessary methods. The first result tells us that a Legendrian foliation determines the germ of the contact structure in its neighborhood.

Theorem 2.2. *Let N be a compact manifold (possibly with boundary) and let (M_1,ξ_1) and (M_2,ξ_2) be contact manifolds. Assume that two embeddings $\iota_1\colon N \hookrightarrow M_1$ and $\iota_2\colon N \hookrightarrow M_2$ are given such that ξ_1 and ξ_2 induce on N the same cooriented Legendrian foliation $\mathcal{F}$. Then we find neighborhoods $U_1 \subset M_1$ of $\iota_1(N)$ and $U_2 \subset M_2$ of $\iota_2(N)$ together with a contactomorphism*

$$\Phi\colon (U_1,\xi_1) \to (U_2,\xi_2)$$

that preserves N, that means, $\Phi \circ \iota_1 = \iota_2$.

Another useful fact is the following theorem that tells us that the singular foliations that can be realized as Legendrian ones are exactly those that admit a regular equation (using the convention from the introduction). This result generalizes the 3-dimensional situation [15], where this property was called a foliation without "*isochore singularities*".

Theorem 2.3. *Let N be a manifold with a singular codimension-1 foliation $\mathcal{F}$ given by a regular equation β. Then we can find an (open) cooriented contact manifold (M,ξ) that contains N as a submanifold such that ξ induces $\mathcal{F}$ as Legendrian foliation on N.*

2.2. Singularities of the Legendrian Foliation

The singular set of a Legendrian foliation $\mathcal{F}$ can be extremely complicated. We will only discuss briefly a few general properties of such points, before we specialize all considerations to two simple situations.

Let N have a singular foliation $\mathcal{F}$ given by a regular equation β, and let $p \in \operatorname{Sing}(\mathcal{F})$ be a singular point of $\mathcal{F}$. Choose a Kupka chart U with coordinates $(s, t, x_1, \ldots, x_{n-1})$ centered at p. In this chart β is represented by

$$a(s,t)\,ds + b(s,t)\,dt$$

with two smooth functions $a, b\colon U \to \mathbb{R}$ that only depend on the s- and t-coordinates, and that vanish at the origin.

To understand the shape of the foliation depending on the functions a and b, we might study trajectories of the vector field

$$X = b(s,t)\frac{\partial}{\partial s} - a(s,t)\frac{\partial}{\partial t}$$

that spans the intersection of the foliation with the (s,t)-slices. Its divergence $\operatorname{div} X = \partial b/\partial s - \partial a/\partial t$ does not vanish, since $d\beta \neq 0$. Up to a genericity condition, we know by the Grobman-Hartman theorem that the flow of X is C^0-equivalent to the flow of its linearization (see [32]). In dimension 2, the Grobman-Hartman theorem even yields a C^1-equivalence, but this does not suffice for our purposes. For one, we would like to stick to a smooth model for all singularities, but in fact it even suffices for our goals to only look at singularities whose leaves are all radial, so we will use below a more hands-on approach.

2.2.1. Elliptic Singularities. The first type of singularities we allow for the foliation $\mathcal{F}$ on N are called **elliptic**: In this case, the point $p \in \operatorname{Sing}(\mathcal{F})$ admits a Kupka chart diffeomorphic to $\mathbb{R}^2 \times \mathbb{R}^n$ with coordinates $\{(s, t, x_1, \ldots, x_n)\}$ in which the foliation is given as the kernel of the 1-form

$$s\,dt - t\,ds$$

that means, the leaves are just the radial rays in each (s,t)-slice.

We will always assume that the elliptic singularities of a foliation $\mathcal{F}$ are closed isolated codimension 2 submanifolds S in the interior of N with trivial normal bundle, so that the tubular neighborhood of S is diffeomorphic to $\mathbb{D}^2_\varepsilon \times S$. We assume additionally that the foliation $\mathcal{F}$ in this model

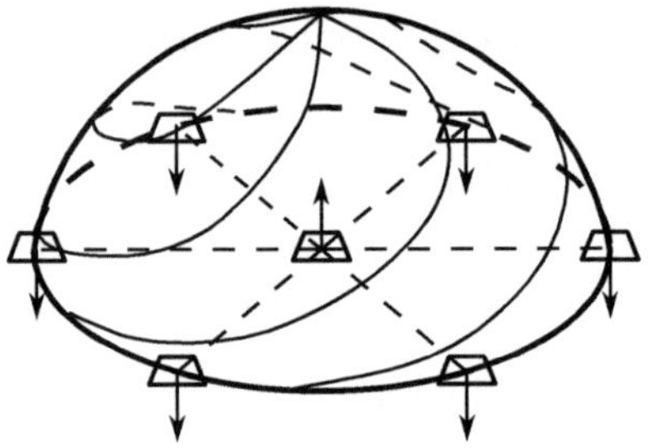

Fig. 2. In dimension 3 it is well-known that we can get rid of 1-dimensional singular sets of a Legendrian foliation by slightly tilting the surface along the singular set. The picture represents how to produce an overtwisted disk whose boundary is a regular compact leaf of the foliation

neighborhood is given by the points with constant angular coordinate on the $\mathbb{D}^2_\varepsilon$-factor.

2.2.2. Singularities of Codimension 1. Singular sets of codimension 1 are extremely ungeneric, but can be often found through explicit constructions (as in Example 2.7). We will show in this section that by slightly deforming the foliated submanifold one can sometimes modify the foliation in a controlled way so that the singular set turns into a regular compact leaf (see Figure 2).

We will treat this situation in detail to illustrate what type of methods are needed for the proofs in this section.

Lemma 2.4. *Let N be a compact manifold with a singular codimension 1 foliation $\mathcal{F}$ given by a regular equation β. Assume that the singular set $\operatorname{Sing}(\mathcal{F})$ of the foliation contains a closed codimension 1 submanifold $S \hookrightarrow N$ that is cooriented.*

Then we can find a tubular neighborhood of S diffeomorphic to $(-\varepsilon, \varepsilon) \times S$ such that β pulls back to

$$s \cdot \widetilde{\beta},$$

where s denotes the coordinate on $(-\varepsilon, \varepsilon)$, and $\widetilde{\beta}$ is a non-vanishing 1-form on S that defines a regular codimension 1 foliation on S.

Proof. Choose a coorientation for S. We first find a vector field X on a neighborhood of S that is transverse to S and lies in the kernel of β. Study the local situation in a Kupka chart U around a point $p \in S$ with coordinates

$(s, t, x_1, \ldots, x_{n-1})$. Assume that β restricts to

$$a(s,t)\,ds + b(s,t)\,dt,$$

such that $S \cap U$ corresponds to the subset $\{s = 0\}$, and such that s increases in direction of the chosen coorientation.

Since a and b vanish along $S \cap U$, we may write this form also as

$$s\,a_s(s,t)\,ds + s\,b_s(s,t)\,dt = s\big(a_s(s,t)\,ds + b_s(s,t)\,dt\big)$$

with smooth functions a_s and b_s that satisfy the conditions

$$a_s(0,t) = \frac{\partial a}{\partial s}(0,t) \quad \text{and} \quad b_s(0,t) = \frac{\partial b}{\partial s}(0,t).$$

The function b_s does not vanish in a small neighborhood of $S \cap U$, because $0 \neq d\beta = \partial_s b\, ds \wedge dt$. Choose then on the Kupka chart U the smooth vector field

$$X_U(s, t, x_1, \ldots, x_{n-1}) = \partial_s - \frac{a(s,t)}{b(s,t)}\partial_t = \partial_s - \frac{a_s(s,t)}{b_s(s,t)}\partial_t.$$

This field lies in $\mathcal{F}$, and is positively transverse to $S \cap U$.

Cover the singular set S with a finite number of Kupka charts $U_1, \ldots, U_N$, construct vector fields X_{U_j} according to the method described above, and glue them together to obtain the desired vector field X by using a partition of unity subordinate to the cover. We can use the flow of X to obtain a tubular neighborhood of S that is diffeomorphic to $(-\varepsilon, \varepsilon) \times S$, where $\{0\} \times S$ corresponds to the submanifold S, and X corresponds to the field ∂_s, where s is the coordinate on the interval $(-\varepsilon, \varepsilon)$, and since $\beta(X) \equiv 0$, it follows that β does not contain any ds-terms.

Let γ be the 1-form given by $\iota_X d\beta$. This form does not vanish on a neighborhood of the singular set S, because $d\beta \neq 0$ while $\beta|_{TS} \equiv 0$, and so we can write

$$0 \equiv \iota_X(\beta \wedge d\beta) = \beta(X)\,d\beta - \beta \wedge (\iota_X d\beta) = -\beta \wedge \gamma.$$

This means that there is a smooth function $F\colon (-\varepsilon, \varepsilon) \times S \to \mathbb{R}$ with $F|_S = 0$ such that $\beta = F\gamma$. Furthermore, we get that

$$\gamma = \iota_X d\beta = dF(X)\gamma + F\iota_X d\gamma$$

does not vanish along S, but F does, so we obtain on S that $dF(X) = 1$, and it follows that S is a regular zero level set of the function F. In fact, we can also easily see from

$$0 \equiv \beta \wedge d\beta = F^2 \gamma \wedge d\gamma$$

that $\gamma \wedge d\gamma$ vanishes everywhere so that $\ker \gamma$ defines a regular foliation $\widetilde{\mathcal{F}}$ that agrees with the initial foliation outside $\operatorname{Sing}(\mathcal{F})$.

Finally, we have $\iota_X \gamma \equiv 0$, and using a similar argument as before, we see

$$0 \equiv \iota_X(\gamma \wedge d\gamma) = -\gamma \wedge \iota_X d\gamma$$

so that there is a smooth function $f\colon (-\varepsilon, \varepsilon) \times S \to \mathbb{R}$ such that $\mathcal{L}_X \gamma = \iota_X d\gamma = f\gamma$. The flow in s-direction possibly rescales the 1-form γ, but it leaves its kernel invariant, thus the foliation $\widetilde{\mathcal{F}}$ is tangent to the s-direction and s-invariant. We can hence represent $\widetilde{\mathcal{F}}$ on $(-\varepsilon, \varepsilon) \times S$ as the kernel of the 1-form $\widetilde{\beta} = \gamma|_{TS}$ that does not depend on the s-coordinate, and does not have any ds-terms. It follows that γ is equal to $\widetilde{F}\gamma|_{TS}$ for a function $\widetilde{F}$ that restricts on S to 1.

For the initial 1-form β this means that $\beta = (F\widetilde{F})\widetilde{\beta}$, and $F\widetilde{F}$ is a smooth function and $\{0\} \times S$ is the (regular) level set of 0. We can redefine the model $(-\varepsilon, \varepsilon) \times S$ by using the flow of a vector field $G^{-1}\partial_s$ with $G = \partial_s(F\widetilde{F})$ to achieve that β reduces on this new model to $s\widetilde{\beta}$. ■

Suppose from now on that the singular foliation is of the form described in Lemma 2.4, that means, we have a closed manifold S with a regular codimension 1 foliation $\mathcal{F}_S$ given as the kernel of a 1-form $\widetilde{\beta}$, and N is diffeomorphic to $(-\varepsilon, \varepsilon) \times S$ with a singular foliation $\mathcal{F}$ given as the kernel of the 1-form $s\widetilde{\beta}$.

Remember that a 1-form σ on S defines a section in T^*S with the property that $\sigma^*\lambda_{\mathrm{can}} = \sigma$. We may realize $\mathcal{F}$ as a Legendrian foliation, by embedding $(-\varepsilon, \varepsilon) \times S$ into the 1-jet space $(\mathbb{R} \times T^*S, dz + \lambda_{\mathrm{can}})$ via the map

$$(s, p) \mapsto (0, s\widetilde{\beta}).$$

The foliations agree, and according to Theorem 2.2 this model describes a small neighborhood of $(N, s\widetilde{\beta})$ embedded into an arbitrary contact manifold.

Assume from now on additionally that $\widetilde{\beta}$ is a *closed* 1-form on S (by a result of Tischler, S fibers over the circle [34]). Choose a smooth odd function $f\colon (-\varepsilon, \varepsilon) \to \mathbb{R}$ with compact support such that the derivative $f'(0) = -1$. The section

$$(-\varepsilon, \varepsilon) \times S \hookrightarrow \mathbb{R} \times T^*S, \quad (s, p) \mapsto \big(\delta f(s), s\widetilde{\beta}\big)$$

describes for small $\delta > 0$ a C^∞-small deformation of N that agrees away from S with N. The perturbed submanifold N' also carries a Legendrian foliation induced by $\ker(ds + \lambda_{\mathrm{can}})$, because the pull-back form $\beta' = f'\,ds + s\widetilde{\beta}$ gives

$$\beta' \wedge d\beta' = \left(f'\,ds + s\widetilde{\beta}\right) \wedge \left(ds \wedge \widetilde{\beta} + s\,d\widetilde{\beta}\right) = s\left(f'\,ds + s\widetilde{\beta}\right) \wedge d\widetilde{\beta} = s f'\,ds \wedge d\widetilde{\beta},$$

which vanishes, so that β' satisfies the Frobenius condition. Furthermore, since β' itself does not vanish anywhere, it is easy to check that $\ker\beta'$ defines a regular foliation $\mathcal{F}'$, and that $\{0\} \times S$ is a closed leaf of $\mathcal{F}'$.

As a conclusion, we obtain

Corollary 2.5. *Let (M,ξ) be a contact manifold containing a submanifold N with an induced Legendrian foliation $\mathcal{F}$. Assume that the singular set of $\mathcal{F}$ contains a cooriented closed codimension 1-submanifold $S \subset N$, and that there is a regular foliation $\mathcal{F}$ that agrees outside N with $\mathcal{F}$, and that corresponds on S with a fibration over the circle. Using an arbitrary small C^∞ perturbation of N close to S, we obtain a new Legendrian foliation for which S has become a regular closed leaf.*

2.3. Examples of Legendrian Foliations

The following example relates Legendrian foliations to Lagrangian submanifolds. It is not important by itself, but it may help understanding the construction of the bLobs in blown down Giroux domains given in [21], and I believe that it might pave the way to other applications.

Example 2.6. Let P be a principal circle bundle over a base manifold B, and suppose that ξ is a contact structure on P that is transverse to the $\mathbb{S}^1$-fibers and invariant under the action. It is well-known that by averaging, we can choose an $\mathbb{S}^1$-invariant contact form α for ξ and that there exists a symplectic form ω on B such that $\pi^*\omega = d\alpha$, where π is the bundle projection $\pi\colon P \to B$. The symplectic form ω represents the image of the Euler class $e(P)$ in $H^2(B,\mathbb{R})$, and hence P cannot be a trivial bundle (see [5]). The manifold (P_L,α) is usually called the **pre-quantization of the symplectic manifold** (B,ω) (or the **Boothby-Wang manifold**).

Let L be a Lagrangian submanifold in (B,ω), and let $P_L := \pi^{-1}(L)$ be the fibration over L. Note first that in this situation, we have $\omega|_{TL} = 0$, so that $e(P_L) = e(P)|_L$ will automatically either vanish or be a torsion class. We assume that $e(P_L) = 0$, so that the fibration P_L will be trivial, and we can find a section $\sigma\colon L \to P_L$.

We have $(\alpha \wedge d\alpha)|_{TP_L} = (\alpha \wedge \pi^*\omega)|_{TP_L} \equiv 0$, so that ξ induces a Legendrian foliation $\mathcal{F}$ on P_L. Furthermore, since the infinitesimal generator X_φ of the circle action satisfies $\alpha(X_\varphi) \equiv 1$, it follows that $\mathcal{F}$ is everywhere regular. Using the section σ, we can identify P_L with $\mathbb{S}^1 \times L$, and write $\alpha|_{TP_L}$ as

$$d\varphi + \beta,$$

where φ is the coordinate on the circle and β is a closed 1-form on L. The leaves of the foliation $\mathcal{F}$ are local sections, but they need not be global ones, and usually these leaves will not even be compact. Instead the proper way to think of them is as the horizontal lift of the flat connection 1-form $\alpha|_{TP_L}$.

Choose any loop $\gamma \subset L$ based at a point $p_0 \in L$. We want to lift $\gamma(t)$ to a path $\widetilde{\gamma}(t) = (e^{i\varphi(t)}, \gamma(t))$ in $P_L \cong \mathbb{S}^1 \times L$ that is always tangent to a leaf of $\mathcal{F}$, so that

$$\widetilde{\gamma}'(t) = \left(-\beta(\gamma'(t)), \gamma'(t)\right).$$

In particular start and end point of $\widetilde{\gamma}$ are related by the monodromy

$$C_\gamma := -\int_\gamma \beta,$$

that means, if $\widetilde{\gamma}$ starts at $(e^{i\varphi_0}, p_0) \in \mathbb{S}^1 \times L$, then its end point will be $(e^{i(\varphi_0 + C_\gamma)}, p_0)$.

Note that since the connection is flat, that means, β is closed, two homologous paths from p_0 to p_1 will lift the end point in the same way. Thus we have a well-defined map

$$H_1(L, \mathbb{Z}) \to \mathbb{S}^1.$$

The leaves of the Legendrian foliation will only be compact, if the image of this map is discrete.

Note that the embedding of $H^1(L, \mathbb{Q}) \to H^1(L, \mathbb{R})$ is dense, and so we find a 1-form β' arbitrarily close to β such that the monodromy for every loop in L will be a rational number. Clearly, we can extend $\delta = \beta' - \beta$ to a 1-form defined on the whole bundle P, and suppose that δ is sufficiently small so that $\alpha' = \alpha + \delta$ determines a contact structure that is isotopic to the initial one. We may hence suppose that after a small perturbation of α that the Legendrian foliation on P_L is given by $d\phi + \beta'$.

In fact, since $H_1(L, \mathbb{Z})$ is finitely generated, we find a number $c \in \mathbb{Q}$ such that all possible values of the monodromy are a multiple of c, and by slightly perturbing α we obtain a regular Legendrian foliation on P_L, with compact leaves.

The second example gives a Legendrian foliation with a codimension 1 singular set.

Example 2.7. Let L be any smooth $(n+1)$-dimensional manifold with a Riemannian metric g. It is well-known that the unit cotangent bundle $\mathbb{S}(T^*L)$ carries a contact structure given as the kernel of the canonical 1-form λ_{can}. The fibers of this bundle are Legendrian spheres, hence if we choose any smooth regular loop $\gamma\colon \mathbb{S}^1 \to L$, and if we study the fibers lying over this path, we obtain the submanifold $N_\gamma := \pi^{-1}(\gamma)$ that has a singular Legendrian foliation.

In fact, we can naturally decompose $T^*L|_\gamma$ into the two subsets U_+ and U_- defined as

$$U_\pm = \{\nu \in N_\gamma \mid \pm\nu(\gamma') \geq 0\}.$$

These sets correspond in each fiber of N_γ to opposite hemispheres. The singular set of the Legendrian foliation on N_γ is $U_+ \cap U_-$, and that the regular leaves correspond to the intersection of each fiber of N_γ with the interior of U_+ and U_-. In particular, if N_γ is orientable, we obtain that it can be written as

$$(\mathbb{S}^1 \times \mathbb{S}^n, x_0\, d\varphi),$$

where φ is the coordinate on $\mathbb{S}^1$, and $(x_0, \ldots, x_n)$ are the coordinates on $\mathbb{S}^n$.

Using the results of Section 2.2.2, we can perturb N_γ to a submanifold with a regular Legendrian foliation composed of two Reeb components.

2.4. Legendrian Open Books

Even though we discussed Legendrian foliations quite generally, we will only be interested in two special types: *Legendrian open books* introduced in [29] and *bordered Legendrian open books* introduced in [21]. Both objects were defined with the aim of generalizings results from 3-dimensional contact topology that hold for the 2-sphere with standard foliation and the overtwisted disk respectively [4, 9, 17, 18].

Definition. Let N be a closed manifold. An **open book** on N is a pair (B, ϑ) where:

- The **binding** B is a nonempty codimension 2 submanifold in the interior of N with trivial normal bundle.
- $\vartheta \colon N \setminus B \to \mathbb{S}^1$ is a fibration, which coincides in a neighborhood $B \times \mathbb{D}^2$ of $B = B \times \{0\}$ with the normal angular coordinate.

Definition. If N is a compact manifold with nonempty boundary, then a **relative open book** on N is a pair (B, ϑ) where:

- The **binding** B is a nonempty codimension 2 submanifold in the interior of N with trivial normal bundle.
- $\vartheta \colon N \setminus B \to \mathbb{S}^1$ is a fibration whose fibers are transverse to ∂N, and which coincides in a neighborhood $B \times \mathbb{D}^2$ of $B = B \times \{0\}$ with the normal angular coordinate.

We are interested in studying contact manifolds with submanifolds with a Legendrian foliation that either define an open book or a relative open book.

Definition. A closed submanifold N carrying a Legendrian foliation $\mathcal{F}$ in a contact manifold (M, ξ) is a **Legendrian open book** (abbreviated **Lob**), if N admits an open book (B, ϑ), whose fibers are the regular leaves of the Legendrian foliation (the binding is the singular set of $\mathcal{F}$).

Definition. A compact submanifold N with boundary in a contact manifold (M, ξ) is called a **bordered Legendrian open book** (abbreviated **bLob**), if N carries a Legendrian foliation $\mathcal{F}$ and if it has a relative open book (B, ϑ) such that:

(i) the regular leaves of $\mathcal{F}$ lie in the fibers of θ,

(ii) $\operatorname{Sing}(\mathcal{F}) = \partial N \cup B$.

A contact manifold that contains a bLob is called ***PS*-overtwisted**.

Example 2.8.

(i) Every Lob in a contact 3-manifold is diffeomorphic to a 2-sphere with the binding consisting of the north and south poles, and the fibers being the longitudes. This special type of Lob has been studied extensively and has given several important applications, see for example [4, 9, 17, 18]. It is easy to find such Lobs locally, for example, the unit sphere in $\mathbb{R}^3$ with the standard contact structure $\xi = \ker(dz + x\,dy - y\,dx)$.

(ii) A bLob in a 3-dimensional contact manifold is an overtwisted disk (with singular boundary).

(iii) In higher dimensions, the plastikstufe had been introduced as a filling obstruction [27], but note that a plastikstufe is just a specific bLob that is diffeomorphic to $\mathbb{D}^2 \times B$, where the fibration is the one of an overtwisted disk (with singular boundary) on the $\mathbb{D}^2$-factor, extended by a product with a closed manifold B. Topologically a bLob might be *much* more general than the initial definition of the plastikstufe. For example, a plastikstufe in dimension 5 is always diffeomorphic to a solid torus $\mathbb{D}^2 \times \mathbb{S}^1$ while a 3-manifold admits a relative open book if and only if its boundary is a nonempty union of tori.

The importance of the previous definitions lie in the following two theorems, which will be proved in Section 4.

Theorem A ([21, 27]). *Let (M, ξ) be a contact manifold that contains a* bLob *N, then M does not admit any semi-positive weak symplectic filling (W, ω) for which $\omega|_{TN}$ is exact.*

The statement above is a generalization of the analogous statement found first for the overtwisted disk in [9, 17].

Remark 2.9. A bLob obstructs always (semi-positive) *strong* symplectic filling, because in that case the restriction of ω to N is exact.

Remark 2.10. In dimension 4 and 6, every symplectic manifold is automatically semi-positive.

Theorem B ([29]). *Let (M, ξ) be a contact manifold of dimension $(2n+1)$ that contains a* Lob *N. If M has a weak symplectic filling (W, ω) that is symplectically aspherical, and for which $\omega|_{TN}$ is exact, then it follows that N represents a trivial class in $H_{n+1}(W, \mathbb{Z}_2)$. If the first and second Stiefel-Whitney classes $w_1(N)$ and $w_2(N)$ vanish, then we obtain that N must be a trivial class in $H_{n+1}(W, \mathbb{Z})$.*

Remark 2.11. The methods from [18] can be generalized for Theorem A, see [2], and for Theorem B, see [29], to find closed contractible Reeb orbits.

2.5. Examples of bLobs

The most important result of these notes is the construction of non-fillable manifolds in higher dimensions. The first such manifolds were obtained by Presas in [33], and modifying his examples it was soon possible to show that every contact structure can be converted into one that is PS-overtwisted [35].

This result was reproved and generalized in [11], where it was shown that we may modify a contact structure into one that is PS-overtwisted without changing the homotopy class of the underlying almost contact structure.

A very nice explicit construction in dimension 5 that is similar to the 3-dimensional Lutz twist was given in [26]. In [21] the construction was extended and produced examples that are not PS-overtwisted but share many properties with 3-manifold that have positive Giroux torsion.

The following unpublished construction is due to Francisco Presas who explained it to me during a stay in Madrid. It is probably the easiest way to produce a closed PS-overtwisted manifolds of arbitrary dimensions.

Theorem 2.12 (Fran Presas). *Let (M_1, ξ_1) and (M_2, ξ_2) be contact manifolds of dimension $2n+1$ that both contain a PS-overtwisted submanifold (N, ξ_N) of codimension 2 with trivial normal bundle. The **fiber sum** of M_1 and M_2 along N is a PS-overtwisted $(2n+1)$-manifold.*

Proof. Let α_N be a contact form for ξ_N. The manifold N has neighborhoods $U_1 \subset M_1$ and $U_2 \subset M_2$ that are contactomorphic to

$$\mathbb{D}^2_{\sqrt{\varepsilon}} \times N$$

with contact structure given as the kernel of the 1-form $\alpha_N + r^2\, d\varphi$ [12, Theorem 2.5.15].

We can remove the submanifold $\{0\} \times N$ in this model, and do a reparametrization of the r-coordinate by $s = r^2$ to bring the neighborhood into the form

$$(0, \varepsilon) \times \mathbb{S}^1 \times N$$

with contact form $\alpha_N + s\, d\varphi$. We extend $M_1 \setminus N$ and $M_2 \setminus N$ by attaching the negative s-direction to the model collar, so that we obtain a neighborhood

$$\big((-\varepsilon, \varepsilon) \times \mathbb{S}^1 \times N,\ \alpha_N + s\, d\varphi\big).$$

Denote these extended manifolds by $(\widetilde{M}_1, \widetilde{\xi}_1)$ and $(\widetilde{M}_2, \widetilde{\xi}_2)$, and glue them together using the contactomorphism

$$(-\varepsilon, \varepsilon) \times \mathbb{S}^1 \times N \to (-\varepsilon, \varepsilon) \times \mathbb{S}^1 \times N$$
$$(s, \varphi, p) \mapsto (-s, -\varphi, p).$$

We call the contact manifold (M', ξ') that we have obtained this way the **fiber sum** of M_1 and M_2 along N.

If S is a bLob in N, then it is easy to see that $\{0\} \times \mathbb{S}^1 \times S$ is a bLob in the model neighborhood $(-\varepsilon, \varepsilon) \times \mathbb{S}^1 \times N$. ∎

With this proposition, we can now construct non-fillable contact manifolds of arbitrary dimension. Every oriented 3-manifold admits an overtwisted contact structure in every homotopy class of almost contact structures.

Let (M, ξ) be a compact manifold, let α_M be a contact form for ξ. A fundamental result due to Emmanuel Giroux gives the existence of a compatible open book decomposition for M [16]. Using this open book decomposition, it is easy to find functions $f, g\colon M \to \mathbb{R}$ such that

$$\big(M \times \mathbb{T}^2, \ker(\alpha_M + f\, dx + g\, dy)\big)$$

is a contact structure, see [6], where (x, y) denotes the coordinates on the 2-torus. The fibers $M \times \{z\}$ are contact submanifold with trivial normal bundle, so that in particular if (M, ξ) is PS-overtwisted, we can apply the construction above to glue two copies of $M \times \mathbb{T}^2$ along a fiber $M \times \{z\}$. This way, we obtain a PS-overtwisted contact structure on $M \times \Sigma_2$, where Σ_2 is a genus 2 surface.

Using this process inductively, we find closed PS-overtwisted contact manifolds of any dimension ≥ 3.

Note that in dimension 5, we can find more easily examples to which we can apply Theorem 2.12, so that it is not necessary to rely on [6]. Let (M, ξ) be an overtwisted 3-manifold with contact form α. After normalizing α with respect to a Riemannian metric, it describes a section

$$\sigma_\alpha\colon M \to \mathbb{S}\big(T^*M\big)$$

in the unit cotangent bundle. It satisfies the fundamental relation $\sigma_\alpha^* \lambda_{\mathrm{can}} = \alpha$, hence it gives a contact embedding of (M, ξ) into $(\mathbb{S}(T^*M), \ker \lambda_{\mathrm{can}})$.

For trivial normal bundle, this allows us to glue with Theorem 2.12 two copies together and obtain a PS-overtwisted 5-manifold.

3. Behavior of J-Holomorphic Disks Imposed by Convexity

The following section only fixes notation, and explains some well-known facts about J-convexity. With some basic knowledge on J-holomorphic curves, one can safely skip it and continue directly to Section 3.2, which describes the local models around the binding and the boundary of the Lobs and bLobs and the behavior of holomorphic disks that lie nearby. The next two sections include a description about moduli spaces and their basic properties, but most results are only explained in an intuitive way without giving any proofs. The fifth section deals with the Gromov compactness of the considered moduli spaces, and the chapter finishes proving the two applications that relate a Lob or a bLob to the topology of a symplectic filling.

3.1. Almost Complex Structures and Maximally Foliated Submanifolds

3.1.1. Preliminaries: J-Convexity.

The Maximum Principle. One of the basic ingredients in the theory of J-holomorphic curves with boundary is the maximum principle, which we will now briefly describe in the special case of Riemann surfaces. We assume in this section that (Σ, j) is a Riemann surface that does not need to be compact and may or may not have boundary. We define the differential operator d^j that associates to every smooth function $f\colon \Sigma \to \mathbb{R}$ a 1-form given by

$$(d^j f)(v) := -df(jv)$$

for $v \in T\Sigma$.

Definition. We say that a function $f\colon (\Sigma, j) \to \mathbb{R}$ is

(a) **harmonic** if the 2-form $dd^j f$ vanishes everywhere,

(b) it is **subharmonic** if the 2-form $dd^j f$ is a positive volume form with respect to the orientation defined by (v, jv) for any non-vanishing vector $v \in T\Sigma$.

(c) If f only satisfies

$$dd^j f(v, jv) \geq 0$$

then we call it **weakly subharmonic**.

In particular, if we choose a complex chart $(U \subset \mathbb{C}, \phi)$ for Σ with coordinate $z = x + iy$, we can represent f by $f_U := f \circ \phi^{-1} \colon U \to \mathbb{R}$. The 2-form $dd^j f$ simplifies on this chart to $dd^i f_U$, because ϕ is holomorphic with respect to j and i, and we can write $dd^i f_U$ in the form $(\triangle f_U)\, dx \wedge dy$, where the Laplacian is defined as

$$\triangle f_U = \frac{\partial^2 f_U}{\partial x^2} + \frac{\partial^2 f_U}{\partial y^2}.$$

Note that f_U is subharmonic, if and only if $dd^i f_U(\partial_x, \partial_y) > 0$, that means, $\triangle f_U > 0$.

For strictly subharmonic functions, it is obvious that they may not have any interior maxima, because the Hessian needs to be negative definite at any such point. We really need to consider both weakly subharmonic functions and the behavior at boundary points. To prove the maximum principle in this more general setup, we use the following technical result.

Lemma 3.1. *Let $f \colon \mathbb{D}^2 \subset \mathbb{C} \to \mathbb{R}$ be a function that is C^1 on the closed unit disk, and both C^2 and weakly subharmonic on the interior of the disk. Assume that f takes its maximum at a boundary point $z_0 \in \partial\mathbb{D}^2$ and is everywhere else strictly smaller than $f(z_0)$. Choose an arbitrary vector $X \in T_{z_0}\mathbb{C}$ at z_0 pointing transversely out of $\overline{\mathbb{D}}^2$.*

Then the derivative $\mathcal{L}_X f(z_0)$ in X-direction needs to be strictly *positive.*

Proof. We will perturb f to a *strictly* subharmonic function making use of the auxiliary function $g \colon \overline{\mathbb{D}}^2 \to \mathbb{R}$ defined by (see Figure 3)

$$g(r) = r^4 - \frac{9}{4} r^2 + \frac{5}{4}.$$

The function g vanishes along the boundary $\partial\mathbb{D}^2$, and its derivative in any direction v that is positively transverse to the boundary $\partial\mathbb{D}^2$ is strictly negative, because $\partial_\varphi g = 0$ and because

$$r\partial_r g = \frac{1}{2} r^2 \left(8r^2 - 9\right).$$

Finally, we also see that g is strictly subharmonic on the open annulus $\mathbb{A} = \{z \in \mathbb{C} \mid 3/4 < |z| < 1\}$ as

$$\triangle g = \frac{\partial^2 g}{\partial x^2} + \frac{\partial^2 g}{\partial y^2} = 16r^2 - 9.$$

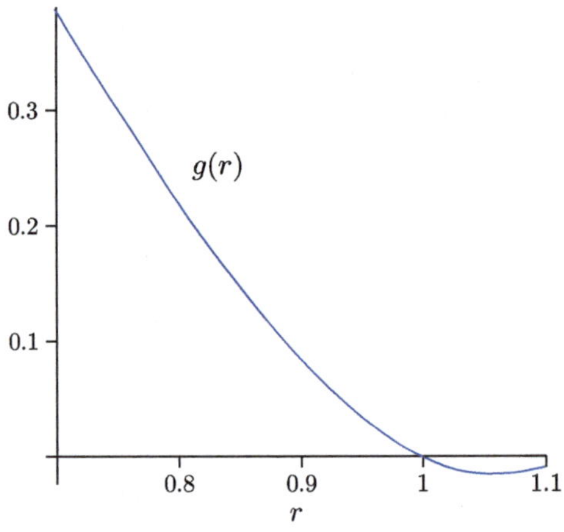

Fig. 3. The function $g(r)$ is subharmonic, vanishes on the boundary, and has negative radial derivative

We slightly perturb f by setting $f_\varepsilon = f + \varepsilon g$ for small $\varepsilon > 0$, and we additionally restrict f_ε to the closure of the annulus $\mathbb{A}$. Note in particular that f_ε must take its maximum on $\partial\mathbb{A}$, because f_ε is *strictly* subharmonic on the interior of $\mathbb{A}$ so that one of $\frac{\partial^2 f_\varepsilon}{\partial x^2}$ or $\frac{\partial^2 f_\varepsilon}{\partial y^2}$ must be strictly positive. This contradicts existence of possible interior maximum points. The functions f_ε are equal to f along the outer boundary of $\mathbb{A}$ so that the maximum of f_ε will either lie in z_0 or on the inner boundary of $\mathbb{A}$.

The initial function f is by assumption strictly smaller than $f(z_0)$ on the inner boundary of the annulus and by choosing ε sufficiently small, it follows that the perturbed function f_ε will still be strictly smaller than $f_\varepsilon(z_0) = f(z_0)$. Thus z_0 will also be the maximum of f_ε. Let X be a vector at z_0 that points transversely out of $\overline{\mathbb{D}}^2$. The derivative $\mathcal{L}_X f_\varepsilon$ at z_0 cannot be strictly negative, because z_0 is a maximum, and so since

$$0 \le \mathcal{L}_X f_\varepsilon = \mathcal{L}_X f + \varepsilon \mathcal{L}_X g,$$

the derivative of f in X-direction has to be *strictly* positive, yielding the desired result. ■

Now we are prepared to state and prove the maximum principle.

Theorem 3.2 (Weak maximum principle). *Let (Σ, j) be a connected compact Riemann surface. A weakly subharmonic function $f\colon \Sigma \to \mathbb{R}$ that attains its maximum at an interior point $z_0 \in \Sigma \setminus \partial\Sigma$ must be constant.*

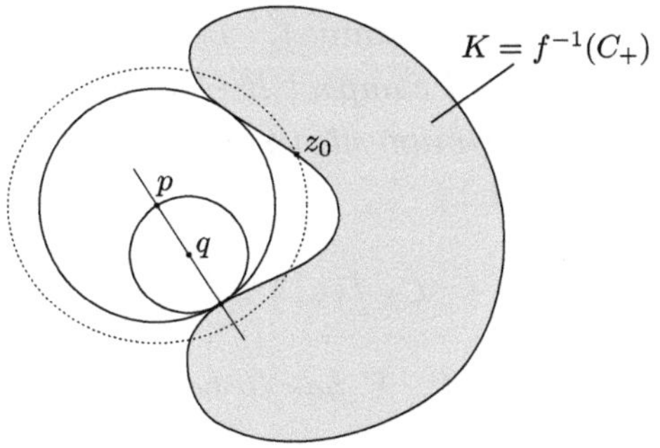

Fig. 4. Constructing a disk that has a single maximum on its boundary

Proof. The proof is classical and holds in much greater generality (see for example [14]). Nonetheless we will explain it in the special case needed by us to show that it only uses elementary techniques. The strategy is simply to find a closed disk in the interior of the Riemann surface with the properties required by Lemma 3.1. Then the function f increases in radial direction further, so that the maximum point was not really a maximum.

More precisely, assume f not to be constant, and to have a maximum at an interior point $z_+ \in \Sigma \setminus \partial\Sigma$ with $C_+ := f(z_+)$. The subset $K := f^{-1}(C_+) \cap \mathring{\Sigma}$ is closed in $\mathring{\Sigma}$. For every point $z \in K$, we find an $R_z > 0$ such that the open disk $D_{R_z}(z)$ is contained in some complex chart. There must be a point $z_0 \in K$ for which the half sized disk $D_{R_{z_0}/2}(z_0)$ intersects $\mathring{\Sigma} \setminus K$, for otherwise K would be open and hence as $\mathring{\Sigma}$ is connected, $K = \mathring{\Sigma}$.

Let p be a point in $D_{R_{z_0}/2}(z_0) \setminus K$ (see Figure 4). It lies so close to z_0 that the entire closed disk of radius $|p - z_0|$ lies in the chart U, and then we can choose first a disk $\overline{\mathbb{D}_R(p)}$ centered at p, where R is the largest number for which the *open* disk does not intersect $f^{-1}(C_+)$. We are interested in finding a closed disk that intersects $f^{-1}(C_+)$ at a *single* boundary point: For this let q be the mid point between p and one of the boundary points in $\partial\mathbb{D}^2_R(p) \cap f^{-1}(C_+)$. The disk $\mathbb{D}^2_{R/2}(q)$ touches $f^{-1}(C_+)$ at exactly one point.

This smaller disk satisfies the conditions of Lemma 3.1, and so it follows that the derivative of f at the maximum is strictly positive in radial direction. But since this point lies in the interior of Σ, it follows that f still increases in that direction and hence this point cannot be the maximum. Of course, the whole existence of the disk was based on the assumption that f was not constant, so we obtain the statement of the theorem. ■

If Σ has boundary, we also get the following refinement.

Theorem 3.3 (Boundary point lemma). *Let $f\colon \Sigma \to \mathbb{R}$ be a weakly subharmonic function on a connected compact Riemann surface (Σ, j) with boundary. Assume f takes its maximum at a point $z_+ \in \partial\Sigma$, then f will either be constant or the derivative at z_+*

$$\mathcal{L}_X f(z_+) > 0$$

in any outward direction $X \in T_{z_+}\Sigma$ has to be strictly positive.

Proof. Denote the maximum $f(z_+)$ by C_+. By the maximum principle, Theorem 3.2, we know that f will be constant if there is a point $z \in \Sigma \setminus \partial\Sigma$ for which $f(z) = C_+$. We can thus assume that for all $z \notin \partial\Sigma$, we have $f < C_+$. Using a chart U around the point z_+, that represents an open set in $\mathbb{H} := \{z \in \mathbb{C} \,|\, \operatorname{Im} z \geq 0\}$, such that z_+ corresponds to the origin, we can easily find a small disk in $\mathbb{H}$ that touches $\partial\mathbb{H}$ only in 0, and hence allows us to directly apply Lemma 3.1 to complete the proof. ■

Plurisubharmonic Functions. We will now explain the connection between the previous section and contact topology.

Let (W, J) be an almost complex manifold, that means that J is a section of the endomorphism bundle $\operatorname{End}(TM)$ with $J^2 = -\mathbf{1}$. Define the differential $d^J f$ of a smooth function $f\colon W \to \mathbb{R}$ as before by

$$(d^J f)(v) := -df(J \cdot v)$$

for any vector $v \in TW$.

Definition. We say that a function $h\colon W \to \mathbb{R}$ is ***J*-plurisubharmonic**, if the 2-form

$$\omega_h := dd^J h$$

evaluates positively on J-complex lines, that means that $\omega_h(v, Jv)$ is strictly positive for every non-vanishing vector $v \in TW$.

If ω_h vanishes, then we say that h is ***J*-harmonic**.

Remark 3.4.

(1) If h is J-plurisubharmonic, then ω_h is an exact symplectic form that tames J.

(2) If ω_h is only non-negative, then we say that h is **weakly J-plurisubharmonic**. This notion might be for example interesting in the context of confoliations.

Let (Σ, j) be a Riemann surface that does not need to be compact, and may or may not have boundary. We say that a smooth map $u\colon \Sigma \to W$ is **J-holomorphic**, if its differential commutes with the pair (j, J), that means, at every $z \in \Sigma$ we have

$$J \cdot Du = Du \cdot j.$$

Using the commutation relation, we easily check for every J-holomorphic map u and every smooth function $f\colon U \to \mathbb{R}$ the formula

$$u^* d^J f = -df \cdot J \cdot Du = -df \cdot Du \cdot j = -d(f \circ u) \cdot j = d^j (f \circ u) = d^j u^* f. \tag{3.1}$$

Corollary 3.5. *If $u\colon (\Sigma, j) \to (W, J)$ is J-holomorphic and $h\colon W \to \mathbb{R}$ is a J-plurisubharmonic function, then $h \circ u$ will be weakly subharmonic, because*

$$dd^j (h \circ u) = d\, u^* d^J h = u^* dd^J h$$

and because the differential Du commutes with the complex structures, so that

$$dd^j (h \circ u)(v, jv) = dd^J h(Du \cdot v, J \cdot Du \cdot v) \geq 0$$

for every vector $v \in T\Sigma$. The function is strictly positive precisely at points $z \in U$, where Du_z does not vanish.

The maximum principle restricts severely the behavior of holomorphic maps:

Corollary 3.6. *Let $u\colon (\Sigma, j) \to (W, J)$ be a J-holomorphic map and $h\colon W \to \mathbb{R}$ be a J-plurisubharmonic function. If u is not a constant map then $h \circ u\colon \Sigma \to \mathbb{R}$ will never take its maximum on the interior of Σ.*

Proof. Since $h \circ u$ is weakly subharmonic, it follows immediately from the maximum principle (Theorem 3.2) that $h \circ u$ must be constant if it takes its maximum in the interior of Σ, and hence $d(h \circ u) = 0$. On the other hand, we know that if there were a point $z \in \Sigma$ with $D_z u \neq 0$, then $\omega_h(Du \cdot v, Du \cdot jv)$ would need to be strictly positive for non-vanishing vectors. This is not possible though, because $u^* \omega_h = dd^j (h \circ u) = 0$. ■

Corollary 3.7. *Let (Σ, j) be a Riemann surface with boundary, $u\colon (\Sigma, j) \to (W, J)$ a J-holomorphic map and $h\colon W \to \mathbb{R}$ be a J-plurisubharmonic function. If $h \circ u\colon \Sigma \to \mathbb{R}$ takes its maximum at $z_0 \in \partial\Sigma$ then it follows either that $d(h \circ u)(v) > 0$ for every vector $v \in T_{z_0}\Sigma$ pointing transversely out of the surface, or u will be constant.*

Proof. The proof is analogous to the previous one, but uses the boundary point lemma (Theorem 3.3) instead of the simple maximum principle. ■

Remark 3.8. Note that if h is only *weakly* plurisubharmonic, then we can only deduce in the two corollaries above that u has to lie in a level set of h, and not that u itself must be constant.

Contact Structures as Convex Boundaries. Now we will finally explain the relation between plurisubharmonic functions and contact manifolds.

Definition. Let (W, J) be an almost complex manifold with boundary. We say that W has **J-convex boundary**, if there exists a smooth function $h\colon W \to (-\infty, 0]$ with the properties

- h is J-plurisubharmonic on a *neighborhood* of ∂W,
- h is a regular equation for ∂W, that means, 0 is a regular value of h and $\partial W = h^{-1}(0)$.

Note that the function h in the definition takes its maximum on ∂W, so that it must be strictly increasing in outward direction.

We will show that the boundary of an almost complex manifold is J-convex if and only if it carries a natural cooriented contact structure (whose conformal symplectic structure tames J). Remember that we are always assuming our contact manifolds to be cooriented. Hence the manifold is oriented, and its contact structure will have a natural conformal symplectic structure.

Definition. Let M be a codimension 1 submanifold in an almost complex manifold (W, J). The **subbundle of complex tangencies** of M is the J-complex subbundle

$$\xi := TM \cap (J \cdot TM).$$

Proposition 3.9. *Let (W, J) be an almost complex manifold with boundary $M := \partial W$ and let ξ be the subbundle of complex tangencies of M. We have the following equivalence:*

(1) *The boundary M is J-convex.*

(2) *The subbundle ξ is a cooriented contact structure whose natural orientation is compatible with the boundary orientation of M, and whose natural conformal symplectic structure tames $J|_\xi$.*

Proof. To prove the direction "(1) $\Rightarrow$ (2)", let h be the J-plurisubharmonic equation of M that exists by assumption. A straight forward calculation shows that the kernel of the 1-form $\alpha := d^J h|_{TM}$ is precisely ξ, and in particular that α does not vanish. Furthermore $d\alpha|_{TM} = \omega_h|_{TM}$ is a symplectic structure on ξ that tames $J|_\xi$, so that α is a contact form. To check that $\alpha \wedge d\alpha^{n-1}$ is a positive volume form with respect to the boundary orientation induced on M by (W, J), let R_α be the Reeb field of α, and define a vector field $Y = -JR_\alpha$. The field Y is positively transverse to ∂W, because $\mathcal{L}_Y h = dh(Y) = d^J h(R_\alpha) = \alpha(R_\alpha) = 1$ is positive. Choosing a basis $(v_1, \dots, v_{2n-2})$ for ξ at a point $p \in M$, we compute

$$\alpha \wedge d\alpha^{n-1}(R_\alpha, v_1, \dots, v_{2n-2}) = d\alpha^{n-1}(v_1, \dots, v_{2n-2}) = \omega_h^{n-1}(v_1, \dots, v_{2n-2}).$$

Similarly, we obtain

$$\begin{aligned} \omega_h^n(Y, R_\alpha, v_1, \dots, v_{2n-2}) &= n\omega_h(Y, R_\alpha) \cdot \omega_h^{n-1}(v_1, \dots, v_{2n-2}) \\ &= n\omega_h(R_\alpha, JR_\alpha) \cdot \omega_h^{n-1}(v_1, \dots, v_{2n-2}), \end{aligned}$$

where we have used that $\omega_h(R_\alpha, v_j) = d\alpha(R_\alpha, v_j) = 0$ for all $j \in \{1, \dots, n-1\}$. The first term $\omega_h(R_\alpha, JR_\alpha)$ is positive, and hence $\alpha \wedge d\alpha^{n-1}$ and $\iota_Y \omega_h^n$ induce identical orientations on M.

To prove the direction "(2) $\Rightarrow$ (1)", choose any collar neighborhood $(-\varepsilon, 0] \times M$ for the boundary, and let t be the coordinate on $(-\varepsilon, 0]$. First note that $\alpha = d^J t|_{TM}$ is a non-vanishing 1-form with kernel ξ, so in particular it will be contact. Let R_α be the Reeb field of α, and set $Y := -JR_\alpha$. As before, the field Y is positively transverse to M, because of $\mathcal{L}_Y t = -dt(JR_\alpha) = \alpha(R_\alpha) = 1$.

Let C be a large constant, whose size will be determined below, and set $h(t, p) := e^{Ct} - 1$. Clearly, h is a regular equation for M, and we claim that for sufficiently large C, h will be a J-plurisubharmonic function.

Let $v \in T_pW$ be any non-vanishing vector at $p \in M$ and represent it as

$$v = aY + bR_\alpha + cZ,$$

where Y and R_α were defined above, and $Z \in \xi$ is a vector in the contact structure that has been normalized such that $d\alpha(Z, JZ) = \omega_t(Z, JZ) = 1$. Note that the 1-form $\alpha_C = d^J h|_{TM} = Ce^{Ct}\alpha$ is a contact form that represents the same coorientation as α.

We compute $\omega_h = dd^J h = Ce^{Ct}(\omega_t + C\,dt \wedge d^J t)$, which simplifies for $t = 0$ further to $\omega_h = C(\omega_t + C\,dt \wedge d^J t)$ and so we have

$$\omega_h(R_\alpha, \cdot) = C\big(\omega_t(R_\alpha, \cdot) - C\,dt\big) \quad \text{and} \quad \omega_h(Y, \cdot) = C\big(\omega_t(\cdot, JR_\alpha) + C\,d^J t\big).$$

This implies $\omega_h(R_\alpha, Z) = \omega_h(R_\alpha, JZ) = 0$ for all $Z \in \xi$, and $\omega_h(Y, R_\alpha) = C^2 + C\omega_t(R_\alpha, JR_\alpha)$ can be made arbitrarily large by increasing the size of C. With these relations we obtain

$$\begin{aligned}
&\omega_h(v, Jv)\\
&\quad= \omega_h(aY + bR_\alpha + cZ, aR_\alpha - bY + cJZ)\\
&\quad= \big(a^2 + b^2\big)\omega_h(Y, R_\alpha) + c^2\omega_h(Z, JZ) + ac\omega_h(Y, JZ) + bc\omega_h(Y, Z)\\
&\quad= \big(a^2 + b^2\big)\big(C^2 + O(C)\big) + C\big(c^2\omega_t(Z, JZ) + ac\omega_t(Y, JZ) + bc\omega_t(Y, Z)\big)
\end{aligned}$$

and setting $A_a = \omega_t(Y, JZ)$ and $A_b = \omega_t(Y, Z)$ and using that $\omega_t(Z, JZ) = 1$

$$\begin{aligned}
&= \big(a^2 + b^2\big)\big(C^2 + O(C)\big) + C\big(c^2 + A_a ac + A_b bc\big)\\
&= \big(a^2 + b^2\big)\big(C^2 + O(C)\big) + \frac{C}{2}\big((c + aA_a)^2 - a^2A_a^2 + (c + bA_b)^2 - b^2A_b^2\big)\\
&= a^2\big(C^2 + O(C)\big) + b^2\big(C^2 + O(C)\big) + \frac{C}{2}\big((c + aA_a)^2 + (c + bA_b)^2\big).
\end{aligned}$$

By choosing C large enough, we can ensure that the a^2- and b^2-coefficients are both positive. Then it is obvious from the computation above that ω_h tames J, and hence h is J-plurisubharmonic. ∎

Legendrian Foliations in Convex Boundaries.

Definition. A **totally real submanifold** N of an almost complex manifold (W, J) is a submanifold of dimension $\dim N = \frac{1}{2}\dim W$ that is not tangent to any J-complex line, that means, $TN \cap (JTN) = \{0\}$, which is equivalent to requiring

$$TW|_N = TN \oplus (JTN).$$

Proposition 3.10. *Let (W, J) be an almost complex manifold with J-convex boundary (M, ξ). Assume N is a submanifold of M for which the complex tangencies ξ induce the Legendrian foliation $\mathcal{F} = TN \cap \xi$. Then it is easy to check that $N \setminus \operatorname{Sing}(\mathcal{F})$ is totally real.*

Proof. If $X \in TN$ is a non-vanishing vector with JX also in TN, then in particular

$$X \in TN \cap (JTN) \subset TM \cap (JTM) = \xi,$$

so that X and JX have to lie in $\mathcal{F}$. The 2-form $d\alpha$ tames $J|_\xi$ so that $d\alpha(X, JX) > 0$, but $d\alpha|_{\mathcal{F}}$ vanishes at regular points of the foliation, and hence X must be 0. ■

We will next study the restrictions imposed by a Legendrian foliation on J-holomorphic curves. Let (Σ, j) be a compact Riemann surface with boundary, and let A be a subset of an almost complex manifold (W, J). We introduce for J-holomorphic maps $u\colon \Sigma \to W$ with $u(\partial\Sigma) \subset A$ the notation

$$u\colon (\Sigma, \partial\Sigma, j) \to (W, A, J).$$

Note that we are always supposing that u is at least C^1 along the boundary.

Corollary 3.11. *Let (W, J) be an almost complex manifold with convex boundary (M, ξ). Let $N \hookrightarrow M$ be a submanifold with an induced Legendrian foliation $\mathcal{F}$, and let u be a J-holomorphic map*

$$u\colon (\Sigma, \partial\Sigma, j) \to \big(W, N \setminus \operatorname{Sing}(\mathcal{F}), J\big).$$

If there is an interior point $z_0 \in \Sigma \setminus \partial\Sigma$ at which u touches M, or if ∂u is not positively transverse to $\mathcal{F}$, then u is a constant map.

Proof. Choose a J-plurisubharmonic function $h\colon W \to \mathbb{R}$ that is a regular equation for M. The first implication follows directly from Corollary 3.6, because z_0 would be an interior maximum for $h \circ u$.

For the second implication note first that $h \circ u$ takes its maximum on $\partial\Sigma$ so that if u is not constant, we have by Corollary 3.7 that the derivative $\mathcal{L}_v(h \circ u)$ is strictly positive for every point $z_1 \in \partial\Sigma$ and every vector $v \in T_{z_1}\Sigma$ pointing out of Σ. Now if $w \in T\Sigma$ is a vector that is tangent to $\partial\Sigma$ such that jw points inward (so that w corresponds to the boundary orientation of $\partial\Sigma$, because $(-jw, w)$ is a positive basis of $T\Sigma$), we obtain

$$\alpha(Du \cdot w) = -dh(JDu \cdot w) = -dh(Du \cdot jw) = -d(h \circ u)(jw) > 0.$$

The boundary of ∂u has thus to be positively transverse to ξ, and so it is in particular positively transverse to the Legendrian foliation $\mathcal{F}$. ■

Note that the result above applies only for holomorphic maps that are C^1 along the boundary.

3.1.2. Preliminaries: ω-Convexity. Above we have explained the notion of J-convexity, and the relevant relationship between contact and almost complex structures. In this section, we want to discuss the notion of ω-convexity, that means the relationship between an (almost) symplectic and a contact structure.

In fact, we are not interested in studying almost complex manifolds for their own sake, but we would like to use the almost complex structure to understand instead a symplectic manifold (W,ω). As initiated by Gromov, we introduce an auxiliary almost complex structure to be able to study J-holomorphic curves in the hope that even though the J-holomorphic curves depend very strongly on the almost complex structure chosen, we'll be able to extract interesting information about the initial symplectic structure.

For this strategy to work, we need the almost complex structure to be **tamed** by ω, that means, we want

$$\omega(X, JX) > 0$$

for every non-vanishing vector $X \in TW$. This tameness condition is important, because it allows us to control the limit behavior of sequences of holomorphic curves (see Section 4.3).

As explained in the previous section, J-convexity is a property that greatly helps us in understanding holomorphic curves in ambient manifolds that have boundary. When (W,ω) is a symplectic manifold with boundary $M = \partial W$, we would thus like to chose an almost complex structure J that is

- tamed by ω, and
- that makes the boundary J-convex.

In particular, if such a J exists, we know that the boundary admits an induced contact structure

$$\xi = TM \cap (J \cdot TM).$$

From the symplectic or contact topological view point, the opposite setup would be more natural though: given a symplectic manifold (W,ω) with contact boundary (M,ξ), can we choose an almost complex structure J that is tamed by ω, and that makes the boundary J-convex such that ξ is the bundle of J-complex tangencies?

The general answer to that question was given in [21].

Definition. Let (M,ξ) be a cooriented contact manifold of dimension $2n-1$, and let (W,ω) be a symplectic manifold whose boundary is M. Let α be a positive contact form for ξ, and assume that the orientation induced by $\alpha \wedge d\alpha^{n-1}$ on M agrees with the boundary orientation of (W,ω). We call (W,ω) a **weak symplectic filling** of (M,ξ), if

$$\alpha \wedge (T\,d\alpha + \omega)^{n-1} > 0$$

for every $T \in [0,\infty)$.

The proofs of the following statements are very lengthy, hence we will omit the proofs referring instead to the Appendix of [21] for more details.

Theorem 3.12. *Let (M,ξ) be a cooriented contact manifold, and let (W,ω) be a symplectic manifold with boundary $M = \partial W$. The following two statements are equivalent*

- *(W,ω) is a weak symplectic filling of (M,ξ).*
- *There exists an almost complex structure J on W that is tamed by ω and that makes M a J-convex boundary whose J-complex tangencies are ξ.*

Furthermore the space of all almost complex structures that satisfy these conditions is contractible (if non-empty).

A weak filling is a notion that is relatively recent in higher dimensions; traditionally it is the concept of a strong symplectic filling that has been studied for a much longer time. Let (W,ω) be a symplectic manifold. A vector field X_L is called a **Liouville vector field**, if it satisfies the equation

$$\mathcal{L}_{X_L}\omega = \omega.$$

Definition. Let (M,ξ) be a cooriented contact manifold, and let (W,ω) be a symplectic manifold whose boundary is M. We call (W,ω) a **strong**

symplectic filling of (M,ξ), if there exists a Liouville vector field X_L on a neighborhood of M such that $\lambda := (\iota_{X_L}\omega)|_{TM}$ is a positive contact form for ξ.

It is easy to see that a strong filling is in particular a weak filling. Note that the symplectic form of a strong filling becomes always exact when restricted to the boundary, but that this needs not be true for a weak filling; if it is then it will usually still not be a strong symplectic filling, but by Corollary 3.15 it can deformed into one.

Lemma 3.13. *Let (W,ω) be a symplectic manifold and let M be a hypersurface (possibly a boundary component of W) together with a non-vanishing 1-form λ. Assume that the restriction of ω to $\ker\lambda$ is symplectic.*

Then there is a tubular neighborhood of M in W that is symplectomorphic to the model

$$\big((-\varepsilon,\varepsilon)\times M, d(t\lambda)+\omega|_{TM}\big),$$

where t is the coordinate on the interval $(-\varepsilon,\varepsilon)$. The 0-slice $\{0\}\times M$ corresponds in this identification to the hypersurface M. If M is a boundary component of W then of course we need to replace the model by $(-\varepsilon,0]\times M$ or by $[0,\varepsilon)\times M$ depending on whether $\lambda\wedge\omega^{n-1}$ is oriented as the boundary of (W,ω) or not.

For the proof see [21, Lemma 2.6].

Proposition 3.14. *Let (W,ω) be a weak filling of a contact manifold (M,ξ), and let Ω be a 2-form on M that is cohomologous to $\omega|_{TM}$. Choose a positive contact form α for (M,ξ). Then if we allow $C>0$ to be sufficiently large, we can attach a collar $[0,C]\times M$ to W with a symplectic form ω_C that agrees close to $\{C\}\times M$ with $d(t\alpha)+\Omega$, and such that the new manifold is a weak filling of $(\{t_0\}\times M,\xi)$ for every $t_0\in[0,C]$.*

The proof can be found in [21, Lemma 2.10].

Corollary 3.15. *Let (W,ω) be a weak symplectic filling of (M,ξ) and assume that ω restricted to a neighborhood of M is an exact symplectic form. Then we may deform ω on a small neighborhood of M such it becomes a strong symplectic filling.*

Proof. Since $\omega|_{TM}$ is exact, we can apply the proposition above with $\Omega = 0$. Afterwards we can isotope the collar back into the neighborhood of the boundary of W. ∎

Note that two contact structures that are strongly filled by the same symplectic manifold are isotopic, while a symplectic manifold may be a weak filling of two different contact manifolds. This is true even when the restriction of the symplectic structure to the boundary is exact, see [21, Remark 2.11].

3.2. Holomorphic Curves and Legendrian Foliations

Let (W, J) be an almost complex manifold with J-convex boundary (M, ξ), and let $N \subset M$ be a submanifold carrying a Legendrian foliation $\mathcal{F}$. The aim of this section will be to better understand the behavior of J-holomorphic maps

$$u\colon (\Sigma, \partial\Sigma, j) \to (W, N, J),$$

that lie close to a singular point $p \in \mathrm{Sing}(\mathcal{F})$ of the Legendrian foliation. For this we will assume that J is of a very specific form in a neighborhood of the point p.

3.2.1. Existence of J-Convex Functions Close to Totally Real Submanifolds. As a preliminary tool, we will need the following result.

Proposition 3.16. *Let (W, J) be an almost complex structure that contains a closed totally real submanifold L. Then there exists a smooth function $f\colon W \to [0, \infty)$ with $L = f^{-1}(0)$ that is J-plurisubharmonic on a neighborhood of L. In particular, it follows that $df_p = 0$ at every point $p \in L$.*

Proof. We will first show that we find around every point $p \in L$ a chart U with coordinates $\{(x_1, \ldots, x_n; y_1, \ldots, y_n)\} \subset \mathbb{R}^{2n}$ such that $L \cap U = \{y_1 = \cdots = y_n = 0\}$ and

$$J\frac{\partial}{\partial x_j}\bigg|_{L\cap U} = \frac{\partial}{\partial y_j}\bigg|_{L\cap U}.$$

For this, start by choosing coordinates $\{(x_1, \ldots, x_n)\} \subset \mathbb{R}^n$ for the submanifold L around the point p, and consider the associated vector fields

$$Y_1 = J\frac{\partial}{\partial x_1}, \ldots, Y_n = J\frac{\partial}{\partial x_n}$$

along L. These vector fields are everywhere linearly independent and transverse to L, hence, we can define a smooth map from a small ball around 0 in $\mathbb{R}^{2n} = \{(x_1,\dots,x_n;y_1,\dots,y_n)\}$ to W by

$$y_1 Y_1(x_1,\dots,x_n) + \dots + y_n Y_1(x_1,\dots,x_n) \mapsto \exp(y_1 Y_1 + \dots + y_n Y_1),$$

where exp is the exponential map for an arbitrary Riemannian metric on W. If the ball is chosen sufficiently small, the map will be a chart with the desired properties.

For such a chart U, we will choose a function

$$f_U\colon U \to [0,\infty), \quad (x_1,\dots,x_n;y_1,\dots,y_n) \mapsto \frac{1}{2}\big(y_1^2 + \dots + y_n^2\big).$$

It is obvious that both the function itself, and its differential vanish along $L \cap U$. Furthermore f is plurisubharmonic close to $L \cap U$, because

$$\begin{aligned} dd^J f_U &= d\big(y_1\, d^J y_1 + \dots + y_n\, d^J y_n\big) \\ &= dy_1 \wedge d^J y_1 + \dots + dy_n \wedge d^J y_n + y_1\, dd^J y_1 + \dots + y_n\, dd^J y_n \end{aligned}$$

simplifies at $L \cap U$ to

$$dd^J f_U\big|_{L\cap U} = dx_1 \wedge dy_1 + \dots + dx_n \wedge dy_n,$$

where we have used that all y_j vanish, and that $J\frac{\partial}{\partial x_j} = \frac{\partial}{\partial y_j}$ and $J\frac{\partial}{\partial y_j} = J^2\frac{\partial}{\partial x_j} = -\frac{\partial}{\partial x_j}$. It is easy to check that this 2-form evaluates positively on complex lines along $L \cap U$, and hence also in a small neighborhood of p.

Now to obtain a global plurisubharmonic function as stated in the proposition, cover L with finitely many charts $U_1,\dots,U_N$, each with a function $f_1,\dots,f_N$ according to the construction given above. Choose a subordinate partition of unity $\rho_1,\dots,\rho_N$, and define

$$f = \sum_{j=1}^{N} \rho_j \cdot f_j .$$

The function f and its differential $df = \sum_{j=1}^{N} (\rho_j\, df_j + f_j\, d\rho_j)$ vanish along L so that the only term in

$$\begin{aligned} dd^J f &= d\sum_{j=1}^{N} \big(\rho_j\, d^J f_j + f_j\, d^J \rho_j\big) \\ &= \sum_{j=1}^{N} \big(\rho_j\, dd^J f_j + d\rho_j \wedge d^J f_j + f_j\, dd^J \rho_j + df_j \wedge d^J \rho_j\big) \end{aligned}$$

that survives along L is the first one, giving us along L

$$dd^J f = \sum_{j=1}^{N} \rho_j \, dd^J f_j .$$

This 2-form is positive on J-complex lines, and hence there is a small neighborhood of L on which f is plurisubharmonic. Finally, we modify f to be positive outside this small neighborhood so that we have $L = f^{-1}(0)$ as required. ■

Corollary 3.17. *Let (W, J) be an almost complex structure that contains a closed totally real submanifold L. Then we find a small neighborhood U of L for which every J-holomorphic map*

$$u\colon (\Sigma, \partial\Sigma, j) \to (W, L, J)$$

from a compact Riemann surface needs to be constant if $u(\Sigma) \subset U$.

Proof. Let $f\colon W \to [0, \infty)$ be the function constructed in Proposition 3.16, and let $U \subset (W, J)$ be the neighborhood of L, where f is J-plurisubharmonic. Because $u(\Sigma) \subset U$, we obtain from Corollary 3.6 that $f \circ u$ must take its maximum on the boundary of Σ, but because $f \circ u$ is zero on all of $\partial\Sigma$, it follows that $f \circ u$ will vanish on the whole surface Σ. The image $u(\Sigma)$ lies then in the totally real submanifold L, and this implies that the differential of u vanishes everywhere. Hence there is a $\mathbf{q}_0 \in L$ with $u(z) = \mathbf{q}_0$ for all $z \in \Sigma$. ■

3.2.2. *J*-Holomorphic Curves Close to Elliptic Singularities of a Legendrian Foliation. The aim of this section will be to show that for a suitable choice of an almost complex structure, elliptic singularities give birth to a family of holomorphic disks, and that apart from these disks and their branched covers, no other holomorphic disks may get close to the elliptic singularities.

Before studying the higher dimensional case, we will construct a model situation for a 4-dimensional almost complex manifold with convex boundary.

Dimension 4. Consider $\mathbb{C}^2$ with its standard complex structure i. Then it is easy to check that $h(z_1, z_2) = \frac{1}{2}(|z_1|^2 + |z_2|^2)$ is a plurisubharmonic function whose regular level sets are the concentric spheres around the origin. We choose the level set $M = h^{-1}(1/2)$, that is, the boundary of the closed unit ball $W := h^{-1}((-\infty, 1/2])$ that is i-convex and has the induced contact form

$$\alpha_0 = d^i h\big|_{TM} = x_1\,dy_1 - y_1\,dx_1 + x_2\,dy_2 - y_2\,dx_2.$$

We only want to study a neighborhood U of $(0,1)$ in W. Embed a small disk by the map

$$\Phi\colon z \mapsto \left(z, \sqrt{1-|z|^2}\right)$$

into $M \cap U$, and denote the image of Φ by N_0. This submanifold is the intersection of $M = \mathbb{S}^3$ with a hyperplane whose z_2-coordinate is purely real. The restriction of α_0 to N_0 reduces to

$$\alpha_0|_{TN_0} = \Phi^*\alpha_0 = x\,dy - y\,dx, \tag{3.2}$$

so that the Legendrian foliation has at the origin an elliptic singularity (of the type described in Section 2.2.1).

Let U be the subset

$$U = \left\{(z_1, z_2) \in \mathbb{C}^2 \;\middle|\; \mathrm{Re}(z_2) > 1 - \delta\right\} \cap h^{-1}((-\infty, 1/2])$$

for small $\delta > 0$, that means, we take the unit ball and cut off all points under a certain x_2-height.

The following propositions explain that there is essentially a unique holomorphic disk with boundary in N_0 passing through a given point $(z_1, z_2) \in N_0 \cap U$. All other holomorphic curves with the same boundary condition will either be constant or will be (branched) covers of that disk.

Proposition 3.18. *Denote the intersection of U with the complex plane $\mathbb{C} \times \{x\}$ for $x \in (1-\delta, 1)$ by L_x. For every $x_2 \in (1-\delta, 1)$, there exists a unique* injective *holomorphic map*

$$u_{x_2}\colon \left(\mathbb{D}^2, \partial\mathbb{D}^2\right) \to (L_{x_2}, \partial L_{x_2})$$

that satisfies $u_{x_2}(0) = (0, x_2)$ and $u_{x_2}(1) \in \{(x_1, x_2) \in U | x_1 > 0\}$.

The last two conditions only serve to fix a parametrization of a given geometric disk.

Proof. The desired map u_{x_2} can be explicitly written down as

$$u_{x_2}(z) = (Cz, x_2)$$

with $C = \sqrt{1 - x_2^2}$.

To prove uniqueness assume that there were a second holomorphic map

$$\tilde{u}_{x_2}\colon \left(\mathbb{D}^2, \partial\mathbb{D}^2\right) \to \left(L_{x_2}, \partial L_{x_2}\right)$$

with the required properties. It is clear that $L_{x_2} = \{(x+iy, x_2) \in \mathbb{C}^2 \mid x^2 + y^2 \le 1 - x_2^2\}$ is a round disk.

By Corollary 3.11, the restriction $u_{x_2}|_{\partial\mathbb{D}^2}$ of the map to the boundary has non-vanishing derivative, and it is by assumption injective, hence it is a diffeomorphism onto ∂L_{x_2}. This proves that u_{x_2} has to be for topological reasons surjective on L_{x_2} (otherwise we could construct a retract of the disk onto its boundary). Note also that the germ of a holomorphic map around the origin in $\mathbb{C}$ is always biholomorphic to $z \mapsto z^k$ for some integer $k \in \mathbb{N}_0$, so that the differential of u_{x_2} may not vanish anywhere, because otherwise u_{x_2} could not be injective.

Together this allows us to define a biholomorphism

$$\varphi := u_{x_2}^{-1} \circ \tilde{u}_{x_2}\colon \left(\mathbb{D}^2, \partial\mathbb{D}^2\right) \to \left(\mathbb{D}^2, \partial\mathbb{D}^2\right)$$

with $\varphi(0) = 0$ and $\varphi(1) = 1$, but the only automorphism of the disk with these properties is the identity, thus showing that $u_{x_2} = \tilde{u}_{x_2}$. ■

Proposition 3.19. *Let*

$$u\colon (\Sigma, \partial\Sigma; j) \to (U, N_0; i)$$

be any holomorphic map from a connected compact Riemann surface (Σ, j) *to* U *with* $u(\partial\Sigma) \subset N_0$.

Either u *is constant or its image is one of the slices* $L_{x_2} = U \cap (\mathbb{C} \times \{x_2\})$. *If* u *is injective at one of its boundary points, then* Σ *will be a disk, and after a reparametrization by a Möbius transformation,* u *will be equal to the map* u_{x_2} *given in Proposition* 3.18.

Proof. Note that we are supposing that u is at least C^1 on the boundary so that by Corollary 3.11 the map u will be constant if it touches the elliptic singularity in N.

The proof of the proposition will be based on the harmonicity of the coordinate functions x_1, y_1, x_2, and y_2. Let $f\colon U \to \mathbb{R}$ be the function $(z_1, z_2) \mapsto y_2 = \operatorname{Im}(z_2)$. Since Σ is a compact domain, the function $f \circ u$ attains somewhere on Σ its maximum and its minimum, and applying the maximum principle, Corollary 3.6, to $f \circ u$ itself and also to $-f \circ u$, we obtain that both the maximum and the minimum have to lie on $\partial\Sigma$. But since $u(\partial\Sigma) \subset N_0$ has vanishing imaginary z_2-part, it follows that $f \circ u \equiv 0$ on the whole surface. Using now the Cauchy-Riemann equations, it immediately follows that the real part of the z_2-coordinate of u has to be constant everywhere. We can deduce that the image of u has to lie in one of the slices $L_{x_2} = \mathbb{C} \times \{x_2\}$, and in particular the boundary $u(\partial\Sigma)$ lies in the circle $\partial L_{x_2} = \{(x+iy, x_2) \in \mathbb{C}^2 \mid x^2 + y^2 = 1 - x_2^2\}$.

Assume that u is not constant. Since u lies in L_{x_2}, we can use the map u_{x_2} from Proposition 3.18, to define a holomorphic map

$$\varphi := u_{x_2}^{-1} \circ u \colon (\Sigma, \partial\Sigma) \to (\mathbb{D}^2, \partial\mathbb{D}^2).$$

If u were not surjective on L_{x_2}, we could suppose (after a Möbius transformation on the target space) that the image of φ does not contain 0. The function $h(z) = -\ln|z|$ on $\mathbb{D}^2 \setminus \{0\}$ is harmonic, because it is locally the real part of a holomorphic function, and because $h \circ \varphi$ would have its maximum on the interior of Σ, we obtain that $h \circ \varphi$ is constant, so that the image of φ lies in $\partial\mathbb{D}^2$. The image of a non-constant holomorphic map is open, and hence u must be constant.

Assume now that u is injective at one of its boundary points. As we have shown in Proposition 3.18 the restriction $u|_{\partial\Sigma} \colon \partial\Sigma \to \partial L_{x_2}$ will be a diffeomorphism for each component of $\partial\Sigma$ so that $\partial\Sigma$ must be connected. Furthermore, it follows that u will also be injective on a small neighborhood of ∂L_{x_2}, because if we find two sequences $(z_k)_k$ and $(\tilde{z}_k)_k$ coming arbitrarily close to $\partial\Sigma$ with $u(z_k) = u(\tilde{z}_k)$ for every k, then after assuming that they both converge (reducing if necessary to subsequences), we see by continuity that $\lim u(z_k) = \lim u(\tilde{z}_k)$ and $\lim z_k, \lim \tilde{z}_k \in \partial\Sigma$, so that we can conclude that $\lim z_k = \lim \tilde{z}_k$. Using that the differential of u in $\lim z_k$ is not singular, we obtain that for k sufficiently large, we will always have $z_k = \tilde{z}_k$ showing that u is indeed injective on a small neighborhood of $\partial\Sigma$.

Assume $z_0 \in \Sigma$ is a point at which the differential $D\varphi$ vanishes. Then we know that φ can be represented in suitable charts as $z \mapsto z^k$ for some $k \in \mathbb{N}$, but if $k > 1$ this yields a contradiction, because we know that φ is a biholomorphism on a neighborhood of $\partial\Sigma$, and hence its degree must be 1. Since φ is holomorphic, it preserves orientations, so that on the other hand, we would have that the degree would need to be *at least* k, if there were such a critical point.

We obtain that φ has nowhere vanishing differential, and hence it must be a regular cover, but since it is of degree 1, it is in fact a biholomorphism, and Σ must be a disk. ■

The Higher Dimensional Situation. In this section, L will always be a closed manifold, and we will choose for T^*L an almost complex structure J_L for which the 0-section L is totally real, so that there is by Proposition 3.16 a function $f_L\colon T^*L \to [0,\infty)$ that vanishes on L (and only on L) and that is plurisubharmonic on a small neighborhood of L.

As before, we will first describe a very explicit manifold that will serve as a model for the neighborhood of an elliptic singularity. Let $\mathbb{C}^2 \times T^*L$ be the almost complex manifold with almost complex structure $J = i \oplus J_L$, where i is the standard complex structure on $\mathbb{C}^2$. We define a function $f\colon \mathbb{C}^2 \times T^*L \to [0,\infty)$ by

$$f(z_1, z_2, \mathbf{q}, \mathbf{p}) = \frac{1}{2}\big(|z_1|^2 + |z_2|^2\big) + f_L(\mathbf{q}, \mathbf{p}).$$

If we stay in a sufficiently small neighborhood of the 0-section of T^*L, this function is clearly J-plurisubharmonic and we denote its regular level set $f^{-1}(1/2)$ by M; its contact form is given by

$$\alpha := d^J f\big|_{TM} = \big(x_1\,dy_1 - y_1\,dx_1 + x_2\,dy_2 - y_2\,dx_2 + d^{J_L} f_L\big)\big|_{TM}.$$

Now we define a submanifold N in M as the image of the map

$$\Phi\colon \mathbb{D}^2 \times L \hookrightarrow M \subset \mathbb{C}^2 \times T^*L$$

given by $\Phi(z;\mathbf{q}) = (z, \sqrt{1-|z|^2}; \mathbf{q}, \mathbf{0})$, that means, the image of Φ is the product of the 0-section in T^*L and the submanifold N_0 given in the previous section. The submanifold has a Legendrian foliation $\mathcal{F}$ induced by

$$\alpha|_{TN} = \Phi^* d^J f = x\,dy - y\,dx.$$

In particular, the leaves of the foliation are parallel to the L-factor in $\mathbb{D}^2 \times L$ and $\mathcal{F}$ has an elliptic singularity in $\{0\} \times L$.

Note that both the almost complex structure as well as the submanifold N split as a product, thus if we consider a holomorphic map

$$u\colon (\Sigma, \partial\Sigma; j) \to \big(\mathbb{C}^2 \times T^*L, N; J\big),$$

we can decompose it into $u = (u_1, u_2)$ with

$$u_1\colon (\Sigma, \partial\Sigma; j) \to \big(\mathbb{C}^2, N_0; i\big)$$

$$u_2\colon (\Sigma, \partial\Sigma; j) \to \big(T^*L, L; J_L\big).$$

This allows us to treat each factor independently from the other one, and we will easily be able to obtain similar results as in the previous section.

Since we are interested in finding a local model, we will first restrict our situation to the following subset

$$U := \{(z_1, z_2; \mathbf{q}, \mathbf{p}) \mid \operatorname{Re}(z_2) \geq 1 - \delta\} \cap f^{-1}([0, 1/2]) \tag{3.3}$$

that is, for δ sufficiently small, a compact neighborhood of N in $f^{-1}([0,1/2])$, because the points $(z_1, z_2; \mathbf{q}, \mathbf{p})$ in U satisfy

$$0 \leq \frac{1}{2}|z_1|^2 + f_L(\mathbf{q}, \mathbf{p}) \leq \frac{1}{2}\left(1 - |z_2|^2\right) \leq \delta - \frac{1}{2}\delta^2 \leq \delta$$

so that all coordinates are bounded. Note in particular, that this bound on the $\mathbf{p}$-coordinates guarantees that f will be J-plurisubharmonic on U.

The submanifold $N \cap U$ can also be written in the following easy form

$$\{(z, x_2; \mathbf{q}, \mathbf{0}) \mid x_2 \geq 1 - \delta \text{ and } |z|^2 = 1 - x_2^2\} \times L.$$

Corollary 3.20. *Let*

$$u\colon (\Sigma, \partial\Sigma, j) \to (U, N \cap U; J)$$

be any holomorphic map from a connected compact Riemann surface (Σ, j) *to* U *with* $u(\partial\Sigma) \subset N$.

Either u *is constant or its image is one of the slices* $L_{x_2,\mathbf{q}_0} = (\mathbb{C} \times \{x_2\} \times \{\mathbf{q}_0\}) \cap U$ *with* $x_2 \in [1-\delta, 1)$ *and* $\mathbf{q}_0$ *a point on the* 0*-section of* T^*L*. If* u *is injective at one of its boundary points, then* Σ *will be a disk, and* u *is equal to*

$$u(z) = \left(u_{x_2} \circ \varphi(z); \mathbf{q}_0, \mathbf{0}\right),$$

where u_{x_2} *is the map given in Proposition* 3.18*, and* φ *is a Möbius transformation of the unit disk.*

Proof. Let u be a J-holomorphic map as in the statement. We will study u by decomposing it into $u = (u_{\mathbb{C}^2}, u_{T^*L})$ with

$$u_{\mathbb{C}^2}\colon (\Sigma, \partial\Sigma, j) \to \left(\mathbb{C}^2, N, i\right)$$
$$u_{T^*L}\colon (\Sigma, \partial\Sigma, j) \to \left(T^*L, L, J_L\right).$$

Using that f_L is J_L-plurisubharmonic on the considered neighborhood of the 0-section contained in U, it follows from Corollary 3.17 that u_{T^*L} is constant.

Once we know that u_{T^*L} is constant, the situation for $u_{\mathbb{C}^2}$ is identical to the one in Proposition 3.19, so that we obtain the desired result. ■

The results obtained so far only explain the behavior of holomorphic curves that are completely contained in the model neighborhood U. Next we will extend this result to show that a holomorphic curve is either disjoint from the subset U or is lies completely inside U.

Assume (W, J) is a compact almost complex manifold with convex boundary $M = \partial W$. Let N be a submanifold of M, and assume that there is a compact subset U in W such that U is diffeomorphic to the model above, with $M \cap U$, $N \cap U$ and $J|_U$ all being equal to the corresponding objects in our model neighborhood.

Proposition 3.21. *Let*

$$u\colon (\Sigma, \partial\Sigma; j) \to (W, N; J)$$

be a holomorphic map, and let U be a compact subset of W that agrees with the model described above.

If $u(\Sigma)$ intersects U, then it has to lie entirely in U, and it will be consequently of the form given by Corollary 3.20.

Proof. Assume u to be a holomorphic map whose image lies partially in U. The set U is a compact manifold with corners, and we write $\partial U = \partial_M U \cup \partial_W U$ (see Figure 5), where

$$\partial_M U = U \cap M$$
$$\partial_W U = \big\{(z_1, z_2; \mathbf{q}, \mathbf{p}) \,|\, \mathrm{Re}(z_2) \geq 1 - \delta\big\} \cap f^{-1}\big([0, 1/2]\big).$$

We will show that the real part of the z_2-coordinate of u needs to be constant. This then proves the proposition, because it prevents u from leaving U.

Thus assume instead that the real part of z_2 does vary on u. Slightly decreasing the cut-off level δ in (3.3) using Sard's theorem, the holomorphic map u will intersect $\partial_W U$ transversely, so that $u^{-1}(\partial_W U)$ will be a properly embedded submanifold of Σ. We will restrict u to the compact subset $G = u^{-1}(U)$, and denote the boundary components of this domain

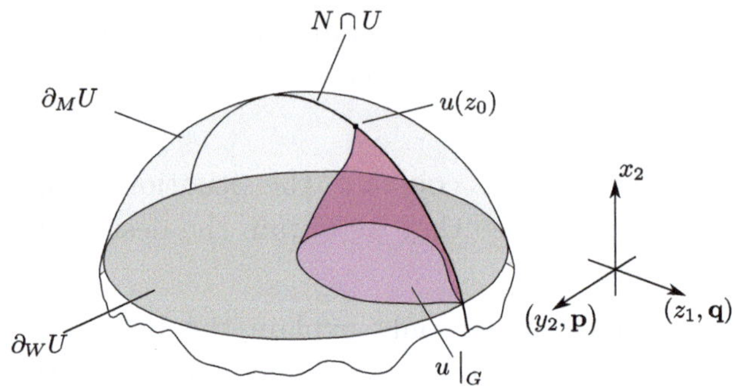

Fig. 5. Sketch of the symplectic model neighborhood of an elliptic singularity. A holomorphic curve lying only partially in this neighborhood would have two types of boundary, one part $u(\partial_M G)$ that lies in $N \cap U$, and a second one $u(\partial_W G)$ where the curve leaves the model neighborhood

by $\partial_M G = u^{-1}(N \cap U)$ and $\partial_W G = u^{-1}(\partial_W U)$. We thus have a holomorphic map

$$u|_G \colon (G, \partial G; j) \to (U, \partial U; J)$$

with $u(\partial_M G) \subset N \cap U$ and $u(\partial_W G) \subset \partial_W U$.

The coordinate maps $h_x \colon (z_1, z_2; \mathbf{q}, \mathbf{p}) \mapsto \operatorname{Re}(z_2)$ and $h_y \colon (z_1, z_2; \mathbf{q}, \mathbf{p}) \mapsto \operatorname{Im}(z_2)$ are harmonic, and it follows by the maximum principle that the maximum of $h_x \circ u|_G$ will lie for each component of G on the boundary of that component.

Furthermore the maximum of $h_x \circ u|_G$ cannot lie on $\partial_W G$, because by our assumption $u|_G$ is transverse to $\partial_W U$. It follows that the maximum of $h_x \circ u|_G$ will be a point $z_0 \in \partial_M G$; in particular z_0 does not lie on one of the edges of G. By the boundary point lemma, either $h_x \circ u|_G$ is constant or the outward derivative of this function at z_0 must be strictly positive. On the other hand, the function $h_y \circ u|_G$ is equal to 0 all along the boundary $\partial_M G$ so that the derivatives of $h_x \circ u|_G$ and $h_y \circ u|_G$ vanish at z_0 in directions that are tangent to the boundary. Using the Cauchy-Riemann equation we see that this implies that the derivatives of these two functions at z_0 vanish in *every direction*, in particular this implies that the function $h_x \circ u|_G$ needs to be constant.

In either case, we have proved that the image of u lies completely inside U. ■

The conclusion of the results in this section is that every curve that intersects a certain neighborhood of the elliptic singularities lies completely in this neighborhood and can be explicitly determined.

3.2.3. J-Holomorphic Curves Close to Codimension 1 Singularities. Let $(N, \mathcal{F})$ be a submanifold with Legendrian foliation and with non-empty boundary. We will show in this section that a boundary component of N lying in the singular set of $\mathcal{F}$ can sometimes exclude that any holomorphic curve gets close to this component. This way, the boundary may block any holomorphic disks from escaping the submanifold N.

The argument is similar to that of the previous section, where we constructed an almost complex manifold that served as a model for the neighborhood of the singular set.

Remark 3.22. We will only be dealing here with the easiest type of singular sets: Products of a closed manifold with $\mathbb{S}^1$. A more general situation has been considered in [21], where the singular set is allowed to be a fiber bundle over the circle.

Let T^*F be the cotangent bundle of a closed manifold F, choose an almost complex structure J_F on T^*F for which F is a totally real submanifold, and let $f_F\colon T^*F \to [0,\infty)$ be the function constructed in Proposition 3.16 that only vanishes along the 0-section of T^*F and that is J_F-plurisubharmonic close to the 0-section F.

Define (W, J) as

$$W := \mathbb{C} \times T^*\mathbb{S}^1 \times T^*F = \big\{(x+iy; \varphi, r; \mathbf{q}, \mathbf{p})\big\},$$

and let J be the almost complex structure $i \oplus i \oplus J_F$, where the complex structure on $T^*\mathbb{S}^1$ is the one induced from the identification of $T^*\mathbb{S}^1$ and $\mathbb{C}/(2\pi\mathbb{Z})$ with $\varphi + ir \sim \varphi + 2\pi + ir$. The function

$$f\colon W \to [0,\infty), (x+iy; \varphi, r; \mathbf{q}, \mathbf{p}) \mapsto \frac{1}{2}\big(x^2+y^2\big) + \frac{1}{2}r^2 + f_F(\mathbf{q}, \mathbf{p})$$

is J-plurisubharmonic on a neighborhood where the values of $\mathbf{p}$ are small. We denote the level set $f^{-1}(1/2)$ by M, and note that for small values of $\mathbf{p}$, it is a smooth contact manifold with contact form

$$\alpha_M := \big(x\,dy - y\,dx - r\,d\varphi + d^{J_F} f_F\big)\big|_{TM}.$$

Let N be the submanifold of M given as the image of the map

$$\Phi\colon \mathbb{S}^1 \times [0,\varepsilon) \times F, (\varphi, r; \mathbf{q}) \mapsto \left(\sqrt{1-r^2}; \varphi, r; \mathbf{q}, \mathbf{0}\right).$$

It has a Legendrian foliation $\mathcal{F}$, because $\Phi^*\alpha_M = -r\,d\varphi$ that becomes singular exactly at the boundary $\partial N = \{1\} \times \mathbb{S}^1 \times F$.

Our local model will be the subset

$$U = \left\{(x+iy; \varphi, r; \mathbf{q}, \mathbf{p}) \mid x \geq 1-\delta\right\} \cap f^{-1}\left([0, 1/2]\right)$$

for sufficiently small $\delta > 0$. Clearly U contains $\partial N = \operatorname{Sing}(\ker(-r\,d\varphi))$. Furthermore U is compact, because all coordinates are bounded: Points $(x+iy; \varphi, r; \mathbf{q}, \mathbf{p})$ in U satisfy

$$0 \leq \frac{1}{2}y^2 + \frac{1}{2}r^2 + f_F(\mathbf{q}, \mathbf{p}) = f(x+iy; \varphi, r; \mathbf{q}, \mathbf{p}) - \frac{1}{2}x^2 \leq 1/2\left(1-x^2\right) \leq \delta.$$

We also obtain that if δ has been chosen small enough, f is everywhere J-plurisubharmonic on U.

Remark 3.23. Note that the construction of the local model also applies in the case of contact 3-manifolds, because F may be just a point.

We will first exclude existence of holomorphic curves that are entirely contained in U.

Proposition 3.24. *A J-holomorphic map*

$$u\colon (\Sigma, \partial\Sigma, j) \to (U, N \cap U, J)$$

from a compact Riemann surface into U, whose boundary is mapped into $N \cap U$, must be constant.

Proof. As in the previous section, we can decompose u as $(u_{\mathbb{C}\times T^*\mathbb{S}^1}, u_{T^*F})$ with

$$u_{\mathbb{C}\times T^*\mathbb{S}^1}\colon (\Sigma, \partial\Sigma, j) \to \left(\mathbb{C} \times T^*\mathbb{S}^1, \left\{\left(\sqrt{1-r^2}; \varphi, r\right) \mid \varphi \in \mathbb{S}^1, r \in [0,\varepsilon)\right\}, i \oplus i\right)$$

$$u_{T^*F}\colon (\Sigma, \partial\Sigma, j) \to \left(T^*F, F, J_F\right).$$

Note in particular that the boundary conditions also split in this decomposition, so that we obtain two completely uncoupled problems. Furthermore, using Corollary 3.17, it follows that the second map is constant, because f_F is a J_F plurisubharmonic function on the considered neighborhood.

To show that $u_{\mathbb{C}\times T^*\mathbb{S}^1}$ is constant, use the harmonic function $g(z;\varphi,r) = \operatorname{Im}(z)$. Since $g\circ u_{\mathbb{C}\times T^*\mathbb{S}^1}$ vanishes along $\partial\Sigma$, it follows that $g\circ u_{\mathbb{C}\times T^*\mathbb{S}^1}$ has to be zero on the whole Riemann surface, and combining this with the Cauchy-Riemann equation, it follows that the real part of the z-coordinate of $u_{\mathbb{C}\times T^*\mathbb{S}^1}$ is equal to a constant $C\in[1-\delta,1]$. Now that we know that the first coordinate of $u_{\mathbb{C}\times T^*\mathbb{S}^1}$ is constant, we see that the boundary of $u_{\mathbb{C}\times T^*\mathbb{S}^1}$ has to lie in the circle $\{(C;\varphi,+\sqrt{1-C^2})\mid\varphi\in\mathbb{S}^1\}\subset\mathbb{C}\times T^*\mathbb{S}^1$.

This allows us to study only the second coordinate of $u_{\mathbb{C}\times T^*\mathbb{S}^1}$ reducing our map to the form

$$u_{T^*\mathbb{S}^1}\colon(\Sigma,\partial\Sigma,j)\to\big(T^*\mathbb{S}^1,S,i\big),$$

where $S=\{r=+\sqrt{1-C^2}\}$. Using that the map $(r,\varphi)\mapsto r$ is harmonic, and that it is constant along the boundary of Σ, we obtain that the whole image of the surface has to lie in the corresponding circle, implying with the Cauchy-Riemann equation that $u_{T^*\mathbb{S}^1}$ needs to be constant. ■

Next we will show that holomorphic curves may not enter the domain U even partially. Let (W,J) be now a compact almost complex manifold with convex boundary $M=\partial W$, and let N be a submanifold of M with $\partial N\neq\emptyset$. Assume that W contains a compact subset U that is identical to the model neighborhood constructed above such that $M\cap U$, $N\cap U$ and $J|_U$ all agree with the corresponding objects in the model.

Proposition 3.25. *If the image of a J-holomorphic map*

$$u\colon(\Sigma,\partial\Sigma,j)\to(W,N,J)$$

intersects the neighborhood U, then it will be constant.

Proof. It suffices to show that the image of u lies inside U, because we can then apply Proposition 3.24. Following the same line of arguments as in the proof of Proposition 3.21, one can show that the real part of the first coordinate of u needs to be constant. We recommend the reader to work out the details as an exercise. ■

Remark 3.26. Note that when the codimension 1 singular set lies in the interior of the maximally foliated submanifold, one can find under additional conditions a family of holomorphic annuli with one boundary component on each side of the singular set (see [30]). The reason why these curves do not appear in the results of this section are that we are assuming that all boundary components of the holomorphic curves lie locally on one side of the singular set.

4. Moduli Spaces of Disks and Filling Obstructions

4.1. The Moduli Space of Holomorphic Disks

Let us assume again that (W, J) is an almost complex manifold, and that $N \subset W$ is a totally real submanifold. We want to study the space of maps

$$u\colon (\mathbb{D}^2, \partial\mathbb{D}^2) \to (W, N; J)$$

that are J-holomorphic (strictly speaking they are (i, J)-holomorphic), meaning that we want the differential of u to be complex linear, so that it satisfies at every $z \in \Sigma$ the equation

$$Du_z \cdot i = J\bigl(u(z)\bigr) \cdot Du_z.$$

Note that J depends on the point $u(z)$!

A different way to state this equation is by introducing the Cauchy-Riemann operator

$$\bar\partial_J u = J(u) \cdot Du - Du \cdot i,$$

and writing $\bar\partial_J u = 0$, so that the space of J-holomorphic maps, we are interested in then becomes

$$\widetilde{\mathcal{M}}\bigl(\mathbb{D}^2, N; J\bigr) = \bigl\{u\colon \mathbb{D}^2 \to W \bigm| \bar\partial_J u = 0 \text{ and } u\bigl(\partial\mathbb{D}^2\bigr) \subset N\bigr\}.$$

Remark 4.1. The situation of holomorphic disks is a bit special compared to the one of general holomorphic maps, because all complex structures on the disk are equivalent. If Σ were a smooth compact surface of higher genus, we would usually need to study the space of pairs (u, j), where j is a complex structure on Σ, and u is a map $u\colon (\Sigma, \partial\Sigma) \to (W, N)$ that should be (j, J)-holomorphic, that means, $J(u) \cdot Du - Du \cdot j = 0$.

To be a bit more precise, we do not choose pairs (u, j) with arbitrary complex structures j on Σ, but we only allow for j a single element in each equivalence class of complex structures: If $\varphi\colon \Sigma \to \Sigma$ is a diffeomorphism, and j is some complex structure, then of course $\varphi^* j$ will generally be a complex structure different from j, but we usually identify all complex structures up to isotopy, and use that the space of equivalence classes of complex structures can be represented as a smooth finite dimensional manifold (see [19] for a nice introduction to this theory).

Fortunately, these complications are not necessary for holomorphic disks (or spheres), and it is sufficient for us to work with the standard complex structure i on $\mathbb{D}^2$.

In this section, we want to explain the topological structure of the space $\widetilde{\mathcal{M}}(\mathbb{D}^2, N; J)$ without entering into too many technical details. Instead of starting directly with our particular case, we will try to argue on an intuitive level by considering a finite dimensional situation that has strong analogies with the problem we are dealing with.

Let us consider a vector bundle E of rank r over a smooth n-manifold B. Choose a section $\sigma\colon B \to E$, and let $M = \sigma^{-1}(0)$ be the set of points at which σ intersects the 0-section. We would "expect" M to be a smooth submanifold of dimension $\dim M = n - r$ (if $n - r < 0$, we could hope not to have any intersections at all); unfortunately, this intuitive expectation might very well be false. A sufficient condition under which it holds, is when σ is transverse to the 0-section, that means, for every $x \in M$, the tangent space to the 0-section T_xB in T_xE spans together with the image $D\sigma \cdot T_xB$ the whole tangent space T_xE. It is well-known that when the transversality condition is initially not true, it can be achieved by slightly perturbing the section σ.

Let us now come again to the Cauchy-Riemann problem. The role of B will be taken by the space of all maps $u\colon (\mathbb{D}^2, \partial\mathbb{D}^2) \to (W, N)$, which we will denote by $\mathcal{B}(\mathbb{D}^2; N)$. We do not want to spend any time thinking about the regularity of the maps and point instead to [25] as reference. It is sufficient for us to observe that the space $\mathcal{B}(\mathbb{D}^2; N)$ is a Banach manifold, that means, an infinite dimensional manifold modeled on a Banach space.

The section σ will be replaced by the Cauchy-Riemann operator $\bar\partial_J$, and before pursuing this analogy further, we want first to specify the target space of this operator. In fact, $\bar\partial_J$ associates to every map $u \in \mathcal{B}(\mathbb{D}^2; N)$ a 1-form on Σ with values in TW. The formal way to state this is that we have for every map u a vector bundle u^*TW over $\mathbb{D}^2$, which allows us to construct

$$\operatorname{Hom}\bigl(T\mathbb{D}^2, u^*TW\bigr).$$

The sections in $\operatorname{Hom}(T\mathbb{D}^2, u^*TW)$ form a vector space, and if we look at all sections for *all* maps u, we obtain a vector bundle over $\mathcal{B}(\mathbb{D}^2; N)$, whose fiber over a point u are all sections in $\mathrm{Hom}(T\mathbb{D}^2, u^*TW)$. We denote this bundle by $\mathcal{E}(\mathbb{D}^2; N)$.

The operator $\bar{\partial}_J$ associates to every u, that means, to every point of $\mathcal{B}(\mathbb{D}^2; N)$ an element in $\mathcal{E}(\mathbb{D}^2; N)$ so that we can think of $\bar{\partial}_J$ as a section in the bundle $\mathcal{E}(\mathbb{D}^2; N)$. The J-holomorphic maps are the points of $\mathcal{B}(\mathbb{D}^2; N)$ where the section $\bar{\partial}_J$ intersects the 0-section. In fact, $\bar{\partial}_J u$ is always anti-holomorphic, because

$$J(u) \cdot \bar{\partial}_J u = -Du - J(u) \cdot Du \cdot i = \big(Du \cdot i - J(u) \cdot Du\big) \cdot i = -(\bar{\partial}_J u) \cdot i,$$

and for analytical reasons we will only consider sections in $\mathrm{Hom}(T\Sigma, u^*TW)$ taking values in the subbundle $\overline{\mathrm{Hom}_{\mathbb{C}}}(T\Sigma, u^*TW)$ of anti-holomorphic homomorphisms. We denote the subbundle of sections taking values in $\overline{\mathrm{Hom}_{\mathbb{C}}}(T\Sigma, u^*TW)$ by $\bar{\mathcal{E}}_{\mathbb{C}}(\mathbb{D}^2; N)$.

4.1.1. The Expected Dimension of $\widetilde{\mathcal{M}}(\mathbb{D}^2, N; J)$. The rank of $\bar{\mathcal{E}}_{\mathbb{C}}(\mathbb{D}^2; N)$ and the dimension of $\mathcal{B}(\mathbb{D}^2; N)$ are both infinite, hence we cannot compute the expected dimension of the solution space $\widetilde{\mathcal{M}}(\mathbb{D}^2, N; J)$ as in the finite dimensional case, where it was just the difference $\dim M - \operatorname{rank} E$. Nonetheless we can associate a so called Fredholm index to a Cauchy-Riemann problem. We will later give some more details about how the index is actually defined, for now we just note that it is an integer that determines the expected dimension of the space $\widetilde{\mathcal{M}}(\mathbb{D}^2, N; J)$.

For a Cauchy-Riemann problem with totally real boundary condition the index has an easy explicit formula (see for example [25, Theorem C.1.10]) that simplifies in our specific case of holomorphic disks to

$$\operatorname{ind}_u \bar{\partial}_J = \frac{1}{2} \dim W + \mu\big(u^*TW, u^*TN\big), \tag{4.1}$$

where we have used that the Euler characteristic of a disk is $\chi(\mathbb{D}^2) = 1$.

Remark 4.2. We would like to warn the reader that the dimension of a moduli space of holomorphic disks or holomorphic spheres tends to increase, if we increase the dimension of the symplectic ambient manifold. Unfortunately, the opposite is true for a higher genus curve Σ: The formula above becomes

$$\operatorname{ind}_u \bar{\partial}_J = \frac{1}{2} \chi(\Sigma) \dim W + \mu\big(u^*TW, u^*TN\big),$$

and since the Euler characteristic is negative, and it is harder to find curves with genus in high dimensional spaces than in lower dimensional ones.

The Maslov index μ is an integer that classifies loops of totally real subspaces up to homotopy:

Definition. Let $E_{\mathbb{C}}$ be a complex vector bundle over the closed 2-disk $\mathbb{D}^2$ and let $E_{\mathbb{R}}$ be a totally real subbundle of $E_{\mathbb{C}}|_{\partial\mathbb{D}^2}$ defined only over the boundary of the disk. The **Maslov index** $\mu(E_{\mathbb{C}}, E_{\mathbb{R}})$ is an integer that is computed by trivializing $E_{\mathbb{C}}$ over the disk, and choosing a continuous frame $A(e^{i\phi}) \in \mathrm{GL}(n,\mathbb{C})$ over the boundary $\partial\mathbb{D}^2$ representing $E_{\mathbb{R}}$. We then set

$$\mu(E_{\mathbb{C}}, E_{\mathbb{R}}) := \deg \frac{\det A^2}{\det(A^* A)},$$

where $\deg(f)$ is the degree of a continuous map $f\colon \mathbb{S}^1 \to \mathbb{S}^1$.

In these notes, we will compute the Maslov index only once, in Section 4.1.3, but note that the index $\mathrm{ind}_u \bar{\partial}_J$ depends on the holomorphic disk u in $\widetilde{\mathcal{M}}(\mathbb{D}^2, N; J)$, we are considering; this should not confuse us however, because it only means that the space of disks may have different components and the expected dimensions of the different components do not need to agree.

We will now briefly explain how the index of $\bar{\partial}_J$ is defined. We have a map $\bar{\partial}_J\colon \mathcal{B}(\mathbb{D}^2; N) \to \bar{\mathcal{E}}_{\mathbb{C}}(\mathbb{D}^2; N)$, and we need to compute the linearization of $\bar{\partial}_J$ at a point of $u \in \mathcal{B}(\mathbb{D}^2; N)$, that means, we have to compute the differential

$$\bar{D}_J(u)\colon T_u\mathcal{B}\big(\mathbb{D}^2; N\big) \to T_{\bar{\partial}_J u}\bar{\mathcal{E}}_{\mathbb{C}}\big(\mathbb{D}^2; N\big).$$

To find $\bar{D}_J(u)$, choose a smooth path u_t of maps in $\mathcal{B}(\mathbb{D}^2; N)$ with $u_0 = u$, then we can regard the image $\bar{\partial}_J u_t$, and take its derivative with respect to t in $t = 0$. If we set $\dot{u}_0 = \frac{d}{dt}|_{t=0} u_t$, this allows us to obtain a linear operator $\bar{D}_J(u)$ by

$$\bar{D}_J(u) \cdot \dot{u}_0 = \frac{d}{dt}\bigg|_{t=0} \bar{\partial}_J u_t.$$

It is a good exercise to determine the domain and target space of this operator, and find a way to describe them.

The index of $\bar{\partial}_J$ at u is defined as

$$\mathrm{ind}_u \bar{\partial}_J := \dim\ker \bar{D}_J(u) - \dim\operatorname{coker} \bar{D}_J(u).$$

It is a remarkable fact that the index is finite and determined by formula (4.1) above. Also note that the index is constant on each connected component of $\mathcal{B}(\mathbb{D}^2; N)$.

4.1.2. Transversality of the Cauchy-Riemann Problem. Just as in the finite dimensional analogue, it may happen that the formal dimension we have computed does not correspond to the dimension we are observing in an actual situation. In fact, if the section σ (or in our infinite dimensional case, $\bar{\partial}_J$) are not transverse to the 0-section, there is no reason why M or $\widetilde{\mathcal{M}}(\mathbb{D}^2, N; J)$ would need to be smooth manifolds at all.

On the other hand, if σ is transverse to the 0-section, then $M = \sigma^{-1}(0)$ is a smooth submanifold of dimension $\dim M - \operatorname{rank} E$, and the analogue result is also true for the Cauchy-Riemann problem: If $\bar{\partial}_J$ is at every point of $\widetilde{\mathcal{M}}(\mathbb{D}^2, N; J)$ transverse to 0 (or said equivalently, if the cokernel of the linearized operator is trivial for every holomorphic disk), then $\widetilde{\mathcal{M}}(\mathbb{D}^2, N; J)$ will be a smooth manifold whose dimension is given by the index of $\bar{\partial}_J$.

In the finite dimensional situation, we can often achieve transversality by a small perturbation of σ, but of course, this might require a subtle analysis of the situation, when we want to perturb σ only within a space of sections satisfying certain prescribed properties.

Definition. Let $u\colon \Sigma \to W$ be a holomorphic map from a Riemann surface with or without boundary. We call u **somewhere injective**, if there exists a point $z \in \Sigma$ with $du_z \neq 0$, and such that z is the only point that is mapped by u to $u(z)$, that means,

$$u^{-1}\big(u(z)\big) = \{z\}.$$

We call a holomorphic curve that is not the multiple cover of any other holomorphic curve a **simple holomorphic curve**. Closed simple holomorphic curves are somewhere injective, [25, Proposition 2.5.1].

It is a non-trivial result that by perturbing the almost complex structure J, we can achieve transversality of the Cauchy-Riemann operator for every disk in W whose boundary is injective in a totally real submanifold N. We could hope that this theoretical result would be sufficient for us, because the considered disks are injective along their boundaries, but we have chosen a very specific almost complex structure in Section 3.2, and perturbing this J would destroy the results obtained in that section. Below, we will prove by hand that $\bar{\partial}_J$ is transverse to 0 for the holomorphic disks in our model neighborhood.

Remark 4.3. Note that often it is not possible to work only with somewhere injective holomorphic curves, and perturbing J will in that case not be sufficient to obtain transversality for holomorphic curves. Sometimes one can work around this problem by requiring that W is semi-positive, see Section 4.3. Unfortunately, there are many situations where this approach won't work either, as is the case of SFT, where transversality has been one of the most important outstanding technical problems.

4.1.3. The Bishop Family. In this section, we will show that the disks that we have found in Section 3.2.2, lying in the model neighborhood are regular solutions of the Cauchy-Riemann problem.

Before starting the actual proof of our claim, we will briefly recapitulate the situation described in Section 3.2.2. Let (W, J) be an almost complex manifold of dimension $2n$ with boundary that contains a model neighborhood U of the desired form. Remember that U was a subset of $\mathbb{C}^2 \times T^*L$ with almost complex structure $i \oplus J_L$, that we had a function $f\colon \mathbb{C}^2 \times T^*L \to [0,\infty)$ given by

$$f(z_1, z_2, \mathbf{q}, \mathbf{p}) = \frac{1}{2}\left(|z_1|^2 + |z_2|^2\right) + f_L(\mathbf{q}, \mathbf{p}),$$

and that the model neighborhood U was the subset

$$U := \left\{(z_1, z_2; \mathbf{q}, \mathbf{p}) \mid \operatorname{Re}(z_2) \geq 1 - \delta\right\} \cap f^{-1}\left([0, 1/2]\right).$$

The totally real manifold N is the image of the map

$$(z; \mathbf{q}) \in \mathbb{D}^2_\varepsilon \times L \mapsto \left(z, \sqrt{1 - |z|^2}; \mathbf{q}, \mathbf{0}\right) \subset \partial U.$$

For every pair $(s, \mathbf{q}) \in [1 - \delta, 1) \times L$, we find a holomorphic map of the form

$$\begin{aligned} u_{s,\mathbf{q}}\colon \left(\mathbb{D}^2, \partial\mathbb{D}^2\right) &\to U \\ z &\mapsto (C_s z, s; \mathbf{q}, \mathbf{0}) \end{aligned}$$

with $C_s = \sqrt{1 - s^2}$. We call this map a **(parametrized) Bishop disk**, and we call the collection of these disks, the **Bishop family**. Sometimes we will not be precise about whether the disks are parametrized or not, and whether we speak about disks with or without a marked point (see Section 4.2), but we hope that in each situation it will be clear what is meant.

To check that a given Bishop disk $u_{s,\mathbf{q}}$ is regular, we will first compute the index of the linearized Cauchy-Riemann operator that gives us the expected dimension for the space of holomorphic disks containing the Bishop family. Note that the observed dimension is $1 + \dim L + 3 = 1 + (n - 2) + 3 = n + 2$.

The first part, $1+\dim L$ corresponds to the s- and $\mathbf{q}$-parameters of the family; the three corresponds to the dimension of the group of Möbius transformations acting on the complex unit disk: If $u_{s,\mathbf{q}}$ is a Bishop disk, and if $\varphi\colon \mathbb{D}^2 \to \mathbb{D}^2$ is a Möbius transformation, then of course $u_{s,\mathbf{q}} \circ \varphi$ will also be a holomorphic map with admissible boundary condition. On the other hand we showed in Corollary 3.20 that every holomorphic disk that lies in U is up to a Möbius transformation one of the Bishop disks.

For the index computations, it suffices by Section 4.1.1 to trivialize the bundle $E_{\mathbb{C}} := u_{s,\mathbf{q}}^* TW$ over $\mathbb{D}^2$, and study the topology of the totally real subbundle $E_{\mathbb{R}} = u_{s,\mathbf{q}}^* TN$ over $\partial\mathbb{D}^2$.

Before starting any concrete computations, we will significantly simplify the setup by choosing a particular chart: Note that the T^*L-part of a Bishop disk $u_{s,\mathbf{q}}$ is constant, we can hence choose a chart diffeomorphic to $\mathbb{R}^{2n-4} = \{(x_1,\dots,x_{n-2};y_1,\dots,y_{n-2})\}$ for T^*L with the properties

- the point $(\mathbf{q},\mathbf{0})$ corresponds to the origin,
- the almost complex structure J_L is represented at the origin by the standard i,
- the intersections of the 0-section L with the chart corresponds to the subspace $(x_1,\dots,x_{n-2};0,\dots,0)$.

In the chosen chart, we write $u_{s,\mathbf{q}}$ as

$$u_{s,\mathbf{q}}(z) = (C_s z, s; 0,\dots,0) \in \mathbb{C}^2 \times \mathbb{R}^{2n-4}$$

with $C_s = \sqrt{1-s^2}$. By our assumption, the complex structure on the second factor is at the origin of $\mathbb{R}^{2n-4}$ equal to i, and there is then a direct identification of $u_{s,\mathbf{q}}^* TW$ with $\mathbb{C}^2 \times \mathbb{C}^{n-2}$. The submanifold N corresponds in the chart to

$$\left\{(z_1,z_2;x_1,\dots,x_{n-2},0,\dots,0) \in \mathbb{C}^2 \times \mathbb{R}^{2n-4} \,\middle|\, \operatorname{Im} z_2 = 0, |z_1|^2 + |z_2|^2 = 1\right\}.$$

The boundary of $u_{s,\mathbf{q}}$ is given by $e^{i\varphi} \mapsto (\sqrt{1-s^2}e^{i\varphi}, s; 0,\dots,0)$, and the tangent space of TN over this loop is spanned over $\mathbb{R}$ by the vector fields

$$(ie^{i\varphi},0;0,\dots,0), \left(-\frac{s}{\sqrt{1-s^2}}e^{i\varphi},1;0,\dots,0\right), (0,0;1,0,\dots,0),\dots,$$
$$(0,0;0,\dots,0,1,0,\dots,0).$$

We can now easily compute the Maslov index $\mu(E_{\mathbb{C}}, E_{\mathbb{R}})$ as

$$\deg \frac{\det A^2}{\det(A^*A)} = \deg \frac{-e^{2i\varphi}}{1} = 2,$$

where A is the matrix composed by the vector fields given above. Hence we obtain for the index

$$\operatorname{ind}_u \bar{\partial}_J = \frac{1}{2} \dim W + \mu\big(u_{s,\mathbf{q}}^* TW, u_{s.\mathbf{q}}^* TN\big) = n+2,$$

which corresponds to the observed dimension computed above.

We will now show that the linearized operator $\bar{D}_J$ is surjective. We do not do this directly, but we compute instead the dimension of its kernel, and show that it is equal (and not larger than) the Fredholm index. From the definition of the index

$$\operatorname{ind}_u \bar{\partial}_J := \ker \bar{D}_J(u) - \operatorname{coker} \bar{D}_J(u),$$

we see that the cokernel needs to be trivial, and this way the surjectivity result follows.

We now compute the linearized Cauchy-Riemann operator at a Bishop disk $u_{s,\mathbf{q}}$. Let v_t be a smooth family of maps

$$v_t \colon \big(\mathbb{D}^2, \partial \mathbb{D}^2\big) \to (U, N)$$

with $v_0 = u_{s,\mathbf{q}}$ (think of each v_t as a smooth map, but for an analytically correct study, we would need to allow here for Sobolev maps).

In this chart, we can write the family v_t as

$$v_t(z) = \big(z_1(z,t), z_2(z,t); \mathbf{x}(z,t), \mathbf{y}(z,t)\big) \in \mathbb{C}^2 \times \mathbb{R}^{2n-4},$$

where we have set $\mathbf{x}(z,t) = (x_1(z,t), \dots, x_{n-2}(z,t))$ and $\mathbf{y}(z,t) = (y_1(z,t), \dots, y_{n-2}(z,t))$, and we require that the boundary of each of the v_t has to lie in N. When we now take the derivative of v_t with respect to t at $t=0$, we obtain a vector in $T_{u_{s,\mathbf{q}}}\mathcal{B}$ that is represented by a map

$$\dot{v}_0 \colon \mathbb{D}^2 \to \mathbb{C}^2 \times \mathbb{R}^{2(n-2)}, \quad z \mapsto \big(\dot{z}_1(z), \dot{z}_2(z); \dot{\mathbf{x}}(z), \dot{\mathbf{y}}(z)\big)$$

with boundary conditions $\dot{\mathbf{y}}(z) = \mathbf{0}$ and $\operatorname{Im} \dot{z}_2(z) = 0$ for every $z \in \partial \mathbb{D}^2$. Furthermore taking the derivative of $|z_1(z,t)|^2 + |z_2(z,t)|^2 = 1$ for every $z \in \partial \mathbb{D}^2$ with respect to t, we obtain the condition

$$\bar{z}_1(z,0) \cdot \dot{z}_1(z) + z_1(z,0) \cdot \dot{\bar{z}}_1(z) + \bar{z}_2(z,0) \cdot \dot{z}_2(z) + z_2(z,0) \cdot \dot{\bar{z}}_2(z) = 0,$$

which simplifies by using the explicit form of $(z_1(z,0), z_2(z,0))$ to

$$C_s\bar{z}\cdot \dot{z}_1(z) + C_s z\cdot \dot{\bar{z}}_1(z) + s\dot{z}_2(z) + s\dot{\bar{z}}_2(z) = 0$$

for every $z\in\partial\mathbb{D}^2$.

The linearization of the Cauchy-Riemann operator $\bar\partial_J$ at $u_{s,\mathbf{q}}$ given by

$$\bar{D}_J\cdot \dot{v}_0 := \left.\frac{d}{dt}\right|_{t=0} \bar\partial_J v_s$$

decomposes into the $\mathbb{C}^2$-part

$$(id\dot{z}_1 - d\dot{z}_1 i, id\dot{z}_2 - d\dot{z}_2 i)$$

and the $\mathbb{R}^{2(n-2)}$-part

$$\left.\frac{d}{dt}\right|_{t=0} \big(J_L\big(\mathbf{x}(z,t),\mathbf{y}(z,t)\big)\cdot\big(d\mathbf{x}(z,t), d\mathbf{y}(z,t)\big) - \big(d\mathbf{x}(z,t)\cdot i, d\mathbf{y}(z,t)\cdot i\big)\big).$$

The second part can be significantly simplified by using first the product rule, and applying then that $\mathbf{x}(z,0)=\mathbf{0}$ and $\mathbf{y}(z,0)=\mathbf{0}$ are constant so that their differentials vanish. We obtain then

$$J_L(\mathbf{0},\mathbf{0})\cdot(d\dot{\mathbf{x}}, d\dot{\mathbf{y}}) - (d\dot{\mathbf{x}}\cdot i, d\dot{\mathbf{y}}\cdot i),$$

and using that $J_L(\mathbf{0},\mathbf{0}) = i$, it finally reduces to

$$(d\dot{\mathbf{y}} - d\dot{\mathbf{x}}\cdot i, -d\dot{\mathbf{x}} - d\dot{\mathbf{y}}\cdot i).$$

We have shown that linearized Cauchy-Riemann operator simplifies for all coordinates to the standard Cauchy-Riemann operator, so that if $\dot{v}_0(z) = (\dot{z}_1(z), \dot{z}_2(z); \dot{\mathbf{x}}(z), \dot{\mathbf{y}}(z))$ lies in the kernel of $\bar{D}_J$ then the coordinate functions $\dot{z}_1(z), \dot{z}_2(z)$ and $\dot{\mathbf{x}}(z) + i\dot{\mathbf{y}}(z)$ need all to be holomorphic in the classical sense.

Now using the boundary conditions, we easily deduce that $\dot{\mathbf{y}}(z)$ needs to vanish, because it is a harmonic function, and it takes both maximum and minimum on $\partial\mathbb{D}^2$. A direct consequence of $\dot{\mathbf{y}}\equiv\mathbf{0}$ and the Cauchy-Riemann equation is that $\dot{\mathbf{x}}(z)$ will be everywhere constant. We get the analogous result for the function $\dot{z}_2(z)$, so that we can write

$$\dot{v}_0(z) = \big(\dot{z}_1(z), \dot{s}; \dot{\mathbf{q}}_0, \mathbf{0}\big),$$

where $\dot{s}$ is a real constant, and $\dot{\mathbf{q}}_0$ is a fixed vector in $\mathbb{R}^{2(n-2)}$, and we only need to still understand the holomorphic function $\dot{z}_1(z)$.

The boundary condition for $\dot{z}_1(z)$ is $\bar{z} \cdot \dot{z}_1(z) + z \cdot \dot{\bar{z}}_1(z) = -\frac{2s\dot{s}}{C_s}$ for every $z \in \partial\mathbb{D}^2$. Using that the function $\dot{z}_1(z)$ is holomorphic, we can write it as power series in the form

$$\dot{z}_1(z) = \sum_{k=0}^{\infty} a_k z^k$$

and we get at $e^{i\varphi} \in \partial\mathbb{D}^2$

$$\dot{z}_1\left(e^{i\varphi}\right) = \sum_{k=0}^{\infty} a_k\, e^{ik\varphi}.$$

Plugging these series into the equation of the boundary condition, we find

$$e^{-i\varphi} \cdot \sum_{k=0}^{\infty} a_k\, e^{ik\varphi} + e^{i\varphi} \cdot \sum_{k=0}^{\infty} \bar{a}_k\, e^{-ik\varphi} = -\frac{2s\dot{s}}{C_s}$$

so that

$$\sum_{k=0}^{\infty} \left(a_k\, e^{(k-1)\, i\varphi} + \bar{a}_k\, e^{-(k-1)\, i\varphi}\right) = -\frac{2s\dot{s}}{C_s}$$

and by comparing coefficients we see that

$$a_1 + \bar{a}_1 = -\frac{2s\dot{s}}{C_s}, \qquad a_0 + \bar{a}_2 = 0, \qquad a_k = 0 \quad \text{for all } k \geq 3.$$

This means that the three (real) parameters we can choose freely are z_0 and $\operatorname{Im} z_1$.

Concluding, we have found that the dimension of the kernel of $\bar{D}_J$ is equal to $3 + 1 + n - 2 = n + 2$ which corresponds to the Fredholm index of our problem. Thus there is no need to perturb J on the neighborhood of the Bishop family to obtain regularity.

Corollary 4.4. *Let (W, ω) be a compact symplectic manifold that is a weak symplectic filling of a contact manifold (M, ξ). Suppose that N is either a* Lob *or a* bLob *in M, then we can choose close to the binding and to the boundary of N the almost complex structure described in the previous sections, and extend it to an almost complex structure J that is tamed by ω, whose bundle of complex tangencies along M is ξ and that makes M J-convex. By a generic perturbation away from the binding and the boundary of N, we can achieve that all somewhere injective holomorphic curves become regular.*

We call a J with these properties an ***almost complex structure adapted to*** *N.*

The argument in the proof of the corollary above is that the Bishop disks are already regular, and that all other simple holomorphic curves have to lie outside the neighborhood where we require an explicit form for J. Thus it suffices to perturb outside these domains to obtain regularity for every other simple curve.

4.2. The Moduli Space of Holomorphic Disks with a Marked Point

Until now, we only have studied the space of certain J-holomorphic *maps*

$$\widetilde{\mathcal{M}}\big(\mathbb{D}^2, N; J\big) = \big\{u\colon \mathbb{D}^2 \to W \mid \bar\partial_J u = 0 \text{ and } u\big(\partial\mathbb{D}^2\big) \subset N\big\},$$

but many maps correspond to different parametrizations of the same geometric disk. To get rid of this ambiguity (and to obtain compactness), we quotient the space of maps by the biholomorphic reparametrizations of the unit disk, that means, by the Möbius transformations, but we will also add a marked point $z_0 \in \mathbb{D}^2$ to preserve the structure of the geometric disk. To simplify the notation, we will also omit the almost complex structure J in $\widetilde{\mathcal{M}}(\mathbb{D}^2, N)$.

From now on let

$$\begin{aligned}\widetilde{\mathcal{M}}\big(\mathbb{D}^2, N; z_0\big) &= \big\{(u, z_0) \mid z_0 \in \mathbb{D}^2,\ \bar\partial_J u = 0 \text{ and } u\big(\partial\mathbb{D}^2\big) \subset N\big\} \\ &= \widetilde{\mathcal{M}}\big(\mathbb{D}^2, N\big) \times \mathbb{D}^2\end{aligned}$$

be the space of holomorphic maps together with a special point $z_0 \in \mathbb{D}^2$ that will be called the **marked point**. The **moduli space** we are interested in is the space of equivalence classes

$$\mathcal{M}\big(\mathbb{D}^2, N; z_0\big) = \widetilde{\mathcal{M}}\big(\mathbb{D}^2, N; z_0\big)/\sim$$

where we identify two elements (u, z_0) and (u', z_0'), if and only if there is a biholomorphism $\varphi\colon \mathbb{D}^2 \to \mathbb{D}^2$ such that $u = u' \circ \varphi^{-1}$ and $z_0 = \varphi(z_0')$. The map $(u, z) \mapsto u(z)$ descends to a well defined map

$$\begin{aligned}\mathrm{ev}\colon \mathcal{M}\big(\mathbb{D}^2, N; z_0\big) &\to W \\ [u, z_0] &\mapsto u(z_0)\end{aligned}$$

on the moduli space, which we call the **evaluation map**.

Let N be a Lob or a bLob, and assume that B_0 is one of the components of the binding of N. Since this is the only situation, we are really interested in these notes, we introduce the notation $\widetilde{\mathcal{M}}_0(\mathbb{D}^2, N)$ for the connected component in $\widetilde{\mathcal{M}}(\mathbb{D}^2, N)$ that contains the Bishop family around B_0. When adding a marked point, we write $\widetilde{\mathcal{M}}_0(\mathbb{D}^2, N; z_0)$ and $\mathcal{M}_0(\mathbb{D}^2, N; z_0)$ for the corresponding subspaces.

It is easy to see that $\mathcal{M}_0(\mathbb{D}^2, N; z_0)$ is a smooth (non-compact) manifold with boundary. Note first that $\widetilde{\mathcal{M}}_0(\mathbb{D}^2, N; z_0)$ is also a smooth and non-compact manifold with boundary: If J is regular, we know that $\widetilde{\mathcal{M}}_0(\mathbb{D}^2, N)$ is a smooth manifold, and so the boundary of the product manifold $\widetilde{\mathcal{M}}_0(\mathbb{D}^2, N; z_0)$ is

$$\partial\widetilde{\mathcal{M}}_0\big(\mathbb{D}^2, N; z_0\big) = \widetilde{\mathcal{M}}_0\big(\mathbb{D}^2, N\big) \times \partial\mathbb{D}^2.$$

Passing to the quotient preserves this structure, because the boundary of the maps in $\widetilde{\mathcal{M}}_0(\mathbb{D}^2, N)$ intersects each of the pages of the open book exactly once (this is a consequence of Corollary 3.11 and Section 3.2.2), and hence each of the disks is injective along its boundary. The only Möbius transformation that preserves the boundary pointwise is the identity, hence it follows that the group of Möbius transformations acts smoothly, freely and properly on $\widetilde{\mathcal{M}}_0(\mathbb{D}^2, N; z_0)$, and hence the quotient will be a smooth manifold of dimension

$$\dim \mathcal{M}_0\big(\mathbb{D}^2, N; z_0\big) = \dim \widetilde{\mathcal{M}}_0\big(\mathbb{D}^2, N; z_0\big) - 3 = \operatorname{ind}_u \bar{\partial}_J + 2 - 3 = n + 1.$$

As before the points on the boundary of $\mathcal{M}_0(\mathbb{D}^2, N; z_0)$ are the classes $[u, z]$ with $z \in \partial\mathbb{D}^2$. It is also clear that the evaluation map $\mathrm{ev}_{z_0} : \mathcal{M}_0(\mathbb{D}^2, N; z_0) \to W$ is smooth.

Remember that the Bishop disks contract to points as they approach the binding B_0. We will show that we incorporate B_0 into the moduli space $\mathcal{M}_0(\mathbb{D}^2, N; z_0)$ and that the resulting space carries a natural smooth structure that corresponds to the intuitive picture of disks collapsing to one point.

The neighborhood of the binding B_0 in W is diffeomorphic to the model

$$U = \big\{(z_1, z_2; \mathbf{q}, \mathbf{p}) \in \mathbb{C}^2 \times T^*B_0 \mid \operatorname{Re}(z_2) > 1 - \delta\big\} \cap h^{-1}\big((-\infty, 1/2]\big)$$

for small $\delta > 0$ with the function

$$h(z_1, z_2) = \frac{1}{2}\big(|z_1|^2 + |z_2|^2\big) + f_{B_0}(\mathbf{q}, \mathbf{p}),$$

see Section 3.2.2.

The content of Proposition 3.21 and of Corollary 3.20 is that for every point

$$(z, s; \mathbf{q}_0, \mathbf{0}) \in U$$

with $s \in (1 - \delta, 1)$ and $\mathbf{q}_0$ in the 0-section of T^*B_0,

- there is up to a Möbius transformation a unique holomorphic map $u \in \widetilde{\mathcal{M}}_0(\mathbb{D}^2, N)$ containing that point in its image, and
- $\widetilde{\mathcal{M}}(\mathbb{D}^2, N)$ does not contain any holomorphic maps whose image is not entirely contained in $U \cap (\mathbb{C} \times \mathbb{R} \times B_0)$.

As a result, it follows that $V = \mathrm{ev}_{z_0}^{-1}(U)$ is an open subset of $\mathcal{M}_0(\mathbb{D}^2, N; z_0)$, and that the restriction of the evaluation map

$$\mathrm{ev}_{z_0}|_V : V \to U$$

is a diffeomorphism onto $U \cap (\mathbb{C} \times (1 - \delta, 1) \times B_0)$. The closure of this subset is the smooth submanifold

$$U \cap (\mathbb{C} \times \mathbb{R} \times B_0),$$

which we obtain by including the binding $\{0\} \times \{1\} \times B_0$ of N.

Using the evaluation map, we can identify V with its image in U, and this way glue B_0 to the moduli space $\mathcal{M}_0(\mathbb{D}^2, N; z_0)$. The new space is also a smooth manifold with boundary, and the evaluation map extends to it, and is a diffeomorphism onto its image in U so that we can effectively identify U with a subset of the moduli space. In particular, it follows that B_0 is a submanifold that is of codimension 2 in the boundary of the moduli space.

The aim of the next section will consist in studying the Gromov compactification of $\mathcal{M}_0(\mathbb{D}^2, N; z_0)$.

4.3. Compactness

Gromov compactness is a result that describes the possible limits of a sequence of holomorphic curves, and ensures under certain conditions that every such sequence contains a converging subsequence. In the limit, a given sequence of holomorphic curves may break into several components, called **bubbles**, each of which is again a holomorphic curve. We will not describe in detail what "convergence" in this sense really means, but we only sketch the idea: The holomorphic curves in a moduli space can be represented by holomorphic maps, and in the optimal case, one could hope that by choosing for each curve in the given sequence a suitable representative, we might have uniform convergence of the maps, and this way we would find the limit of the sequence as a proper holomorphic curve. Unfortunately, this is usually wrong, but it might be true that for the correct choice of parametrization we have convergence on subdomains. Choosing different reparametrizations, we then obtain convergence on different domains, and each such domain gives then rise to a bubble, that means, a holomorphic curve that represents one component of the Gromov limit.

Theorem 4.5 (Gromov compactness). *Let (W, J) be a compact almost complex manifold (with or without boundary), and assume that J is tamed by a symplectic form ω. Let L be a compact totally real submanifold. Choose a sequence of J-holomorphic maps $u_k \colon (\mathbb{D}^2, \partial\mathbb{D}^2) \to (W, L)$ whose ω-**energy***

$$E(u_k) := \int_{\mathbb{D}^2} u_k^* \omega$$

is bounded by a constant $C > 0$.

Then there is a subsequence of $(u_{k_l})_l$ that converges in the Gromov sense to a bubble tree composed of a finite family of non-constant holomorphic disks $u_\infty^{(1)}, \dots, u_\infty^{(K)}$ whose boundary lies in L, and a finite family of non-constant holomorphic spheres $v_\infty^{(1)}, \dots, v_\infty^{(K')}$. The total energy is preserved so that

$$\lim_{l \to \infty} E(u_{k_l}) = \sum_{j=1}^{K} E\big(u_\infty^{(j)}\big) + \sum_{j=1}^{K'} E\big(v_\infty^{(j)}\big).$$

If each of the disks u_k is equipped with a marked point $z_k \in \mathbb{D}^2$, then after possibly reducing to a another subsequence, there is a marked point z_∞ on one of the components of the bubble tree such that $\lim_k z_k = z_\infty$ in a suitable sense.

The ω-energy is fundamental in the proof of the compactness theorem to limit the number of possible bubbles: By [25, Proposition 4.1.4], there exists in the situation of Theorem 4.5 a constant $\hbar > 0$ that bounds the energy of every holomorphic sphere or every holomorphic disk $u_k \colon (\mathbb{D}^2, \partial\mathbb{D}^2) \to (W, L)$ from below. Since every bubble needs to have at least an $\hbar$-quantum of energy, and since the total energy of the curves in the sequence is bounded by C, the limit curve will never break into more than $C/\hbar$ bubbles (the upper bound of the energy is also used to make sure that each bubble is a compact surface).

We will show in the rest of this section that we can apply Gromov compactness to sequences of holomorphic disks lying in the moduli space $\mathcal{M}_0(\mathbb{D}^2, N)$ studied in the previous section, and how we can incorporate these limits into $\mathcal{M}_0(\mathbb{D}^2, N; z_0)$ to construct the compactification $\overline{\mathcal{M}}_0(\mathbb{D}^2, N; z_0)$.

Proposition 4.6. *Let N be a* Lob *or a* bLob *in the contact boundary (M,ξ) of a symplectic filling (W,ω), and assume that we find a contact form α for ξ such that $\omega|_{TN} = d\alpha|_{TN}$.*

There is a global energy bound $C > 0$ for all holomorphic disks in $\widetilde{\mathcal{M}}_0(\mathbb{D}^2, N)$.

Proof. There is a slight complication in our proof, because we may not assume that ω is globally exact, which would allow us to obtain the energy of a holomorphic disk by integrating over the boundary of the disk. To prove the desired statement, proceed as follows: Let $u\colon (\mathbb{D}^2, \partial\mathbb{D}^2) \to (W, N)$ be any element in $\widetilde{\mathcal{M}}_0(\mathbb{D}^2, N)$. By our assumption, there exists a smooth path of maps u_t that starts at the constant map $u_0(z) \equiv b_0 \in B_0$ in the binding and ends at the chosen map $u_1 = u$. This family of disks may be interpreted as a map from the 3-ball into W. The boundary consists of the image of u_1, and the union of the boundary of all disks $u_t|_{\partial\mathbb{D}^2}$.

Using Stokes' theorem, we get

$$0 = \int_{[0,1]\times\mathbb{D}^2} u_t^* d\omega = \int_{\mathbb{D}^2} u_1^*\omega + \int_{[0,1]\times\partial\mathbb{D}^2} u_t^*\omega$$

so that $E(u) = -\int_{[0,1]\times\partial\mathbb{D}^2} u_t^*\omega$.

By our assumption, we have a contact form on the contact boundary M for which $\omega|_{TN} = d\alpha|_{TN}$, so that using Stokes' theorem a second time (and that $u_0(z) = b_0$) we get

$$E(u) = \int_{\partial\mathbb{D}^2} u^*\alpha.$$

The Legendrian foliation on N is an open book whose pages are fibers of a fibration $\vartheta\colon N \setminus B \to \mathbb{S}^1$. Hence the 1-form $d\vartheta$ and $\alpha|_{TN}$ have the same kernel, and it follows that there exists a smooth function $f\colon N \to [0,\infty)$ such that

$$\alpha|_{TN} = f\, d\vartheta.$$

The function f vanishes on the binding and on the boundary of a bLob, and f is hence bounded on N so that we define $C := 2\pi \max_{x\in N} |f(x)|$.

Using that the boundary of u intersects every leaf of the open book exactly once, we obtain for the energy of u the estimate

$$E(u) = \int_{\partial\mathbb{D}^2} u^*\alpha \le \max_{x\in N} |f(x)| \int_{\partial\mathbb{D}^2} u^* d\vartheta \le 2\pi \max_{x\in N} |f(x)| = C. \quad \blacksquare$$

With the given energy bound, we obtain now Gromov compactness in form of the following corollary.

Corollary 4.7. *Let N be a* Lob *or a* bLob *in the contact boundary (M,ξ) of a symplectic filling (W,ω), and assume that we find a contact form α for ξ such that $\omega|_{TN} = d\alpha|_{TN}$. Let $(u_k)_k$ be a sequence of holomorphic maps in $\widetilde{\mathcal{M}}_0(\mathbb{D}^2,N)$.*

There exists a subsequence $(u_{k_l})_l$ that converges either

- *uniformly up to reparametrizations of the domain to a J-holomorphic map $u_\infty \in \widetilde{\mathcal{M}}_0(\mathbb{D}^2,N)$,*
- *to a constant disk $u_\infty(z) \equiv b_0$ lying in the binding of N,*
- *or to a bubble tree composed of a single holomorphic disk $u_\infty : (\mathbb{D}^2,\partial\mathbb{D}^2) \to (W,N)$ and a finite family of non-constant holomorphic spheres $v_1,\dots,v_j$ with $j \ge 1$.*

Proof. We will apply Theorem 4.5. The submanifold N is not totally real along the binding B and ∂N, but we simply remove a small open neighborhood of both sets. By Proposition 3.24, none of the holomorphic disks u_k may get close to ∂N, and by Proposition 3.21 we know precisely how the curves look like that intersect a neighborhood of B. If we find disks in $(u_k)_k$ that get arbitrarily close to the binding of N, then using that B is compact, we may choose a subsequence that converges to a single point in the binding. If $(u_k)_k$ stays at finite distance from B, we may assume that the neighborhood, we have removed from N is so small that the holomorphic disks we are studying all lie inside.

If the sequence $(u_k)_k$ does not contain any subsequence that can be reparametrized in such a way that it converges to a single non-constant disk u_∞, we use Gromov compactness to obtain a subsequence that splits into a finite collection of holomorphic spheres and disks. But as a consequence from Corollary 3.11, we see that non-constant holomorphic disks attached to N need to intersect the pages of the open book transversely in positive direction. A sequence of holomorphic disks that intersects every page of the open book exactly once, cannot split into several disks intersecting pages several times. In particular possible bubble trees contain by this argument a single disk in its limit. ■

Above, we have obtained compactness for a sequence of disks, but we would like to understand how these limits can be incorporated into the moduli space. Adding the bubble trees to the space of parametrized maps does not give rise to a valid topology, because the bubbling phenomenon can only be understood by using different reparametrizations of the disk to recover all components of the bubble tree.

We will denote the compactification of $\mathcal{M}_0(\mathbb{D}^2, N; z_0)$ by $\overline{\mathcal{M}}_0(\mathbb{D}^2, N; z_0)$. For us, it is not necessary to understand the topology of $\overline{\mathcal{M}}_0(\mathbb{D}^2, N; z_0)$ in detail, but it will be sufficient to see that bubbling is a "codimension 2 phenomenon". In fact, it is not the topology of the moduli space itself we are interested in, but our aim is to obtain information about the symplectic manifold. For this we want to make sure that the image under the evaluation map of all bubble trees that appear in the limit, that means, of $\overline{\mathcal{M}}_0(\mathbb{D}^2, N; z_0) \setminus \mathcal{M}_0(\mathbb{D}^2, N; z_0)$ is contained in the image of a smooth map defined on a finite union of manifolds each of dimension at most

$$\dim \mathcal{M}_0\big(\mathbb{D}^2, N; z_0\big) - 2.$$

For this to be true, we need to impose additional conditions for (W, ω).

Definition. A $2n$-dimensional symplectic manifold (M, ω) is called

- **symplectically aspherical**, if $\omega([A])$ vanishes for every $A \in \pi_2(M)$.
- It is called **semipositive** if every $A \in \pi_2(M)$ with $\omega([A]) > 0$ and $c_1(A) \geq 3 - n$ has non-negative Chern number.

Note that every symplectic 4- or 6-manifold is obviously semipositive.

In a symplectically aspherical manifold no J-holomorphic spheres exist, because their energy would be zero. So in particular they may not appear in any bubble tree and Corollary 4.7 implies in our situation that every sequence of holomorphic disks contains a subsequence that either collapses into the binding or that converges to a single disk in $\mathcal{M}_0(\mathbb{D}^2, N; z_0)$. Using the results of Section 4.2, we obtain the following corollary.

Corollary 4.8. *Let (W, ω) be a compact symplectically aspherical manifold that is a weak filling of a contact manifold (M, ξ). Let N be a* Lob *or a* bLob *in* M, *and assume that we find a contact form for* ξ *such that* $\omega|_{TN} = d\alpha|_{TN}$. *Choose an almost complex structure* J *that is adapted to* N *(as in Corollary* 4.4*).*

Then the compactification of the moduli space $\mathcal{M}_0(\mathbb{D}^2, N; z_0)$ *is a smooth compact manifold*

$$\overline{\mathcal{M}}_0(\mathbb{D}^2, N; z_0) = \mathcal{M}_0(\mathbb{D}^2, N; z_0) \cup (\textit{binding of } N)$$

with boundary. The binding of N *is a submanifold of codimension* 2 *in the boundary* $\partial\overline{\mathcal{M}}_0(\mathbb{D}^2, N; z_0)$.

The condition of asphericity is very strong, and we will obtain more general results by studying instead semipositive manifolds. The important point here is that a generic almost complex structure only ensure transversality for somewhere injective holomorphic curves, see Section 4.1.2. Even though the holomorphic disks in $\mathcal{M}_0(\mathbb{D}^2, N; z_0)$ are simple, it could happen that once the disks bubble, there appear spheres that are multiple covers. For these, we cannot guarantee transversality, and hence we cannot directly predict if the compactification of $\mathcal{M}_0(\mathbb{D}^2, N; z_0)$ consists of adding "codimension 2 strata" or if we will be forced to include too many bubble trees.

Still, we know that every sphere that is not simple is the multiple cover of a simple one (by the Riemann-Hurwitz formula a sphere can only multiply cover a sphere), we can hence compute the dimension of the moduli space of the underlying simple spheres, and use this information as an upper bound for the dimension of the spheres that appear in the bubble tree.

Let $v\colon \mathbb{S}^2 \to W$ be a holomorphic sphere that is a k-fold cover of a sphere $\widetilde{v}$ representing a homology class $[v]$ and $[\widetilde{v}] \in H_2(W, \mathbb{Z})$ respectively with $[v] = k[\widetilde{v}]$ and with $\omega([\widetilde{v}]) > 0$. The expected dimension of the space of maps containing v is by an index formula

$$\operatorname{ind}_v \bar{\partial}_J = 2n + 2c_1\big([v]\big) = 2n + 2kc_1\big([\widetilde{v}]\big).$$

The space of biholomorphisms of $\mathbb{S}^2$ has dimension 6, and hence the expected dimension of the moduli space of unparametrized spheres that contain $[v]$ is $\operatorname{ind}_v \bar{\partial}_J - 6 = 2(n-3) + 2kc_1([\widetilde{v}])$.

As we explained above and in Section 4.1.2, this expected dimension does not correspond in general to the observed dimension of the bubble trees, instead we study the expected dimension of the underlying simple spheres. The dimension of the space containing $\widetilde{v}$ is given by $\operatorname{ind}_{\widetilde{v}} \bar{\partial}_J - 6 =$

$2(n-3)+2c_1([\widetilde{v}])$. If $c_1([\widetilde{v}])<n-3$, then the expected dimension will be negative, and since we obtain regularity of all simple holomorphic curves by choosing a generic almost complex structure, it follows that the moduli space containing $\widetilde{v}$ is generically empty. As a consequence bubble trees appearing as limits do not contain any component that is the k-fold cover of a simple sphere representing the homology class $[\widetilde{v}]$.

If $c_1([\widetilde{v}])\geq n-3$, the definition of semipositivity implies that $c_1([\widetilde{v}])\geq 0$. When we compare the expected dimension of the moduli space containing v with the one of the underlying disk $\widetilde{v}$, we observe that $\operatorname{ind}_v\bar{\partial}_J-6=2(n-3)+2kc_1([\widetilde{v}])\geq 2(n-3)+2c_1([\widetilde{v}])=\operatorname{ind}_{\widetilde{v}}\bar{\partial}_J-6$.

Consider now the image in W of all spheres in the moduli space of v that are k-fold multiple covers of some simple sphere. Their image is contained in the image of the simple spheres lying in the same moduli space as $\widetilde{v}$. The dimension of this second moduli space is smaller or equal than the expected dimension of the initial moduli space containing v, and even though we cannot ensure regularity for v, we have an estimate on the dimension of the subset containing all singular spheres.

The following result allows us to find the desired bound for the dimension of the image of complete bubble trees.

Proposition 4.9. *Assume that (W,ω) is semipositive. To compactify the moduli space $\mathcal{M}_0(W,N,z_0)$, one has to add bubbled curves. We find a finite set of manifolds $X_1,\dots,X_N$ with $\dim X_j\leq\dim\mathcal{M}_0(W,N,z_0)-2$ and smooth maps $f_j\colon X_j\to W$ such that the image of the bubbled curves under the evaluation map ev_{z_0} is contained in*

$$\bigcup f_j(X_j).$$

When we consider instead the compactification of the boundary $\partial\mathcal{M}_0(W,N,z_0)$, that means the space of holomorphic disks with a marked point on the boundary of the disk only, then we obtain the analogue result, only that the manifolds $X_1,\dots,X_N$ have dimension $\dim X_j\leq\dim\partial\mathcal{M}_0(W,N,z_0)-2=\dim\mathcal{M}_0(W,N,z_0)-3$.

Proof. The standard way to treat bubbled curves consists in considering them as elements in a bubble tree: Here such a tree is composed by a simple holomorphic disk $u_0\colon(\mathbb{D}^2,\mathbb{S}^1)\to(W,N)$ and holomorphic spheres $u_1,\dots,u_{k'}\colon\mathbb{S}^2\to W$. These holomorphic curves are connected to each other in a certain way. We formalize this relation by saying that the holomorphic curves are vertices in a tree, i.e. in a connected graph without cycles. We denote the edges of this graph by $\{u_i, u_j\}$, $0 \le i < j \le k'$.

Now we assign to any edge two nodal points z_{ij} and z_{ji}, the first one in the domain of the bubble u_i, the other one in the domain of u_j, and we require that $\mathrm{ev}_{z_{ij}}(u_i) = \mathrm{ev}_{z_{ji}}(u_j)$. For technical reasons, we also require nodal points on each holomorphic curve to be pairwise distinct. To include into the theory, trees with more than one bubble connected at the same point to a holomorphic curve, we add "ghost bubbles". These are constant holomorphic spheres inserted at the point where several bubbles are joined to a single curve. Now all the links at that point are opened and reattached at the ghost bubble. Ghost bubbles are the only constant holomorphic spheres we allow in a bubble tree.

The aim is to give a manifold structure to these bubble trees, but unfortunately this is not always possible, when multiply covered spheres appear in the bubble tree.

Instead, we note that the image of every bubble tree is equal to the image of a simple bubble tree, that means, to a tree, where every holomorphic sphere is simple and any two spheres have different image. Since we are only interested in the image of the evaluation map on the bubble trees, it is for our purposes equivalent to consider the simple bubble tree instead of the original one. The disk u_0 is always simple, and does not need to be replaced by another simple curve.

Let $u_0, u_1, \ldots, u_{k'}$ be the holomorphic curves composing the original bubble tree, and let $A_i \in H_2(W)$ be the homology class represented by the holomorphic sphere u_i. The simple tree is composed by $u_0, v_1, \ldots, v_k$ such that for every u_j there is a bubble sphere v_{i_j} with equal image

$$u_j(\mathbb{S}^2) = v_{i_j}(\mathbb{S}^2)$$

and in particular $A_j = m_j B_{i_j}$, where $B_{i_j} = [v_{i_j}] \in H_2(W)$ and $m_j \ge 1$ is an integer. Write also $\mathbf{A}$ for the sum $\sum_{j=1}^{k'} A_j$ and $\mathbf{B}$ for the sum $\sum_{i=1}^{k} B_i$. Below we will compute the dimension of this simple bubble tree.

The initial bubble tree $u_0, u_1, \ldots, u_{k'}$ is the limit of a sequence in the moduli space $\mathcal{M}_0(W, N, z_0)$. Hence the connected sum $u_\infty := u_0 \# \cdots \# u_{k'}$ is, as element of $\pi_2(W, N)$, homotopic to a disk u in the bishop family, and the Maslov indeces

$$\mu(u) := \mu\big(u^*TW, u^*TN\big) \quad \text{and} \quad \mu(u_\infty) := \mu\big(u_\infty^*TW, u_\infty^*TN\big)$$

have to be equal. With the standard rules for the Maslov index (see for example [25, Appendix C.3]), we obtain

$$2 = \mu(u) = \mu(u_\infty) = \mu(u_0) + \sum_{j=1}^{k'} 2c_1([u_j]) = \mu(u_0) + 2c_1(\mathbf{A}).$$

The dimension of the unconnected set of holomorphic curves $\widetilde{\mathcal{M}}_{[u_0]}(W,N,z_0) \times \prod_{j=1}^k \widetilde{\mathcal{M}}_{B_j}(W)$ for the simple bubble tree is

$$\begin{aligned}(n + \mu(u_0)) + \sum_{j=1}^{k} 2(n + c_1(B_j)) &= n + 2 - 2c_1(A) + 2nk + \sum_{j=1}^{k} 2c_1(B_j)\\ &= n + 2 + 2nk + 2(c_1(\mathbf{B}) - c_1(\mathbf{A})).\end{aligned}$$

In the next step, we want to consider the subset of connected bubbles, i.e. we choose a total of k pairs of nodal points, which then have to be pairwise equal under the evaluation map. The nodal points span a manifold

$$Z(2k) \subset \{(1,\dots,2k) \to \mathbb{D}^2 \amalg \mathbb{S}^2 \amalg \cdots \amalg \mathbb{S}^2\}$$

of dimension $4k$. The dimension reduction comes from requiring that the evaluation map

$$\mathrm{ev}\colon \widetilde{\mathcal{M}}_{[u_0]}(W,N,z_0) \times \prod_{j=1}^{k} \widetilde{\mathcal{M}}_{B_j}(W) \times Z(2k) \to W^{2k}$$

sends pairs of nodal points to the same image in the symplectic manifold. By regularity and transversality of the evaluation map to the diagonal submanifold $\triangle(k) \hookrightarrow W^{2k}$, the dimension of the space of holomorphic curves is reduced by the codimension of $\triangle(k)$, which is $2nk$.

As a last step, we have to add the marked point z_0 used for the evaluation map ev_{z_0}, this way increasing the dimension by 2, and then we take the quotient by the automorphism group to obtain the moduli space. The dimension of the automorphism group is $6k+3$. Hence the dimension of the total moduli space is

$$\begin{aligned}&n + 2 + 2nk + 2(c_1(\mathbf{B}) - c_1(\mathbf{A})) + 4k - 2nk + 2 - (6k+3)\\ &= n + 1 - 2k + 2(c_1(\mathbf{B}) - c_1(\mathbf{A})) \le n + 1 - 2k.\end{aligned}$$

The inequality holds because by the assumption of semipositivity, all the Chern classes are non-negative on holomorphic spheres, and all coefficients n_j

in the difference $c_1(\mathbf{B}) - c_1(\mathbf{A}) = \sum_j c_1(B_j) - \sum_i c_1(A_i) = \sum_j c_1(B_j) - \sum_i m_i c_1(B_{j_i}) = \sum_j n_j c_1(B_j)$ are non-positive integers.

The computations for the disks in $\partial\mathcal{M}_0(\mathbb{D}^2, N; z_0)$ only differs by the requirement that the marked point needs to lie on the boundary of the disk u_0 instead of moving freely on the bubble tree. Instead of having two degrees of freedom for this choice, we thus only add one extra dimension. ■

4.4. Proof of the Non-fillability Theorem A

Theorem A. *Let (M, ξ) be a contact manifold that contains a* bLob *N, then M does not admit any semi-positive weak symplectic filling (W, ω) for which $\omega|_{TN}$ is exact.*

Assume there were a semi-positive symplectic filling (W, ω) for which $\omega|_{TN}$ is exact. Let α be a positive contact form for ξ. By Proposition 3.14, we can extend (W, ω) with a collar in such a way that we have $\omega|_{TN} = d\alpha|_{TN}$, which will allow us to use the energy estimates of the previous section. Now we choose an almost complex structure that is adapted to the bLob N as in Corollary 4.4, and we will study the moduli space $\mathcal{M}_0(\mathbb{D}^2, N; z_0)$ defined in Section 4.2 of holomorphic disks with one marked point lying in the same component as the Bishop family around a chosen component B_0 of the binding of N.

Trace a smooth path $\gamma\colon [0,1] \to N$ that starts at $\gamma(0) \in B_0$ and ends on the boundary ∂N. Assume further that γ is a regular curve, and that it intersects the binding and ∂N only on the endpoints of $[0,1]$. We want to select a 1-dimensional moduli space in $\mathcal{M}_0(\mathbb{D}^2, N; z_0)$ by only considering

$$\mathcal{M}^\gamma := \mathrm{ev}_{z_0}^{-1}\big(\gamma(I)\big).$$

It will be important for us that $\gamma(I)$ does not intersect the image of any bubble trees in $\overline{\mathcal{M}}_0(\mathbb{D}^2, N; z_0) \setminus \mathcal{M}_0(\mathbb{D}^2, N; z_0)$.

By Proposition 4.9, we have that the bubble trees in $\overline{\partial\mathcal{M}}_0(\mathbb{D}^2, N; z_0)$ lie in the image of a finite union of smooth maps defined on manifolds of dimension $\dim \partial\mathcal{M}_0(\mathbb{D}^2, N; z_0) - 2 = \dim N - 2$. The subset $N \setminus \mathrm{ev}_{z_0}(\text{bubble trees})$ is connected and we can deform γ keeping the endpoints fixed so that it does not intersect any of the bubble trees.

For a small perturbation of J (away from the binding and the boundary of N), we can make sure that the evaluation map ev_{z_0} is transverse to the path $\gamma(I)$. If the perturbed J lies sufficiently close to the old one, then γ will also not intersect any bubble trees for this new J, for otherwise we could choose a sequence of almost complex structures J_k converging to the unperturbed J such that for everyone there existed a bubble tree v_k intersecting γ. We would find a converging subsequence of v_k yielding a bubble tree v_∞ for the unperturbed almost complex structure intersecting γ, which contradicts our assumption.

It follows that $\mathcal{M}^\gamma$ is a collection of compact 1-dimensional submanifolds of $\partial\mathcal{M}_0(\mathbb{D}^2, N; z_0)$. There is one component in $\mathcal{M}^\gamma$, which we will denote by $\mathcal{M}_0^\gamma$ that contains the Bishop disks that intersect $\gamma([0,\varepsilon))$. We know that the Bishop disks are the only disks close to the binding, and hence it follows that $\mathcal{M}_0^\gamma$ cannot be a loop that closes, but must be instead a closed interval.

The first endpoint of $\mathcal{M}_0^\gamma$ is the constant disk with image $\gamma(0) \in B_0$, and we will deduce a contradiction by showing that no holomorphic disk can be the second endpoint of $\mathcal{M}_0^\gamma$.

By Proposition 3.24, there is a small neighborhood of ∂N that cannot be entered by any holomorphic disk. By our construction the endpoint of $\mathcal{M}_0^\gamma$ cannot be any bubble tree either. It follows that the endpoint needs to be a regular disk $[u, z_0] \in \partial\mathcal{M}_0(\mathbb{D}^2, N; z_0)$ for which the boundary of u lies in $N \setminus (\partial N \cup B)$ and whose interior points cannot touch ∂W either, because we are assuming that the boundary of W is convex.

It follows that this regular disk cannot really be the endpoint of $\mathcal{M}_0^\gamma$, because the evaluation map ev_{z_0} will also be transverse to γ at $[u, z_0]$ so that we can extend $\mathcal{M}_0^\gamma$ further.

This leads to a contradiction that shows that the assumption that the boundary of W is everywhere convex cannot hold.

4.5. Proof of Theorem B

For the proof, we first recall the definition of the degree of a map.

Definition. Let X and Y be closed oriented n-manifolds. The **degree** of a map $f\colon X \to Y$ is the integer $d = \deg(f)$ such that

$$f_\#[X] = d \cdot [Y],$$

where $[X] \in H_n(X, \mathbb{Z})$ and $[Y] \in H_n(Y, \mathbb{Z})$ are the fundamental classes of the corresponding manifolds. When the manifolds X and Y are not orientable, we define the degree to be an element of $\mathbb{Z}_2$ using the same formula, where the fundamental classes are elements in $H_n(X,\mathbb{Z}_2)$ and $H_n(Y,\mathbb{Z}_2)$.

Note that we can easily compute the degree of a smooth map f between smooth manifolds by considering a regular value $y_0 \in Y$ of f (which by Sard's theorem exist in abundance), and adding

$$\deg f = \sum_{x \in f^{-1}(y_0)} \operatorname{sign} Df_x,$$

where the point x contributes to the sum with $+1$, whenever Df_x is orientation preserving, and contributes with -1 otherwise. In case the manifolds are not orientable, we can always add $+1$ in the above formula, but need to take sum over $\mathbb{Z}_2$.

Theorem B. *Let (M,ξ) be a contact manifold of dimension $(2n+1)$ that contains a* Lob *N. If M has a weak symplectic filling (W,ω) that is symplectically aspherical, and for which $\omega|_{TN}$ is exact, then it follows that N represents a trivial class in $H_{n+1}(W,\mathbb{Z}_2)$. If the first and second Stiefel-Whitney classes $w_1(N)$ and $w_2(N)$ vanish, then we obtain that $[N]$ must be a trivial class in $H_{n+1}(W,\mathbb{Z})$.*

Using Proposition 3.14 we can assume that $\omega|_{TN} = d\alpha|_{TN}$ for a chosen contact form α. Choose an almost complex structure J on W that is adapted to the Lob N, and let $\mathcal{M}_0(\mathbb{D}^2,N;z_0)$ be the moduli space of holomorphic disks with one marked point lying in the same component as the Bishop family around a chosen component of the binding of N.

Since W is symplectically aspherical, we obtain by Corollary 4.8 that $\overline{\mathcal{M}}_0(\mathbb{D}^2,N;z_0)$ is a compact smooth manifold with boundary. It was shown in [13] that $\mathcal{M}_0(\mathbb{D}^2,N;z_0)$ is orientable if the first and second Stiefel-Whitney classes of $N \setminus B$ vanish. With our assumptions this is the case, because $w_j(N\setminus B) = w_j(N)|_{(N\setminus B)}$. If $\mathcal{M}_0(\mathbb{D}^2,N;z_0)$ is orientable then $\overline{\mathcal{M}}_0(\mathbb{D}^2,N;z_0)$ will also be orientable: If there were an orientation reversing loop γ in the compactified moduli space (which is obtained from $\mathcal{M}_0(\mathbb{D}^2,N;z_0)$ by gluing in B as codimension 3 submanifold), then due to the large codimension we could easily push γ completely into the regular part of the moduli space, where it would still need to be orientation reversing.

It follows that the boundary $\partial\overline{\mathcal{M}}_0(\mathbb{D}^2,N;z_0)$ is also homologically a boundary (either with $\mathbb{Z}$- or $\mathbb{Z}_2$-coefficients depending on the orientability of the considered spaces).

Denote the restriction of the evaluation map

$$\operatorname{ev}_{z_0}|_{\partial\overline{\mathcal{M}}_0(\mathbb{D}^2,N;z_0)} : \partial\overline{\mathcal{M}}_0\big(\mathbb{D}^2,N;z_0\big) \to N,$$

by f. We know that close to the binding every point is covered by a unique Bishop disk, this implies by the remarks made above that the degree $\deg(f)$ needs to be ± 1.

We have the following obvious equation

$$\operatorname{ev}_{z_0} \circ \iota_{\partial\overline{\mathcal{M}}} = \iota_N \circ f,$$

where $\iota_{\partial\overline{\mathcal{M}}}$ denotes the embedding of $\partial\overline{\mathcal{M}}_0(\mathbb{D}^2,N;z_0)$ in $\overline{\mathcal{M}}_0(\mathbb{D}^2,N;z_0)$ and ι_N the embedding of N in W. The homomorphism induced by $\iota_{\partial\overline{\mathcal{M}}}$ is the trivial map on the $(n+1)$-st homology group, so that the left side of the equation gives rise to the 0-map

$$H_{n+1}\big(\partial\overline{\mathcal{M}}_0\big(\mathbb{D}^2,N;z_0\big),R\big) \to H_{n+1}(W,R)$$

with R being either $\mathbb{Z}$ or $\mathbb{Z}_2$. Since $f_\#$ is $\pm$ identity, it follows that ι_N has to induce the trivial map on homology, which implies that N is homologically trivial in W.

References

[1] M. Abreu, Topology of symplectomorphism groups of $S^2 \times S^2$. Invent. Math. **131**(1), 1–23 (1998)

[2] P. Albers, H. Hofer, On the Weinstein conjecture in higher dimensions. Comment. Math. Helv. **84**(2), 429–436 (2009)

[3] M. Audin, J. Lafontaine (eds.), *Holomorphic Curves in Symplectic Geometry*. Progress in Mathematics, vol. 117 (Birkhäuser, Basel, 1994)

[4] E. Bedford, B. Gaveau, Envelopes of holomorphy of certain 2-spheres in $\mathbf{C}^2$. Am. J. Math. **105**(4), 975–1009 (1983)

[5] W.M. Boothby, H.C. Wang, On contact manifolds. Ann. Math. **68**, 721–734 (1958)

[6] F. Bourgeois, Odd dimensional tori are contact manifolds. Int. Math. Res. Not. **2002**(30), 1571–1574 (2002)

[7] S.K. Donaldson, Symplectic submanifolds and almost-complex geometry. J. Differ. Geom. **44**(4), 666–705 (1996)

[8] S.K. Donaldson, Lefschetz pencils on symplectic manifolds. J. Differ. Geom. **53**(2), 205–236 (1999)

[9] Y. Eliashberg, Filling by holomorphic discs and its applications, in *Geometry of Low-Dimensional Manifolds, 2*, Durham, 1989. London Math. Soc. Lecture Note Ser., vol. 151 (Cambridge Univ. Press, Cambridge, 1990), pp. 45–67

[10] Y. Eliashberg, Unique holomorphically fillable contact structure on the 3-torus. Int. Math. Res. Not. (2), 77–82 (1996)

[11] J. Etnyre, D. Pancholi, On generalizing Lutz twists. J. Lond. Math. Soc. **84**(3), 670–688 (2011)

[12] H. Geiges, *An Introduction to Contact Topology*. Cambridge Studies in Advanced Mathematics, vol. 109 (Cambridge University Press, Cambridge, 2008)

[13] P. Georgieva, Orientability of moduli spaces and open Gromov-Witten invariants. Ph.D. thesis, Stanford University, 2011

[14] D. Gilbarg, N. Trudinger, *Elliptic Partial Differential Equations of Second Order*. Classics in Mathematics (Springer, Berlin, 2001). Reprint of the 1998 ed.

[15] E. Giroux, Convexité en topologie de contact. Comment. Math. Helv. **66**(4), 637–677 (1991)

[16] E. Giroux, Talk given at ICM Beijing 2002 on contact structures and open books (2002)

[17] M. Gromov, Pseudo holomorphic curves in symplectic manifolds. Invent. Math. **82**, 307–347 (1985)

[18] H. Hofer, Pseudoholomorphic curves in symplectizations with applications to the Weinstein conjecture in dimension three. Invent. Math. **114**(3), 515–563 (1993)

[19] C. Hummel, *Gromov's Compactness Theorem for Pseudo-Holomorphic Curves*. Progress in Mathematics, vol. 151 (Birkhäuser, Basel, 1997)

[20] I. Kupka, The singularities of integrable structurally stable Pfaffian forms. Proc. Natl. Acad. Sci. USA **52**, 1431–1432 (1964)

[21] P. Massot, K. Niederkrüger, C. Wendl, Weak fillability of higher dimensional contact manifolds. Invent. Math. **192**(2), 287–373 (2013)

[22] D. McDuff, The structure of rational and ruled symplectic 4-manifolds. J. Am. Math. Soc. **3**(3), 679–712 (1990)

[23] D. McDuff, Symplectic manifolds with contact type boundaries. Invent. Math. **103**(3), 651–671 (1991)

[24] D. McDuff, D. Salamon, *Introduction to Symplectic Topology*, 2nd edn. Oxford Mathematical Monographs (Oxford University Press, New York, 1998)

[25] D. McDuff, D. Salamon, *J-Holomorphic Curves and Symplectic Topology*. American Mathematical Society Colloquium Publications, vol. 52 (American Mathematical Society, Providence, 2004)

[26] A. Mori, Reeb foliations on S^5 and contact 5–manifolds violating the Thurston-Bennequin inequality. Preprint (2009)

[27] K. Niederkrüger, The plastikstufe—a generalization of the overtwisted disk to higher dimensions. Algebr. Geom. Topol. **6**, 2473–2508 (2006)

[28] K. Niederkrüger, On fillability of contact manifolds. Habilitation à diriger des recherches, Toulouse (2013), oai:tel.archivesouvertes.fr:tel-00922320

[29] K. Niederkrüger, A. Rechtman, The Weinstein conjecture in the presence of submanifolds having a Legendrian foliation. J. Topol. Anal. **3**(4), 405–421 (2011)

[30] K. Niederkrüger, C. Wendl, Weak symplectic fillings and holomorphic curves. Ann. Sci. Éc. Norm. Super. **44**(5), 801–853 (2011)

[31] A. Oancea, C. Viterbo, On the topology of fillings of contact manifolds and applications. Comment. Math. Helv. **87**(1), 41–69 (2012)

[32] J. Palis, W. de Melo, *Geometric Theory of Dynamical Systems* (Springer, New York, 1982)

[33] F. Presas, A class of non-fillable contact structures. Geom. Topol. **11**, 2203–2225 (2007)

[34] D. Tischler, On fibering certain foliated manifolds over $\mathbb{S}^1$. Topology **9**, 153–154 (1970)

[35] O. van Koert, K. Niederkrüger, Every contact manifolds can be given a nonfillable contact structure. Int. Math. Res. Not. **2007**(23), rnm115 (2007), 22

[36] C. Wendl, Strongly fillable contact manifolds and J-holomorphic foliations. Duke Math. J. **151**(3), 337–384 (2010)

K. Niederkrüger

Université de Toulouse
Toulouse
France

BOLYAI SOCIETY
MATHEMATICAL STUDIES, 26

Contact and Symplectic Topology
pp. 245–273.

Contact Invariants in Floer Homology

GORDANA MATIĆ

1. Heegaard Floer Homology—A Very Quick Introduction

In a pair of seminal papers [24, 25] Peter Ozsváth and Zoltan Szabó defined a collection of homology groups they named Heegaard-Floer homologies. To define these groups, they first associate to a 3-manifold M a Heegaard diagram, use it to define a chain complex, and then show that the associated homology groups do not depend on choices made.

A Heegaard decomposition of a 3-manifold is a decomposition $M = H_1 \cup H_2$ into two handlebodies with $\Sigma_g = \partial H_1 = -\partial H_2 = H_1 \cap H_2$. We can define a genus g handlebody as a 3-manifold obtained by gluing g "handles" $D^2 \times [0,1]$ to B^3 by attaching $D^2 \times 0$ and $D^2 \times 1$ along disjoint pairs of discs in S^2 and smoothing. Every 3-manifold has a Heegaard decomposition. The easiest way to "visualize" one is to take a triangulation of M and take the neighborhood of the 1-skeleton as H_1 and the complement as H_2, It is clear that H_1 is a handlebody—if we take a maximal tree T in the 1-skeleton and cut the neighborhood along the discs perpendicular to the edges not in T, we have a ball. The complement is a neighborhood of the dual 1-skeleton, so is a handlebody by the same argument.

Another way to obtain a Heegaard decomposition is to look at Morse functions on M. Assume that a Morse function $f : M \to [0,3]$ is self-indexing, i.e. that it has critical points of index i at critical values $i = 0, \ldots, 3$. When there are g critical points of index 1 and 2, the mid-level surface $\Sigma = f^{-1}(3/2)$

F. Bourgeois et al. (eds.), *Contact and Symplectic Topology*,
Bolyai Society Mathematical Studies 26, DOI 10.1007/978-3-319-02036-5_6,

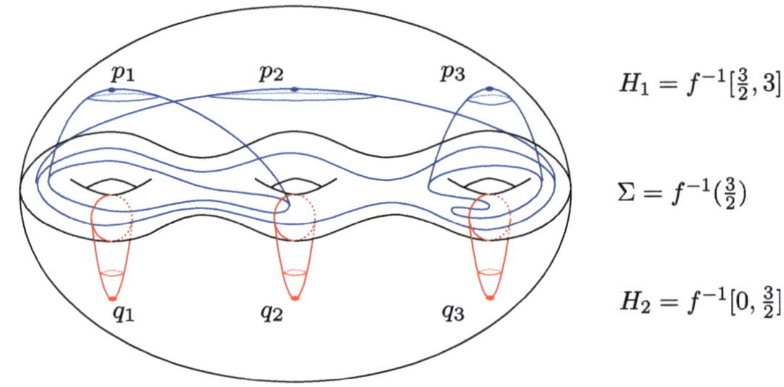

Fig. 1. Self-indexing Morse function $f : M \to [0, 3]$ with critical points of index 1 and their ascending discs in red, and critical points of index 2 and their descending discs in blue tracing the Heegaard diagram on $\Sigma = f^{-1}(\frac{3}{2})$

has genus g. The descending discs D_{p_i} of the index 2 critical points $\{p_i | i = 1, \ldots, g\}$ and the ascending discs D_{q_i} of the index 1 critical points $\{q_i | i = 1, \ldots, g\}$ cut the handlebodies $H_1 = f^{-1}[3/2, 3]$ and $H_2 = f^{-1}[0, 3/2]$ into 3-balls (see Figure 1).

Collections of their boundary curves $\{\alpha_i | \alpha_i = \partial D_{p_i}\}$ (correspondingly $\{\beta_i | \beta_i = \partial D_{q_i}\}$) cut the Heegaard surface Σ_g into a $2g$-times punctured sphere. The *Heegaard diagram* $(\Sigma, \{\alpha_i\}, \{\beta_i\})$ determines M—it can be obtained from $\Sigma \times [-1, 1]$ by gluing compressing discs along $\alpha_i \times \{-1\}$ and along $\beta_i \times \{1\}$, thickening the discs and then finally gluing in two copies of B^3. It is a theorem of Reidemeister and Singer [28, 31] that any two Heegaard diagrams for the same 3-manifold can be related to each other by three moves: (1) stabilization, in which genus of Σ is increased by adding a one handle and a pair of dual curves α_0, β_0 is added to $\{\alpha_i\}$, $\{\beta_i\}$ (and destabilization), (2) isotopy of multicurves $\{\alpha_i\}$, $\{\beta_i\}$ and (3) handle slides, where $\{\alpha_1, \alpha_2, \ldots\}$ is replaced by $\{\alpha_1 + \alpha_2, \alpha_2 \ldots\}$.

Heegaard Floer homology is a variant of Lagrangian Floer homology applied to the two Lagrangian submanifolds $\mathbb{T}_\alpha = \alpha_1 \times \cdots \times \alpha_g$ and $\mathbb{T}_\beta = \beta_1 \times \cdots \times \beta_g$ in the singular symplectic manifold $Sym^g(\Sigma_g)$, the symmetric product of g copies of Σ_g. In [24, 25] Ozsváth and Szabó define several versions of this invariant. To describe the simplest of them, $\widehat{HF}(M)$ with $\mathbb{Z}/2$ coefficients, we define the chain group $\widehat{CF}(\Sigma, \alpha, \beta) = \mathrm{Span}_{Z/2}\{\mathbf{x} | \mathbf{x} \in \mathbb{T}_\alpha \cap \mathbb{T}_\beta\}$ to be the free $\mathbb{Z}/2$-module generated by the points $\mathbf{x} = (x_1, \ldots, x_g)$ in $\mathbb{T}_\alpha \cap \mathbb{T}_\beta$. Note that such a g-tuple $\mathbf{x} = (x_1, \ldots, x_g)$ contains one point on each α_i and one point on each β_i curve.

Grading and the boundary maps are defined by considering pseudoholomorphic discs in $Sym^g(\Sigma_g)$ with boundary on $\mathbb{T}_\alpha$ and $\mathbb{T}_\beta$. More precisely, we look at maps ϕ from the unit disc $D^2 \subset \mathbb{C}$ to $Sym^g(\Sigma)$ that map $-i \mapsto \mathbf{x}$, $i \mapsto \mathbf{y}$, $S^1 \cap \{\text{Re } z \geq 0\}$ to $\mathbb{T}_\alpha$ and $S^1 \cap \{\text{Re } z \leq 0\}$ to $\mathbb{T}_\beta$. We denote the set of homotopy classes of such maps by $\pi_2(\mathbf{x}, \mathbf{y})$. With a generic choice of almost complex structure on $Sym^g(\Sigma_g)$, the space of pseudoholomorphic discs in the homotopy class of $\phi \in \pi_2(\mathbf{x}, \mathbf{y})$ is a smooth manifold $\mathcal{M}(\phi)$. These spaces carry a free $\mathbb{R}$ action by translation. To see this translation action we think of a biholomorphism between $D^2 \setminus \{i, -i\}$ and $[-1, 1] \times \mathbb{R}$ taking $\pm i$ to $\pm\infty$ and translate in the $\mathbb{R}$ direction. The dimension of the space $\mathcal{M}(\phi)$ of holomorphic maps in the homotopy class of ϕ is calculated via the Maslov index $\mu(\phi)$, which defines a relative grading on $\widehat{CF}(\Sigma, \alpha, \beta)$. When the dimension of $\mathcal{M}(\phi)$ is one, compactness arguments give finiteness of $\widehat{\mathcal{M}}_\phi = \mathcal{M}(\phi)/\mathbb{R}$, and we define the boundary operator on $\widehat{CF}(\Sigma, \alpha, \beta)$ by counting (mod 2) the number of points in the 0-dimensional space $\widehat{\mathcal{M}}_\phi$. If we fix a marked point z in the complement of the α and β curves, the function $n_z(\phi)$ on $\pi_2(\mathbf{x}, \mathbf{y})$ is given by the intersection number of $\phi(D^2)$ with $z \times Sym^{g-1}(\Sigma)$. If the image of ϕ misses $z \times Sym^{g-1}(\Sigma)$ then $n_z(\phi) = 0$. The differential

$$\partial \mathbf{x} = \sum_{\mathbf{y}} \sum_{\substack{\phi \in \pi_2(\mathbf{x}, \mathbf{y}) \\ \mu(\phi)=1, n_z(\phi)=0}} \#(\widehat{\mathcal{M}}_\phi) \mathbf{y}$$

has grading -1 and satisfies $\partial \circ \partial = 0$.

To understand better what we are summing over, note that the map $\phi : D^2 \to Sym^g(\Sigma)$ corresponds to a map $\widetilde{\phi} : \widetilde{D} \to \Sigma$ from a g-fold branched cover $\widetilde{D}$ of D^2 to Σ (branching is over the preimage of the intersection of $\phi(D^2)$ with the diagonal in $Sym^g(\Sigma)$). The two marked points i and $-i$ on the boundary of D^2 lift to $2g$ marked points on the boundary of $\widetilde{D}$, and under $\widetilde{\phi}$ these are mapped in alternating fashion to the coordinates x_i and y_i of $\mathbf{x}$ and $\mathbf{y}$. The segments on the boundary of $\widetilde{D}$ between these $2g$ points are mapped in alternating fashion to segments on α and β curves connecting x_i and y_j. The image of the map $\widetilde{\phi}$ is a union of domains D_i in the complement of the α and β curves in Σ, and the Maslov index can be computed from this picture according to a combinatorial formula of Lipshitz [20]. Analysis of the moduli spaces corresponding to Maslov index 2 maps shows that the boundary map satisfies $\partial^2 = 0$. The resulting homology is shown in [24] to be independent of the choice of a marked Heegaard diagram $(\Sigma, \alpha_i, \beta_i, z)$ for M, and therefore an invariant of the 3-manifold M.

Given a marked point z, each intersection point $\mathbf{x} = (x_1, \ldots, x_g)$ in $\mathbb{T}_\alpha \cap \mathbb{T}_\beta$ defines a spin^c structure $\mathbf{s}(\mathbf{x})$ on the 3-manifold M determined by

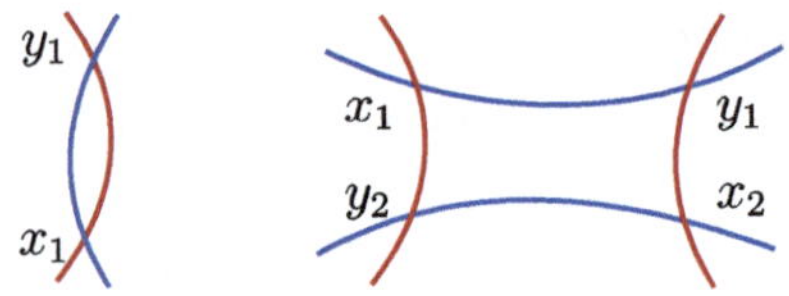

Fig. 2. A bigon and a rectangular domain

the Heegaard diagram. If there is a topological disc from $\mathbf{x}=(x_1,\ldots,x_g)$ to $\mathbf{y}=(y_1,\ldots,y_g)$ with boundary contained in the union of two tori $\mathbb{T}_\alpha$ and $\mathbb{T}_\beta$ the two spin^c structures $\mathbf{s}(\mathbf{x})$ and $\mathbf{s}(\mathbf{y})$ agree. Therefore $\widehat{CF}(\Sigma,\alpha,\beta,z)$ splits as the sum of $\widehat{CF}((\Sigma,\alpha,\beta,z),\mathbf{s})$ and, as is clear from the definition, the boundary map preserves this splitting. Hence the complex, and the homology of the complex as well, split according to spin^c structures:

$$\widehat{HF}(M)=\sum_{\mathbf{s}\in Spin^c(M)}\widehat{HF}(M,\mathbf{s})$$

Other versions of Heegaard Floer homology, HF^∞ and HF^+, are defined by considering the free $\mathbb{Z}/2[U,U^{-1}]$ or $\mathbb{Z}/2[U]$ modules, respectively, generated by points in $\mathbb{T}_\alpha\cap\mathbb{T}_\beta$ and counting also the pseudoholomorphic curves that cross the marked point z in the differential, by recording $n_z(\phi)$:

$$\partial\mathbf{x}=\sum_{\mathbf{y}}\sum_{\substack{\phi\in\pi_2(\mathbf{x},\mathbf{y})\\ \mu(\phi)=1}}\#(\widehat{\mathcal{M}}_\phi)U^{n_z(\phi)}\mathbf{y}$$

Defining orientations on the moduli spaces $\widehat{\mathcal{M}}_\phi$ makes it possible to count the number of points in the 0-dimensional moduli spaces $\widehat{\mathcal{M}}_\phi/\mathbb{R}$ with sign in order to work over $\mathbb{Z}$.

It is highly nontrivial to count the number of points in the moduli spaces $\widehat{\mathcal{M}}_\phi$ for a general ϕ, as that count depends on the choice of almost complex structure on $Sym^g(\Sigma)$. However, if the image of $\widetilde{\phi}$ is a topological disc connecting two intersection points $\mathbf{x}=(x_1,x_2,x_3\ldots x_g)$ and $\mathbf{y}=(y_1,x_2,x_3\ldots x_g)$ that differ in just one coordinate, called a bigon domain (see Figure 2), or between two intersection points $\mathbf{x}=(x_1,x_2,x_3\ldots x_g)$ and $\mathbf{y}=(y_1,y_2,x_3\ldots x_g)$ that differ in exactly two coordinates, called a rectangular domain (and constant maps on other components of the cover), there is a unique holomorphic disc in that homotopy class.

This makes it possible to have a combinatorial calculation of the chain complex and the homology. Sarkar and Wang [30] described a method to

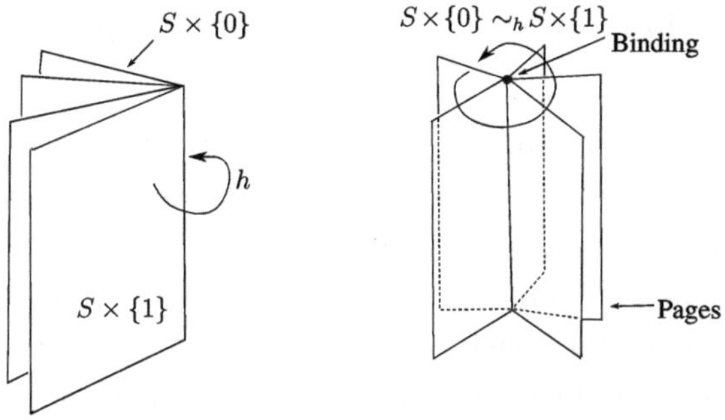

Fig. 3. A neighborhood of the binding in an open book decomposition

produce "nice Heegaard diagrams", i.e. diagrams for which all but one distinguished domain are bigons and rectangles and when the base point z is placed in that distinguished domain, the count involved in calculating $\widehat{HF}(M)$ becomes combinatorial. There are other combinatorial approaches to calculating the Heegaard Floer homology and a variant associated to a knot $K \subset M$, the knot Floer homology $\widehat{HFK}(M,K)$, notably the gird diagrams and convenient diagrams ([22, 27] and many other sources).

2. Open Book Decompositions, Contact Structures and Convex Surfaces

Let S be a surface with boundary and let $Aut(S, \partial S) = \{h \in \mathrm{Diff}(S, \partial S) | h|_{\partial S} = id\}$. Moding $S \times [0,1]$ out by the equivalence relation $(x,1) \sim_h (h(x), 0)$ produces a manifold $M(S,h)$ with a torus boundary component for each component of ∂S. Moding out the boundary by further identifying $(x,t) \sim_h (x,t')$ for all $t,t' \in [0,1]$ and $x \in \partial S$ reduces each torus boundary component to a knot in a closed 3 manifold $M_{(S,h)} = S \times [0,1]/\sim_h$ (Figure 3). Denote by $B_{(S,h)} \subset M_{(S,h)}$ the image of $\partial S \times [0,1]$ under the quotient projection. The complement of a small neighborhood of $B_{(S,h)}$ in $M_{(S,h)}$ is diffeomorphic to $M(S,h)$ and fibers over $S^1 = [0,1]/0 \sim 1$. The link $B_{(S,h)}$ is called the *binding* of the open book and the image of each $S \times \{t\}$ under the quotient projection is a *page*. The projection $\pi : \mathrm{Int}(S) \times [0,1]/\sim_h \to [0,1]/_{0\sim 1}$ induces a bundle structure on the complement of the binding with the preimage of a point in $t \in S^1 = [0,1]/_{0\sim 1}$ being the interior of the page S_t.

An open book decomposition for a 3-manifold M with page S, binding B and monodromy $h \in Aut(S, \partial S)$ is a projection $p : M \setminus B \to S^1$ and a homeomorphism $\Phi : (M,B) \to (M_{(S,h)}, B_{(S,h)})$ that commutes with projections p

and π. We say in this case that (S,h) is a *formal open book* decomposition for (M,B,p).

It is a theorem of Alexander that every 3-manifold has an open book decomposition. Different open book decompositions of the same manifold are related by sequences of positive/negative stabilizations, where stabilization changes S by adding a 1-handle to S and changes h by composing it with a positive/negative Dehn twist about a curve γ dual to the handle. More precisely, we attach a handle $H=[0,1]\times[-\epsilon,\epsilon]$ to S along $\{0,1\}\times[-\epsilon,\epsilon]$ to obtain S' and let $h\cup id$ be extension of h by identity on $[0,1]\times[-\epsilon,\epsilon]$. Let γ be any simple closed curve in S' that intersects the co-core $1/2\times[-\epsilon,\epsilon]$ of the handle H once, and let R_γ (R_γ^{-1}) be the right (left) handed Dehn twist. If $h'=R_\gamma^{\pm1}\circ(h\cup id)$ we say that (S',h') is a positive (negative) stabilization of (S,h). It is not hard to see that if (S',h') is a positive (negative) stabilization of (S,h), and (S,h) is a formal open book decomposition of (M,B,p), then (S',h') gives a formal open book decomposition of (M,B',p') where the binding B' is obtained by a Murasugi sum along γ of the binding B with a positive (negative) Hopf link. For a nice description see [29].

Open book decompositions of 3-manifolds are intimately related to contact topology. A contact structure ξ on a 3-manifold M is a 2-plane filed which is the kernel of a nondegenerate 1-form α, the *contact form.* (Nondegeneracy means that $\alpha\wedge d\alpha = d\,vol$.) The contact structure ξ is said to be *supported by the open book decomposition* (M,B,p) if:

1. B is transverse to ξ,
2. there is a contact 1-form α for which $d\alpha$ induces a symplectic form on each fiber $S_t=p^{-1}(t)$ of $p:M\setminus B\to S^1$,
3. the orientation on B given by α is the same as the boundary orientation induced from S_t oriented by the symplectic structure induced by $d\alpha$.

Thurston and Winkelnkemper [32] showed that any open book decomposition (S,h) of M supports a contact structure ξ by constructing a contact form $\xi_{(S,h)}$ on $M_{(S,h)}$ with these properties. Torisu [33] and Giroux [10] proved that the converse is true, namely that every contact manifold is supported by (has a compatible) open book decomposition. Giroux established the following correspondence:

Theorem 2.1 (Giroux). *Any contact structure (M,ξ) on a closed 3-manifold M is supported by an open book decomposition (S,h). Moreover, any two open book decompositions (S,h) and (S',h') which support the same*

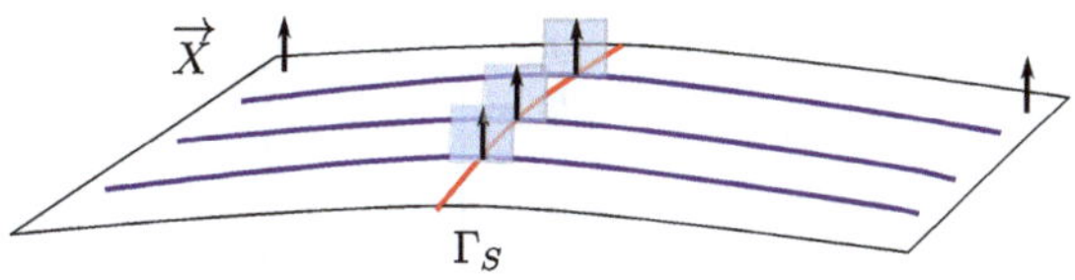

Fig. 4. Convex surface and its characteristic foliation and dividing set

contact structure (M,ξ) are equivalent under a sequence of positive stabilizations/destabilizations.

To give an indication of how a compatible open book decomposition can be found for a given a contact manifold (M,ξ), we need to first talk about convex decomposition methods in contact topology. In the early 90's Emanuel Giroux introduced an important tool in contact topology—convex surfaces. A vector field $\vec{v}$ in a contact manifold (M,ξ) is called *contact* if its flow preserves ξ. A surface $S \subset M$ is called *convex* if there exists a contact vector field $\vec{v}$ transverse to S. Note that a convex surface S can be transverse to many different contact vector fields. To a convex surface S and a contact vector field $\vec{v}$ transverse to S one can associate

$$\Gamma_S = \{x \in S | \vec{v}(x) \in \xi(x)\}$$

its *dividing set* (see Figure 4). It is not hard to show that Γ_S is a smooth multicurve, and that the isotopy class of Γ_S is independent of $\vec{v}$.

If $g(\,,)$ is a Riemannian metric on M and we denote by $\vec{n}_\xi$ a normal vector field to ξ, the dividing set on a convex surface $S \subset M$ determined by a contact vector field $\vec{v}(p)$ can be described as $\Gamma_S = \{p \in S \mid g(\vec{v}(p), \vec{n}_\xi(p)) = 0\}$. The dividing set Γ clearly divides the convex surface S into two regions $R_+(S)$ and $R_-(S)$ where $R_+ = \{p \in S \mid g(\vec{v}(p), \vec{n}_\xi(p)) \geq 0\}$ and $R_- = \{p \in S \mid g(\vec{v}(p), \vec{n}_\xi(p)) \leq 0\}$. It is not hard to see that, if S is a closed convex surface in the contact manifold (M,ξ), the contact class of ξ evaluates on S as $\chi(\xi)[S] = \chi(R_+(S)) - \chi(R_-(S))$.

A knot is *Legendrian* if it is everywhere tangent to contact planes. Contact planes (cooriented by $\vec{n}_\xi$) induce, on a surface S oriented by $\vec{v}$, an oriented singular foliation $\mathcal{F}$ by Legendrian curves called the *characteristic foliation.* The regions $R_\pm$ correspond to "source" and "sink" regions in this picture. Legendrian curves of the foliation intersect the dividing set transversally as in Figure 4. If a convex surface S has Legendrian boundary, then the dividing set intersects the boundary in an even number of points $2n$, and n describes the number of twists that the contact plane makes relative to the framing for the Legendrian boundary curve coming from the convex surface.

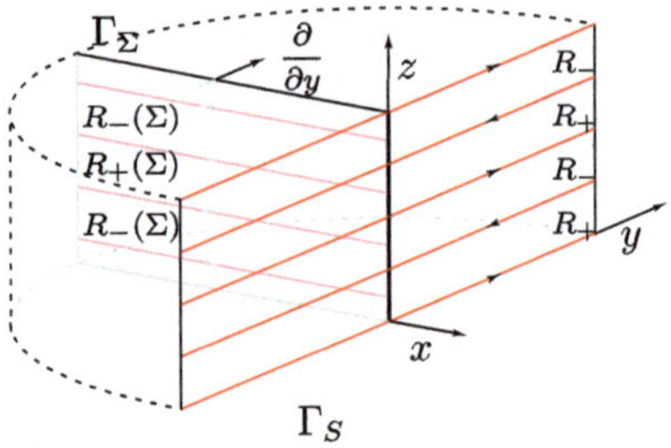

Fig. 5. Standard neighborhood of a Legendrian curve at the intersection of two perpendicular convex surfaces

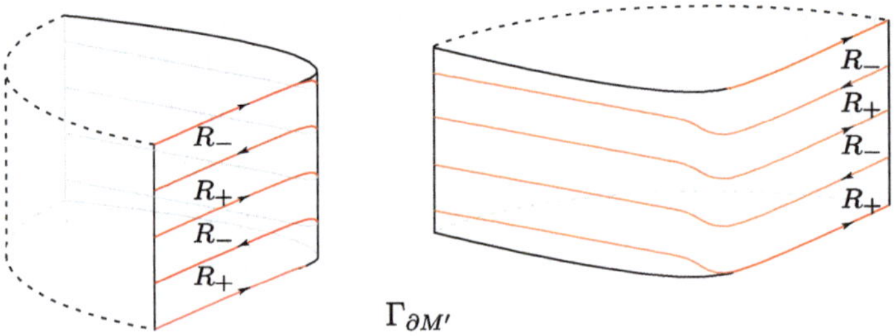

Fig. 6. Smoothing of the dividing set after cutting along convex surface Σ and rounding corners

We can decompose a contact manifold (M,ξ) by cutting it along properly embedded convex surfaces Σ. For a gentler introduction see [12]. When (M,ξ) is a contact manifold with boundary we assume that the boundary ∂M is convex, and that the convex cutting surface Σ has Legendrian boundary $L=\partial\Sigma\subset S$. A standard neighborhood theorem says that if L intersects the dividing set Γ_Σ geometrically $2n$ times, there is a neighborhood $N_\varepsilon(L)$ of L in M and local coordinates (x,y,z) on it so that $N_\varepsilon(L)=\{(x,y,z)|x^2+y^2<\varepsilon, x\leq 0\}$ in $\mathbb{R}^2\times(\mathbb{R}/\mathbb{Z})$ (see Figure 5).

Here L lies on the z axis, the set of points with $x=0$ corresponds to an annular neighborhood of L in Σ, the set of points with $y=0$ corresponds to a neighborhood of L in S, and the contact structure ξ is given as the kernel of $\alpha=\sin(2\pi nz)dx+\cos(2\pi nz)dy$, $n\neq 0$. If we choose the contact vector fields for S and Σ to be $v_S=\frac{\partial}{\partial x}$ and $v_\Sigma=\frac{\partial}{\partial y}$ it is not hard to calculate that the dividing sets are $\Gamma_S=\{(0,y,\frac{k}{2n})|0\leq k<2n\}$ and $\Gamma_\Sigma=\{(x,0,\frac{1+2k}{4n})|0\leq k<2n\}$.

After cutting M along Σ and rounding corners to obtain M', transverse contact vector fields on the new boundary components of $\partial M'=S'$ can be chosen to be $v_S=a\frac{\partial}{\partial x}+b\frac{\partial}{\partial y}$ so that they rotate between $\frac{\partial}{\partial x}$ and $\frac{\partial}{\partial y}$. A simple

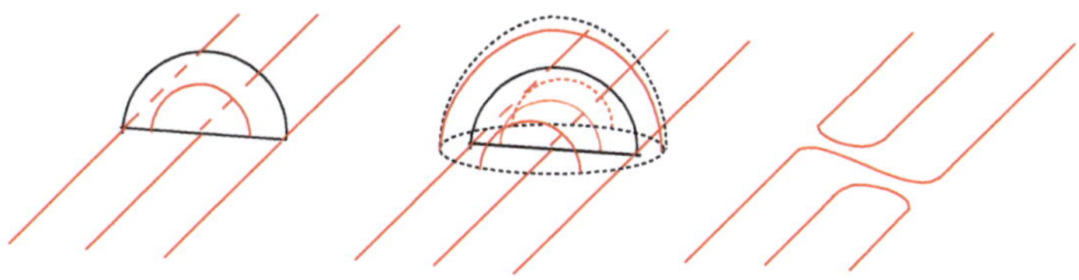

Fig. 7. Adding a bypass

calculation shows that the new dividing set on $\partial M'$ is obtained by the "turn to the right" method illustrated in Figure 6.

Giroux and Honda showed that isotoping a convex surface S through a contact manifold preserves the isotopy class of the dividing set except for at finitely many levels of the isotopy. Honda described how at these levels the product contact structure changes by adding a "bypass" to the product structure on $S \times I$, i.e. engulfing a "half overtwisted disc" B bounded by a Legendrian arc α in S connecting three dividing curves (possibly not globally different), and an Legendrian arc β along which the contact planes coincide with the tangent planes of B. Thickening the bypass disc to $B \times [-\varepsilon, \varepsilon]$ attached along $\alpha \times [-\varepsilon, \varepsilon]$, looking at the dividing set and rounding to obtain S' we see that the change in the dividing set after adding a bypass is as pictured in Figure 7.

To describe briefly how to find an open book compatible with a given contact structure ξ on M we start with a triangulation of M that is fine enough so that each 3-simplex is contained in a standard contact chart for ξ. The 1-skeleton can then be perturbed to be Legendrian in such a way that the relative twisting of the contact planes along the boundaries of the discs in the 2-skeleton is such that these discs can be made convex with the given Legendrian boundary. By adding extra pieces to the 1-skeleton to divide the discs in the 2-skeleton in such a way that each contains exactly one arc in its dividing set we can achieve the following:

(1) There is a handlebody decomposition of the 3-manifold into $H_1 = \nu(M^{(1)})$, the neighborhood of the 1-skeleton, and the complementary handlebody $H_2 = \nu(\bar{M}^{(1)})$ which is the nighbourhood of the dual 1-skeleton.

(2) The common boundary Σ is a convex surface.

(3) The contact handlebodies H_1 and H_2 with convex boundary are *disc decomposable*, i.e. there is a family of compressing discs with Legendrian boundary such that the boundary of each disc intersects the dividing set in exactly two points.

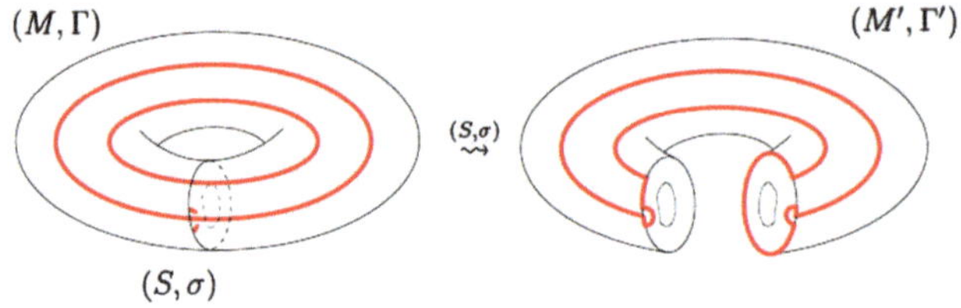

Fig. 8. Cutting a genus one handlebody along a convex compressing disc

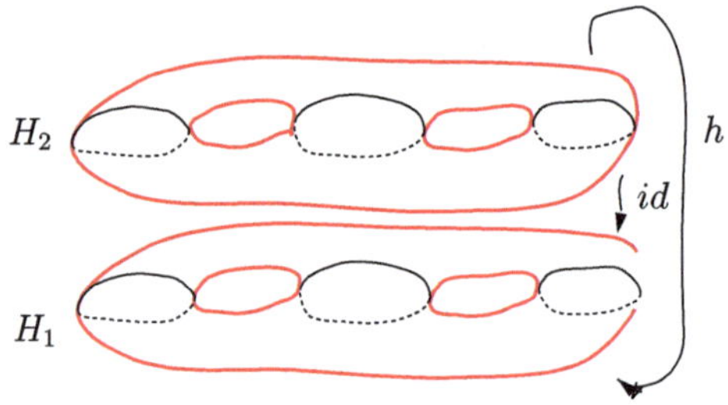

Fig. 9. Gluing disc decomposable contact structures on H_1 and H_2

Note that the convex discs in (3) are just small discs transverse to the 1-cells of the 1-skeleton in H_1 and the 2-cells of the subdivided complex for H_2. By cutting along the compressing discs until we obtain B^3, we can see that contact handlebodies H_i are contactomorphic to product contact manifolds $R \times I$ where R is homeomorphic to Σ_+ (and Σ_-), as we think of obtaining H_i by gluing contact 1-handles onto a standard ball with convex boundary S^2 and dividing set S^1. The simplest case, of a genus one surface with two parallel longitudinal dividing curves, is shown in Figure 8.

Note that when we consider Σ as the boundary of H_2 the roles of Σ_+ and Σ_- are reversed in comparison to what they are when we consider Σ as the boundary of H_1 (due to the change of orientation on Σ). We identify H_1 with $R \times [0, 1/2]$ in such a way that Σ_- corresponds to $R \times \{0\}$ and Σ_+ corresponds to $R \times \{1/2\}$, and identify H_2 with $R \times [1/2, 1]$ in such a way that Σ_- corresponds to $R \times \{1/2\}$ and Σ_+ to $R \times \{1\}$. Then the identification of $R \times \{0\}$ and $R \times \{1\}$ coming form the way H_1 and H_2 are glued inside M defines the monodromy map $h : R \to R$ which realizes M as corresponding to the open book (R, h), and the open book inside M is compatible with ξ (see Figure 9).

We say that a contact structure ξ on a 3-manifold M is *overtwisted* if there is an embedded D^2 in M such that the tangent plane $T_P D^2$ and the contact plane ξ_P agree at all points $P \in \partial D^2$. If ξ is not overtwisted we say it is *tight*. There are familiar examples of tight contact structures. It is a theorem of Bennequin [4] that the standard contact structure on $\mathbb{R}^3$ is tight. Giroux showed that a product neighborhood of a convex surface is tight if and only if it has no contractible dividing curve, or is S^2 with $\Gamma = S^1$.

A theorem of Eliashberg [5] says that overtwisted contact structures are classified up to isotopy through contactomorphisms (contact isotopy) by the homotopy class of their plane fields, i.e. two overtwisted contact structures that are homotopic as plane fields are isotopic through a family of contact structures. This is not true in the case of tight contact structures, which are more closely related to the finer topology of the manifold. A central question in contact topology on 3-manifolds is to construct, recognize and classify tight contact structures on a given M^3.

There is a rich source of tight examples; the fillable contact structures. A contact manifold (M,ξ) is *Stein fillable* if there is a compact complex Stein manifold W with convex boundary such that $M = \partial W$ and contact planes are complex lines in TM. (M,ξ) is *strongly symplectically fillable* if M is the boundary of a symplectic manifold (W,ω) which looks like a Stein manifold near the boundary. Finally, (M,ξ) is *weakly symplectically fillable* if M is boundary of a symplectic manifold (W,ω) and $\omega|_\xi \geq 0$. It is a theorem of Gromov and Eliashberg that fillable contact structures are tight [7, 11].

To describe tightness in the framework of open book decompositions we use the notion of "right-veering" homeomorphisms [13]. We say that a homeomorphism $h \in Aut(S,\partial S)$ is right-veering if every properly embedded oriented arc α in S is mapped "to the right" of α as in Figure 10. For two properly embedded arcs α and β with the same initial point $\alpha(0) = \beta(0)$ which are isotoped rel boundary to intersect transversally in a minimal number of points, we say that β is to the right of α if the tangent vectors $\{\beta'(0), \alpha'(0)\}$ define the orientation of S.

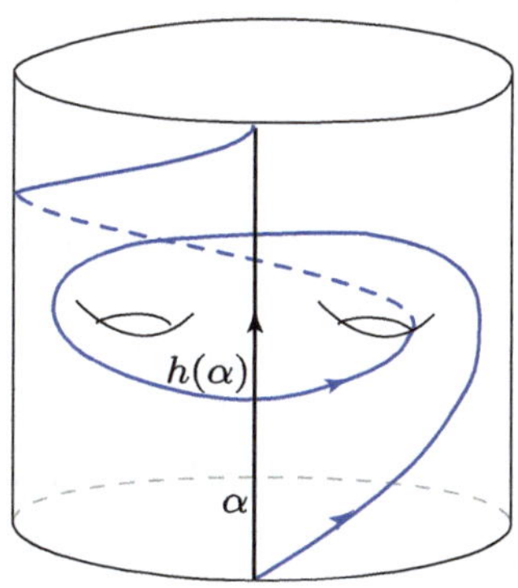

Fig. 10. The image arc $h(\alpha)$ is to the right of α

Theorem 2.2 (Honda-Kazez-Matić). *A contact 3-manifold (M,ξ) is tight if and only if all of its adapted open book decompositions have right-veering monodromy.*

Right-veering diffeomorphisms of S form a monoid $Veer(S,\partial S)$ in $Aut(S,\partial S)$. Monodromies in the submonoid $Dehn^+(S,\partial S) \subset Veer(S,\partial S)$ consisting of diffeomorphisms that are products of positive Dehn twists give rise to Stein fillable structures by the work of Eliashberg [6]. That every Stein fillable structure has a monodromy in $Dehn^+(S,\partial S)$ is a theorem by Loi-Piergallini and Akbulut-Ozbagci [1, 21]. John Baldwin [3] and Baker, Etnyre and Van-Horn Morris [2] have shown that there are open books for Stein fillable contact structures that are not in $Dehn^+(S,\partial S)$. They have also shown that the open book monodromies corresponding to Stein, strongly and weekly fillable contact structures form monoids. The natural inclusions $Dehn^+(S,\partial S) \subsetneq Stein(S,\partial S) \subsetneq Strong(S,\partial S) \subsetneq Week(S,\partial S) \subsetneq Tight(S,\partial S) \subsetneq Veer(S,\partial S)$ are all proper according to work of, in order, Baker-Etnyre-VanHorn Morris and independently Wand, Ghiggini, Eliashberg, Ghiggini and Honda-Kazez-Matić and Goodman. However, it is not clear that $Tight(S,\partial S)$ is a monoid, or how to describe the tightness in terms of one open book. It is easy to show that any open book can be stabilized to be right veering, hence there are right veering open books supporting overtwisted contact structures. Even if we know that an open book has right veering monodromy and that it is not the result of a stabilization, this does not guarantee tightness, as first shown by Lekili [19].

3. From Open Books to Contact Invariants

Ozsváth and Szabó [26] used the one-to-one correspondence between equivalence classes of open books for M under positive stabilization and isotopy classes of contact structures on M to define an invariant $c(\xi)$ of contact structures that lives in the Heegaard Floer homology of $-M$. To do this they associated to an open book decomposition (M,B,p) compatible with ξ the fibered manifold $M_0(B)$ obtained by performing 0-framed surgery with respect to the page framing on the binding B. Heegaard Floer homology for a fibered 3-manifold is special—it is one dimensional in the spin^c structure corresponding to the fibration. Ozsváth and Szabó defined $c(\xi)$ to be the image of the generator of this group under the map induced by the cobordism defined by the 0-handle attachment from $\widehat{HF}(-M_0(B))$ to $\widehat{HF}(-M)$. It has the property that $c(\xi)=0$ if ξ is overtwisted, and that $c(\xi)\neq 0$ when ξ is fillable.

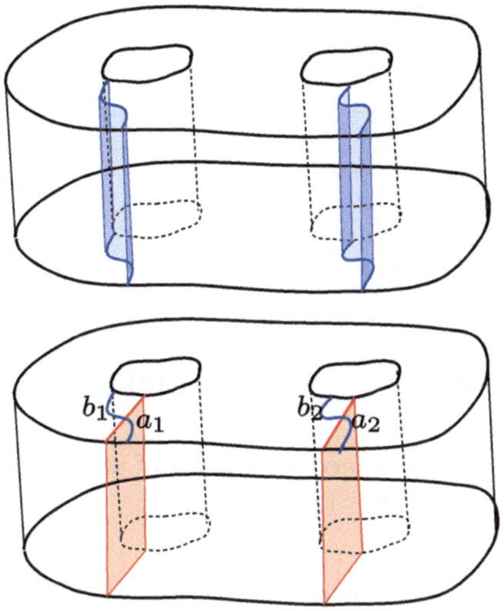

Fig. 11. The Heegaard decomposition and the diagram determined by a basis of arcs for the page of an open book decomposition

To see a concrete generator of $c(\xi)$ we will use the open book decomposition to construct a Heegaard diagram in which this generator sits in a distinguished way. An open book decomposition adapted to ξ gives rise to a Heegaard decomposition into two handlebodies $H_1 = S \times [0, 1/2]/_{\sim_h}$ and $H_2 = S \times [1/2, 1]/_{\sim_h}$.

A *basis of arcs* for the surface with boundary S is a collection of properly embedded arcs $\{a_i | i = 1, \ldots, r\}$ that cuts the surface into a disc, $S \setminus (\bigcup a_i) = D^2$. Given a basis of arcs we can construct the family of compressing discs $D_{a_i} = a_i \times [0, 1/2]$ for H_1 which cut H_1 into a ball. We perturb the a_i slightly to obtain a basis of arcs b_i such that each b_i intersects a_i transversely at exactly one point (positively) and consider discs $D_{b_i} = b_i \times [1/2, 1]$ which are a set of compressing discs that cut H_2 down to a ball.

Taking the boundary curves of these discs $\alpha_i = \partial(D_{a_i})$ and $\beta_i = \partial(D_{b_i})$ and looking at them as curves in the separating surface $\Sigma = S \times \{1/2\} \cup -S \times \{0\} = S \times \{1/2\} \cup -S \times \{1\}$ (here Σ is the common boundary of H_1 and H_2, and $S \times \{1\}$ is identified with $S \times \{0\}$ via the monodromy h), we obtain a Heegaard diagram $(\Sigma, \alpha_i, \beta_i, z)$ (see Figure 11). There is a distinguished generator $\mathbf{x} = (x_1, \ldots, x_i, \ldots)$ for $\widehat{CF}(\Sigma, \alpha_i, \beta_i, z)$ that is contained in $\Sigma = S \times \{1/2\}$ where $x_i = \alpha_i \cap \beta_i$. If we choose the marked point z to lie in $\Sigma = S \times \{1/2\}$ and outside the thin strips bounded by α_i and β_i curves, it is easy to see that $\mathbf{x}$ is a cycle, and in fact:

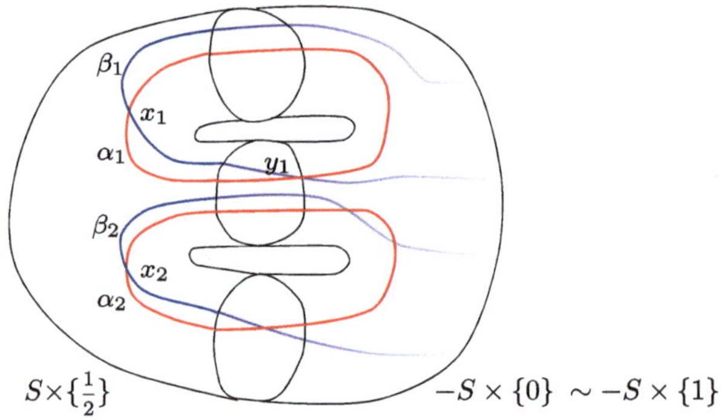

Fig. 12. The Heegaard diagram determined by a basis of arcs and the contact element $\mathbf{x}=(x_1,x_2)$

Theorem 3.1 [14]. *The generator* $\mathbf{x}=(x_1,x_2,\ldots,x_r)$ *is a cycle in* $\widehat{CF}(\Sigma,\beta_i,\alpha_1,z)$ *and its homology class is the Ozsváth-Szabó contact class,* $\mathbf{x}=c(\xi)\in\widehat{HF}(-M)$.

In particular, for a different open book decomposition compatible with the same contact structure ξ, we get the same class $c(\xi)\in\widehat{HF}(-M)$. Note that we needed to switch the role of α and β curves, hence the orientation of M in order for $\mathbf{x}$ to be a cycle.

From this description it is easy to see that for an overtwisted contact structure $c(\xi)=0$. Specifically, an overtwisted contact structure has a compatible open book decomposition and a basis of arcs such that the holonomy takes at least one of the arcs, we can name it α_1, to the left of itself. Then in the corresponding Heegaard diagram, there is a bigon connecting $\mathbf{y}=(y_1,x_2,\ldots,x_r)$ to $\mathbf{x}=(x_1,x_2,\ldots,x_r)$, which gives $\partial\mathbf{y}=\mathbf{x}$, hence $\mathbf{x}=0$ in homology. When looking at the Figure 12, it might seem that α_1 is in fact mapped to the right, but that is an artifact of reversal of orientation on $-S\times\{0\}$.

Given that $c(\xi)=0$ for overtwisted contact structure, and that $c(\xi)\neq 0$ when ξ is fillable (by a theorem of Gromov and Eliashberg [7, 11]), it was a natural question to ask whether $c(\xi)\neq 0$ is a characterization of tightness. Ghiggini [8] showed that there are tight contact manifolds with $c(\xi)=0$. The examples he found contain Giroux torsion, i.e. a contact embedding of $T^2\times[0,1]=\mathbb{R}^2/\mathbb{Z}^2\times[0,1]$ with contact structure given by $\xi_{tor}=ker(\cos(2\pi z)dx-\sin(2\pi z)dy)$.

The question then was reformulated to ask if $c(\xi)=0$ is equivalent to the requirement that ξ contains Giroux torsion. The answer was obtained with the use of a contact invariant defined for contact manifolds with boundary, the sutured contact invariant, which we will discuss in the next section. On one hand, the invariant is used in [9] to prove that any contact manifold containing Giroux torsion has $c(\xi)=0$. On the other hand, it is used in [16] to show that for a general surface with boundary S there are S^1-invariant contact structures on $S\times S^1$ generalizing Giroux torsion that have the property that any contact manifold that contains them has $c(\xi)=0$, and is hence not fillable.

4. Sutured Manifolds and Partial Open Books

When a 3-manifold M has nonempty boundary, we will study contact structures on M for which the boundary ∂M is a convex surface with a prescribed dividing set $\Gamma\subset\partial M$ dividing ∂M into R_+ and R_- regions. A pair (M,Γ) of a manifold and a dividing set Γ on the boundary (which divides every boundary component) is called a *sutured manifold.* Sutured manifolds were first defined by Gabai for use in the study of foliations. When $\chi(R_+)=\chi(R_-)$ we call the sutured manifold (M,Γ) balanced (Figure 13).

We say that the balanced sutured manifold (M,Γ) carries a compatible contact structure ξ with convex boundary if the suture Γ agrees with the dividing set Γ_ξ on the boundary. Recall that on a closed convex surface Σ in a contact manifold (M,ξ) the contact class evaluates as $c(\xi)[\Sigma]=\chi(R_+(\Sigma))-\chi(R_-(\Sigma))$. Since $\partial M=\Sigma$ is zero in homology, $c(\xi)[\Sigma]=0$ and $\chi(R_+(\Sigma))=\chi(R_-(\Sigma))$. Hence a sutured manifold that supports a contact structure with convex boundary is balanced.

We want to define an analogue of a Heegaard diagram in the case of a manifold with boundary, and sutured manifolds provide the right framework. A *sutured Heegaard diagram* consists of a surface with boundary Σ of genus g and two families of attaching curves $\{\alpha_i|i=1,\dots,k\}$ and $\{\beta_i|i=1,\dots,l\}$ with $k,l\leq g$. When we attach 2-handles to $\Sigma\times[0,1]$ along those curves

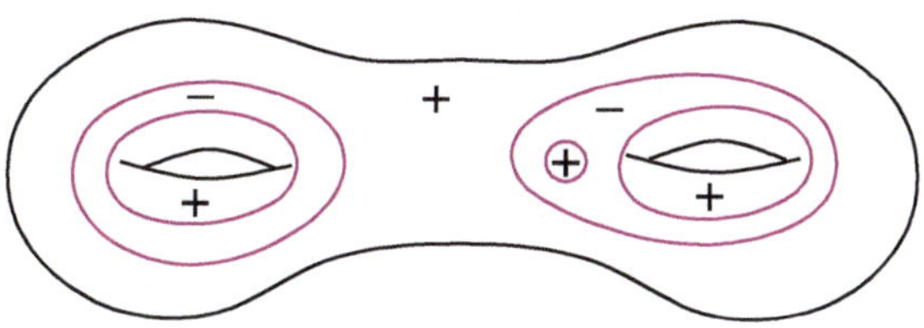

Fig. 13. A balanced sutured manifold

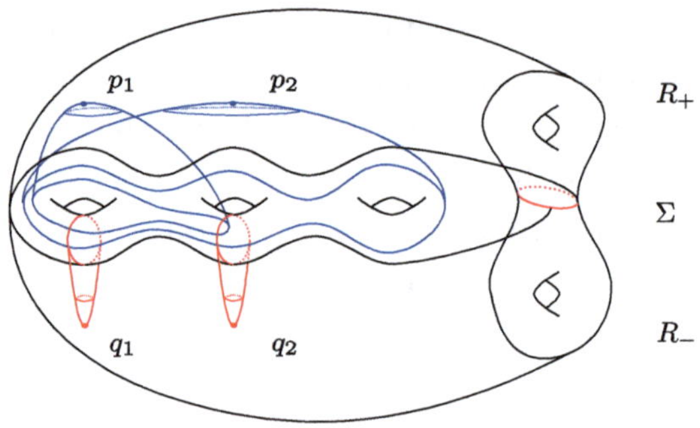

Fig. 14. A Morse function picture of a balanced sutured manifold

we will build a 3-manifold with boundary. It will have a suture $\Gamma = \partial\Sigma \times \{\frac{1}{2}\}$ which divides the boundary into two regions: R_+ which is obtained by compressing $\Sigma \times \{1\}$ along α_i curves (cutting open along α and filling in by attaching pairs of discs), and R_- obtained by compressing $\Sigma \times \{0\}$ along β_i curves. On the 3-manifold with boundary we obtain this way there is clearly a Morse function picture, analogous to the closed case, that has the centers of the attached 2-handles as critical points (see Figure 14). Note that in the case of a closed manifold and a classical Heegaard diagram of genus g and α and β curves cutting it to a $2g$ times punctured sphere, adding 2-handles to the Heegaard surface along the α and β curves produces first a manifold with two S^2 boundary components to which, in the end, we add two 3-handles. A way to build a closed manifold from the sutured manifold is to attach enough handles until we obtain a union of sutured spheres on the boundary, one for each of the boundary components of the Heegaard surface, and then add 3-balls. It is easy to see that in the sutured Heegaard diagram where $k = l$, i.e. when we attach the same number of compressing discs to $\Sigma \times \{1\}$ as to $\Sigma \times \{0\}$, the sutured manifold we build is balanced.

Andras Juhasz [17] defined Sutured Floer Homology for a balanced sutured manifold in analogy to the Heegaard Floer Homology $\widehat{HF}$. As is done in the case of a closed manifold, he associated to a balanced Heegaard diagram $(\Sigma, \alpha_i, \beta_i)$ a chain complex generated by the intersection points of the two tori $\mathbb{T}_\alpha = \alpha_1 \times \cdots \times \alpha_k$ and $\mathbb{T}_\beta = \beta_1 \times \cdots \times \beta_k$ in $Sym^k(\Sigma)$ and defined the boundary operator by counting holomorphic disc. The role played by the base-point z in the closed case is played by the boundary $\partial\Sigma$, namely we consider only domains that do not go out to the boundary. The homology of this complex is denoted by $SFH(M, \Gamma)$ and Juhasz proved it does not depend on the choice of the sutured Heegaard diagram for (M, Γ).

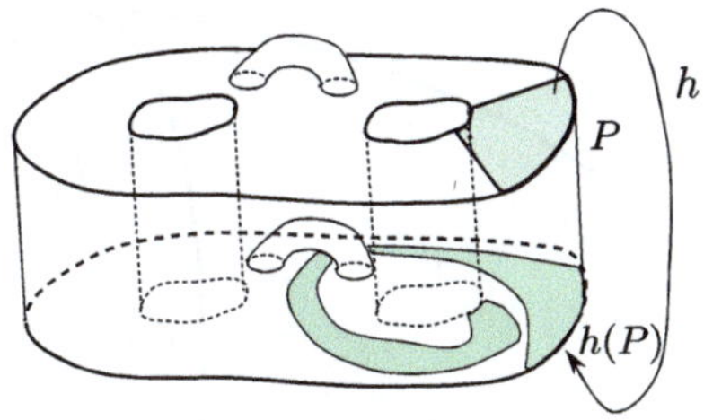

Fig. 15. Gluing to obtain a partial open book

We want to associate to a contact structure ξ with convex boundary supported by (M,Γ) an element in the Sutured Heegaard Floer homology $SFH(-M,-\Gamma)$. The change of orientation is parallel to the fact that contact invariant in the closed case lives in $\widehat{HF}(-M)$. In order to define $c(\xi)\in SFH(-M,-\Gamma)$, we first need to define an analogue of open book decompositions for manifolds with sutured boundary.

Definition 4.1. A *partial open book* (S,P,h) consists of the following data: a compact, oriented surface S with nonempty boundary, and a "partial" monodromy map $h: P\to S$ defined on a subset $P\subset S$ such that $\partial P\cap\partial S\neq 0$ and $h|_{\partial P\cap\partial S}=id$.

To obtain a sutured manifold associated to a partial open book decomposition (S,P,h) define an equivalence relation $\sim_h$ on $S\times[0,1]$ by setting $(x,1)\sim_h(h(x),0)$ for $x\in P$, and $(x,t)\sim_h(x,t')$ for $t,t'\in[0,1]$ $x\in\partial S$ (see Figure 15). It is not difficult to see that the glued-up space can be smoothed out to a manifold with boundary, where $R_+=(S\setminus P)\times\{1\}$ and $R_-=(S\setminus h(P))\times\{0\}$ and the dividing set is $\Gamma=\partial(S\setminus int\, P)$.

To motivate this definition let us think about the construction of the open book compatible with a contact structure ξ in the closed case that we described in Section 2. We will adapt this construction to the case of balanced sutured (M,Γ). We again take a cell decomposition of M by cells small enough so that the contact structure on them is standard, make the 1-skeleton Legendrian and 2-cells convex (keeping the boundary fixed), and this time consider a Legendrian 1-complex L that consists of the portion of the Legendrian 1-skeleton that is in the interior of the manifold together with enough of the Legendrian 1-simplices that come out and meet the boundary in points on the dividing set to meet every component of it in at least 2 points. If we take the cell decomposition to be fine enough, we can ensure that the neighborhood of L is a disc-decomposable contact handlebody H_2 with

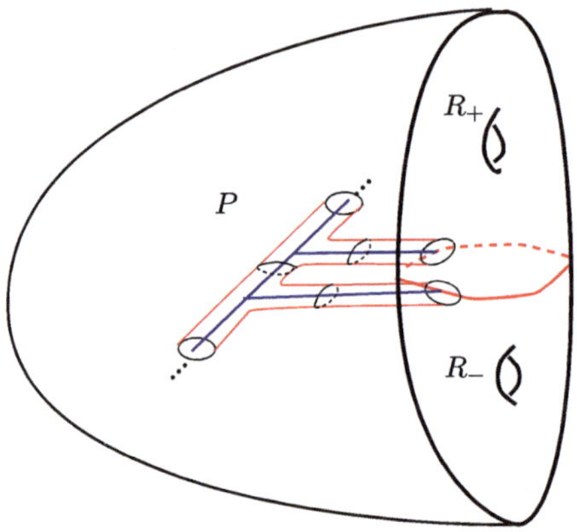

Fig. 16. Neighborhood of the Legendrian skeleton L near the boundary

convex boundary, and that its complement H_1 is also a disc-decomposable handlebody [15].

The boundary of $H_2 = \nu(L)$ consists of a "tube region" P and some discs that are part of the boundary and intersect the dividing set Γ in one segment each (see Figure 16). The tube P is divided by its dividing set into two regions $P_\pm$ and the boundary of the complement H_1 consists of $S_+ = R_+ \cup P_+$ and $S_- = R_- \cup P_-$ (we are being a bit sloppy and identifying $R_\pm$ with $R_\pm \setminus \partial(\nu(L))$. The choice of a fine enough decomposition ensures that the two contact handlebodies H_1 and H_2 are disc-decomposable and we can identify $H_1 = S \times [0, 1/2]$ and $H_2 = P \times [1/2, 1]$, where $S = R_+ \cup P_+ = R_- \cup P_-$ (see Figure 17). For simplicity of notation we ignore in this picture the fact that in the product handlebodies we mod out by $(x, t) \sim (x, t')$ to get to the real disc decomposable handlebody picture. By looking at the gluing of H_2 to H_1 inside M we see that we can think of M as obtained first by gluing along P_- to obtain the glued up $S \times [0, 1/2] \cup_{P \times \{1/2\}} P \times [1/2, 1]$ as homeomorphic to $S \times [0, 1]$, and of the final gluing along P_+ as gluing by the partial monodromy (after we identify $S \times [0, 1/2] \cup_{P \times \{1/2\}} P \times [1/2, 1]$ with $S \times [0, 1]$ in an obvious way).

We now want to associate a sutured Heegaard diagram to this decomposition. Define a *basis of arcs* in P to be a collection $\{a_i | i = 1, \ldots, k\}$ of disjoint properly embedded arcs in P with boundary on $\partial P \cap \partial S$ such that $S \setminus \bigcup \{a_i\}_{i=1,\ldots,k}$ deformation retracts onto $R_+ = S - P$. In our example in Figures 17 and 18, the basis consists of just one arc a_1. Let b_i, $i = 1, \ldots, k$, be pushoffs of a_i in the direction of ∂S so that a_i and b_i intersect exactly once at a point x_i. It is not hard to see that if we set $\Sigma = (S \times \{0\}) \cup (P \times \{\frac{1}{2}\})$,

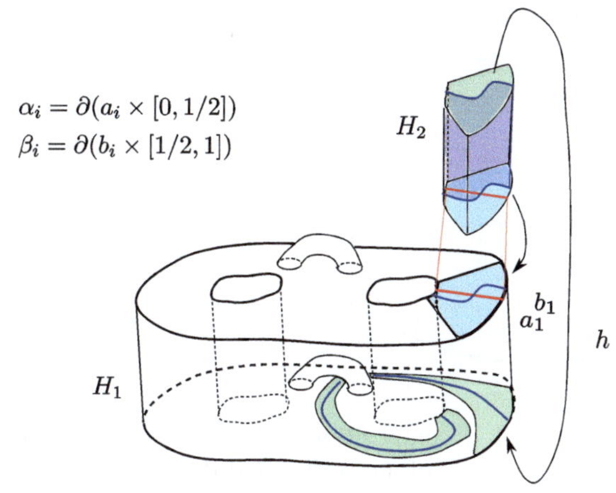

Fig. 17. The partial open book and the handlebody decomposition

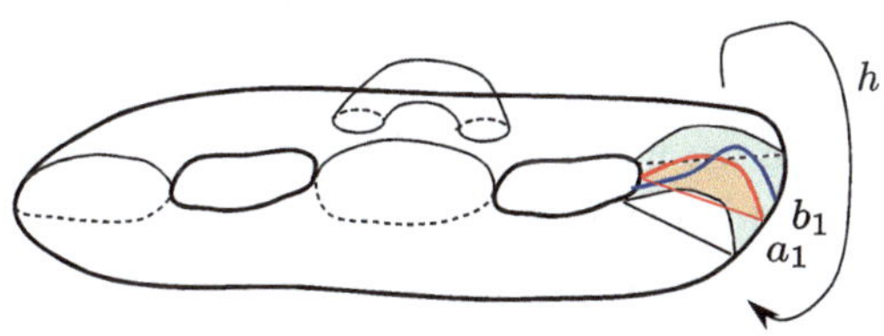

Fig. 18. The partial open book and the Heegaard diagram

$\alpha_i = \partial(a_i \times [0, \frac{1}{2}])$ and $\beta_i = (b_i \times \{\frac{1}{2}\}) \cup (h(b_i) \times \{0\})$, then (Σ, β, α) is a Heegaard diagram for $(-M, -\Gamma)$. See Figure 19.

We can again look at the special generator $\mathbf{x} = (x_1, \ldots, x_i, \ldots)$ for $SFH(\Sigma, \beta, \alpha)$ and, as in the case of the closed manifold, $\mathbf{x} = [(x_1, \ldots, x_k)]$ is a cycle. It is shown in [15] that it defines a contact invariant.

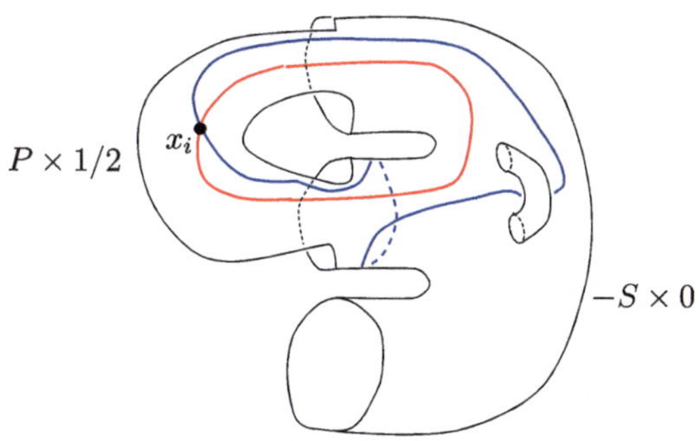

Fig. 19. The Heegaard diagram determined by a basis of arcs a_1 and the contact element $\mathbf{x} = (x_1)$

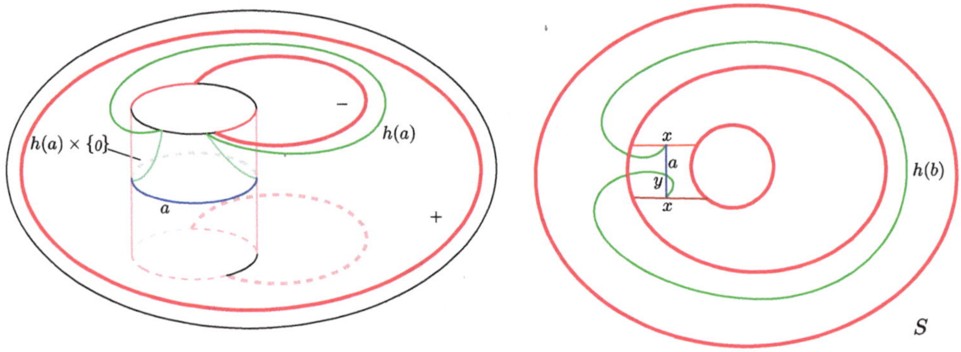

Fig. 20. A partial open book decomposition and a Heegaard diagram for a neighborhood of an overtwisted disc

Theorem 4.2. *The point* $\mathbf{x} = [(x_1, \ldots, x_k)]$ *is independent of choices up to* ± 1 *and generates the sutured contact invariant* $c(\xi) \in SFH(-M, -\Gamma)$.

A concrete example we will look at here is a partial open book decomposition for a neighborhood of an overtwisted disc. The corresponding sutured manifold is (B^3, Γ) with the dividing set consisting of 3 parallel curves. It is enough to take just one segment connecting two nonadjacent components of the dividing set for the Legendrian 1-complex L, to get the complement to be a disc-decomposable handlebody. The segment comprising L is the core of the cylinder in Figure 20, while $P_\pm$ are two halves of the cylinder, and the basis of arcs consists of a single arc a on P. In the left-hand diagram of Figure 20 arc a is shown on $P \times \{1/2\} \subset S \times \{1/2\}$ (thus might more properly be denoted by $a \times \{1/2\}$). Isotoping a (rel endpoints) through $N(L)$ where we use the homeomorphism $N(L) = P \times [1/2, 1]$ produces, by the definition of the monodromy h, the arc $h(a) \times \{0\} \subset S \times \{0\}$. Finally pushing $h(a) \times \{0\}$ (rel endpoints) through the fibration $M - N(L)$ that identifies $M - N(L) = S \times [0, 1/2]$ results in $h(a) \times \{1/2\} \subset S \times \{1/2\}$, and this is denoted simply by $h(a)$. The right side of the figure shows a and $h(a)$ in $S = S \times \{1/2\}$.

5. Gluing Theorem for Sutured Manifolds

In [13] we work to understand the effect of cutting and gluing of contact manifolds along convex surfaces in sutured manifolds in the context of the contact invariant. We say that one balanced sutured manifold (M', Γ') is a *sutured submanifold* of another balanced sutured manifold (M, Γ) if M' is a submanifold with boundary of M and $M' \subset int(M)$. A contact structure ξ defined on $M \subset int(M')$ is *compatible* with the sutured manifold structures of M and M' if the dividing set of ξ on the boundary of $M - int(M')$ is $\Gamma \cup -\Gamma'$. In this section we will define a map on sutured Floer homology induced by the inclusion of (M', Γ') into (M, Γ) in the presence of a compatible contact structure in the complement. We will see how triviality of the contact invariant on a sutured submanifold implies the triviality of the contact invariant of the manifold itself. In the next section we will use the gluing theorem to calculate the contact invariant in some examples, and obtain some interesting obstructions to fillability. For simplicity, we can think of all constructions as done over $\mathbb{Z}/2\mathbb{Z}$, so we do not have to worry about the sign ambiguity.

If a connected component N of $M \setminus int(M')$ contains no components of ∂M we say that N is *isolated*. When $M \setminus int(M')$ has no isolated components we have the following:

Theorem 5.1 [13]. *Let (M', Γ') be a sutured submanifold of (M, Γ), and let ξ be a compatible contact structure on $M \setminus int(M')$. Assume that $M - int(M')$ has no isolated components. Then ξ induces a natural map:*

$$\Phi_\xi : SFH(-M', -\Gamma') \to SFH(-M, -\Gamma)$$

Moreover, if ξ' is any contact structure on M' compatible with Γ' then

$$\Phi_\xi(c(M', \Gamma', \xi')) = c(M, \Gamma, \xi' \cup \xi)$$

where $\xi' \cup \xi$ is a contact structure on M that restricts to ξ on $M \setminus int(M')$ and to ξ' on M'.

There is a more complicated statement in the case of existence of isolated components which involves considering multi-pointed Heegaard diagrams and tensoring with $\widehat{HF}(S^1 \times S^2)$, see [13].

Brief description of Φ_ξ. To define this map we have to carefully extend a sutured Heegaard diagram for (M', Γ') to a diagram for (M, Γ). For details of this construction look at [13]. Here is just a very quick idea. We use, in an essential way, the contact structure ξ on $M \setminus int(M')$ compatible with the sutures Γ and Γ' to define the map. We start from a Heegaard surface for M'. If we are given ξ' on M' take $(\Sigma', \beta', \alpha')$ to be defined by a partial open book compatible with ξ'. If we are not given a ξ', let $(\Sigma', \beta', \alpha')$ be a Heegaard diagram arising from a partial open book decomposition of some contact structure ζ which has dividing set Γ' on $\partial M'$. We would like to join this Heegaard diagram to one generated by ξ on $M \setminus int(M')$. However, that is in general not precise enough, as it does not provide enough compressing discs for the union. To connect the two sides, we need to start with N, a contact product neigbourhood of $\partial M'$ in $M \setminus int(M')$, and let $M'' = M \setminus int(M' \cup N)$. We carefully choose a Heegaard surface Σ_N which is compatible with the $[0,1]$-invariant contact structure $\xi|_N$, as well as a basis of arcs $\{a_i^N\}$ for it. We then extend Σ_N to a Heegaard surface $\Sigma_{M''}$ and denote a basis of arcs on the union extending $\{a_i^N\}$ in a way necessary to obtain sutured Heegaard diagram for $(M \setminus int\, M', \Gamma \cup \Gamma')$ by $\{\alpha_i''\}$. This needs to be done in a way compatible with $\xi \cup \zeta$, i.e. so that after $\{a_i''\}$ and their perturbations $\{b_i''\}$ are chosen, and we look at boundaries $\{\alpha_i''\}$ and $\{\beta_i''\}$ of corresponding compressing discs, the special point $\mathbf{x}'' = (\ldots, \mathbf{x}_i'', \ldots)$, consisting of $x_i'' = a_i'' \cap b_i''$, is the contact class of $\xi \cup \zeta$. After gluing we get a Heegaard diagram for (M, Γ) by taking $\alpha = \alpha' \cup \alpha''$ and $\beta' = \beta' \cup \beta''$. We then define

$$\Phi_\xi : CF(\Sigma', \beta', \alpha') \to CF(\Sigma, \beta' \cup \beta'', \alpha' \cup \alpha''),$$
$$\mathbf{y} \mapsto (\mathbf{y}, \mathbf{x}''). \quad \blacksquare$$

Above theorem has as an immediate consequence:

Theorem 5.2 [13]. *Let $i : (M', \Gamma', \xi') \to (M, \Gamma, \xi)$ be an inclusion such that $\xi|_{M'} = \xi'$. If $c(M, \Gamma, \xi) \neq 0$, then $c(M', \Gamma', \xi') \neq 0$.*

Juhász [17] showed that we can recover the Heegaard Floer Homology of a closed manifold M by calculating the sutured Floer homology of the manifold with sutured boundary $(M \setminus B^3, \Gamma)$ obtained by removing a solid ball B^3 from M and letting the suture Γ be $S^1 \subset S^2$ on the boundary $S^2 = \partial(M \setminus B^3)$. This isomorphism is one to one on generators when we consider the sutured Heegaard diagram on $M \setminus B^3$ and the corresponding Heegaard diagram on M obtained by closing the Heegaard surface by adding a disc along its boundary and making its center a marked point. This isomorphism takes in a natural way the sutured contact invariant of $(M \setminus B^3, S^1)$ to the contact invariant of the closed manifold. Since fillable structures have nonzero contact invariants we have:

Theorem 5.3 [13]. *If $c(M, \Gamma, \xi) = 0$, then (M, Γ, ξ) does not embed into any fillable contact structure.*

It is clear from the construction, as we had remarked after defining the contact invariant generator in the closed case, that an arc that is taken to the left by the monodromy produces a holomorphic disc that kills the contact invariant. The partial open book we constructed in Figure 20 shows that $c(\xi) = 0$ for the sutured manifold which is the neighborhood of the overtwisted disc (there is a disc from y to x making $\partial y = x$). From this, coupled with the embedding theorem, we see another proof that $c(\xi) = 0$ for any overtwisted contact structure.

6. A TQFT Aspect of $c(\xi)$ and Fillability Obstructions

In this section we study S^1 invariant contact structures ξ on $\Sigma \times S^1$, such that $\Sigma \times \{t\}$ is convex surface with Legendrian boundary, and for all the components of $\partial\Sigma$ there is twisting of ξ with respect to the framing determined by Σ. These contact structures are classified by their dividing sets Γ_ξ which are properly embedded multicurves (disjoint union of curves and arcs) that intersect every component of the boundary of Σ in an even number of points, and divide Σ and hence $\partial\Sigma$ into positive and negative regions. We say that a properly embedded multicurve $K \subset \Sigma$ is *isolating* if $\Sigma \setminus K$ contains a component that does not intersect $\partial\Sigma$ (see Figure 21). Examples of contact structures with $c(M, \Gamma, \xi) = 0$ can now be obtained quite easily from:

Theorem 6.1 [13]. *Let ξ_K be the S^1 invariant contact structures on $\Sigma \times S^1$, such that $\Sigma \times \{t\}$ is convex with dividing set $\Gamma_\Sigma = K$. If K is isolating then* $c(\Sigma \times S^1, \xi_K) = 0$.

To prove this theorem we use some TQFT-like properties of contact invariants for S^1 invariant contact structures on $\Sigma \times S^1$. Consider a "bordered"

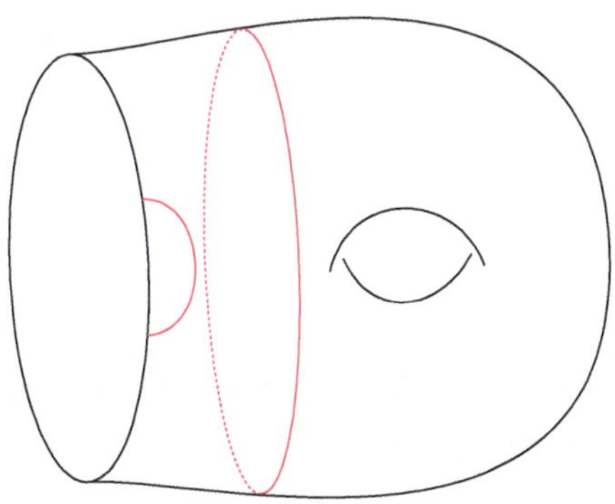

Fig. 21. An isolating dividing set on a punctured torus

surface with boundary (Σ, F), i.e. a surface Σ with a finite subset $F \subset \partial\Sigma$ consisting of 2n points that divide the boundary $\gamma = \partial\Sigma$ into alternating positive and negative regions $\gamma_\pm$, with $\gamma \setminus F = \gamma_- \sqcup \gamma_+$. We say that a union K of closed curves and properly embedded arcs in Σ with $\partial K = F$ is a dividing set on (Σ, F), or that K divides Σ, if $\Sigma \setminus K$ is a disjoint union of positive and negative regions $R_\pm$, and $\partial R_\pm = K \cup \gamma_\pm$. Denote by $\mathcal{D}(\Sigma, F)$ the family of all dividing sets for (Σ, F). We will use Sutured Floer Homology and sutured contact invariants to define a map that assigns a vector space $V(\Sigma, F)$ to each (Σ, F) and an element in that vector space to each $K \in \mathcal{D}(\Sigma, F)$ with following TQFT-like properties:

1. If Σ is connected, then

$$V(\Sigma, F) = \mathbb{F}^2 \otimes \cdots \otimes \mathbb{F}^2,$$

   where $\mathbb{F} = \mathbb{Z}/2\mathbb{Z}$, the number of copies of $\mathbb{F}^2$ is $r = n - \chi(\Sigma)$, and $\mathbb{F}^2 = \mathbb{F} \oplus \mathbb{F}$ is a graded $\mathbb{F}$-module whose first summand has grading 1 and the second summand has grading -1. If (Σ, F) is the disjoint union of (Σ_1, F_1) and (Σ_2, F_2), then

$$V(\Sigma_1 \sqcup \Sigma_2, F_1 \sqcup F_2) \simeq V(\Sigma_1, F_1) \otimes V(\Sigma_2, F_2).$$

2. To each $K \in \mathcal{D}(\Sigma, F)$ it assigns $c(K) \in V(\Sigma, F)$. If K has a homotopically trivial closed component, then $c(K) = 0$.
3. Given (Σ, F), possibly disconnected, let $\delta, \delta' \subset \partial\Sigma$ be mutually disjoint submanifolds of $\partial\Sigma$, such that their endpoints do not lie in F, and let τ be a diffeomorphism $\tau : \delta \xrightarrow{\sim} \delta'$ which identifies $\delta \cap F \xrightarrow{\sim} \delta' \cap F$ and preserves the $\pm$ labeling and reverses the orientation on δ, δ' inherited from $\partial\Sigma$. Denote by (Σ', F') the result of identifying γ and γ' via τ. For every $K \in \mathcal{D}(\Sigma, F)$ denote by $\overline{K}$ the dividing set obtained from K by gluing $K|_\gamma$ and $K|_{\gamma'}$. Then there exists a map

$$\Phi_\tau : V(\Sigma, F) \to V(\Sigma', F'),$$

   which satisfies

$$c(K) \mapsto c(\overline{K}).$$

Figure 22 shows the case when Σ is a union of disjoint surfaces Σ'' and Σ''', and hence Σ' is obtained by gluing Σ'' and Σ'''.

To define this assignment for every $(\Sigma, F, \gamma_\pm)$, we first perturb F by moving it slightly in the direction opposite to the one defined by the orientation

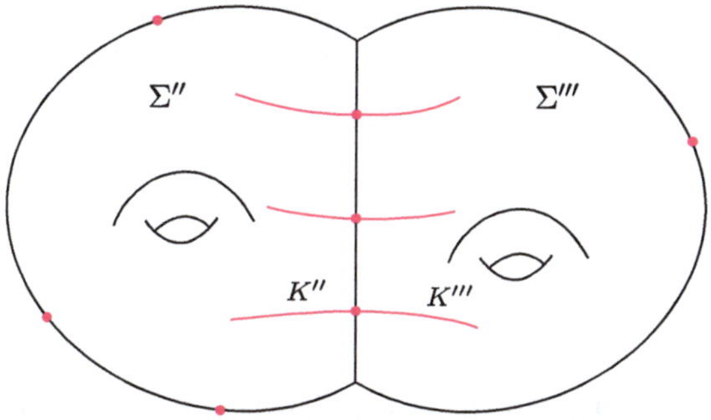

Fig. 22. Gluing (Σ'', K'') and (Σ''', K''')

on $\gamma = \partial\Sigma$ to obtain F^0, and shifting $\gamma_\pm$ in the same way to obtain $\gamma^0_\pm$. We then consider the sutured 3-manifold $(\Sigma \times S^1, \Gamma)$ where $\Gamma = F^0 \times S^1$ and $R_\pm = \gamma^0_\pm \times S^1$. Denote by $V(\Sigma, F)$ the sutured Floer homology of $(-\Sigma \times S^1, -\Gamma)$. A dividing set $K \in \mathcal{D}(\Sigma, F)$ defines an S^1-invariant contact structure ξ_K on $(\Sigma \times S^1, \Gamma)$, and hence the corresponding contact invariant $c(K) = c(\xi) \in V(\Sigma, F)$. We use F^0 instead of F since the two convex surfaces $\Sigma \times \{t\}$ and $\partial(\Sigma \times S^1)$ are transverse, so the two dividing sets have to mark "interlocking points" along the Legendrian intersection curve, i.e. points in $F^0 = \Gamma \cap \partial\Sigma$ must lie between the endpoints F of K (see Figure 5).

To prove that $V(\Sigma, F) = \mathbb{F}^2 \otimes \cdots \otimes \mathbb{F}^2$ with appropriate number of factors, we first need to calculate $SFH(D^2 \times S^1, \Gamma_2)$ for a solid torus with dividing set Γ_2 made up of 4 longitudinal curves. It is shown in [13, Section 5, Example 3] that the result is $SFH(D^2 \times S^1, \Gamma_2) = \mathbb{F}^2 = \mathbb{F}_{(1)} \oplus \mathbb{F}_{(-1)}$. There are exactly two S^1 invariant contact structures compatible with these sutures, each determined by one of the two dividing sets on D^2 consisting of two arcs, and each generating one of the $\mathbb{F}_{(\pm 1)}$. We cut the surface Σ repeatedly by properly embedded arcs that connect + and − regions of $\partial\Sigma$ until we get a disc. Every time we do a cut, each of the two new arcs in the boundary that correspond to the cut gets one marked point in F, thus each cut adds two points to the boundary, increasing n by one. A tensor product formula by Juhász [18, Proposition 8.10] that holds for splitting sutured manifolds along product annuli applies. The annulus here is product of a cutting arc with S^1. The number of summands, $r = n - \chi(\Sigma)$ corresponds to the number of discs, each with 4 marked points on the boundary, that Σ is finally cut into, and the $\mathbb{F}^2$ factor corresponds to the contribution of each such disc to SFH.

If K has a homotopically trivial closed component, then ξ_K is overtwisted and hence $c(K) = 0$ since we defined $c(K) \in V(\Sigma, F)$ to be the value of the contact invariant for the S^1-invariant contact structure ξ_K determined by K.

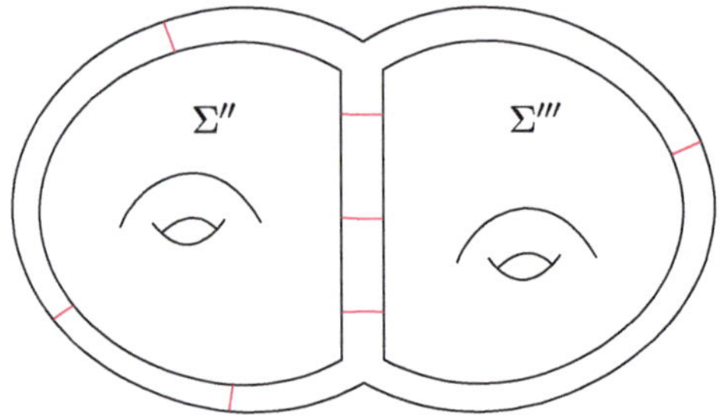

Fig. 23. The dividing set K_0 on $\Sigma' \setminus \Sigma$ is given in red

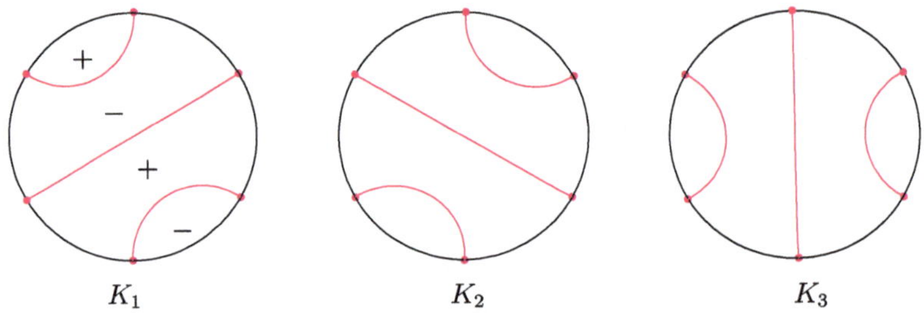

Fig. 24. Dividing sets on $\Sigma = D^2$ for $|F| = 6$

The Gluing Theorem 5.1 applied to $\Sigma \times S^1 \subset \Sigma' \times S^1$ gives us the map Φ_τ. We think of Σ as a subset of Σ' as in Figure 23, where $\Sigma = \Sigma'' \sqcup \Sigma'''$ and Σ'' and Σ''' are identified with slightly shrunk copies inside σ'.

The contact structure on $\Sigma' \times S^1 \setminus \Sigma \times S^1$ is the S^1 invariant structure determined by the dividing set K_τ on $\Sigma' \setminus \Sigma$ determined by F'', F''' and the identification τ.

By studying what we have in the case of $\Sigma = D^2$ and $|F| = 6$ we see in [16] that for dividing sets K_1, K_2, K_3 as in Figure 24 we have that the corresponding $c(K_1), c(K_2), c(K_3)$ are nonzero and distinct, and satisfy $c(K_1) = c(K_2) + c(K_3)$. Note that these three configurations are related by bypass addition. In fact, K_2 is obtained by adding a bypass to the front of K_1 along an arc connecting the three dividing curves, and K_3 is obtained by adding a bypass from the back of K_1 (digging a bypass).

When we combine this with the gluing theorem, we obtain the same relationship for any three dividing sets $\{K_i,\ i = 1, 2, 3\}$ related by bypass addition on a general Σ. We will quickly argue that $c(\xi_K) = 0$ for our example in Figure 21.

It is not hard to see that by adding and digging bypasses along the bypass arc δ given in blue in K_1 in Figure 25, we obtain dividing curves as in K_2

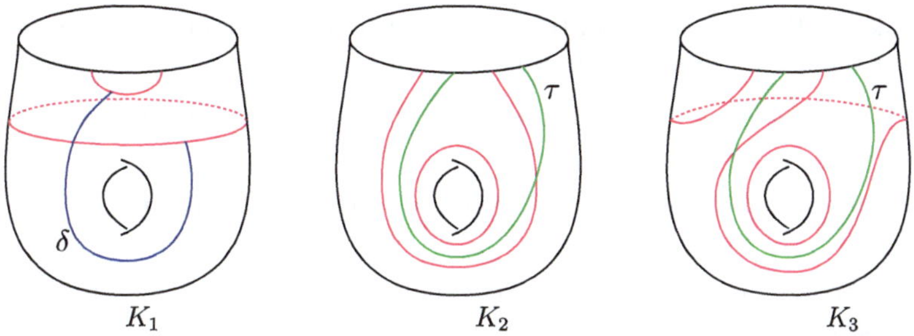

Fig. 25. Bypass adition in the proof of $c(\xi_K) = 0$ for an isolating dividin set

and K_3. Cutting Σ along the green curve τ and looking at the dividing sets K_2' and K_3' resulting from K_2 and K_3 on the resulting annulus, we see that $c(K_2') = c(K_3')$. Juhász' annulus theorem says gluing along τ induces an isomorphism, so that we get $c(K_2) = c(K_3)$. This completes the proof that $c(\xi_{K_1}) = 0$ since $c(K_1) = c(K_2) + c(K_3) = 0$ as we are working over $\mathbb{F} = \mathbb{Z}/2\mathbb{Z}$.

More general isolating dividing sets on surfaces of higher genus can be dealt with by similar methods or reduced to this case, thus proving Theorem 6.1. For full explanation see [16]. The fact that the contact invariant vanishes for isolating dividing sets was proved over $\mathbb{Z}$ coefficients by Patrick Massot [23].

Theorem 6.1 together with Theorem 5.3 shows that $(\Sigma \times S^1, \xi_K)$ with contact structures corresponding to an isolating K form a vast family of universally tight contact structures that do not embed into fillable structures and are thus generalizing Giroux torsion. Similar results were obtained by looking at holomorphic curves and contact homology by Chris Wendl [34].

Acknowledgements. I want to thank the organizers for inviting me to give this lecture series, and to Laura Starkston for sharing her notes from the lectures. Thanks go to my collaborators Ko Honda and Will Kazez for all the interesting mathematics we did together that resulted in many results mentioned in these notes, and for originally drawing many figures that appear here, as well as to Whitney George and David Gay for help with some of the figures. Finally, thanks to Margaret Symington for reading the original draft and suggesting many improvements.

References

[1] S. Akbulut, B. Ozbagci, Lefschetz fibrations on compact Stein surfaces. Geom. Topol. **5**, 319–334 (2001)

[2] K. Baker, J.B. Etnyre, J. Van Horn-Morris, Cabling, contact structures and mapping class monoids. J. Differ. Geom. **90**, 1–80 (2012)

[3] J.A. Baldwin, Contact monoids and Stein cobordisms. Math. Res. Lett. **19**(1), 31–40 (2012)

[4] D. Bennequin, Entrelacements et equations de Pfaff. Astérisque **107–108**, 87–161 (1983)

[5] Y. Eliashberg, Classification of overtwisted contact structures on 3-manifolds. Invent. Math. **98**, 623–637 (1989)

[6] Y. Eliashberg, Topological characterization of Stein manifolds of dimension > 2. Int. J. Math. **1**, 29–46 (1990)

[7] Y. Eliashberg, Filling by holomorphic discs and its applications. Lond. Math. Soc. Lect. Note Ser. **151**, 45–67 (1991)

[8] P. Ghiggini, Infinitely many universally tight contact manifolds with trivial Ozsváth-Szabó contact invariants. Geom. Topol. **10**, 335–357 (2006)

[9] P. Ghiggini, K. Honda, J. Van Horn-Morris, The vanishing of the contact invariant in the presence of torsion. arXiv:0706.1602

[10] E. Giroux, Géométrie de contact: de la dimension trois vers les dimensions supérieures, in *Proceedings of the International Congress of Mathematicians*, vol. II, Beijing, 2002 (Higher Ed. Press, Beijing, 2002), pp. 405–414

[11] M. Gromov, Pseudo holomorphic curves in symplectic manifolds. Invent. Math. **82**, 307–347 (1985)

[12] K. Honda, W. Kazez, G. Matić, Convex decomposition theory. Int. Math. Res. Not. **2002**, 55–88 (2002)

[13] K. Honda, W. Kazez, G. Matić, Right veering automorphisms of compact surfaces with boundary I. Invent. Math. **169** (2007)

[14] K. Honda, W. Kazez, G. Matić, On the contact class in Heegaard Floer homology. J. Differ. Geom. **83**(2), 289–311 (2009)

[15] K. Honda, W. Kazez, G. Matić, The contact invariant in sutured Floer homology. Invent. Math. **176**(3), 637–676 (2009)

[16] K. Honda, W. Kazez, G. Matić, Contact structures, sutured Floer homology and TQFT. Preprint, arXiv:0807.2431

[17] A. Juhász, Holomorphic discs and sutured manifolds. Algebr. Geom. Topol. **6**, 1429–1457 (2006) (electronic)

[18] A. Juhász, Floer homology and surface decompositions. Geom. Topol. **12**, 299–350 (2008)

[19] Y. Lekili, Planar open books with four binding components. Algebr. Geom. Topol. **11**, 909–928 (2011)

[20] R. Lipshitz, A cylindrical reformulation of Heegaard Floer homology. Geom. Topol. **10**, 955–1097 (2006) (electronic)

[21] A. Loi, R. Piergallini, Compact Stein surfaces with boundary as branched covers of B4. Invent. Math. **143**, 325–348 (2001)

[22] C. Manolescu, P. Ozsváth, Z. Szabó, D. Thurston, On combinatorial link Floer homology. Geom. Topol. **11**, 2339–2412 (2007)

[23] P. Massot, Infinitely many universally tight torsion free contact structures with vanishing Ozsváth–Szabó contact invariants. Math. Ann. **353**(4), 1351–1376 (2012)

[24] P. Ozsváth, Z. Szabó, Holomorphic disks and topological invariants for closed three-manifolds. Ann. Math. **159**, 1027–1158 (2004)

[25] P. Ozsváth, Z. Szabó, Holomorphic disks and three-manifold invariants: properties and applications. Ann. Math. **159**, 1159–1245 (2004)

[26] P. Ozsváth, Z. Szabó, Heegaard Floer homology and contact structures. Duke Math. J. **129**, 39–61 (2005)

[27] P. Ozsváth, A. Stipsicz, Z. Szabó, Combinatorial Heegaard Floer homology and nice Heegaard diagrams. Adv. Math. **231**, 102–171 (2012)

[28] K. Reidemeister, Zur dreidimensionalen Topologie. Abh. Math. Semin. Univ. Hamb. **9**, 189–194 (1933)

[29] D. Rolfsen, *Knots and Links* (Publish or Perish Inc., Houston, 1990). Corrected revision of the 1976 original

[30] S. Sarkar, J. Wang, An algorithm for computing some Heegaard Floer homologies. Ann. Math. **171**(2), 1213–1236 (2010)

[31] J. Singer, Three-dimensional manifolds and their Heegaard diagrams. Trans. Am. Math. Soc. **35**, 88–111 (1933)

[32] W. Thurston, H. Winkelnkemper, On the existence of contact forms. Proc. Am. Math. Soc. **52**, 345–347 (1975)

[33] I. Torisu, Convex contact structures and fibered links in 3-manifolds. Int. Math. Res. Not. **2000**, 441–454 (2000)

[34] C. Wendl, A hierarchy of local symplectic filling obstructions for contact 3-manifolds. Duke Math. J. **162**(12), 2197–2283 (2013)

G. Matić (✉)

University of Georgia
Athens, GA 30602
USA

e-mail: gordana@math.uga.edu

url: http://www.math.uga.edu/~gordana

BOLYAI SOCIETY
MATHEMATICAL STUDIES, 26

Contact and Symplectic Topology
pp. 275–355.

Notes on Bordered Floer Homology

ROBERT LIPSHITZ, PETER OZSVÁTH, and DYLAN P. THURSTON

1. Introduction

Heegaard Floer homology is a kind of $(3+1)$-dimensional topological field theory defined by the second author and Z. Szabó. More precisely, one variant of Heegaard Floer homology associates to each connected, oriented 3-manifold Y an abelian group $\widehat{HF}(Y)$ [46] (see also [16]), and to each smooth, connected, 4-dimensional cobordism W from Y_1 to Y_2 a group homomorphism $\hat{F}\colon \widehat{HF}(Y_1) \to \widehat{HF}(Y_2)$ [50]. This assignment is functorial: composition of cobordisms corresponds to composition of maps. As the name suggests, the Heegaard Floer homology groups are the homologies of chain complexes $\widehat{CF}(Y)$, defined via Lagrangian-intersection Floer homology[1]. The invariant is also multiplicative: the chain complex $\widehat{CF}(Y_1 \# Y_2)$ associated to the connected sum of Y_1 and Y_2 is the tensor product $\widehat{CF}(Y_1) \otimes \widehat{CF}(Y_2)$ of the chain complexes associated to Y_1 and Y_2. The other variants of Heegaard Floer homology—$HF^+(Y)$, $HF^-(Y)$ and $HF^\infty(Y)$—are modules over $\mathbb{Z}[U]$, but otherwise behave fairly similarly to $\widehat{HF}(Y)$ (but see point (4) below).

Heegaard Floer homology has received widespread attention largely because of its striking topological applications. Many of these applications draw on the remarkable geometric content of the Heegaard Floer invariants:

[1]Strictly speaking, in the original definition the manifolds were only totally-real, not Lagrangian. It was shown in [54] that a Kähler form can be chosen making the relevant submanifolds Lagrangian.

F. Bourgeois et al. (eds.), *Contact and Symplectic Topology*,
Bolyai Society Mathematical Studies 26, DOI 10.1007/978-3-319-02036-5_7,

(1) The group $\widehat{HF}(Y)$ detects the Thurston norm of Y; similarly, the variant of Heegaard Floer homology $\widehat{HFK}(Y,K)$ associated to a null-homologous knot K, called knot Floer homology [44, 56], detects the genus of K [43].

(2) The group $\widehat{HF}(Y)$ detects whether and how Y fibers over S^1; similarly, $\widehat{HFK}(Y,K)$ detects whether K is fibered [9, 41].

(3) The two previous properties are reminiscent of the Alexander polynomial, which gives partial information in each case. There is a precise relationship between $\widehat{HFK}$ and the Alexander polynomial. Specifically, if K is a knot in S^3, then $\widehat{HFK}(K)$ is endowed with an integral bigrading $\widehat{HFK}(K) = \bigoplus_{d,s\in\mathbb{Z}} \widehat{HFK}_d(K,s)$, and

$$\sum_d (-1)^d T^s \operatorname{rank} \widehat{HFK}_d(K,s) = \Delta_K(T)$$

[44, 56].

(4) The Heegaard Floer homology groups of closed 3-manifolds are now known to agree with the Seiberg-Witten Floer homology groups [6–8, 19–23, 59–63]. Moreover, one can use Heegaard Floer homology to define an invariant of smooth, closed 4-manifolds [50], with similar properties to the Seiberg-Witten invariant [14, 47, 57]; it is expected that the two invariants agree. Note, however, that to capture the analogue of the Seiberg-Witten invariant one needs to work with the HF^+ and HF^- variants of Heegaard Floer homology.

As mentioned above, Heegaard Floer homology is defined using Lagrangian-intersection Floer homology, i.e., by counting holomorphic curves. Consequently, it is in general hard to compute—though there are now several algorithms for doing so; see particularly [37–40, 58]. With the goal of computing and better understanding Heegaard Floer homology in mind, we have been developing bordered Heegaard Floer homology, a tool for understanding the behavior of the Heegaard Floer homology group $\widehat{HF}(Y)$ under cutting and gluing of Y along surfaces. Roughly, bordered Floer homology is a $(2+1+1)$-dimensional field theory. That is, roughly, it assigns to each connected, oriented surface F a differential graded algebra $\mathcal{A}(F)$ and to a cobordism Y from F_1 to F_2 an $(\mathcal{A}(F_1), \mathcal{A}(F_2))$-bimodule $\widehat{CFDA}(Y)$. Composition of cobordisms corresponds to tensor product of bimodules.

More precisely, like in Heegaard Floer homology, in bordered Floer homology, the invariants are not associated directly to the topological objects of interest—manifolds of dimensions 2 through 4—but rather to certain combinatorial representations for these objects, which we describe next.

The combinatorial representations of oriented surfaces which appear in bordered Floer homology, the *pointed matched circles*, which we denote by $\mathcal{Z}$, consist essentially of a handle-decomposition of the surface. (See Definition 2.1 below for a more precise formulation.) We will let $F(\mathcal{Z})$ denote the surface underlying $\mathcal{Z}$. Bordered Floer homology associates to such a pointed matched circle a differential-graded (*dg*) algebra $\mathcal{A}(\mathcal{Z})$; the definition of $\mathcal{A}(\mathcal{Z})$ is purely combinatorial.

The three-dimensional objects studied in the bordered theory are cobordisms, i.e., three-manifolds with parameterized boundary. More precisely, a *bordered* 3-*manifold* consists of a compact, oriented 3-manifold-with-boundary Y and a homeomorphism $\phi : F(\mathcal{Z}) \to \partial Y$, where $\mathcal{Z}$ is some pointed matched circle.

Bordered Floer homology associates to a bordered 3-manifold $(Y, \phi : F(\mathcal{Z}) \to \partial Y)$ a left *dg* $\mathcal{A}(-\mathcal{Z})$-module, which we denote $\widehat{CFD}(Y)$. (The minus sign in front of $\mathcal{Z}$ denotes a reversal of orientation.) Explicitly, $\widehat{CFD}(Y)$ is a left module over the *dg* algebra $\mathcal{A}(-\mathcal{Z})$; and $\widehat{CFD}(Y)$ is equipped with a differential which satisfies the Leibniz rule[2] with respect to the action by the algebra;

$$\partial_{\widehat{CFD}(Y)}(a \cdot x) = d_{\mathcal{A}(-\mathcal{Z})}(a) \cdot x + a \cdot \partial_{\widehat{CFD}(Y)}(x).$$

Like the algebras, the modules $\widehat{CFD}$ are also associated to combinatorial representations of the underlying structure. In this case, the combinatorial structure is called a *bordered Heegaard diagram* (Definition 2.5 below). Unlike the algebras, the definition of $\widehat{CFD}$ then depends on further analytic choices (specifically, a family of complex structures on the underlying Heegaard surface); but the quasi-isomorphism type of the module does not depend on these further choices.

The modules $\widehat{CFD}$ can be used to reconstruct the Heegaard Floer homology $\widehat{HF}$ via *pairing theorems*, which come in several variants. For example, recall that if M_1 and M_2 are two *dg*-modules over some algebra $\mathcal{A}$, we can consider their chain complex of morphisms $\mathrm{Mor}_{\mathcal{A}}(M_1, M_2)$, which is to be

[2]The ground ring for bordered Floer homology is $\mathbb{Z}/2\mathbb{Z}$; hence the signs usually appearing in the differential graded Leibniz rule become irrelevant.

thought of as the space of $\mathcal{A}$-linear maps $\phi: M_1 \to M_2$, equipped with a differential

$$d_{\mathrm{Mor}}(\phi) = d_{M_2} \circ \phi + \phi \circ d_{M_1}.$$

Theorem 1.1. *Let Y_1 and Y_2 be two $\mathcal{Z}$-bordered three-manifolds. Then there is an isomorphism between the homology of the morphism space $\mathrm{Mor}_{\mathcal{A}(-\mathcal{Z})}(\widehat{CFD}(Y_1), \widehat{CFD}(Y_2))$ and the Heegaard Floer homology $\widehat{HF}(Y)$ of the three-manifold $Y = -Y_1 \cup_{F(\mathcal{Z})} Y_2$ obtained by gluing $-Y_1$ and Y_2 along their common boundary $F(\mathcal{Z})$ (according to the identifications specified by their borderings).*

(This was not the original formulation of the pairing theorem; rather it is a re-formulation appearing first in [1]; see also [32].)

The discussion above naturally raises the following questions:

(1) To what extent is the algebra of a pointed matched surface an invariant of the underlying surface?
(2) In what way does the bordered invariant $\widehat{CFD}(Y)$ depend on the parameterization of the boundary of Y?

Perhaps not too surprisingly, the answers to both of these questions are governed by certain bimodules.

Given a homeomorphism $\psi: F(-\mathcal{Z}_1) \to F(-\mathcal{Z}_2)$, there is an $\mathcal{A}(\mathcal{Z}_1)$-$\mathcal{A}(\mathcal{Z}_2)$-bimodule $\widehat{CFDD}(\psi)$ which allows one to change the framing of a bordered three-manifold. There is a mild technical point which becomes important when discussing these bimodules: as we will see, $F(\mathcal{Z})$ contains a distinguished disk, and the homeomorphism ψ is required to fix this disk pointwise.

We can now state the dependence of the modules on the parameterization in terms of these bimodules. To state the dependence, recall that if $\mathcal{A}_1$ and $\mathcal{A}_2$ are two *dg* algebras, B is an $\mathcal{A}_1$-$\mathcal{A}_2$-bimodule and M is a *dg* $\mathcal{A}_1$-module, then the space $\mathrm{Mor}_{\mathcal{A}_1}(B, M)$ is naturally a left *dg* $\mathcal{A}_2$-module.

Theorem 1.2. *If $(Y, \phi: F(-\mathcal{Z}_2) \to \partial Y)$ is a bordered three-manifold and $\psi: F(-\mathcal{Z}_1) \to F(-\mathcal{Z}_2)$ is a homeomorphism then there is a quasi-isomorphism:*

$$\widehat{CFD}(Y, \phi \circ \psi) \simeq \mathrm{Mor}_{\mathcal{A}(\mathcal{Z}_1)}\big(\widehat{CFDD}(\psi), \widehat{CFD}(Y, \phi)\big).$$

Theorem 1.2 can be thought of as a kind of pairing theorem, as well. The bimodule $\widehat{CFDD}(\psi)$ appearing above is the invariant associated to a very simple bordered three-manifold with two boundary components: the underlying three-manifold here is the product of an interval with the surface $F(\mathcal{Z}_2)$. It is best to think of this as the special case of a more general construction, involving bordered three-manifolds with two boundary components. It turns out that these three-manifolds need to be equipped with some additional structure, giving the *arced cobordisms* of Definition 2.10 below. Theorem 1.2 then becomes a special case of a pairing theorem for gluing bordered three-manifolds to arced cobordisms (Theorem 2.23, below); see Example 2.24.

Theorem 1.2 answers Question (2) above. The bimodules associated to mapping classes also answer Question (1): while $\mathcal{A}(\mathcal{Z})$ is not an invariant of $F(\mathcal{Z})$, the (equivalence class of the) derived category of modules over $\mathcal{A}(\mathcal{Z})$ is an invariant of (the homeomorphism type of) $F(\mathcal{Z})$. For more details, see [29, Theorem 1].

Arguably more excitingly, Theorems 1.1 and 1.2 are an effective tool for computing Heegaard Floer homology. They can be used to give an algorithm for computing $\widehat{HF}(Y)$ for an arbitrary closed, oriented three-manifold Y [31]; the map $\hat{F}_W$ associated to any smooth cobordism W [36]; and the spectral sequence [49] from Khovanov homology to $\widehat{HF}$ of the branched double cover [30, 35]. (We sketch the algorithm for computing $\widehat{HF}(Y)$ in Section 6.) In a different direction, the torus boundary case of bordered Floer homology has been particularly useful for practical computations; see Section 5.

Bordered Floer homology also associates another kind of module, denoted $\widehat{CFA}(Y)$, to a bordered 3-manifold $(Y, \phi : F(\mathcal{Z}) \to \partial Y)$. The module $\widehat{CFA}(Y)$ is a right A_∞-module over $\mathcal{A}(\mathcal{Z})$. To avoid digressing into A_∞-algebra, we have suppressed $\widehat{CFA}(Y)$, and will continue to do so throughout these notes to the extent possible. (Another drawback of $\widehat{CFA}(Y)$ is that its definition requires counting more holomorphic curves than $\widehat{CFD}(Y)$, making $\widehat{CFA}(Y)$ typically harder to compute.) There is one place that $\widehat{CFA}(Y)$ seems unavoidable: in the proof of the pairing theorem, which we sketch in Section 4.4.

These notes are organized into five lectures. The first of these (Section 2) focuses primarily on the combinatorial representations for manifolds (pointed matched circles and Heegaard diagrams for bordered and arced three-manifolds) which are used in the definitions of the modules. After a sufficient amount of the background is laid out, we give a second, more detailed overview of the theory during the middle of the first lecture. Finally, Section 2 concludes by defining the algebra $\mathcal{A}(\mathcal{Z})$ associated to a pointed matched circle $\mathcal{Z}$.

The second lecture is devoted to defining the module $\widehat{CFD}(Y)$ associated to a bordered 3-manifold Y, as well as its generalization $\widehat{CFDD}(Y)$ to an arced cobordism. That lecture starts by reviewing both the original definition and the cylindrical reformulation of the invariant $\widehat{HF}(Y)$ for a closed 3-manifold. The lecture then turns to $\widehat{CFD}(Y)$ and the moduli spaces used to define it, proves the surgery exact triangle for $\widehat{HF}$ (originally proved in [45]) and concludes by briefly defining the extension $\widehat{CFDD}(Y)$.

In the third lecture, we describe the analysis which underpins the theory. This allows us to sketch the proof that the differential on $\widehat{CFD}$ is, in fact, a differential. It also allows us to sketch a proof of the pairing theorem; in the process, the invariant $\widehat{CFA}(Y)$, elsewhere absent from these notes, arises naturally.

The last two lectures are computational. The fourth lecture is devoted to the torus-boundary case. After recalling some terminology about knot Floer homology, it explains how one can recover the knot Floer homology group $\widehat{HFK}(Y,K)$ from the bordered Floer homology of $Y \setminus K$; indeed, this process also allows one to obtain, with a little more work, the knot Floer homology of any satellite of K. The lecture then discusses the other direction: for a knot K in S^3, one can recover the bordered Floer homology $S^3 \setminus K$ from the knot Floer complex $CFK^-(K)$. Combining these results, one obtains a theorem about the behavior of knot Floer homology under taking satellites.

Finally, the last lecture describes an algorithm coming from bordered Floer homology for computing $\widehat{HF}(Y)$ for closed three-manifolds Y.

There are a number of important aspects of the theory which are missing from these notes. These include:

- Any discussion of the grading on bordered Floer homology. The grading takes a somewhat complicated form—the algebras are graded by a non-commutative group $G(\mathcal{Z})$ and the modules by $G(\mathcal{Z})$-sets—and we refer the reader to [27, Chapter 10] for this part of the story.
- A more thorough treatment of $\widehat{CFA}$. This would involve a lengthy algebraic digression which might distract from the underlying geometry in the theory. Again, we refer the reader to [27] to fill in this omission.
- A discussion of the proof of invariance of the bordered modules (Theorem 2.16). Most of the ideas in the proof of invariance, however, are present in the proof that $\partial^2 = 0$ on $\widehat{CFD}$ and the proof of invariance in the closed case [46].

- A proof of the Mor versions of the pairing theorem (Theorems 2.21 and 2.23). We refer the reader to [32] for these proofs.
- The connection between bordered Floer homology and Juhász's *sutured Floer homology* [15]. This connection is given by Zarev's *bordered sutured theory* [64].

There are two other expository articles on bordered Heegaard Floer homology, with somewhat different focuses, in which the reader might be interested: [28, 33]. The paper [34] is also intended to be partly expository.

2. Combinatorial Representations of Surfaces and 3-Manifolds with Boundary. Formal Structure of Bordered Floer Homology. The Algebra Associated to a Surface

Much of this lecture lays out in detail the combinatorial representations of the topological objects used in the definition of bordered Floer homology. We start with surfaces (encoded by pointed matched circles), and then move on to bordered three-manifolds (encoded by Heegaard diagrams). With this material in place, we give a more detailed overview of the formal structure of bordered Floer homology. The lecture concludes with the definition of the algebra associated to a pointed matched circle.

2.1. Arc Diagrams and Surfaces

Definition 2.1. A *pointed matched circle* consists of an oriented circle Z, a point $z \in Z$, a finite set of points $\mathbf{a} \subset Z$ disjoint from z, and a fixed-point free involution $M : \mathbf{a} \to \mathbf{a}$. The map M matches the points $\mathbf{a}$ in pairs; that is, we can view $\mathbf{a}$ as a union of S^0s. We require that the result Z' of doing surgery on $(Z, \mathbf{a})$ according to M be connected. See Figure 1.

A pointed matched circle specifies a surface. There are a few essentially equivalent constructions; here is one:

Construction 2.2. Fix a pointed matched circle $\mathcal{Z} = (Z, \mathbf{a}, M, z)$. Build an oriented surface-with-boundary $F^\circ(\mathcal{Z})$ as follows. Start with $[0,1] \times Z$. Attach a strip (2-dimensional 1-handle) to each pair of matched points in $\mathbf{a} \times \{0\}$. The result has boundary $(Z \times \{1\}) \amalg Z'$. Fill in Z' with a copy of $\mathbb{D}^2$. The result is $F^\circ(\mathcal{Z})$. Again, see Figure 1.

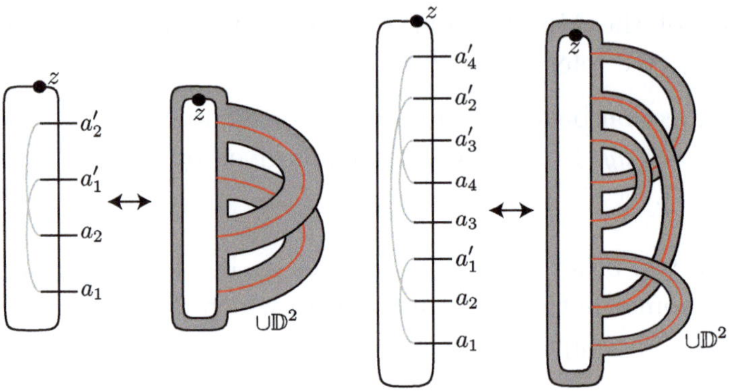

Fig. 1. Pointed matched circles and surfaces. Left: a pointed matched circle specifying a once-punctured torus. Right: a pointed matched circle specifying a once-punctured genus 2 surface. In both cases, the involution M exchanges a_i and a'_i

As a slight variant, we could fill in the boundary of $F^\circ(\mathcal{Z})$ with a disk. This gives a surface $F(\mathcal{Z})$ with a distinguished disk in it—the disk $F(\mathcal{Z}) \setminus F^\circ(\mathcal{Z})$—and a distinguished basepoint on the boundary of this disk. That is, $F(\mathcal{Z})$ is a *strongly based surface.* (Papers in the subject sometimes treat a pointed matched circle as specifying a surface with boundary, and sometimes as specifying a closed, strongly based surface; it makes no essential difference.)

Remark 2.3. Pointed matched circles are a special case of Zarev's *arc diagrams*; any orientable surface with non-empty boundary can be represented by an arc diagram, and there is an associated algebra similar to the one we will describe in Section 2.4.3. Arc diagrams are, in turn, closely related to fat graphs and chord diagrams.

2.2. Bordered Heegaard Diagrams for 3-Manifolds

We start with 3-manifolds with one boundary component:

Definition 2.4. A *bordered* 3*-manifold* consists of a compact, oriented 3-manifold-with-boundary Y and a homeomorphism $\phi : F(\mathcal{Z}) \to \partial Y$ for some pointed matched circle $\mathcal{Z}$.

Call two bordered 3-manifolds $(Y_1, \phi_1 : F(\mathcal{Z}) \to \partial Y_1)$ and $(Y_2, \phi_2 : F(\mathcal{Z}) \to \partial Y_2)$ *equivalent* if there is a homeomorphism $\psi : Y_1 \to Y_2$ so that

$\phi_2 = \psi \circ \phi_1$, i.e.,

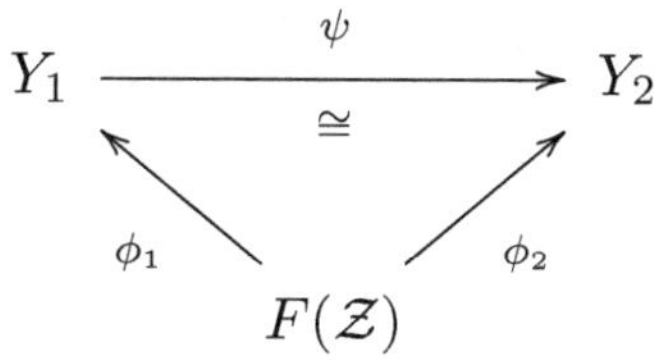

commutes.

We often drop the parametrization ϕ from the notation, writing Y to denote a bordered 3-manifold, i.e., $Y = (Y, \phi)$.

We can represent bordered 3-manifolds combinatorially, as follows:

Definition 2.5. Let $\mathcal{Z}$ be a pointed matched circle representing a surface of genus k. A *bordered Heegaard diagram with boundary* $\mathcal{Z}$ is a tuple

$$\mathcal{H} = \big(\Sigma_g, \overbrace{\overbrace{\alpha_1^a, \dots, \alpha_{2k}^a}^{\boldsymbol{\alpha}^a}, \overbrace{\alpha_1^c, \dots, \alpha_{g-k}^c}^{\boldsymbol{\alpha}^c}}^{\boldsymbol{\alpha}}, \overbrace{\beta_1, \dots, \beta_g}^{\boldsymbol{\beta}}, z\big)$$

where

- Σ_g is a compact, oriented surface of genus g with one boundary component.
- $\boldsymbol{\beta}$ is a g-tuple of pairwise disjoint circles in the interior of Σ.
- $\boldsymbol{\alpha}^c$ is a $(g-k)$-tuple of pairwise disjoint circles in the interior of Σ.
- $\boldsymbol{\alpha}^a$ is a $(2k)$-tuple of pairwise disjoint arcs in Σ with boundary in $\partial\Sigma$.
- z is a basepoint in $\partial\Sigma \setminus \boldsymbol{\alpha}^a$.
- $\boldsymbol{\alpha}^a \cap \boldsymbol{\alpha}^c = \emptyset$.
- $\Sigma \setminus (\boldsymbol{\alpha}^c \cup \boldsymbol{\alpha}^a)$ and $\Sigma \setminus \boldsymbol{\beta}$ are both connected.
- $\mathcal{Z} = (\partial\Sigma, \boldsymbol{\alpha}^a \cap \partial\Sigma, M, z)$. Here, M matches (exchanges) the two endpoints of each α_i^a.

Especially when we are considering holomorphic curves, we will abuse notation and also use Σ to denote $\Sigma \setminus \partial\Sigma$; and similarly for the α-arcs.

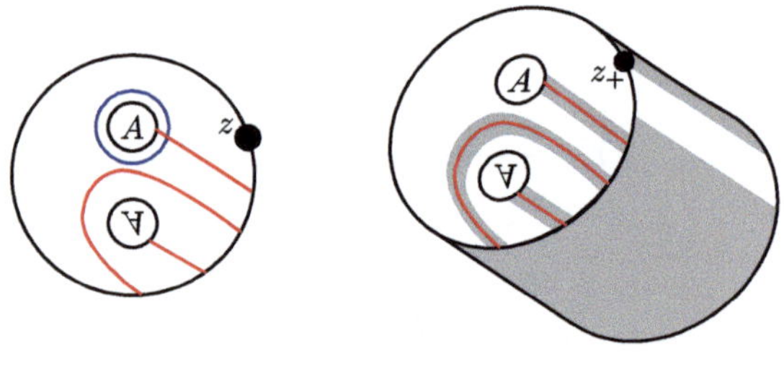

Fig. 2. A bordered Heegaard diagram and the associated 3-manifold. The picture on the left is a Heegaard diagram for the bordered solid torus shown on the right. (The labels A indicate a handle between the corresponding circles.) The shaded part of the boundary is $F^\circ(\mathcal{Z})$. This figure is adapted from [29, Figure 12]

Construction 2.6. Let $\mathcal{H}=(\Sigma,\boldsymbol{\alpha},\boldsymbol{\beta},z)$ be a bordered Heegaard diagram with boundary $\mathcal{Z}$. There is a corresponding bordered 3-manifold $Y(\mathcal{H})$ constructed as follows.

(1) Thicken Σ to $\Sigma\times[0,1]$.
(2) Attach three-dimensional two-handles along the α-circles in $\Sigma\times\{0\}$.
(3) Attach three-dimensional two-handles along the β-circles in $\Sigma\times\{1\}$.

A parameterization of the boundary is specified as follows. Consider the graph

$$\big(\boldsymbol{\alpha}^a\cup\big(\partial\Sigma\setminus\mathrm{nbd}(z)\big)\big)\times\{0\}\subset\Sigma\times\{0\},$$

thought of as a subset of ∂Y. The closure F° of a neighborhood of this graph is naturally identified with $F^\circ(\mathcal{Z})$. The complement of F° in ∂Y is a disk, and is identified with $F(\mathcal{Z})\setminus F^\circ(\mathcal{Z})$. See Figure 2.

The orientations in Construction 2.6 are confusing; see [29, Construction 5.3] for a discussion of this point.

Example 2.7. Figure 2 shows a Heegaard diagram for a solid torus. This is one of many Heegaard diagrams for bordered solid tori; see Section 3.4 for more Heegaard diagrams for solid tori.

Example 2.8. Figure 26 (page 345) shows a Heegaard diagram for a genus 2 handlebody. Again, this is one among many.

Example 2.9. Fix an oriented surface Σ, equipped with a g-tuple of pairwise disjoint, homologically independent curves $\boldsymbol{\beta}$ and a $(g-1)$-tuple of pairwise disjoint, homologically independent curves $\boldsymbol{\alpha}^c=\{\alpha_1^c,\dots,\alpha_1^{g-1}\}$. Then

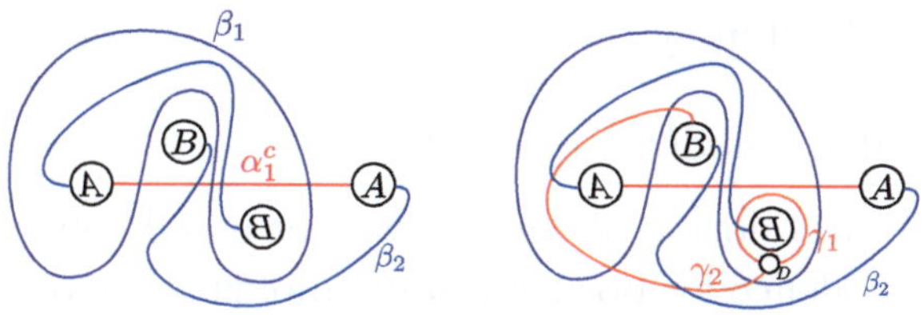

Fig. 3. A bordered Heegaard diagram for the trefoil complement. Left: a Heegaard diagram for the complement of the trefoil. The circles labeled A (respectively B) denote a handle attached to the plane. Right: a bordered Heegaard diagram, obtained by adding the curves γ_1 and γ_2 and deleting a disk. It may be instructive to compare this diagram with Figure 16

$(\Sigma, \boldsymbol{\alpha}^c, \boldsymbol{\beta})$ is a Heegaard diagram for a three-manifold with torus boundary, and indeed any such three-manifold Y can be described by a Heegaard diagram of this type. To turn such a diagram into a bordered Heegaard diagram, we proceed as follows. Fix an additional pair of circles γ_1 and γ_2 in Σ so that:

- γ_1 and γ_2 are disjoint from $\alpha_1^c, \dots, \alpha_{g-1}^c$,
- γ_1 and γ_2 intersect, transversally, in a single point p and
- both of the homology classes $[\gamma_1]$ and $[\gamma_2]$ are homologically independent from $[\alpha_1^c], \dots, [\alpha_{g-1}^c]$.

Let D be a disk around p which is disjoint from all the above curves, except for γ_1 and γ_2, each of which it meets in a single arc. Then, the complement of D specifies a bordered Heegaard diagram for Y, for some parametrization of ∂Y. A bordered Heegaard diagram for the trefoil complement is illustrated in Figure 3.

(This example is drawn from [27, Section 4.2]. See also the discussion around [27, Figure 11.8].)

We also consider 3-dimensional cobordisms:

Definition 2.10. Fix pointed matched circles $\mathcal{Z}_L = (Z_L, \mathbf{a}_L, M_L, z_L)$ and $\mathcal{Z}_R = (Z_R, \mathbf{a}_R, M_R, z_R)$. An *arced cobordism from $\mathcal{Z}_L$ to $\mathcal{Z}_R$* consists of:

- A compact, oriented 3-manifold-with-boundary Y,
- an injection $\phi : (-F^\circ(\mathcal{Z}_L)) \amalg F^\circ(\mathcal{Z}_R) \to \partial Y$ (where $-$ denotes orientation reversal) and

- a path γ in $\partial Y \setminus \mathrm{Im}(\phi)$

such that $Y \setminus (\mathrm{Im}(\phi) \cup \mathrm{nbd}(\gamma))$ is a disk.

There is a natural notion of equivalence for arced cobordisms, similar to the notion of equivalence for bordered 3-manifolds; we leave it as an exercise.

As for bordered 3-manifolds, we will typically denote all of the data of an arced cobordism simply by Y. Also as with bordered 3-manifolds, there are several other essentially equivalent ways to formulate the notion of an arced cobordism; see for instance [29, Section 5] and [32, Section 3].

Again, a combinatorial representation of arced cobordisms will be important to us:

Definition 2.11. An *arced Heegaard diagram* is a tuple

$$\mathcal{H} = \Big(\Sigma_g, \overbrace{\overbrace{\alpha_1^{a,L}, \dots, \alpha_{2k_L}^{a,L}}^{\boldsymbol{\alpha}^{a,L}}, \overbrace{\alpha_1^{a,R}, \dots, \alpha_{2k_R}^{a,R}}^{\boldsymbol{\alpha}^{a,R}}, \overbrace{\alpha_1^{c}, \dots, \alpha_{g-k_L-k_R}^{c}}^{\boldsymbol{\alpha}^{c}}}^{\boldsymbol{\alpha}}, \overbrace{\beta_1, \dots, \beta_g}^{\boldsymbol{\beta}}, \mathbf{z}\Big)$$

where

- Σ_g is a compact, oriented surface of genus g with two boundary components, $\partial_L \Sigma$ and $\partial_R \Sigma$;
- $\boldsymbol{\beta}$ is a g-tuple of pairwise disjoint curves in the interior of Σ;
- $\boldsymbol{\alpha}^{a,L}$ is a collection of pairwise-disjoint embedded arcs with boundary on $\partial_L \Sigma$;
- $\boldsymbol{\alpha}^{a,R}$ is a collection of pairwise-disjoint embedded arcs with boundary on $\partial_R \Sigma$;
- $\boldsymbol{\alpha}^{c}$ is a collection of pairwise-disjoint circles in the interior of Σ; and
- $\mathbf{z}$ is a path in $\Sigma \setminus (\boldsymbol{\alpha}^{a,L} \cup \boldsymbol{\alpha}^{a,R} \cup \boldsymbol{\alpha}^{c} \cup \boldsymbol{\beta})$ between $\partial_L \Sigma$ and $\partial_R \Sigma$.

These are required to satisfy:

- $\boldsymbol{\alpha}^{a,L}$, $\boldsymbol{\alpha}^{a,R}$ and $\boldsymbol{\alpha}$ are all disjoint,
- $\Sigma \setminus \boldsymbol{\alpha}$ and $\Sigma \setminus \boldsymbol{\beta}$ are connected and
- $\boldsymbol{\alpha}$ intersects $\boldsymbol{\beta}$ transversely.

(Compare [29, Definition 5.4].)

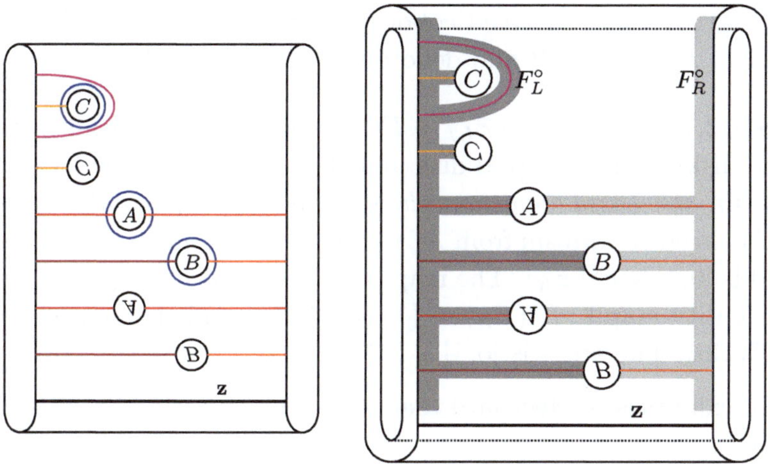

Fig. 4. Constructing a bordered 3-manifold with two boundary components from an arced bordered Heegaard diagram. The Heegaard diagram on the left represents an elementary cobordism from the genus two surface to the genus one surface. On the right is a (somewhat schematic) depiction of the resulting 3-manifold. The inside part of the boundary, which corresponds to $\Sigma \times \{1\}$, is a cylinder, since the β-circles caused the handles to be filled in. The outside part of the boundary, corresponding to $\Sigma \times \{0\}$, is a surface of genus 3 with two boundary components. The region F_L° (respectively F_R°) is darkly (respectively lightly) shaded

Observe that each boundary component of an arced Heegaard diagram is a pointed matched circle.

Construction 2.12. Fix an arced Heegaard diagram $\mathcal{H} = (\Sigma_g, \boldsymbol{\alpha}^{a,L}, \boldsymbol{\alpha}^{a,R}, \boldsymbol{\alpha}^c, \boldsymbol{\beta}, \mathbf{z})$ with boundary $\mathcal{Z}_L \amalg \mathcal{Z}_R$. Build a 3-manifold-with-boundary Y as follows:

(1) Thicken Σ to $\Sigma \times [0,1]$.

(2) Attach three-dimensional two-handles along the α-circles in $\Sigma \times \{0\}$.

(3) Attach three-dimensional two-handles along the β-circles in $\Sigma \times \{1\}$.

Consider the graphs

$$\Gamma_L = \big(\boldsymbol{\alpha}^{a,L} \cup \big(\partial_L \Sigma \setminus \mathrm{nbd}(\mathbf{z})\big)\big) \times \{0\} \subset \Sigma \times \{0\}$$

$$\Gamma_R = \big(\boldsymbol{\alpha}^{a,R} \cup \big(\partial_R \Sigma \setminus \mathrm{nbd}(\mathbf{z})\big)\big) \times \{0\} \subset \Sigma \times \{0\}$$

thought of as subsets of ∂Y. The closure F_L° (respectively F_R°) of a neighborhood of Γ_L (respectively Γ_R) is naturally identified with $F^\circ(\mathcal{Z}_L)$ (respectively $F^\circ(\mathcal{Z}_R)$). Let ϕ denote this identification $F^\circ(\mathcal{Z}_L) \amalg F^\circ(\mathcal{Z}_R) \to F_L^\circ \amalg F_R^\circ$. The arc $\gamma_{\mathbf{z}} = \mathbf{z} \times \{0\}$ connects F_L° and F_R°, and $\partial Y \setminus (F_L^\circ \cup F_R^\circ \cup \mathrm{nbd}(\gamma_{\mathbf{z}}))$ is a disk.

The data $(Y, \phi, \gamma_{\mathbf{z}})$ is an arced cobordism; we call this cobordism the *arced cobordism associated to* $\mathcal{H}$ and denote it by $Y(\mathcal{H})$. See Figure 4.

Example 2.13. Let $\psi : F(\mathcal{Z}_L) \to F(\mathcal{Z}_R)$ be a homeomorphism taking the preferred disk to the preferred disk and the basepoint to the basepoint; that is, ψ is a *strongly based homeomorphism.* The *mapping cylinder of* ψ, denoted M_ψ, is the arced cobordism from $\mathcal{Z}_L$ to $\mathcal{Z}_R$ given as follows. The underlying 3-manifold is $[0,1] \times F^\circ(\mathcal{Z}_R)$. The map $\phi : -F^\circ(\mathcal{Z}_L) \amalg F^\circ(\mathcal{Z}_R) \to \partial M_\psi$ is given by the identity map $\mathbb{I} : F^\circ(\mathcal{Z}_R) \to \{1\} \times F^\circ(\mathcal{Z}_R)$ and the map $\psi : F^\circ(\mathcal{Z}_L) \to \{0\} \times F^\circ(\mathcal{Z}_R)$. The arc γ is $[0,1] \times \{z\}$.

Some examples of Heegaard diagrams for mapping cylinders are shown in Figure 9.

Gluing the mapping cylinder for ψ to a bordered 3-manifold (Y, ϕ) in the sense of Exercise 2.2 gives $(Y, \phi \circ \psi)$.

As in the closed case, the key properties of bordered Heegaard diagrams are that every bordered 3-manifold can be represented by a bordered Heegaard diagram, and any two such diagrams can be related by certain elementary moves:

Theorem 2.14. *Let* $(Y, \phi : F(\mathcal{Z}) \to \partial Y)$ *be a bordered* 3*-manifold. Then* Y *is represented by some bordered Heegaard diagram* $\mathcal{H}$*. Similarly, let* $(Y, \phi : F^\circ(\mathcal{Z}_L) \amalg F^\circ(\mathcal{Z}_R) \to \partial Y, \gamma)$ *be an arced cobordisms. Then* Y *is represented by some arced Heegaard diagram* $\mathcal{H}$*.*

The case of bordered Heegaard diagrams is [27, Lemma 4.9] while the arced Heegaard diagram case is [29, Proposition 5.10].

Theorem 2.15. *Suppose that* $\mathcal{H}$ *and* $\mathcal{H}'$ *are bordered Heegaard diagrams representing equivalent bordered* 3*-manifolds* $Y(\mathcal{H}) \cong Y(\mathcal{H}')$*. Then* $\mathcal{H}$ *and* $\mathcal{H}'$ *can be made diffeomorphic by a sequence of the following moves:*

- *Isotopies of the* α*- and/or* β*-curves.*
- *Handleslides or* α*-circles over* α*-circles,* α*-arcs over* α*-circles, and* β*-circles over* β*-circles.*
- *Stabilizations and destabilizations of the diagram, i.e., taking connected sums with the standard Heegaard diagram for* S^3*.*

(See Figure 5.*)*

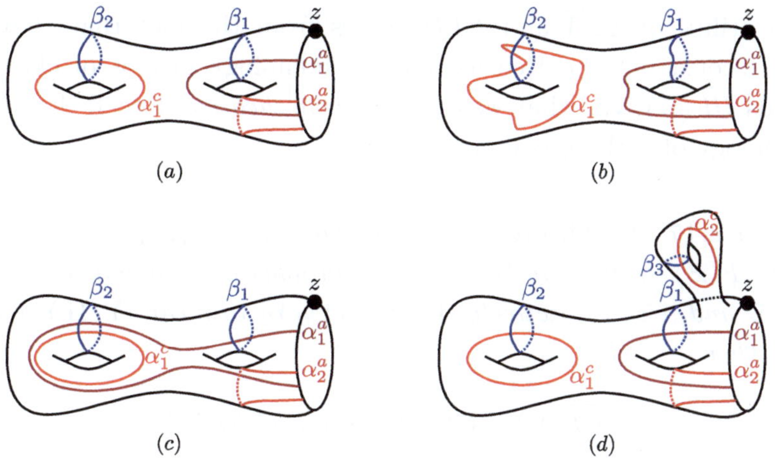

Fig. 5. Heegaard moves. (a) A genus 2 bordered Heegaard diagram for a solid torus. (b) The result of applying some isotopies to the α- and β-curves. (c) The result of a handleslide of α_1^a over α_1^c. (d) The result of a stabilization

An exactly analogous statement holds for arced Heegaard diagrams and arced cobordisms.

The case of bordered Heegaard diagrams is [27, Proposition 4.10] while the arced Heegaard diagram case is [29, Proposition 5.11].

2.3. The Structure of Bordered Floer Homology

2.3.1. The Connected Boundary Case. For simplicity, we begin with the connected boundary case. Bordered Floer homology assigns:

Pointed matched circle $\mathcal{Z}$	*dg* algebra $\mathcal{A}(\mathcal{Z})$
Bordered 3-manifold $(Y, \phi : F(\mathcal{Z}) \to \partial Y)$	Right A_∞ $\mathcal{A}(\mathcal{Z})$-module $\widehat{CFA}(Y)$ Left *dg* $\mathcal{A}(-\mathcal{Z})$-module $\widehat{CFD}(Y)$.

Actually, the modules $\widehat{CFA}(Y)$ and $\widehat{CFD}(Y)$ depend on a choice of bordered Heegaard diagram $\mathcal{H}$ for Y, as well as another auxiliary choice—an almost-complex structure. However:

Theorem 2.16 [27, Theorems 1.1 and 1.2]**.** *Up to quasi-isomorphism, the modules $\widehat{CFA}(Y)$ and $\widehat{CFD}(Y)$ depend only on the equivalence class of bordered 3-manifold Y.*

The utility of $\widehat{CFA}$ and $\widehat{CFD}$ comes from the fact that they can be used to reconstruct the Heegaard Floer homology groups of closed three-manifolds $\widehat{HF}(Y)$, via what we call a *pairing theorem*. Recall that $\widehat{HF}(Y)$ is the homology of a chain complex $\widehat{CF}(Y)$.

Theorem 2.17 [27, Theorem 1.3]. *Suppose that* $(Y_1, \phi_1 : F(\mathcal{Z}) \to \partial Y)$ *and* $(Y_2, \phi_2 : -F(\mathcal{Z}) \to \partial Y)$ *are bordered* 3*-manifolds with boundaries parameterized by* $\mathcal{Z}$ *and* $-\mathcal{Z}$*, respectively. Write* $Y_1 \cup_\partial Y_2$ *to mean* $(Y_1 \amalg Y_2)/(\phi_1(x) \sim \phi_2(x))$*. Then*

$$\widehat{CF}(Y) \simeq \widehat{CFA}(Y_1) \widetilde{\otimes}_{\mathcal{A}(F)} \widehat{CFD}(Y_2).$$

Here, $\widetilde{\otimes}$ denotes the appropriate notion of tensor product given that $\widehat{CFA}$ may be an A_∞-module. In the case that $\widehat{CFA}$ is an ordinary module, this reduces to the derived tensor product—which is good, since $\widehat{CFA}$ is only well-defined up to quasi-isomorphism. But this distinction is not so important: the module $\widehat{CFD}$ is projective, so the derived and ordinary tensor products agree.

The modules $\widehat{CFA}(Y)$ and $\widehat{CFD}(Y)$ are defined using holomorphic curves (though for certain kinds of diagrams the techniques of [58] can be used to compute them combinatorially). By contrast, the algebras $\mathcal{A}(\mathcal{Z})$ are defined combinatorially. A few further properties of the algebras:

- Each $\mathcal{A}(\mathcal{Z})$ is a finite-dimensional algebra over $\mathbb{F}_2$.
- The algebra $\mathcal{A}(\mathcal{Z})$ decomposes as a direct sum of subalgebras

$$\mathcal{A}(\mathcal{Z}) = \bigoplus_{i=-k}^{k} \mathcal{A}(\mathcal{Z}, i).$$

  Here, k is the genus of $F(\mathcal{Z})$. The action of $\mathcal{A}(\mathcal{Z}, i)$ on $\widehat{CFA}(Y)$ and $\widehat{CFD}(Y)$ is trivial for $i \neq 0$, but the other summands come up for the cobordism invariants below.

- The algebra $\mathcal{A}(\mathcal{Z}, -k)$ is isomorphic to $\mathbb{F}_2$ (with trivial differential). In particular, if $\mathcal{Z}$ is the (unique) pointed matched circle for S^2 then $\mathcal{A}(\mathcal{Z}) = \mathbb{F}_2$. The algebra $\mathcal{A}(\mathcal{Z}, k)$ is quasi-isomorphic to $\mathbb{F}_2$.

- If $\mathcal{Z}$ is the unique pointed matched circle for the torus then $\mathcal{A}(\mathcal{Z},0)$ has no differential; in terms of generators and relations, $\mathcal{A}(\mathcal{Z},0)$ is given by

$$\iota_0 \bullet \;\underset{\rho_3}{\overset{\rho_1}{\underset{\xleftarrow{\rho_2}}{\rightleftarrows}}}\; \bullet \iota_1 \;/(\rho_2\rho_1 = \rho_3\rho_2 = 0). \tag{2.1}$$

This algebra is 8-dimensional over $\mathbb{F}_2$. It will appear frequently, so we name the rest of the elements in its standard basis: let $\rho_{12} = \rho_1\rho_2$, $\rho_{23} = \rho_2\rho_3$ and $\rho_{123} = \rho_1\rho_2\rho_3$.

(Our notation for path algebras might be somewhat non-standard. The vertices ι_0 and ι_1 are, of course, idempotents. The arrow ρ_1 indicates that $\iota_0\rho_1\iota_1 = \rho_1$.)

2.3.2. Invariants of Arced Cobordisms. To get a useful theory, we need to generalize to three-manifolds with two boundary components. In fact, the invariants which come up in this two-boundary-component case are associated to three-manifolds equipped with some extra structure: the arced cobordisms of Definition 2.10.

Suppose Y is an arced cobordism from $\mathcal{Z}_1$ to $\mathcal{Z}_2$. Then there are several kinds of bimodules we can associate to Y: we can treat each boundary component of Y in either a "type D" way or a "type A" way. (What this means will be clearer after Sections 3 and 4.) This gives invariants $\widehat{CFDD}(Y)$ (both boundaries viewed in a type D way), $\widehat{CFDA}(Y)$ (one boundary, say $\mathcal{Z}_1$, viewed in a type D way and the other in a type A way), and $\widehat{CFAA}(Y)$ (both boundaries viewed in a type A way). The bimodule $\widehat{CFDD}(Y)$ is an ordinary—indeed, bi-projective—*dg* bimodule; both of $\widehat{CFDA}(Y)$ and $\widehat{CFAA}(Y)$ are typically A_∞-bimodules.

As with the modules associated to bordered 3-manifolds, the bimodules $\widehat{CFDD}(Y)$, $\widehat{CFDA}(Y)$ and $\widehat{CFAA}(Y)$ depend on the choices of Heegaard diagrams and almost-complex structures. Again, up to quasi-isomorphism they are invariants:

Theorem 2.18 [29, Theorem 8]. *Up to quasi-isomorphism, the bimodules $\widehat{CFDD}(Y)$, $\widehat{CFDA}(Y)$ and $\widehat{CFAA}(Y)$ depend only on the equivalence class of arced cobordism Y.*

By convention, we view $\widehat{CFDD}(Y)$ as having commuting left actions by $\mathcal{A}(-\mathcal{Z}_1)$ and $\mathcal{A}(-\mathcal{Z}_2)$; $\widehat{CFDA}(Y)$ as having a left action by $\mathcal{A}(-\mathcal{Z}_1)$ and a right action by $\mathcal{A}(\mathcal{Z}_2)$; and $\widehat{CFAA}(Y)$ as having right actions by $\mathcal{A}(\mathcal{Z}_1)$ and $\mathcal{A}(\mathcal{Z}_2)$. However, $\mathcal{A}(-\mathcal{Z})$ is the opposite algebra to $\mathcal{A}(\mathcal{Z})$ (Exercise 2.13) so we can move actions from one side to the other at the cost of introducing / deleting minus signs. In the literature, we often find it convenient to decorate the invariants with the algebras they are over, writing

$$^{\mathcal{A}(-\mathcal{Z}_1),\mathcal{A}(-\mathcal{Z}_2)}\widehat{CFDD}(Y) \qquad ^{\mathcal{A}(-\mathcal{Z}_1)}\widehat{CFDA}(Y)_{\mathcal{A}(\mathcal{Z}_2)} \qquad \widehat{CFAA}(Y)_{\mathcal{A}(\mathcal{Z}_1),\mathcal{A}(\mathcal{Z}_1)}.$$

The superscripts indicate that the module structure is projective, and subscripts indicate the module structure may be A_∞. This notation leads to a kind of Einstein summation behavior for tensor products in the pairing theorems:

Theorem 2.19 [29, Theorem 11]. *Let Y_1 be a bordered 3-manifold with boundary $\mathcal{Z}_1$ and Y_2 be an arced cobordism from $\mathcal{Z}_1$ to $\mathcal{Z}_2$. Let $Y_1 \cup_{F(\mathcal{Z})} Y_2$ be the bordered 3-manifold obtained by gluing Y_1 to Y_2 (Exercise 2.2). Then there are quasi-isomorphisms*

$$\widehat{CFA}(Y_1)\,\widetilde{\otimes}_{\mathcal{A}(\mathcal{Z}_1)}\,\widehat{CFDA}(Y_2) \simeq \widehat{CFA}(Y_1 \cup_{F(\mathcal{Z}_1)} Y_2)$$
$$\widehat{CFAA}(Y_2)\,\widetilde{\otimes}_{\mathcal{A}(-\mathcal{Z}_1)}\,\widehat{CFD}(Y_1) \simeq \widehat{CFA}(Y_1 \cup_{F(\mathcal{Z}_1)} Y_2)$$
$$\widehat{CFA}(Y_1)\,\widetilde{\otimes}_{\mathcal{A}(\mathcal{Z}_1)}\,\widehat{CFDD}(Y_2) \simeq \widehat{CFD}(Y_1 \cup_{F(\mathcal{Z}_1)} Y_2)$$
$$\widehat{CFDA}(Y_2)\,\widetilde{\otimes}_{\mathcal{A}(-\mathcal{Z}_1)}\,\widehat{CFD}(Y_1) \simeq \widehat{CFD}(Y_1 \cup_{F(\mathcal{Z}_1)} Y_2).$$

Theorem 2.20 [29, Theorem 12]. *Let Y_1 be an arced cobordism from $\mathcal{Z}_1$ to $\mathcal{Z}_2$ and Y_2 an arced cobordism from $\mathcal{Z}_2$ and $\mathcal{Z}_3$. Let $Y_1 \cup_{F(\mathcal{Z}_2)} Y_2$ be the result of gluing Y_1 to Y_2 along $F(\mathcal{Z}_2)$ (Exercise 2.2). Then there are quasi-isomorphisms of bimodules:*

$$\widehat{CFDA}(Y_1)\,\widetilde{\otimes}_{\mathcal{A}(\mathcal{Z}_2)}\,\widehat{CFDA}(Y_2) \simeq \widehat{CFDA}(Y_1 \cup_{F(\mathcal{Z}_2)} Y_2)$$
$$\widehat{CFAA}(Y_1)\,\widetilde{\otimes}_{\mathcal{A}(\mathcal{Z}_2)}\,\widehat{CFDA}(Y_2) \simeq \widehat{CFAA}(Y_1 \cup_{F(\mathcal{Z}_2)} Y_2)$$
$$\widehat{CFDA}(Y_1)\,\widetilde{\otimes}_{\mathcal{A}(\mathcal{Z}_2)}\,\widehat{CFDD}(Y_2) \simeq \widehat{CFDD}(Y_1 \cup_{F(\mathcal{Z}_2)} Y_2)$$
$$\widehat{CFAA}(Y_1)\,\widetilde{\otimes}_{\mathcal{A}(\mathcal{Z}_2)}\,\widehat{CFDD}(Y_2) \simeq \widehat{CFDA}(Y_1 \cup_{F(\mathcal{Z}_2)} Y_2).$$

The compact way of stating Theorems 2.19 and 2.20 is that if you tensor type A boundaries with type D boundaries then you get what you expect.

2.3.3. Pairing Theorems without *A* Modules. To avoid a long detour into A_∞ formalism, in most of these lectures we will avoid $\widehat{CFA}$. (The exception will be the discussion of the pairing theorem in Section 4.) So, it will be useful to have versions of the pairing theorems—Theorems 2.17, 2.19 and 2.20—making use only of type D modules. We can accomplish this using certain dualities of bordered Floer invariants:

Theorem 2.21 [32, Theorem 2]. *Let Y be a bordered 3-manifold with boundary $F(\mathcal{Z})$. Let $-Y$ denote Y with its orientation reversed, which has boundary $F(-\mathcal{Z})$. Then there are quasi-isomorphisms:*

$$\mathrm{Mor}_{\mathcal{A}(-\mathcal{Z})}\big(\widehat{CFD}(Y), \mathcal{A}(-\mathcal{Z})\big) \simeq \widehat{CFA}(-Y) \tag{2.2}$$

$$\mathrm{Mor}_{\mathcal{A}(\mathcal{Z})}\big(\widehat{CFA}(Y), \mathcal{A}(\mathcal{Z})\big) \simeq \widehat{CFD}(-Y). \tag{2.3}$$

In Formula (2.2), Mor denotes the chain complex of module homomorphisms from $\widehat{CFD}(Y)$ to $\mathcal{A}(-\mathcal{Z})$, with differential given by

$$\partial(f) = f \circ \partial_{\widehat{CFD}(Y)} + d_{\mathcal{A}(-\mathcal{Z})} \circ f.$$

So, for instance, the cycles in the Mor complex are the *dg* module homomorphisms, i.e., chain maps which respect the module structure. In Formula (2.3), Mor denotes the chain complex of A_∞-morphisms.

Corollary 2.22 [32, Theorem 1]. *Suppose that Y_1 and Y_2 are bordered 3-manifolds with boundary $F(\mathcal{Z})$. Then*

$$\begin{aligned}\widehat{CF}(-Y_1 \cup_{F(\mathcal{Z})} Y_2) &\simeq \mathrm{Mor}_{\mathcal{A}(-\mathcal{Z})}\big(\widehat{CFD}(Y_1), \widehat{CFD}(Y_2)\big) \\ &\simeq \mathrm{Mor}_{\mathcal{A}(\mathcal{Z})}\big(\widehat{CFA}(Y_1), \widehat{CFA}(Y_2)\big)\end{aligned}$$

so

$$\begin{aligned}\widehat{HF}(-Y_1 \cup_{F(\mathcal{Z})} Y_2) &\simeq \mathrm{Ext}_{\mathcal{A}(-\mathcal{Z})}\big(\widehat{CFD}(Y_1), \widehat{CFD}(Y_2)\big) \\ &\simeq \mathrm{Ext}_{\mathcal{A}(\mathcal{Z})}\big(\widehat{CFA}(Y_1), \widehat{CFA}(Y_2)\big).\end{aligned}$$

For bimodules the situation is somewhat more subtle: there are a few natural notions of "dual", and some versions introduce boundary Dehn twists in the bimodules. The following result will be more than sufficient for these lectures:

Theorem 2.23 [32, Corollary 8]. *If Y_1 is a bordered 3-manifold with boundary $F(\mathcal{Z}_1)$ and Y_2 is an arced cobordism from $-\mathcal{Z}_1$ to $-\mathcal{Z}_2$ then*

$$\begin{aligned}\widehat{CFA}\big(Y_1 \cup_{F(\mathcal{Z}_1)} (-Y_2)\big) &\simeq \mathrm{Mor}_{\mathcal{A}(-\mathcal{Z}_1)}\big(\widehat{CFDD}(Y_2), \widehat{CFD}(Y_1)\big) \\ \widehat{CFD}(-Y_1 \cup_{F(\mathcal{Z}_1)} Y_2) &\simeq \mathrm{Mor}_{\mathcal{A}(-\mathcal{Z}_1)}\big(\widehat{CFD}(Y_1), \widehat{CFDD}(Y_2)\big).\end{aligned} \tag{2.4}$$

Example 2.24. The bimodules $\widehat{CFDD}(\psi)$ discussed in the introduction are defined to be $\widehat{CFDD}(M_\psi)$ associated to the mapping cylinder of ψ (Example 2.13). So, Theorem 1.2 from the introduction is a special case of Theorem 2.23.

For further results like these, including some involving boundary Dehn twists, see the introduction to [32].

2.4. The Algebra Associated to a Pointed Matched Circle

We will define the algebras associated to pointed matched circles in three steps. We start with a warm-up in Section 2.4.1, discussing the group ring of the symmetric group S_n and a deformation of it called the nilCoxeter algebra. In Section 2.4.2 we define a family of algebras $\mathcal{A}(n,k)$ $(n,k \in \mathbb{N})$, which are a kind of directed, distributed version of the nilCoxeter algebra. The algebra $\mathcal{A}(\mathcal{Z})$ associated to a pointed matched circle for a surface of genus k is defined as a subalgebra of $\bigoplus_{i=0}^{2k} \mathcal{A}(2k,i)$; the definition is given in Section 2.4.3. (It is also possible to give a more direct definition of $\mathcal{A}(\mathcal{Z})$; see, for instance, [31, Section 1.1].)

2.4.1. A Graphical Representation of Permutations. Consider the symmetric group S_n on $\underline{n} = \{1,\dots,n\}$. We can represent elements of S_n graphically as homotopy classes of maps

$$\left(\textstyle\coprod_{i=1}^n [0,1], \coprod_{i=1}^n \{0\}, \coprod_{i=1}^n \{1\}\right) \xrightarrow{\phi} \left([0,1]\times[0,n], \coprod_{i=1}^n \{0\}\times \underline{n}, \coprod_{i=1}^n \{1\}\times \underline{n}\right)$$

such that the restrictions $\phi|_{\coprod_{i=1}^n \{0\}}$ and $\phi|_{\coprod_{i=1}^n \{1\}}$ are injective. For example, the permutation $\left(\begin{smallmatrix}1&2&3&4&5\\3&1&2&5&4\end{smallmatrix}\right) \in S_5$ is represented by the diagram

(2.5)

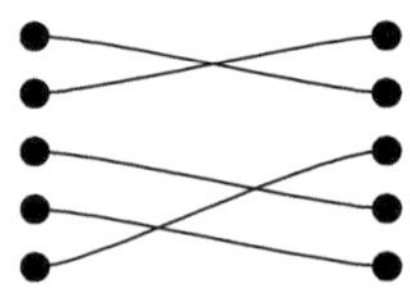

In the graphical notation, multiplication corresponds to juxtaposition. So, the group ring $\mathbb{Z}[S_n]$ of S_n is given by formal linear combinations of diagrams as in (2.5), with product given by juxtaposition. Moreover, notice that essential crossings in diagrams like Formula (2.5) correspond to *inversions*, i.e., pairs $i,j \in \{1,\dots,n\}$ such that $i<j$ but $\sigma(j)<\sigma(i)$.

In $\mathbb{Z}[S_n]$, double-crossings can be undone via Reidemeister II-like moves:

(2.6) 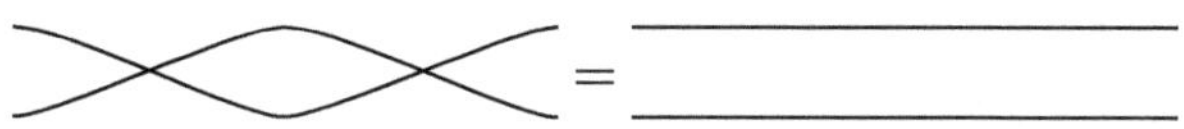

If we replace this relation by the relation that double-crossings are 0,

(2.7) 

we arrive at another algebra, the *nilCoxeter algebra* $\mathcal{N}_n$; see, for instance [17]. Note that even though $\mathcal{N}_n \not\cong \mathbb{Z}[S_n]$, S_n still gives a basis for $\mathcal{N}_n$. Let $\mathrm{Inv}(\sigma)$ denote the set of inversions of σ. An equivalent formulation is that we define

$$\sigma \cdot_{\mathcal{N}} \tau = \begin{cases} \tau \circ \sigma & \text{if } \#\mathrm{Inv}(\tau \circ \sigma) = \#\mathrm{Inv}(\sigma) + \#\mathrm{Inv}(\tau) \\ 0 & \text{else.} \end{cases}$$

If we work over $\mathbb{F}_2$, as is our tendency, we can define a differential on $\mathcal{N}_n$ by declaring that $d(\sigma)$ is the sum of all ways of smoothing a crossing in σ. More formally, let $\tau_{i,j}$ denote the transposition exchanging i and j. Then define

$$d(\sigma) = \sum_{\substack{(i,j)\in\mathrm{Inv}(\sigma) \\ \#\mathrm{Inv}(\tau_{i,j}\sigma)=\#\mathrm{Inv}(\sigma)-1}} \tau_{i,j} \circ \sigma. \tag{2.8}$$

It is straightforward to verify that this makes $\mathcal{N}_n$ into a differential algebra. (If we want to define this differential with signs, we need an odd version of the nilCoxeter algebra; see [18].)

2.4.2. The Algebra $\mathcal{A}(n,k)$. Now, instead of permutations of $\{1,\dots,n\}$, consider *partial permutations*, i.e., triples (S,T,σ) where $S,T \subset \underline{n}$ and $\sigma : S \to T$ is a bijection. Call a partial permutation (S,T,σ) *upward-veering* if $\sigma(i) \geq i$ for all $i \in S$. Let $\mathcal{A}(n)$ denote the $\mathbb{F}_2$-vector space generated by all upward-veering partial permutations. Define a product on $\mathcal{A}(n)$ by

$$(S,T,\phi)\cdot(U,V,\psi) = \begin{cases} 0 & \text{if } T \neq U \\ 0 & \text{if } \#\mathrm{Inv}(\psi\circ\phi) \neq \#\mathrm{Inv}(\psi) + \#\mathrm{Inv}(\phi) \\ (S,V,\psi\circ\phi) & \text{otherwise.} \end{cases} \tag{2.9}$$

Define a differential on $\mathcal{A}(n)$ by setting

$$d(S,T,\phi) = \sum_{\substack{(i,j)\in\mathrm{Inv}(\phi) \\ \#\mathrm{Inv}(\tau_{i,j}\circ\phi)=\#\mathrm{Inv}(\phi)-1}} (S,T,\tau_{i,j}\circ\phi).$$

Graphically, we can still represent generators of $\mathcal{A}(n)$ as strand diagrams; for example, in $n=5$, we draw the partial permutation $(\{1,2,3\},\{3,4,5\},(1\mapsto 5,2\mapsto 4,3\mapsto 3))$ as

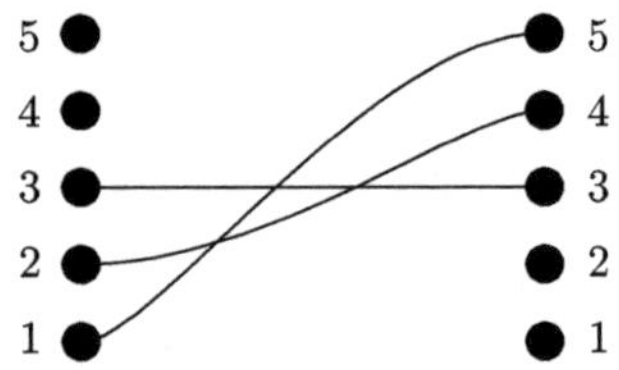

Multiplication is 0 if the endpoints do not match up (the first condition in Equation (2.9)) or if the concatenation contains a double crossing (the second condition in Equation (2.9)); otherwise, the product is just the concatenation. The differential is gotten by summing over all ways of smoothing one crossing, and then throwing away any diagrams involving double crossings.

Proposition 2.25 [27, Lemma 3.1]. *These operations make $\mathcal{A}(n)$ into a differential algebra.*

Proposition 2.25 is not especially difficult, though keeping track of the double-crossing condition adds some complication. The reader is invited to prove it as an extra exercise.

Notice that $\mathcal{A}(n)$ decomposes as a direct sum

$$\mathcal{A}(n)=\bigoplus_{l=0}^{n}\mathcal{A}(n,i) \tag{2.10}$$

where $\mathcal{A}(n,i)$ is generated by partial permutations (S,T,ϕ) with $|S|=|T|=i$.

The algebra $\mathcal{A}(n)$ has an obvious grading by the number of crossings. This grading does not, however, descend in a nice way to the subalgebras associated to pointed matched circles.

2.4.3. The Algebra Associated to a Pointed Matched Circle. Fix a pointed matched circle $\mathcal{Z}=(Z,\mathbf{a},M,z)$ for a surface of genus k, so $|\mathbf{a}|=4k$. The basepoint z and orientation of Z identify $\mathbf{a}$ with $\underline{4k}=\{1,\dots,4k\}$. The algebra $\mathcal{A}(\mathcal{Z})$ is a subalgebra of $\mathcal{A}(4k)$.

Call a generator (S,T,ϕ) of $\mathcal{A}(4k)$ *M-admissible* if $S\cap M(S)=T\cap M(T)=\emptyset$. (This terminology is not used elsewhere in the literature.) Write $\mathrm{Fix}(\phi)=\{i\in S\mid \phi(i)=i\}$. Suppose that ϕ is M-admissible. Then, given $U\subset \mathrm{Fix}(\phi)$ we can define a new element $(S\setminus U\cup M(U),T\setminus U\cup M(U),\phi_U)\in$

$\mathcal{A}(n)$ by replacing the horizontal strands at U by horizontal strands at $M(U)$. That is, ϕ_U is characterized by $\phi_U|_{S\setminus U} = \phi|_{S\setminus U}$ and $\phi_U|_{M(U)} = \mathbb{I}$. Given an M-admissible (S,T,ϕ) define

$$a(S,T,\phi) = \sum_{U \subset \mathrm{Fix}(\phi)} (S \setminus U \cup M(U), T \setminus U \cup M(U), \phi_U).$$

For example,

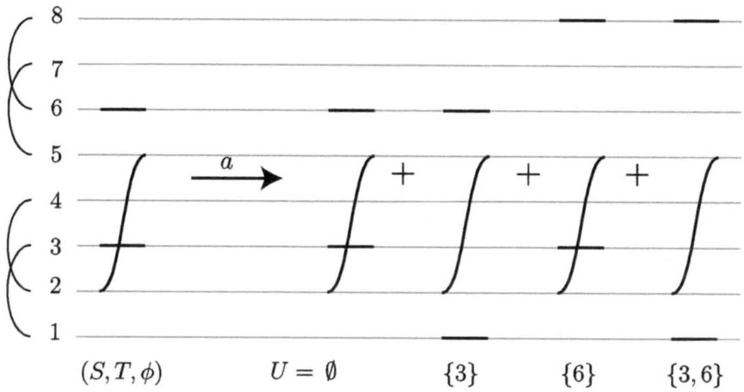

Now, $\mathcal{A}(\mathcal{Z})$ is defined to be the subalgebra of $\mathcal{A}(4k)$ generated by $a(S,T,\phi)$ for M-admissible generators (S,T,ϕ).

The decomposition of $\mathcal{A}(n)$ from Formula (2.10) gives a decomposition of $\mathcal{A}(\mathcal{Z})$. It is convenient to change the indexing slightly: let $\mathcal{A}(\mathcal{Z},i) = \mathcal{A}(\mathcal{Z}) \cap \mathcal{A}(4k, k+i)$, so $\mathcal{A}(\mathcal{Z}) = \bigoplus_{i=-k}^{k} \mathcal{A}(\mathcal{Z},i)$.

2.5. Exercises

Exercise 2.1. Let Y be a closed 3-manifold. How do you go from a pointed Heegaard diagram for Y to a bordered Heegaard diagram for $Y \setminus \mathbb{D}^3$? Vice-versa? (Hint: both directions are easy.)

Exercise 2.2. Let Y_1 be a bordered 3-manifold with boundary $\mathcal{Z}_1$ and Y_2 an arced cobordism from $\mathcal{Z}_1$ to $\mathcal{Z}_2$. There is a natural way to glue Y_1 and Y_2 to get a bordered 3-manifold with boundary $\mathcal{Z}_2$; how?

Similarly, if Y_1 is an arced cobordism from $\mathcal{Z}_1$ to $\mathcal{Z}_2$ and Y_2 is an arced cobordism from $\mathcal{Z}_2$ to $\mathcal{Z}_3$ then there is a natural way to glue Y_1 to Y_2 to obtain an arced cobordism from $\mathcal{Z}_1$ to $\mathcal{Z}_3$; how?

(Both parts are a little awkward with our definition of arced cobordism; the definitions in [29] and [32] make them more obvious.)

Exercise 2.3. Let $\mathcal{H}$ be a bordered Heegaard diagram with no α circles. What is the underlying three-manifold $Y(\mathcal{H})$?

Exercise 2.4. Formulate precisely the notion of equivalence for arced cobordisms.

Exercise 2.5. The bordered Heegaard diagram in Figure 3 represents the trefoil complement with some particular framing. Which one (as an element of $\mathbb{Z}$)?

Exercise 2.6. Draw a bordered Heegaard diagram for the 0-framed complement of the figure eight knot.

Exercise 2.7. Verify that the differential given in Formula (2.8) makes the nilCoxeter algebra into a differential algebra, i.e., that it satisfies $d^2 = 0$ and the Leibniz rule.

Exercise 2.8. Give an example of an element $(S, T, \phi) \in \mathcal{A}(n)$ and a pair $(i, j) \in \mathrm{Inv}(\phi)$ so that $(S, T, \tau_{i,j} \circ \phi)$ is not in $d(S, T, \phi)$.

Exercise 2.9. Verify the path algebra description (Equation (2.1)) for the algebra $\mathcal{A}(T^2, 0)$.

Exercise 2.10. Prove: There is a one-to-one correspondence between indecomposable idempotents in $\mathcal{A}(\mathcal{Z})$ and subsets of the set of matched pairs of $\mathcal{Z}$, i.e., subsets of $\mathbf{a}/M$. (An idempotent I is called *indecomposable* if for any idempotent J, either $I \cdot J = I$ or $I \cdot (1 - J) = I$.) (Hint: this should be easy.)

Exercise 2.11. In this exercise we explain how to produce arced Heegaard diagrams for mapping cylinders. This algorithm is explained in somewhat more detail in [29, Section 5.3].

(1) Show that the arced Heegaard diagram on the left of Figure 6 represents the mapping cylinder of the identity map (of the pointed matched circle for a torus). Generalize this to give a diagram for the identity map of any pointed matched circle. (See Figure 27 for the standard arced Heegaard diagram for the identity map of another pointed matched circle.)

(2) Let $\phi : F(\mathcal{Z}_L) \to F(\mathcal{Z}_R)$ be a strongly based homeomorphism. Recall from Construction 2.12 that a neighborhood F_L° of the graph Γ_L is

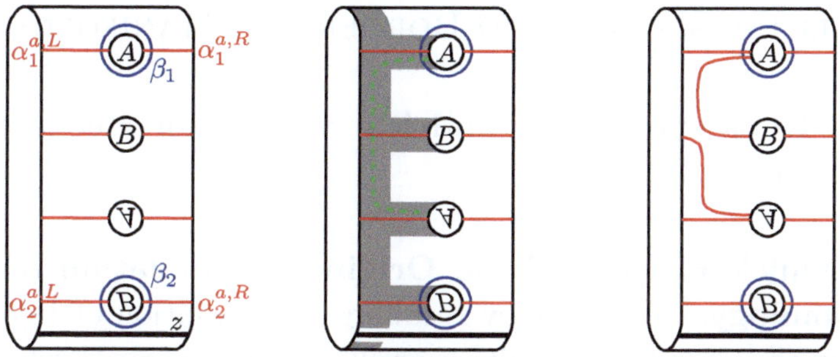

Fig. 6. Building Heegaard diagrams for mapping cylinders. Left: a Heegaard diagram for the identity map of the torus. Center: the sub-surface $F^\circ(\mathcal{Z}_L)$ and a dashed curve γ on $F^\circ(\mathcal{Z}_L)$. Right: a Heegaard diagram for a Dehn twist around γ. This figure is adapted from [29, Figure 15]

identified with $F^\circ(\mathcal{Z}_L)$. Start with the identity Heegaard diagram for $F(\mathcal{Z}_L)$, and apply the homeomorphism ϕ to the $\alpha_i^{a,L} \subset F_L^\circ$. (See Figure 6 for an example.) Prove: the result is an arced Heegaard diagram for ϕ.

Exercise 2.12. There is a unique pointed matched circle representing the once-punctured torus.

(1) List several different pointed matched circles representing the once-punctured genus 2 surface.

(2) Show that the set of matched circles representing the once-punctured genus k surface is in bijection with the set of gluing patterns for the $4k$-gon giving the genus k surface.

Exercise 2.13. Prove that $\mathcal{A}(-\mathcal{Z})$ is the opposite algebra to $\mathcal{A}(\mathcal{Z})$.

Exercise 2.14. Let $\mathcal{Z}$ be the *split pointed matched circle* for a surface of genus k, as illustrated in Figure 21 (page 341). Give a path algebra description of $\mathcal{A}(\mathcal{Z}, -k+1)$, similar to Formula (2.1).

Similarly, let $\mathcal{Z}$ be the *antipodal pointed matched circle* for a surface of genus k, i.e., the pointed matched circle in which a_i is matched to a_{i+2k} $(i = 1, \dots, 2k)$. Give a path algebra description of $\mathcal{A}(\mathcal{Z}, -k+1)$, similar to Formula (2.1). (For a solution to this part, see [34, Example 2.4].)

3. Modules Associated to Bordered 3-Manifolds

3.1. Brief Review of the Cylindrical Setting for Heegaard Floer Homology

3.1.1. A Quick Review of the Original Formulation of Heegaard Floer Homology. We start by recalling the definition of Heegaard Floer homology in the closed setting [46], as well as a "cylindrical" reformulation of the definition [26]; this reformulation will be useful for defining the bordered Floer invariants.

Fix a pointed Heegaard diagram $\mathcal{H} = (\Sigma, \boldsymbol{\alpha}, \boldsymbol{\beta}, z)$ (in the sense of [46]) for a closed 3-manifold Y. Associated to $\mathcal{H}$ are various Heegaard Floer homology groups; as noted in the previous lecture, bordered Floer homology (so far) relates to the technically simplest of these, $\widehat{HF}(Y)$. The group $\widehat{HF}(Y)$ is defined as follows. Suppose Σ has genus g. Choosing a complex structure j_Σ on Σ makes the symmetric product

$$\mathrm{Sym}^g(\Sigma) = \overbrace{\Sigma \times \cdots \times \Sigma}^{g \text{ copies}} / S_g$$

into a smooth—in fact, Kähler—manifold. (This is not obvious.) Writing $\boldsymbol{\alpha} = \{\alpha_1, \dots, \alpha_g\}$ and $\boldsymbol{\beta} = \{\beta_1, \dots, \beta_g\}$, the tori $\alpha_1 \times \cdots \times \alpha_g, \beta_1 \times \cdots \times \beta_g \subset \Sigma^{\times g}$ project to embedded tori T_α and T_β in $\mathrm{Sym}^g(\Sigma)$. Each of T_α and T_β is totally real; in fact, it was shown in [54] that for an appropriate choice of Kähler form the tori T_α and T_β are Lagrangian. Then, $\widehat{HF}(Y)$ is the Lagrangian Floer homology of (T_α, T_β) inside $\mathrm{Sym}^g(\Sigma \setminus \{z\})$.

In a little more detail, $\widehat{HF}(Y)$ is the homology of a chain complex $(\widehat{CF}(Y), \partial)$. $\widehat{CF}(Y)$ is the free $\mathbb{F}_2$-vector space generated by $T_\alpha \cap T_\beta$. The differential $\partial : \widehat{CF}(Y) \to \widehat{CF}(Y)$ is defined by counting holomorphic disks of the following kind. Given $\mathbf{x}, \mathbf{y} \in T_\alpha \cap T_\beta$ we consider the space of maps $\mathbb{D}^2 \to \mathrm{Sym}^g(\Sigma \setminus \{z\})$ such that:

- $-i$ maps to $\mathbf{x}$.
- $+i$ maps to $\mathbf{y}$.
- $\{p \in \partial\mathbb{D}^2 \mid \Re(p) > 0\}$ maps to T_α.
- $\{p \in \partial\mathbb{D}^2 \mid \Re(p) < 0\}$ maps to T_β.

See Figure 7. Such disks are called *Whitney disks.* Let $\mathcal{B}(\mathbf{x}, \mathbf{y})$ denote the space of Whitney disks from $\mathbf{x}$ to $\mathbf{y}$. Further:

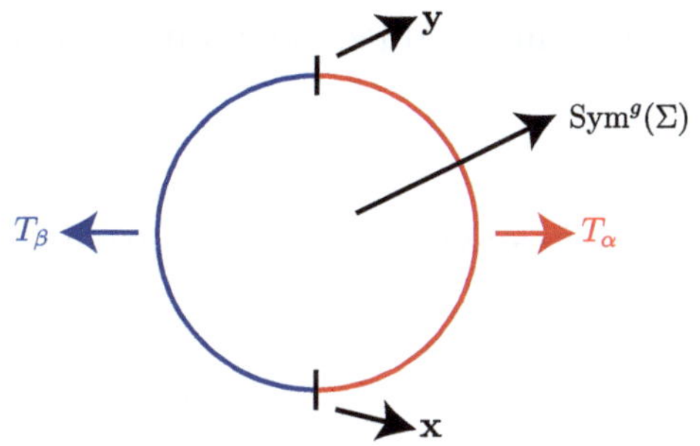

Fig. 7. Boundary conditions for Whitney disks

- Let $\pi_2(\mathbf{x},\mathbf{y})$ denote the set of homotopy classes of Whitney disks, i.e., the set of path components in $\mathcal{B}(\mathbf{x},\mathbf{y})$.
- Let $\widetilde{\mathcal{M}}(\mathbf{x},\mathbf{y}) \subset \mathcal{B}(\mathbf{x},\mathbf{y})$ denote the space of holomorphic Whitney disks.

The space $\widetilde{\mathcal{M}}(\mathbf{x},\mathbf{y})$ decomposes according to elements of $\pi_2(\mathbf{x},\mathbf{y})$:

$$\widetilde{\mathcal{M}}(\mathbf{x},\mathbf{y}) = \coprod_{B\in\pi_2(\mathbf{x},\mathbf{y})} \widetilde{\mathcal{M}}^B(\mathbf{x},\mathbf{y}).$$

If $\widetilde{\mathcal{M}}(\mathbf{x},\mathbf{y})$ is transversally cut-out, each space $\widetilde{\mathcal{M}}^B(\mathbf{x},\mathbf{y})$ is a smooth manifold whose dimension is given by a number $\mu(B)$ called the *Maslov index* of B. There is an $\mathbb{R}$-action on both $\mathcal{B}(\mathbf{x},\mathbf{y})$ and $\widetilde{\mathcal{M}}(\mathbf{x},\mathbf{y})$ by translation in the source (thought of as an infinite strip). Let $\mathcal{M}^B(\mathbf{x},\mathbf{y}) = \widetilde{\mathcal{M}}^B(\mathbf{x},\mathbf{y})/\mathbb{R}$. Finally, the differential on $\widehat{CF}(Y)$ is given by

$$\partial(\mathbf{x}) = \sum_{\mathbf{y}\in T_\alpha\cap T_\beta} \sum_{\substack{B\in\pi_2(\mathbf{x},\mathbf{y})\\ \mu(B)=1}} (\#\mathcal{M}^B(\mathbf{x},\mathbf{y}))\mathbf{y}. \tag{3.1}$$

(Here, # denotes the modulo-2 count of points.) Under certain assumptions on $\mathcal{H}$, called *admissibility*, this count is guaranteed to be finite, so ∂ is well-defined. Moreover:

Theorem 3.1 [46]. *For any suitably generic choice of almost-complex structure, the map ∂ satisfies $\partial^2 = 0$. Moreover, the homology $\widehat{HF}(Y) = H_*(\widehat{CF}(Y),\partial)$ is an invariant of Y.*

3.1.2. The Cylindrical Reformulation. Before proceeding to bordered Floer homology, it will be helpful to have a mild reformulation of the definition of $\widehat{HF}$. It is based on the *tautological correspondence* between maps from $\mathbb{D}^2$ to $\mathrm{Sym}^g(\Sigma)$ and multi-valued functions from $\mathbb{D}^2$ to Σ:

$$\text{Holomorphic maps } \mathbb{D}^2 \to \mathrm{Sym}^g(\Sigma) \longleftrightarrow \text{Diagrams} \quad \begin{array}{ccc} S & \xrightarrow{u_\Sigma} & \Sigma \\ {\scriptstyle u_{\mathbb{D}}}\big\downarrow & & \\ \mathbb{D}^2 & & \end{array}$$

with u_Σ, $u_{\mathbb{D}}$ holomorphic,
$u_{\mathbb{D}}$ a g-fold branched cover.

One direction is easy: given a diagram as on the right, consider the map $\mathbb{D}^2 \to \mathrm{Sym}^g(\Sigma)$ given by mapping p to the g-tuple $u_\Sigma(u_{\mathbb{D}}^{-1}(p))$. The other direction is not hard, either; see, for instance, [26, Section 13].

In light of the tautological correspondence, we can reformulate $\widehat{HF}$ in terms of maps to $\Sigma \times \mathbb{D}^2$. It will be convenient later to view $\mathbb{D}^2 \setminus \{\pm i\}$ as a strip $[0,1] \times \mathbb{R}$. Then:

- Generators of $\widehat{CF}(Y)$ correspond to g-tuples of points $\mathbf{x} = \{x_i\}_{i=1}^g$ with $x_i \in \alpha_i \cap \beta_{\sigma(i)}$ for some $\sigma \in S_g$. These generators can be thought of as g-tuples of chords $\mathbf{x} \times [0,1] \subset \Sigma \times [0,1]$, connecting $\boldsymbol{\alpha} \times \{1\}$ and $\boldsymbol{\beta} \times \{0\}$.
- The differential counts embedded holomorphic maps

  $$u : (S, \partial S) \to \big((\Sigma \setminus \{z\}) \times [0,1] \times \mathbb{R}, (\boldsymbol{\alpha} \times \{1\} \times \mathbb{R}) \cup (\boldsymbol{\beta} \times \{0\} \times \mathbb{R})\big) \tag{3.2}$$

  modulo translation in $\mathbb{R}$. Here, S is a Riemann surface with boundary and punctures on its boundary. The punctures are divided into $+$ punctures and $-$ punctures. Near the $-$ punctures, u is asymptotic to $\mathbf{x} \times [0,1] \times \{-\infty\}$ and near the $+$ punctures u is asymptotic to $\mathbf{y} \times [0,1] \times \{+\infty\}$.

In the cylindrical setting, the set of homotopy classes $\pi_2(\mathbf{x}, \mathbf{y})$ of Whitney disks becomes the set of homology classes (in a suitable sense) of maps as in Formula (3.2). (Philosophically, this is related to the Dold-Thom theorem that $\pi_k(\mathrm{Sym}^\infty(X)) \cong H_k(X)$.)

We have been suppressing almost-complex structures. In order to achieve transversality, one typically perturbs the complex structure $j_\Sigma \times j_{\mathbb{D}}$ on $\Sigma \times [0,1] \times \mathbb{R}$ to a more generic almost-complex structure J. In this cylindrical setting, it is important to ensure that translation in $\mathbb{R}$ remains J-holomorphic. Some other conditions which are necessary or convenient are given in [26, Section 1].

Remark 3.2. It would have been more consistent with conventions in contact homology to consider $\mathbb{R} \times [0,1] \times \Sigma$ rather that $\Sigma \times [0,1] \times \mathbb{R}$.

3.2. Holomorphic Curves and Reeb Chords

Now consider a bordered Heegaard diagram $\mathcal{H} = (\Sigma, \boldsymbol{\alpha}^a, \boldsymbol{\alpha}^c, \boldsymbol{\beta}, z)$. Rather than viewing Σ as a compact surface-with-boundary, attach a cylindrical end $\mathbb{R} \times S^1$ to $\partial\Sigma$; and extend the α-arcs $\boldsymbol{\alpha}^a$ in a translation-invariant way to $\mathbb{R} \times S^1$. (Topologically, this is the same as simply deleting $\partial\Sigma$; but if one is paying attention to the symplectic form and almost-complex structure then there is a difference.) We abuse notation, using the same notation Σ and $\boldsymbol{\alpha}^a$ for the versions with cylindrical ends. We will still consider holomorphic maps as in Formula (3.2); but now there is a third source of non-compactness, $\partial\Sigma$, and these maps can have asymptotics there as well.

We start with the asymptotics at $\pm\infty$. A term for the asymptotics at $\pm\infty$:

Definition 3.3. By a *generator* we mean a g-tuple $\mathbf{x} \subset \boldsymbol{\alpha} \cap \boldsymbol{\beta}$ which has one point on each α-circle, one point on each β-circle, and at most one point on each α-arc.

We consider holomorphic curves disjoint from a neighborhood of z. It follows from this and the fact that only the α-arcs touch $\partial\Sigma$ that the asymptotics at $\partial\Sigma$ are of the form $\rho_i \times (1, t_i)$, where ρ_i is a chord in $\partial\Sigma \setminus \{z\}$ with boundary on $\boldsymbol{\alpha}^a$. We collect these curves into moduli spaces. Let $\widetilde{\mathcal{M}}(\mathbf{x}, \mathbf{y}; \rho_1, \dots, \rho_n)$ denote the moduli space of embedded holomorphic maps as in Formula (3.2) where:

- S is a surface with boundary and punctures on its boundary. Of these punctures, g are labeled $-$, g are labeled $+$, and the rest are labeled e.
- $\mathbf{x}$ and $\mathbf{y}$ are generators.
- at the punctures labeled $-$, u is asymptotic to $\mathbf{x} \times [0,1] \times \{-\infty\}$.
- at the punctures labeled $+$, u is asymptotic to $\mathbf{y} \times [0,1] \times \{+\infty\}$.
- at the punctures labeled e, u is asymptotic to the chords $\rho_i \times (1, t_i) \in \partial\Sigma \times \{1\} \times \mathbb{R}$. Moreover, we require that $t_1 < t_2 < \cdots < t_n$.

There is an $\mathbb{R}$-action on $\widetilde{\mathcal{M}}(\mathbf{x}, \mathbf{y}; \rho_1, \dots, \rho_n)$ by translation in the target; let

$$\mathcal{M}(\mathbf{x}, \mathbf{y}; \rho_1, \dots, \rho_n) = \widetilde{\mathcal{M}}(\mathbf{x}, \mathbf{y}; \rho_1, \dots, \rho_n)/\mathbb{R}.$$

We call the chords ρ *Reeb chords*; they are Reeb chords for the contact structure on $S^1 = \partial\Sigma$. This comes from thinking of the setup as related to a Morse-Bott case of (relative) symplectic field theory. The asymptotic boundary is then $(\partial\Sigma \times [0,1] \times \mathbb{R}, \partial\boldsymbol{\alpha}^a \times \{1\} \times \mathbb{R})$, and we are in the Levi-flat case of, e.g., [5].

As in the closed case, the space of maps of the form just described naturally decomposes into homology classes; see [27, Section 4.3]. To keep notation consistent with the closed case, we let $\pi_2(\mathbf{x},\mathbf{y})$ denote the set of homology classes of maps connecting $\mathbf{x}$ to $\mathbf{y}$; note that we do not specify the Reeb chords here. Then

$$\mathcal{M}(\mathbf{x},\mathbf{y};\rho_1,\dots,\rho_n) = \coprod_{B\in\pi_2(\mathbf{x},\mathbf{y})} \mathcal{M}^B(\mathbf{x},\mathbf{y};\rho_1,\dots,\rho_n).$$

As in the closed case, we have been suppressing the almost-complex structure J from the discussion; the interested reader is referred to [27, Section 5.2]. For a generic choice of J, each of the spaces $\mathcal{M}^B(\mathbf{x},\mathbf{y};\rho_1,\dots,\rho_n)$ is a manifold whose dimension is given by a number $\mathrm{ind}(B;\rho_1,\dots,\rho_n)-1$. The notation ind stands for index: as is usual for holomorphic curves, the dimension is given by the index of the linearized $\overline{\partial}$-operator. One can give an explicit formula for $\mathrm{ind}(B;\rho_1,\dots,\rho_n)$; see [27, Section 5.7].

The next natural thing to talk about, from an analytic perspective, is what the compactifications of $\mathcal{M}^B(\mathbf{x},\mathbf{y};\rho_1,\dots,\rho_n)$ look like. We defer this discussion to Section 4, and instead turn to the definition of the bordered invariant $\widehat{CFD}(Y)$.

3.3. The Definition of $\widehat{CFD}$

3.3.1. Reeb Chords and Algebra Elements. Before defining $\widehat{CFD}(\mathcal{H})$ we need one more piece of notation. Let $\mathcal{Z} = (Z,\mathbf{a},M,z)$ be a pointed matched circle and ρ a chord in $Z\setminus\{z\}$ with boundary in $\mathbf{a}$. Orienting ρ according to the orientation of Z and identifying $\mathbf{a} = \{1,\dots,4k\}$, the chord ρ has an initial point i and a terminal point j. Write

$$a(\rho) = \sum_{\substack{S\subset\underline{4k}\\ i\in S}} (S, S\setminus\{i\}\cup\{j\}, \phi_S) \tag{3.3}$$

where $\phi_S(i) = j$ and $\phi_S|_{S\setminus i} = \mathbb{I}$, and the sum is only over S's so that S and $S\setminus\{i\}\cup\{j\}$ are M-admissible. That is, $a(\rho)$ is the union of a strand from i to

j and any admissible set of horizontal strands. A somewhat trivial example is given by Exercise 2.9.

3.3.2. The Definition of $\widehat{CFD}$. Fix a bordered Heegaard diagram $\mathcal{H} = (\Sigma, \boldsymbol{\alpha}^a, \boldsymbol{\alpha}^c, \boldsymbol{\beta}, z)$ with boundary $\mathcal{Z}$. We will define a left *dg* module $\widehat{CFD}(\mathcal{H})$ over $\mathcal{A}(-\mathcal{Z})$ (where, as usual, $-$ denotes orientation reversal). The module $\widehat{CFD}(\mathcal{H})$ will lie over $\mathcal{A}(-\mathcal{Z},0)$, in the sense that the other summands $\mathcal{A}(-\mathcal{Z},i)$, $i \neq 0$, of $\mathcal{A}(-\mathcal{Z})$ act trivially on $\widehat{CFD}(\mathcal{H})$.

Let $\mathfrak{S}(\mathcal{H})$ denote the set of generators for $\mathcal{H}$. Given a generator $\mathbf{x} \in \mathfrak{S}(\mathcal{H})$, let $I(S)$ denote the set of α-arcs which are disjoint from $\mathbf{x}$.[3] Then $I(S)$ corresponds to a set of matched pairs in $-\mathcal{Z}$, and hence, by Exercise 2.10, to an indecomposable idempotent of $\mathcal{A}(-\mathcal{Z})$. As a (left) module, define

$$\widehat{CFD}(\mathcal{H}) = \bigoplus_{\mathbf{x} \in \mathfrak{S}(\mathcal{H})} \mathcal{A}(-\mathcal{Z}) \cdot I(S).$$

It remains to define the differential on $\widehat{CFD}(\mathcal{H})$. For $\mathbf{x} \in \mathfrak{S}(\mathcal{H})$ define

$$\partial(\mathbf{x}) = \sum_{\substack{\mathbf{y} \in \mathfrak{S}(\mathcal{H}) \\ n \geq 0 \\ (\rho_1,\dots,\rho_n) \\ B \mid \operatorname{ind}(B,\rho_1,\dots,\rho_n)=1}} (\#\mathcal{M}^B(\mathbf{x},\mathbf{y};\rho_1,\dots,\rho_n)) a(-\rho_1)\cdots a(-\rho_n)\mathbf{y}. \tag{3.4}$$

Here, the minus signs are included because $\widehat{CFD}$ is a module over $\mathcal{A}(-\mathcal{Z})$ rather than $\mathcal{A}(\mathcal{Z})$; $-\rho$ is the chord ρ but viewed as running in the opposite direction (i.e., as a chord in $-\mathcal{Z}$).

Extend the differential to the rest of $\widehat{CFD}(Y)$ by the Leibniz rule. This completes the definition of $\widehat{CFD}(Y)$.

Example 3.4. Consider the bordered Heegaard diagram in Figure 8. We have labeled the three length-1 Reeb chords; notice that we have ordered them in the opposite of the order induced by the orientation of $\partial\mathcal{H}$, because we are thinking of the algebra $\mathcal{A}(-\partial\mathcal{H})$. The module $\widehat{CFD}(\mathcal{H})$ has three generators, x, a and b. With notation as in Formula (2.1), the idempotents are given by

$$I(x) = \iota_1 \qquad I(a) = \iota_0 \qquad I(b) = \iota_0.$$

[3]This $I(S)$ was denoted $I_D(S)$ in [27], where $I(S)$ was used for $I_A(S)$ introduced in Section 4.4.

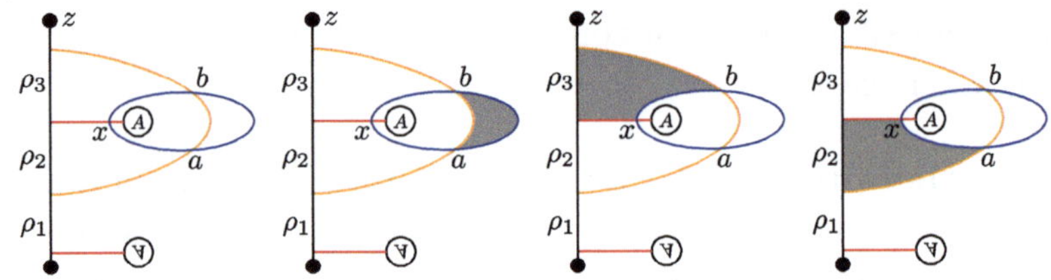

Fig. 8. A Heegaard diagram for a solid torus, and some holomorphic curves in it. The circles labeled A indicate a handle. The shaded regions in the second through fourth figures indicate the domains giving $a \in \partial(b)$, $\rho_3 x \in \partial b$, and $\rho_2 a \in \partial x$, respectively

The differentials are given by

$$\partial(b) = a + \rho_3 x$$
$$\partial(x) = \rho_2 a$$
$$\partial(a) = 0.$$

Each of these differentials comes from a disk mapped to $\Sigma \times [0,1] \times \mathbb{R}$; the projections of these disks to Σ (their *domains*—see Definition 3.5) are indicated in the figure. Since $\rho_3\rho_2 = 0$, $\partial^2 = 0$.

3.3.3. Finiteness Conditions. As in the closed case, the definition of $\widehat{CFD}$ (Formula (3.4)) only makes sense if the sums involved are finite. To ensure finiteness, we add assumptions on the Heegaard diagram $\mathcal{H}$, analogous to admissibility in the closed case:

Definition 3.5. Given a homology class $B \in \pi_2(\mathbf{x}, \mathbf{y})$, the projection of B to Σ defines a cellular 2-chain with respect to the cellulation of Σ given by $\boldsymbol{\alpha} \cup \boldsymbol{\beta}$. This 2-chain is called the *domain of B*, and determines B. A non-trivial class B is called *positive* if its local multiplicities are all non-negative. The domains of homology classes $B \in \pi_2(\mathbf{x}, \mathbf{x})$ are called *periodic domains.* The set of periodic domains does not depend on $\mathbf{x}$.

The Heegaard diagram $\mathcal{H}$ is called *provincially admissible* if it has no positive periodic domains which have multiplicity 0 everywhere along $\partial\Sigma$.

The Heegaard diagram $\mathcal{H}$ is called *admissible* if it has no positive periodic domains.

Lemma 3.6 [27, Lemma 6.5]. *If $\mathcal{H}$ is provincially admissible then the sums in Formula (3.4) are finite. Moreover, if $\mathcal{H}$ is admissible then the operator ∂ is nilpotent in the following sense. Consider sequences of generators*

$(\mathbf{x}_1, \mathbf{x}_2, \ldots, \mathbf{x}_n)$ such that $\mathbf{x}_{i+1}$ occurs in $\partial \mathbf{x}_i$ with nonzero coefficient. If $\mathcal{H}$ is admissible then there is a universal bound on the length of such sequences.

The proof of Lemma 3.6 is not hard; it is an adaptation of the proof of the corresponding fact from the closed case [46, Lemma 4.14]. The nilpotency condition in Lemma 3.6 guarantees that $\widehat{CFD}(\mathcal{H})$ is projective (or rather, $\mathcal{K}$-projective in the sense of, e.g., [4]). It is not particularly relevant until we start taking tensor products, e.g. in the statement of Theorem 2.17.

Theorem 3.7 [27, Proposition 6.7]. *Let $\mathcal{H}$ be a provincially admissible Heegaard diagram. Then $\widehat{CFD}(\mathcal{H})$ is a differential module.*

The only nontrivial thing to check is that $\partial^2 = 0$. The proof involves studying the boundaries of 1-dimensional moduli spaces; we will sketch it in Section 4.

3.4. The Surgery Exact Triangle[4]

Recall that Heegaard Floer homology admits a surgery exact triangle [45]. Specifically, for a pair (M, K) of a 3-manifold M and a framed knot K in M, there is an exact triangle

(3.5)
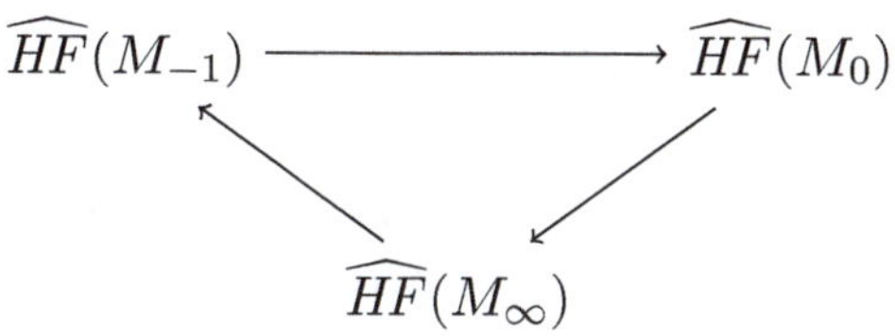

where M_{-1}, M_0, and M_∞ are -1, 0, and ∞ surgery on K, respectively. As a simple application of bordered Floer theory, we reprove this result.

Consider the three diagrams

(3.6)
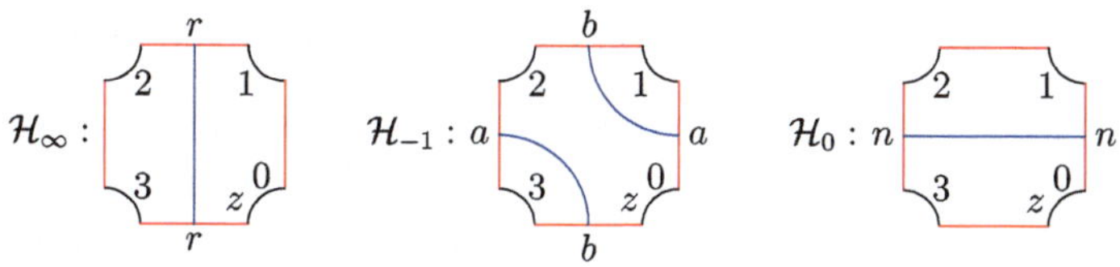

[4]The discussion in this section is taken from [27, Section 11.2].

(Opposite edges are identified, to give $T^2 \setminus \mathbb{D}^2$. Each diagram has two α-arcs and one β-circle. The numbers indicate which chord, in the notation of Formula (2.1), corresponds to which arc in $\partial\mathcal{H}_\bullet$. Note again that the chords are numbered in the opposite of the order induced by the orientation of $\partial\mathcal{H}_\bullet$.) A generator for $\widehat{CFD}(\mathcal{H}_\bullet)$ consists of a single intersection point between the β-circle in $\mathcal{H}_\bullet$ and an α-arc. These intersections are labeled above.

The boundary operators on the $\widehat{CFD}(\mathcal{H}_\bullet)$ (and the relevant domains) are given by

$\partial r = \rho_{23} r$ $\qquad$ $\partial a = \rho_3 b + \rho_1 b$ $\qquad$ $\partial b = 0$ $\qquad$ $\partial n = \rho_{12} n$

There is a short exact sequence

$$0 \longrightarrow \widehat{CFD}(\mathcal{H}_\infty) \xrightarrow{\varphi} \widehat{CFD}(\mathcal{H}_{-1}) \xrightarrow{\psi} \widehat{CFD}(\mathcal{H}_0) \longrightarrow 0$$

where the maps ϕ and ψ are given by

$$\varphi(r) = b + \rho_2 a \qquad \psi(a) = n \qquad \psi(b) = \rho_2 n.$$

Now, the surgery exact triangle follows immediately from the pairing theorem and properties of the derived tensor product.

3.5. The Definition of $\widehat{CFDD}$

Suppose $\mathcal{Z}_L$ and $\mathcal{Z}_R$ are pointed matched circles. We can form their connected sum $\mathcal{Z}_L \# \mathcal{Z}_R$. There are two natural choice of where to put a basepoint in $\mathcal{Z}_L \# \mathcal{Z}_R$; let z be a point in one of these places and w a point in the other. Thinking of z as the basepoint, there is an associated algebra $\mathcal{A}(\mathcal{Z}_L \# \mathcal{Z}_R)$. Moreover, there is an algebra homomorphism

$$p\colon \mathcal{A}(\mathcal{Z}_L \# \mathcal{Z}_R) \to \mathcal{A}(\mathcal{Z}_L) \otimes_{\mathbb{F}_2} \mathcal{A}(\mathcal{Z}_R)$$

given by setting to zero any algebra element crossing the extra basepoint w.

Now, suppose that $\mathcal{H}$ is an arced Heegaard diagram. Performing surgery on $\mathcal{H}$ along the arc $\mathbf{z}$ gives a bordered Heegaard diagram $\mathcal{H}_{dr}$. (Again, there are two choices of where to put the basepoint in $\mathcal{H}_{dr}$; choose either.) If the boundary of $\mathcal{H}$ was $\mathcal{Z}_L \amalg \mathcal{Z}_R$ then the boundary of $\mathcal{H}_{dr}$ is $\mathcal{Z}_L \# \mathcal{Z}_R$.

Associated to $\mathcal{H}_{dr}$ is a bordered module $\widehat{CFD}(\mathcal{H}_{dr})$ over $\mathcal{A}(-(\mathcal{Z}_L \# \mathcal{Z}_R))$.

Definition 3.8. With notation as above, let

$$\widehat{CFDD}(\mathcal{H}) = ((-\mathcal{A}(\mathcal{Z}_L)) \otimes_{\mathbb{F}_2} (-\mathcal{A}(\mathcal{Z}_R))) \otimes_{\mathcal{A}(-(\mathcal{Z}_L \# \mathcal{Z}_R))} \widehat{CFD}(\mathcal{H}_{dr}),$$

be the image of the bordered bimodule $\widehat{CFD}(\mathcal{H}_{dr})$ under the induction functor associated to the homomorphism p. Via the correspondence between left-left $(\mathcal{A}(-\mathcal{Z}_L), \mathcal{A}(-\mathcal{Z}_R))$-bimodules and left $((-\mathcal{A}(\mathcal{Z}_L)) \otimes_{\mathbb{F}_2} (-\mathcal{A}(\mathcal{Z}_R)))$-modules, we view $\widehat{CFDD}(\mathcal{H})$ as a left-left $(\mathcal{A}(-\mathcal{Z}_L), \mathcal{A}(-\mathcal{Z}_R))$-bimodules.

Of course, this definition can be unpacked to define $\widehat{CFDD}(\mathcal{H})$ directly in terms of intersection points and holomorphic curves; doing so is Exercise 3.8.

3.6. Exercises

Exercise 3.1. There is a unique almost-complex structure $\mathrm{Sym}^g(j_\Sigma)$ on $\mathrm{Sym}^g(\Sigma)$ so that the projection map $(\Sigma^{\times g}, j_\Sigma^{\times g}) \to (\mathrm{Sym}^g(\Sigma), \mathrm{Sym}^g(j_\Sigma))$ is holomorphic. In the tautological correspondence of Section 3.1.2, show that if u_Σ and $u_{\mathbb{D}}$ are holomorphic then the map $\mathbb{D}^2 \to \mathrm{Sym}^g(\Sigma)$, $p \mapsto u_\Sigma(u_{\mathbb{D}}^{-1}(p))$ is holomorphic with respect to $\mathrm{Sym}^g(j_\Sigma)$.

Exercise 3.2. Consider the Heegaard diagrams of Section 3.4. Replacing the blue (β) curve in the diagrams $\mathcal{H}_\bullet$ by a circle of slope p/q gives a bordered Heegaard diagram $\mathcal{H}_{p/q}$ for a p/q-framed solid torus. It is fairly easy to compute the invariants $\widehat{CFD}(\mathcal{H}_{p/q})$ for these diagrams; compute some.

For any triple of rational numbers $(p_1/q_1, p_2/q_2, p_3/q_3)$ (with p_i, q_i relatively prime) such that $p_1+p_2+p_3 = q_1+q_2+q_3 = 0$ there is a corresponding surgery triangle; check this for some other examples.

Exercise 3.3. Compute $\mathrm{Mor}(\widehat{CFD}(\mathcal{H}_{p/q}), \widehat{CFD}(\mathcal{H}_{r/s}))$ for a few choices of p, q, r, s. For example, $\mathrm{Mor}(\widehat{CFD}(\mathcal{H}_\infty), \widehat{CFD}(\mathcal{H}_{-1}))$ has generators $(r \mapsto b)$, $(r \mapsto \rho_{23} b)$ and $(r \mapsto \rho_2 a)$. The differentials are given by

$$\partial(r \mapsto b) = (r \mapsto \rho_{23} b)$$
$$\partial(r \mapsto \rho_2 a) = (r \mapsto \rho_{23} b).$$

In particular, the homology of this Mor complex is 1-dimensional.

Recall that $\widehat{HF}(L(p,q)) \cong (\mathbb{F}_2)^p$, and $\widehat{HF}(S^2 \times S^1) \cong (\mathbb{F}_2)^2$; check that your answers are consistent with this.

Exercise 3.4. We explain the type DD bimodule $\widehat{CFDD}(\mathbb{I},0)$ associated to the mapping cylinder for the identity map of $F(\mathcal{Z})$. The notation is somewhat cumbersome, as $\widehat{CFDD}(\mathbb{I},0)$ has two commuting left actions by $\mathcal{A}(T^2,0)$. We write one of these copies of $\mathcal{A}(T^2,0)$ in the notation of Formula (2.1), and the other in the same way but with σ's in place of ρ's and η's in place of ι's. Then, the bimodule $\widehat{CFDD}(\mathbb{I},0)$ has two generators, x and y, with

$$\iota_0 x = \eta_0 x = x \qquad \iota_1 y = \eta_1 y = y$$

and differential given by

$$\begin{aligned} \partial x &= (\rho_1\sigma_3 + \rho_3\sigma_1 + \rho_{123}\sigma_{123}) \otimes y \\ \partial y &= (\rho_2\sigma_2) \otimes x. \end{aligned} \tag{3.7}$$

(Compare [27, Section A.3.3].)

Verify that for the modules $\widehat{CFD}(\mathcal{H}_\bullet)$ of Section 3.4, $\mathrm{Mor}(\widehat{CFDD}(\mathbb{I},0),\cdot)$ acts as the identity. That is, check that

$$\mathrm{Mor}_{\mathcal{A}(T^2,0)}\big(\widehat{CFDD}(\mathbb{I},0), \widehat{CFD}(\mathcal{H}_0)\big) \simeq \widehat{CFD}(\mathcal{H}_0),$$

and similarly for $\mathcal{H}_{-1}$, $\mathcal{H}_\infty$. (You will have to use the equivalence of categories between left $\mathcal{A}(T^2,0)$-modules and right $\mathcal{A}(T^2,0)$-modules coming from the fact that $\mathcal{A}(T^2,0) \cong \mathcal{A}(T^2,0)^{\mathrm{op}}$. Note that this isomorphism exchanges ρ_1 and ρ_3.)

Remark 3.9. There are two non-equivalent notions of the Mor complex above, depending on how one treats the other algebra action on $\widehat{CFDD}(\mathbb{I},0)$. The exercise will be true with either notion. See [32, Theorems 5 and 6] for an example where this distinction matters.

Remark 3.10. It is sometimes convenient to encode the operations in Formula (3.7) by:

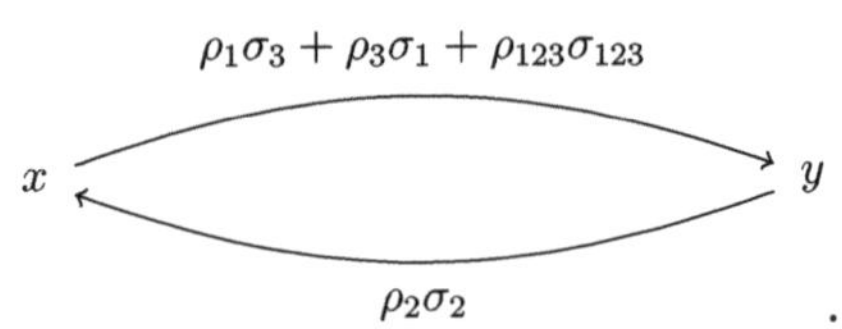

This way of encoding operations on DD bimodules will be used in Exercise 3.6.

Exercise 3.5. Note that the identity for Mor is the $\mathcal{A}$-bimodule $\mathcal{A}$. In spite of the computations in Exercise 3.4, $\widehat{CFDD}(\mathbb{I}) \not\simeq \mathcal{A}(T^2, 0)$. Check this two ways:

- Directly. (Think about the rank of the homologies.)
- By finding a module M over $\mathcal{A}(T^2, 0)$ so that $\widehat{CFDD}(\mathbb{I}) \otimes_{\mathcal{A}(T^2,0)} M \not\simeq M$. (Or, you can use $\mathrm{Mor}(\widehat{CFDD}(\mathbb{I}), M)$ if you prefer.)

Exercise 3.6. Let τ_μ and τ_λ denote the Dehn twists of the torus along a meridian and a longitude, respectively. Heegaard diagrams for the mapping cylinders of τ_μ and τ_λ are shown in Figure 9. With notation as in Remark 3.10, the type *DD* bimodules associated to these Dehn twists and their inverses are given by

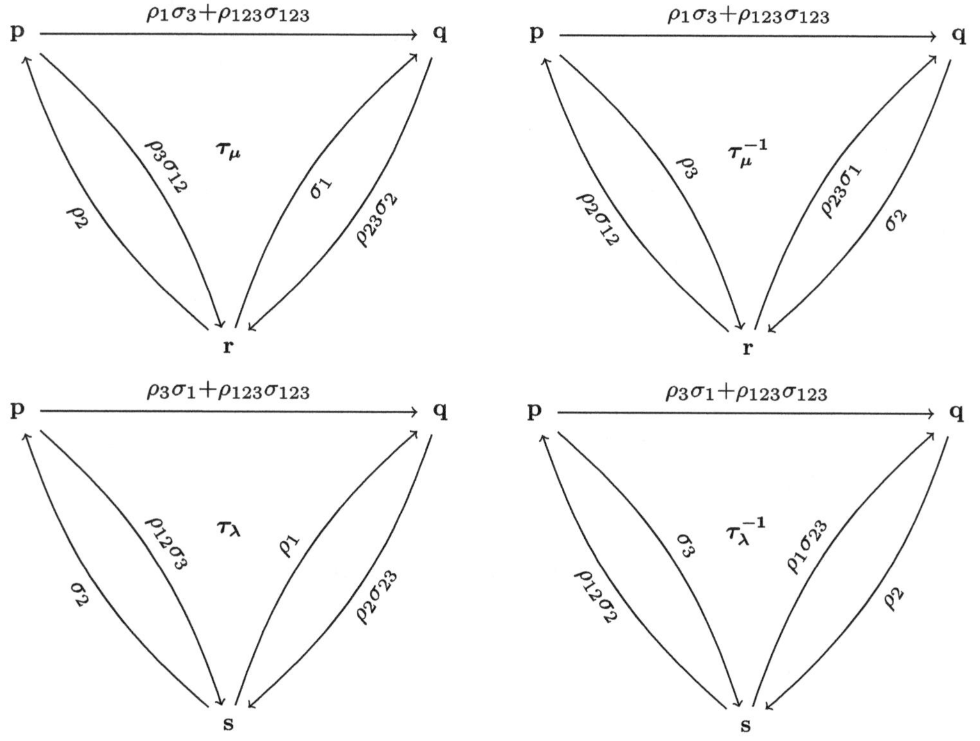

Convince yourself that these bimodules satisfy $\partial^2 = 0$. Compute $\mathrm{Mor}(\widehat{CFDD}(\tau_\mu), \mathcal{H}_0)$ and $\mathrm{Mor}(\widehat{CFDD}(\tau_\lambda), \mathcal{H}_0)$. Compare the results with the answers you computed in Exercise 3.2.

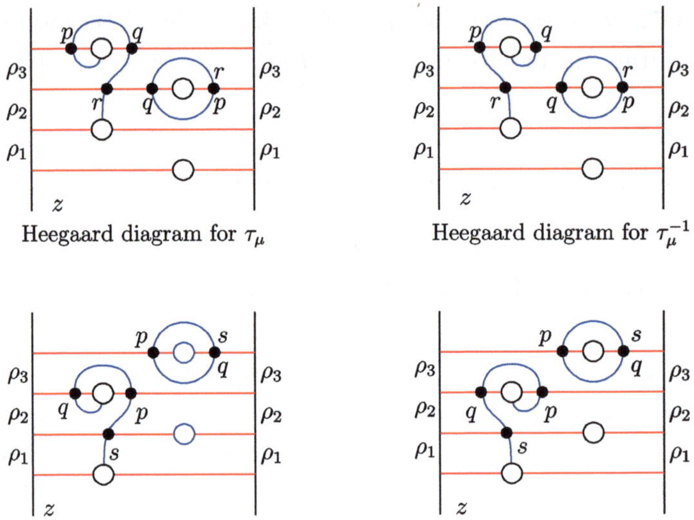

Heegaard diagram for τ_μ Heegaard diagram for τ_μ^{-1}

Heegaard diagram for τ_λ Heegaard diagram for τ_λ^{-1}

Fig. 9. Heegaard diagrams for mapping class group elements. Genus 2 diagrams for τ_μ, τ_μ^{-1}, τ_λ and τ_λ^{-1} are shown. In each of the four diagrams, there are three generators in the $i = 0$ summand. (This figure is drawn from [28, Figure A.2])

Exercise 3.7. Up to Heegaard moves, there are some symmetries relating the diagrams in Figure 9. How are these symmetries reflected in the bimodules in Exercise 3.6?

Exercise 3.8. Unpack the definition of $\widehat{CFDD}$ from Section 3.5 to give a direct definition, avoiding the induction functor.

4. Analysis Underlying the Invariants and the Pairing Theorem

4.1. Broken Flows in the Cylindrical Setting

As a warm-up, we begin this lecture by discussing the proof that $\partial^2 = 0$ for the cylindrical picture for Heegaard Floer homology. We start with an example. Consider the Heegaard diagram for S^3 shown in Figure 10. There are five generators, labeled a, b, c, d and e. The differentials are given by

$$\partial(a) = b + c \qquad \partial(b) = \partial(c) = d \qquad \partial(d) = 0 \qquad \partial(e) = b + c.$$

(Remember that we are working with $\mathbb{F}_2$-coefficients.)

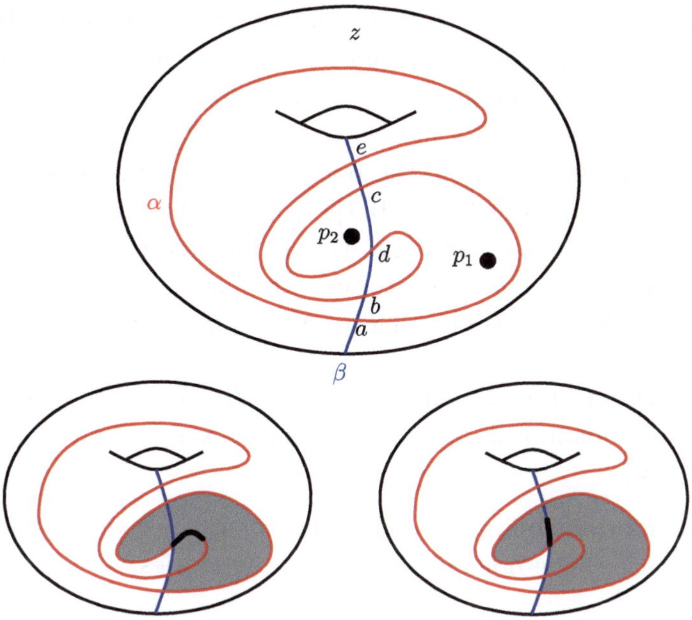

Fig. 10. An unnecessarily complicated diagram for S^3. In the two pictures on the bottom we have indicated the image $\pi_\Sigma(u(\partial\mathbb{D}^2))$ for two typical elements of $\mathcal{M}(a,d)$. The thick black segments indicate cuts

Consider the moduli space $\mathcal{M}(a,d)$ of curves connecting a to d. This moduli space consists of holomorphic maps

$$u\colon \big(\mathbb{D}^2 \setminus \{\pm i\}\big) \to \Sigma \times [0,1] \times \mathbb{R}.$$

Suppose we are working with the almost-complex structure $j_\Sigma \times j_\mathbb{D}$. Then there are projection maps $\pi_\Sigma : \Sigma \times [0,1] \times \mathbb{R} \to \Sigma$ and $\pi_\mathbb{D} : \Sigma \times [0,1] \times \mathbb{R} \to [0,1] \times \mathbb{R}$, and u being holomorphic is equivalent to $\pi_\Sigma \circ u$ and $\pi_\mathbb{D} \circ u$ being holomorphic.

The map $\pi_\mathbb{D} \circ u$ is a 1-fold branched cover, i.e., an isomorphism; up to translation, there is a unique such isomorphism.

A short argument using the Riemann mapping theorem shows that the map $\pi_\Sigma \circ u$ is determined by the image of $\partial\mathbb{D}^2$. Figure 10 shows two possibilities for $\pi_\Sigma(u(\partial\mathbb{D}^2))$. Note the branch point on α_1 or β_1. The whole moduli space is determined by where the branch point lies; so, $\mathcal{M}(a,d)$ is an (open) interval. The ends of $\mathcal{M}(a,d)$ occur when the branch point approaches b or c.

We want to describe the limiting objects. In the ordinary setting for Morse theory, these would be broken flows. In this setting, they are multistory holomorphic buildings. We see this as follows.

Consider a sequence of curves u_i approaching the end of $\mathcal{M}(a,d)$ where the branch point approaches c. Notice the points $p_1, p_2 \in \Sigma$ shown in Figure 10. Consider the points $q_1 = (\pi_\Sigma \circ u_i)^{-1}(p_1)$ and $q_2 = (\pi_\Sigma \circ u_i)^{-1}(p_2)$ in $\mathbb{D}^2$. The points $(\pi_\mathbb{D} \circ u_i)(q_1)$ and $(\pi_\mathbb{D} \circ u_i)(q_2)$ in $[0,1] \times \mathbb{R}$ are getting farther and farther apart. Indeed, from the point of view of q_1, half of the holomorphic curve is heading towards $\Sigma \times [0,1] \times \{+\infty\}$, while from the point of view of q_2, half of the holomorphic curve is heading towards $\Sigma \times [0,1] \times \{-\infty\}$. So, the limiting object has two "stories": the part of the limit containing q_1 and the part of the limit containing q_2. More formally:

Definition 4.1. An *ℓ-story holomorphic building* connecting $\mathbf{x}$ to $\mathbf{y}$ consists of a sequence of holomorphic curves $u_i \in \mathcal{M}(\mathbf{x}_i, \mathbf{x}_{i+1})$, $i = 1, \ldots, \ell$, with $\mathbf{x}_1 = \mathbf{x}$ and $\mathbf{x}_{\ell+1} = \mathbf{y}$.

Each holomorphic building carries a homology class in $\pi_2(\mathbf{x}, \mathbf{y})$, by adding up (concatenating) the homology classes of its stories.

We should now give a topology on the space of holomorphic buildings, to say precisely what it means for a sequence of one-story buildings, i.e., elements of $\mathcal{M}(a,d)$, to converge to a multi-story building. Instead, however, we refer the reader to [5].

The main structural result is:

Theorem 4.2. *Suppose that $B \in \pi_2(\mathbf{x}, \mathbf{y})$ has $\mu(B) = 2$. Let $\overline{\mathcal{M}}^B(\mathbf{x}, \mathbf{y})$ denote the space of 1- or 2-story holomorphic buildings connecting $\mathbf{x}$ to $\mathbf{y}$ in the homology class B. Then for a generic choice of almost-complex structure, $\overline{\mathcal{M}}^B(\mathbf{x}, \mathbf{y})$ is a compact 1-dimensional manifold-with-boundary. The boundary of $\overline{\mathcal{M}}^B(\mathbf{x}, \mathbf{y})$ consists exactly of the 2-story holomorphic buildings connecting $\mathbf{x}$ to $\mathbf{y}$ in the homology class B.*

In the cylindrical formulation, this is [26, Corollary 7.2]; the analogous result for Heegaard Floer homology in the non-cylindrical setting was proved in [46]. (Both proofs are relatively modest adaptations of standard holomorphic curve techniques.)

To conclude the warm-up, we recall that $\partial^2 = 0$ follows from Theorem 4.2 by a standard argument:

Corollary 4.3. *Let $\mathcal{H}$ be an admissible Heegaard diagram for a closed 3-manifold. Then the differential ∂ on $\widehat{CF}(\mathcal{H})$ satisfies $\partial^2 = 0$.*

Proof. The proof involves the usual looking at ends of one-dimensional moduli spaces, as is familiar in Floer homology:

$$\begin{aligned}
\partial(\mathbf{x}) &= \sum_{\mathbf{y}\in\mathfrak{S}(\mathcal{H})} \sum_{\substack{B_1\in\pi_2(\mathbf{x},\mathbf{y})\\ \mu(B_1)=1}} (\#\mathcal{M}^{B_1}(\mathbf{x},\mathbf{y}))\mathbf{y} \\
\partial^2(x) &= \sum_{\mathbf{y}\in\mathfrak{S}(\mathcal{H})} \sum_{\substack{B_1\in\pi_2(\mathbf{x},\mathbf{y})\\ \mu(B_1)=1}} (\#\mathcal{M}^{B_1}(\mathbf{x},\mathbf{y}))\partial(\mathbf{y}) \\
&= \sum_{\mathbf{y},\mathbf{z}\in\mathfrak{S}(\mathcal{H})} \sum_{\substack{B_1\in\pi_2(\mathbf{x},\mathbf{y})\\ \mu(B_1)=1}} \sum_{\substack{B_2\in\pi_2(\mathbf{y},\mathbf{z})\\ \mu(B_2)=1}} (\#\mathcal{M}^{B_1}(\mathbf{x},\mathbf{y}))(\#\mathcal{M}^{B_2}(\mathbf{y},\mathbf{z}))\mathbf{z} \\
&= \sum_{\mathbf{z}\in\mathfrak{S}(\mathcal{H})} \sum_{\substack{B\in\pi_2(\mathbf{x},\mathbf{z})\\ \mu(B)=2}} (\#\partial\mathcal{M}^{B}(\mathbf{x},\mathbf{z}))\mathbf{z} \\
&= 0.
\end{aligned}$$

Most of this is just manipulation of symbols; the key point is the fourth equality, which uses Theorem 4.2. The last equality follows from the fact that a 1-dimensional manifold-with-boundary has an even number of ends. (The assumption about admissibility is used to ensure that the sums involved at each stage are finite.) ■

4.2. The Codimension-One Boundary: Statement

To prove that $\partial^2=0$ for $\widehat{CFD}$ we need to investigate the boundary of the 1-dimensional moduli spaces, analogously to Theorem 4.2. So, fix a bordered Heegaard diagram $\mathcal{H}=(\Sigma,\boldsymbol{\alpha}^c,\boldsymbol{\alpha}^a,\boldsymbol{\beta})$. As above, we can have breaking at $\pm\infty$, giving multi-story holomorphic buildings; but now there are two other sources of non-compactness:

(1) The manifold Σ has a cylindrical end, giving another direction in which curves in $\Sigma\times[0,1]\times\mathbb{R}$ can break.

(2) In the moduli space $\mathcal{M}^B(\mathbf{x},\mathbf{y};\rho_1,\dots,\rho_n)$ we had Reeb chords $\rho_i\times(1,t_i)$ where $t_1<t_2<\cdots<t_n$. This can degenerate when $t_{i+1}-t_i\to 0$.

(There is overlap between the two cases.)

Degenerations of type (1) lead to the analogue of 2-story holomorphic buildings, but in the "horizontal", i.e., Σ, direction. In principle, one can have degenerations in both the vertical ($\mathbb{R}$) and horizontal (Σ) directions at once. We called the resulting objects *holomorphic combs* [27, Definition 5.20]. In codimension 1, the kinds of combs that can appear are quite limited, so rather than giving the general story we will simply explain these cases.

By *east* ∞ we mean $\mathbb{R}\times(\partial\Sigma)\times[0,1]\times\mathbb{R}$; this is the symplectic manifold that one sees at the ("horizontal") end of Σ. Note that there are projection maps

$$\begin{aligned}\pi_\Sigma &: \mathbb{R}\times(\partial\Sigma)\times[0,1]\times\mathbb{R}\to\mathbb{R}\times(\partial\Sigma)\\ \pi_\mathbb{D} &: \mathbb{R}\times(\partial\Sigma)\times[0,1]\times\mathbb{R}\to[0,1]\times\mathbb{R}\\ t &: \mathbb{R}\times(\partial\Sigma)\times[0,1]\times\mathbb{R}\to\mathbb{R},\end{aligned}$$

where t is projection onto the second (last) $\mathbb{R}$-factor. Degenerations of type (1) lead to pairs (u,v) where u is a curve in $\Sigma\times[0,1]\times\mathbb{R}$ of the kind we have been considering and v is a curve at east ∞, i.e., a holomorphic map

$$v:(S,\partial S)\to\big(\mathbb{R}\times(\partial\Sigma)\times[0,1]\times\mathbb{R},\mathbb{R}\times(\boldsymbol{\alpha}\cap\partial\Sigma)\times\{1\}\times\mathbb{R}\big).$$

Here, S is a surface with boundary and punctures on the boundary. Each puncture is labeled either e or w. Near each e puncture, v is asymptotic to some $\{\infty\}\times\rho_i\times(1,t_i)$ where ρ_i is a chord in $\partial\Sigma$ and $t_i\in\mathbb{R}$. Similarly, near each w puncture, v is asymptotic to some $\{-\infty\}\times\rho_i\times(1,t_i)$.

It follows from the boundary conditions and asymptotics that for each component of v, the map $\pi_\mathbb{D}\circ v$ is, in fact, constant. This makes describing holomorphic curves at east ∞ relatively straightforward. Three kinds of curves will play special roles in studying $\widehat{CFD}$:

- A *trivial component* is a disk in $\mathbb{R}\times(\partial\Sigma)\times[0,1]\times\mathbb{R}$ which is invariant under translation in the first $\mathbb{R}$-factor. It follows that a trivial component has one w punctures and one e puncture, and is asymptotic to the same chord ρ at both punctures.
- A *join component* is a disk in $\mathbb{R}\times(\partial\Sigma)\times[0,1]\times\mathbb{R}$ with two w punctures and one e puncture. At the two w punctures the curve is asymptotic to chords ρ_1 and ρ_2 and at the e puncture the curve is asymptotic to a chord ρ. With respect to the cyclic ordering of the punctures (ρ,ρ_1,ρ_2) around the boundary of the disk (see Figure 11), the terminal endpoint of ρ_2 is the initial endpoint of ρ_1; and $\rho=\rho_2\cup\rho_1$.

  A *join curve* is the disjoint union of one join component and finitely many trivial components.

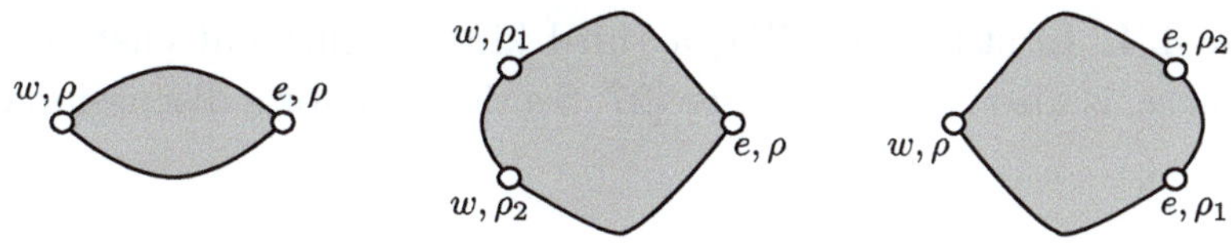

Fig. 11. Sources of curves at east ∞. Left: a trivial component. Center: a join component. Right: a split component. This is [27, Figure 5.3]

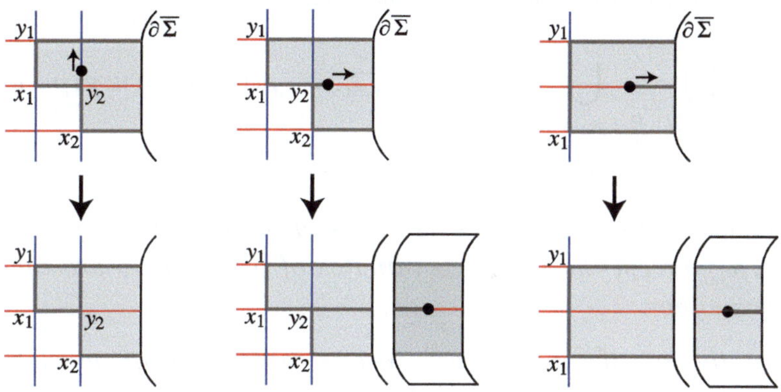

Fig. 12. Examples of three kinds of codimension 1 degenerations. The large black dot represents a boundary branch point of $\pi_\Sigma \circ u$. Left: degenerating into a two-story building. Center: degenerating a join curve. Right: degenerating a split curve. The diagrams show the projection of the curve in $\Sigma \times [0, 1] \times \mathbb{R}$ to Σ. This figure is adapted from [27, Figure 5.1]

- Roughly, a *split component* is the mirror of a join component. In more detail, a split component is a disk in $\mathbb{R} \times (\partial\Sigma) \times [0, 1] \times \mathbb{R}$ with one w punctures and two e puncture. At the two e punctures the curve is asymptotic to chords ρ_1 and ρ_2 and at the w puncture the curve is asymptotic to a chord ρ. With respect to the cyclic ordering of the punctures (ρ, ρ_1, ρ_2) around the boundary of the disk (see Figure 11), the terminal endpoint of ρ_1 is the initial endpoint of ρ_2; and $\rho = \rho_1 \cup \rho_2$.

  For our purposes, a *split curve* is the disjoint union of one split component and finitely many trivial components. (If we were also interested in $\widehat{CFA}$, we would have to allow more than one split component in a split curve.)

Figure 12 gives examples of degenerating a join curve and a split curve at east ∞, as well as breaking into a two-story holomorphic building.

Remark 4.4. In studying $\widehat{CFA}$, a third kind of curve at east ∞, called a *shuffle curve*, is also important. See [27, Section 5.3] for a discussion of shuffle curves.

Theorem 4.5. *Suppose that* $\operatorname{ind}(B;\rho_1,\dots,\rho_n)=2$*. Then the ends of the moduli space* $\mathcal{M}^B(\mathbf{w},\mathbf{y};\rho_1,\dots,\rho_n)$ *consist exactly of the following configurations:*

(1) *Two-story holomorphic buildings, i.e.,*

$$\bigcup_{i=0}^{n}\ \bigcup_{\mathbf{x}\in\mathfrak{S}(\mathcal{H})}\ \bigcup_{\substack{B_1\in\pi_2(\mathbf{w},\mathbf{x})\\ B_2\in\pi_2(\mathbf{x},\mathbf{y})\\ B_1*B_2=B}} \mathcal{M}^{B_1}(\mathbf{w},\mathbf{x};\rho_1,\dots,\rho_i)\times\mathcal{M}^{B_2}(\mathbf{x},\mathbf{y};\rho_{i+1},\dots,\rho_n).$$

(2) *Collapses of levels, i.e., curves u as in the definition of* $\mathcal{M}^B(\mathbf{w},\mathbf{y};\rho_1,\dots,\rho_n)$ *except that the t-coordinates of ρ_i and ρ_{i+1} are equal. Moreover, either:*

 (2a) *the set of (one or two) α-arcs containing $\partial\rho_i$ must be disjoint from the set of (one or two) α-arcs containing $\partial\rho_{i+1}$, or*

 (2b) *the initial endpoint of ρ_i is the same as the final endpoint of ρ_{i+1}.*

(3) *Join curve degenerations, i.e., pairs (u,v) where u is a curve like those in*

$$\mathcal{M}^B(\mathbf{w},\mathbf{y};\rho_1,\dots,\rho_i',\rho_i'',\rho_{i+1},\dots,\rho_n)$$

except that the t-coordinates of ρ_i' and ρ_i'' are equal; and v is a join curve with w asymptotics $\rho_1,\dots,\rho_i',\rho_i'',\dots,\rho_n$ and e asymptotics $\rho_1,\dots,\rho_i,\dots,\rho_n$. In particular, $\rho_i=\rho_i''\cup\rho_i'$. Moreover:

- *The α-arc containing the terminal end of ρ_i'' is distinct from the α-arcs containing the initial and terminal ends of ρ_i.*
- *The t-coordinates of the w asymptotics of v agree with the t-coordinates of the e asymptotics of u.*

(4) *Split curve degenerations, i.e., pairs (u,v) where* $u\in\mathcal{M}^B(\mathbf{w},\mathbf{y};\rho_1,\dots,\rho_i\cup\rho_{i+1},\dots,\rho_n)$ *and v is a split curve with w asymptotics $\rho_1,\dots,(\rho_i\cup\rho_{i+1}),\dots,\rho_n$ and e asymptotics $\rho_1,\dots,\rho_i,\rho_{i+1},\dots,\rho_n$. Moreover, the t-coordinates of the w asymptotics of v agree with the t-coordinates of the e asymptotics of u.*

 In particular, the space of such pairs (u,v) can be canonically identified with $\mathcal{M}^B(\mathbf{w},\mathbf{y};\rho_1,\dots,\rho_i\cup\rho_{i+1},\dots,\rho_n)$.

This is a combination of [27, Theorem 5.55] and [27, Lemma 5.70].

As in most of holomorphic curve theory, the key ingredients in the proof of Theorem 4.5 are:

- A transversality statement: for generic almost-complex structures, the relevant moduli spaces are transversally cut out. For curves in $\Sigma \times [0,1] \times \mathbb{R}$ this is [27, Proposition 5.6]; for curves at east ∞, it is [27, Proposition 5.16]. Because we are not able to perturb the complex structure at east ∞, less transversality holds for curves at east ∞ than one might like. (Specifically, we can not always ensure that the evaluation maps at the punctures are transverse to the diagonal.)
- A compactness statement: sequences of holomorphic curves in $\Sigma \times [0,1] \times \mathbb{R}$ converge to holomorphic combs. This is [27, Proposition 5.23].
- Various gluing statements. Because of the Morse-Bott nature of the asymptotics at east ∞ and transversality issues for curves at east ∞, these statements become somewhat intricate. See [27, Section 5.5].
- An analysis of which of the possible degenerations can occur in codimension-1. See [27, Sections 5.6 and 5.7.3].

There is one more ingredient, because we are working with embedded curves:

- A computation of the index of the $\overline{\partial}$ operator shows that sequences of embedded curves converge to embedded curves. Philosophically, this is related to the adjunction formula. See [27, Section 5.7] for further discussion.

Remark 4.6. The fact that $\pi_{\mathbb{D}}$ is constant on each component of a curve at east ∞ suggests that we have lost some information in our formulation of the limiting objects. One could recover this information by rescaling while taking the limit. Specifically, suppose a sequence of holomorphic curves u_i converges to a pair (u,v), where $v : T \to \mathbb{R} \times (\partial\Sigma) \times [0,1] \times \mathbb{R}$ is a curve at east ∞. Fix a marked point p_i on each u_i converging to a marked point p on u. In taking the limit, rescale the map $\pi_{\mathbb{D}} \circ u_i$ on a neighborhood of p_i so that $d_{p_i}(\pi_{\mathbb{D}} \circ u_i)$ has norm 1. With some work, one thus obtains a rescaled version of $\pi_{\mathbb{D}} \circ v$ in the form of a map $T \to \{x+iy \in \mathbb{C} \mid x \leq 1\}$.

The moduli spaces at east ∞ are sufficiently simple that this refined limiting procedure turns out not to be necessary to construct the bordered invariants; but it seems more relevant to constructing a bordered version of $HF^{\pm}$.

4.3. $\partial^2 = 0$ on $\widehat{CFD}$

With the codimension-1 boundary in hand, we are now ready to prove that $\widehat{CFD}$ is a *dg* module.

Theorem 4.7 [27, Proposition 6.7]. *Fix a provincially admissible bordered Heegaard diagram $\mathcal{H}$. Then for a generic choice of almost-complex structure, the differential ∂ on $\widehat{CFD}(\mathcal{H})$ satisfies $\partial^2 = 0$.*

Sketch of proof. It suffices to show that for each generator $\mathbf{w} \in \mathfrak{S}(\mathcal{H})$, $\partial^2(\mathbf{w}) = 0$. We have

$$\begin{aligned}
\partial^2(\mathbf{w}) &= \partial\Bigg(\sum_{\substack{\mathbf{y}\in\mathfrak{S}(\mathcal{H})\\(\rho_1,\dots,\rho_n)\\B\in\pi_2(\mathbf{w},\mathbf{y})}} (\#\mathcal{M}^B(\mathbf{w},\mathbf{y};\rho_1,\dots,\rho_n))a(-\rho_1)\cdots a(-\rho_n)\mathbf{y}\Bigg)\\
&= \sum_{\substack{\mathbf{x}\in\mathfrak{S}(\mathcal{H})\\(\rho_1,\dots,\rho_i)\\B_1\in\pi_2(\mathbf{w},\mathbf{x})}} \sum_{\substack{\mathbf{y}\in\mathfrak{S}(\mathcal{H})\\(\rho_{i+1},\dots,\rho_n)\\B_2\in\pi_2(\mathbf{w},\mathbf{x})}} (\#\mathcal{M}^{B_1}(\mathbf{w},\mathbf{x};\rho_1,\dots,\rho_i))\\
&\qquad \times (\#\mathcal{M}^{B_2}(\mathbf{w},\mathbf{x};\rho_{i+1},\dots,\rho_n))\\
&\qquad \times a(-\rho_1)\cdots a(-\rho_i)a(-\rho_{i+1})\cdots a(-\rho_n)\mathbf{y}\\
&\quad + \sum_{\substack{\mathbf{x}\in\mathfrak{S}(\mathcal{H})\\(\rho_1,\dots,\rho_n)\\B\in\pi_2(\mathbf{w},\mathbf{x})}} (\#\mathcal{M}^B(\mathbf{w},\mathbf{x};\rho_1,\dots,\rho_n))a(-\rho_1)\cdots d(a_i)\cdots a(-\rho_n)\mathbf{x}.
\end{aligned}$$

(There is some possibly confusing re-indexing: in the second line we have replaced $n \to i$, $\mathbf{y} \to \mathbf{x}$, and $B \to B_1$. In the last line we use the same notation as in the first line, however.)

The sum in the second line corresponds exactly to the 2-story holomorphic buildings, degeneration (1) in Theorem 4.5. The sum in the last line corresponds to the split curve degenerations, degeneration (4) in Theorem 4.5.

It remains to see that the remaining ends of the 1-dimensional moduli spaces cancel in pairs. Indeed, it is easy to see that Case (2a) ends of $\mathcal{M}^B(\mathbf{w},\mathbf{y};\rho_1,\dots,\rho_n)$ correspond to Case (2a) ends of $\mathcal{M}^B(\mathbf{w},\mathbf{y};\rho_1,\dots,\rho_{i+1},\rho_i,\dots,\rho_n)$; and Case (2b) ends of $\mathcal{M}^B(\mathbf{w},\mathbf{y};\rho_1,\dots,\rho_n)$ correspond to join curve ends of $\mathcal{M}^B(\mathbf{w},\mathbf{y};\rho_1,\dots,\rho_i \cup \rho_{i+1},\dots,\rho_n)$. This completes the proof. ∎

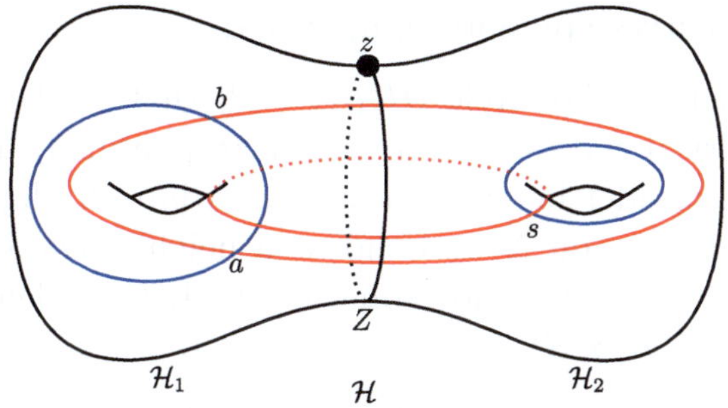

Fig. 13. Splitting a closed Heegaard diagram. The bordered Heegaard diagrams $\mathcal{H}_1$ and $\mathcal{H}_2$ are glued along the circle $Z \subset \mathcal{H}$

4.4. Deforming the Diagonal, $\widehat{CFA}$ and the Pairing Theorem

Our goals for the rest of the lecture are two-fold:

(1) Define the invariant $\widehat{CFA}(Y)$ associated to a bordered 3-manifold.

(2) Prove the pairing theorem, Theorem 2.17.

We will do this in the opposite order: we will start proving Theorem 2.17, and $\widehat{CFA}$ will appear naturally. The material in this section is drawn from [27, Chapter 9], to which we refer the reader for further details.

So, fix bordered Heegaard diagrams $\mathcal{H}_1$, $\mathcal{H}_2$ with $\partial\mathcal{H}_1 = \mathcal{Z} = -\partial\mathcal{H}_2$ and let $\mathcal{H} = \mathcal{H}_1 \cup_\partial \mathcal{H}_2$. (See Figure 13.) We want to understand $\widehat{CF}(\mathcal{H})$ in terms of invariants of $\mathcal{H}_1$ and $\mathcal{H}_2$.

On the level of generators, this is trivial: a generator $\mathbf{x} \in \widehat{CF}(\mathcal{H})$ corresponds to a pair of generators $(\mathbf{x}_1, \mathbf{x}_2)$ for $\mathcal{H}_1$ and $\mathcal{H}_2$ so that the α-arcs occupied by $\mathbf{x}_1$ are complementary to the α-arcs occupied by $\mathbf{x}_2$. So, if we define $I_A(\mathbf{x}_1)$ to be the idempotent in $\mathcal{A}(\mathcal{Z})$ corresponding to the α-arcs occupied by $\mathbf{x}_1$—this is the opposite of $I(\mathbf{x})$ as defined in Section 3.3.2—and let

$$\widehat{CFA}(\mathcal{H}_1) = \mathbb{F}_2\langle\mathfrak{S}(\mathcal{H}_1)\rangle,$$

with $I_A(\mathbf{x}_1)\mathbf{x}_1 = \mathbf{x}_1$, so other indecomposable idempotents kill $\mathbf{x}_1$, then we have

$$\widehat{CFA}(\mathcal{H}_1) \otimes_{\mathcal{A}(\mathcal{Z})} \widehat{CFD}(\mathcal{H}_2) \tag{4.1}$$

as $\mathbb{F}_2$-vector spaces. Note that we have *not* defined an $\mathcal{A}(\mathcal{Z}_1)$-module structure on $\widehat{CFA}(\mathcal{H}_1)$ yet: Equation (4.1) uses only the action of the idempotents and the fact that $\widehat{CFD}(\mathcal{H}_2)$ is a sum of elementary projective modules.

Holomorphic curves are more complicated.

Let $Z \subset \mathcal{H}$ denote the circle $\partial\mathcal{H}_1$. Recall that to define $\widehat{CFD}(\mathcal{H}_2)$ we attached a cylindrical end to $-Z = \partial\Sigma_2$. Correspondingly, to prove the pairing theorem, we consider inserting a long neck into Σ along Z. That is, fix a complex structure j_Σ on Σ and choose a neighborhood U of Z which is biholomorphic to $[-\epsilon,\epsilon] \times S^1$ for some $\epsilon > 0$. Let j_Σ^R denote the result of replacing U by $[-R,R] \times S^1$.

Let $R_i \in \mathbb{R}$ be a sequence with $R_i \to \infty$, and suppose $u_i \in \mathcal{M}^B(\mathbf{x},\mathbf{y})$ is a sequence of holomorphic curves with respect to $j_\Sigma^{R_i} \times j_{\mathbb{D}}$. We are interested in the limit of the sequence $\{u_i\}$. Modulo some technicalities, this is the kind of limit studied in symplectic field theory; the limiting objects have the following form:

Definition 4.8. A *matched holomorphic curve* is a pair of curves

$$(u_1,u_2) \in \mathcal{M}^{B_1}(\mathbf{x}_1,\mathbf{y}_1;\rho_1,\dots,\rho_n) \times \mathcal{M}^{B_2}(\mathbf{x}_2,\mathbf{y}_2;\rho_1,\dots,\rho_n)$$

so that for each $i = 1,\dots,n$, the t-coordinate at which u_1 is asymptotic to ρ_i is equal to the t-coordinate at which u_2 is asymptotic to ρ_i.

Equivalently, there is an evaluation map

$$\mathrm{ev}\colon \mathcal{M}^{B_i}(\mathbf{x}_i,\mathbf{y}_i;\rho_1,\dots,\rho_n) \to \mathbb{R}^{n-1}$$

which takes a curve asymptotic to $\rho_1 \times (1,t_1),\dots,\rho_n \times (1,t_n)$ to $(t_2 - t_1, t_3 - t_2,\dots,t_n - t_{n-1})$. Then a matched holomorphic curve is a pair (u_1,u_2) such that $\mathrm{ev}(u_1) = \mathrm{ev}(u_2)$.

Let $\mathcal{M}^B(\mathbf{x},\mathbf{y};\infty)$ denote the moduli space of matched holomorphic curves in the homology class B. That is,

$$\mathcal{M}^B(\mathbf{x},\mathbf{y};\infty) = \bigcup_{(\rho_1,\dots,\rho_n)} \mathcal{M}^{B_1}(\mathbf{x}_1,\mathbf{y}_1;\rho_1,\dots,\rho_n)\, {}_{\mathrm{ev}}\!\times_{\mathrm{ev}} \mathcal{M}^{B_2}(\mathbf{x}_2,\mathbf{y}_2;\rho_1,\dots,\rho_n). \tag{4.2}$$

Here, $\mathbf{x}$ (respectively $\mathbf{y}$) corresponds to the pair of generators $(\mathbf{x}_1,\mathbf{x}_2)$ (respectively $(\mathbf{y}_1,\mathbf{y}_2)$) and B_i is the intersection of B with $\mathcal{H}_i$.

Proposition 4.9. *Let $\mathcal{M}^B(\mathbf{x},\mathbf{y};R)$ denote the moduli space of holomorphic curves (in $\Sigma \times [0,1] \times \mathbb{R}$, in the homology class B) with respect to an appropriate perturbation*[5] *of the almost-complex structure $j_\Sigma^R \times j_{\mathbb{D}}$. Suppose that*

[5] As usual, we will suppress the fact that one needs to perturb the almost-complex structure in order to achieve transversality from the discussion.

$\mu(B) = 1$. *Then* $\bigcup_{R>0} \mathcal{M}^B(\mathbf{x},\mathbf{y};R)$ *is a* 1*-manifold whose ends as* $R \to \infty$ *are identified with* $\mathcal{M}^B(\mathbf{x},\mathbf{y};\infty)$. *More precisely, let*

$$\mathcal{M}^B(\mathbf{x},\mathbf{y};\geq R_0) = \mathcal{M}^B(\mathbf{x},\mathbf{y};\infty) \cup \bigcup_{R \geq R_0} \mathcal{M}^B(\mathbf{x},\mathbf{y};R).$$

Then there is a there is a topology on $\mathcal{M}^B(\mathbf{x},\mathbf{y};\geq R_0)$ *and an* R_0 *so that* $\mathcal{M}^B(\mathbf{x},\mathbf{y};\geq R_0)$ *is a compact* 1*-manifold with boundary exactly*

$$\mathcal{M}^B(\mathbf{x},\mathbf{y};\infty) \amalg \mathcal{M}^B(\mathbf{x},\mathbf{y};R_0).$$

This follows from compactness and gluing arguments, in a fairly standard way.

Corollary 4.10. *Define* $\partial_1 : \widehat{CF}(\mathcal{H}) \to \widehat{CF}(\mathcal{H})$ *by*

$$\partial_1(\mathbf{x}) = \sum_{\mathbf{y} \in T_\alpha \cap T_\beta} \sum_{\substack{B \in \pi_2(\mathbf{x},\mathbf{y}) \\ \mu(B)=1}} \#\mathcal{M}^B(\mathbf{x},\mathbf{y};\infty)\mathbf{y} \tag{4.3}$$

(cf. Formula (3.1)*). Then* $H_*(\widehat{CF}(\mathcal{H}), \partial_1) \cong \widehat{HF}(Y)$.

Example 4.11. Consider the splitting in Figure 13. The complex $\widehat{CF}(\mathcal{H})$ has two generators, $\mathbf{x} = \{a, s\}$ and $\mathbf{y} = \{b, s\}$; in the notation above, $\mathbf{x}_1 = \{a\}$, $\mathbf{x}_2 = \{s\}$, $\mathbf{y}_1 = \{b\}$ and $\mathbf{y}_2 = \{s\}$. The generator $\mathbf{y}$ occurs twice in $\partial(\mathbf{x})$: once from the small bigon region near the left of the diagram and once from the annular region crossing through the circle Z. We focus on the second of these contributions, the domain of which is shown in Figure 14. (It takes a little work to show that this domain has a holomorphic representative; see Exercise 4.5.)

Now, consider the result of stretching the neck along Z. There are two cases, depending on whether the cut goes through Z or not (which in turn depends on the complex structure on $\mathcal{H}$). If the cut does not go through z, the resulting matched curve (u_1, u_2) has u_1 a disk with one Reeb chord and u_2 an annulus with one Reeb chord. (In fact, this case does not occur in the limit; see Exercise 4.6.)

The more interesting case—and the one which actually occurs—is when the cut does pass through Z. Then both u_1 and u_2 are disks with two Reeb chords on each of their boundaries. The disk u_2 is rigid, but the disk u_1 comes in a 1-parameter family, depending on the length of the cut. There is algebraically one length of cut for which the height difference of the two

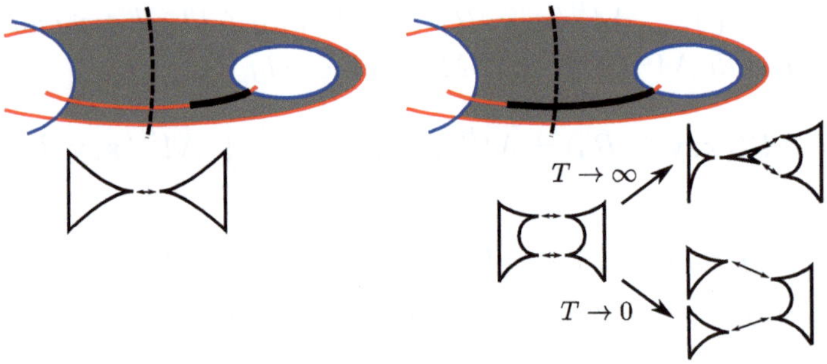

Fig. 14. Splitting an interesting domain. Depending on the complex structure, there are two possible phenomena after splitting: either the cut stays entirely on the right side of the diagram, as in the left picture, or the cut runs through the collapsed circle Z, as in the right picture. We have drawn schematic illustrations of the matched holomorphic curves below the two pictures

Reeb chords in u_1 agrees with the height difference of the Reeb chords in u_2 (Exercise 4.7).

Corollary 4.10 is a step in the direction of a pairing theorem: it gives a definition of $\widehat{HF}$ in terms of holomorphic curves in $\Sigma_1 \times [0,1] \times \mathbb{R}$ and $\Sigma_2 \times [0,1] \times \mathbb{R}$. But as we saw in Example 4.11, the corollary still has two (related) drawbacks:

(1) The moduli spaces we are considering in for $\mathcal{H}_1$ and $\mathcal{H}_2$ are typically high-dimensional. Indeed, in Formula (4.3), we have

$$\dim \mathcal{M}^{B_1}(\mathbf{x}_1, \mathbf{y}_1; \rho_1, \dots, \rho_n) + \dim \mathcal{M}^{B_2}(\mathbf{x}_2, \mathbf{y}_2; \rho_1, \dots, \rho_n) = n - 1.$$

(2) Since we are taking a fiber product of moduli spaces, which curves we want to consider in $\mathcal{H}_1$ depends on $\mathcal{H}_2$. So, it is not yet obvious how to define independent invariants of $\mathcal{H}_1$ and $\mathcal{H}_2$ containing the information needed to compute ∂_1.

To address complaint (2) we could try to formulate an algebra which remembers the chain $\mathrm{ev}_*[\mathcal{M}^{B_1}(\mathbf{x}_1, \mathbf{y}_1; \rho_1, \dots, \rho_n)] \in C_*(\mathbb{R}^{n-1})$. This is a natural way to try to define a bordered Heegaard Floer invariant, and with enough effort it could probably be made to work. This approach would be far from combinatorial, and is also unnecessarily complicated, as we will now show.

The next step is to deform the fiber product in Formula (4.2):

Definition 4.12. A *T-matched holomorphic curve* is a pair

$$(u_1,u_2)\in\mathcal{M}^{B_1}(\mathbf{x}_1,\mathbf{y}_1;\rho_1,\dots,\rho_n)\times\mathcal{M}^{B_2}(\mathbf{x}_2,\mathbf{y}_2;\rho_1,\dots,\rho_n)$$

such that $T\cdot\mathrm{ev}(u_1)=\mathrm{ev}(u_2)$. Let $\mathcal{M}^B_T(\mathbf{x},\mathbf{y};\infty)$ denote the moduli space of T-matched holomorphic curves, i.e.,

$$\mathcal{M}^B_T(\mathbf{x},\mathbf{y};\infty) = \bigcup_{(\rho_1,\dots,\rho_n)} \mathcal{M}^{B_1}(\mathbf{x}_1,\mathbf{y}_1;\rho_1,\dots,\rho_n)\ {}_{T\cdot\mathrm{ev}}\!\times_{\mathrm{ev}} \mathcal{M}^{B_2}(\mathbf{x}_2,\mathbf{y}_2;\rho_1,\dots,\rho_n).$$

So, in particular, a 1-matched holomorphic curve is just a matched holomorphic curve.

A standard continuation-map argument shows:

Proposition 4.13. *Let ∂_T denote the map defined analogously to Formula* (3.1) *(or Formula* (4.3)*) but using the moduli spaces $\mathcal{M}^B_T(\mathbf{x},\mathbf{y};\infty)$. Then $H_*(\widehat{CF}(\mathcal{H}),\partial_T)\cong\widehat{HF}(Y)$.*

Now, of course, we send $T\to\infty$. Consider a sequence of T_i-matched curves (u_1^i,u_2^i) with $T_i\to\infty$. Suppose that $u_1^i\in\mathcal{M}^{B_1}(\mathbf{x}_1,\mathbf{y}_1;(\rho_1,\dots,\rho_n))$. Let s_j^i be the $\mathbb{R}$-coordinate at which u_1^i is asymptotic to ρ_j and let t_j^i be the $\mathbb{R}$-coordinate at which u_2^i is asymptotic to ρ_j. Then, after passing to a subsequence, for each ρ_j, either:

- $(s_{j+1}^i-s_j^i)\in(0,\infty)$ stays bounded away from 0 and $(t_{j+1}^i-t_j^i)\to\infty$ as $i\to\infty$; or
- $(s_{j+1}^i-s_j^i)\to 0$ and $(t_{j+1}^i-t_j^i)$ stays bounded as $i\to\infty$.

So, in the limit:

- On the right we have an ℓ-story holomorphic building (for some ℓ) $U_2^\infty=(v_1,\dots,v_\ell)$, where $v_j\in\mathcal{M}(\mathbf{x}_{1,j},\mathbf{x}_{1,j+1};\rho_{n_j},\dots,\rho_{n_{j+1}})$, $\mathbf{x}_{1,j}=\mathbf{x}_1$, $\mathbf{x}_{1,\ell+1}=\mathbf{y}_1$, $1=n_1\le n_2\le\cdots\le n_{\ell+1}=n$.
- On the left we have a curve u_1^∞ asymptotic to some sets of Reeb chords $\boldsymbol{\rho}_1,\dots,\boldsymbol{\rho}_\ell$ at t-coordinates $t_1<\cdots<t_\ell\in\mathbb{R}$. Let

  $$\mathcal{M}^{B_2}(\mathbf{x}_2,\mathbf{y}_2;\boldsymbol{\rho}_1,\dots,\boldsymbol{\rho}_\ell)$$

  denote the moduli space of such curves.

Importantly, there is no longer a matching condition between the curves u_1^∞ and U_2^∞.

Example 4.14. Continuing with Example 4.11 in the case that the cut goes through the neck, as on the right of Figure 14, as $T \to \infty$ the $\mathbb{R}$-coordinates of the two Reeb chords in u_1 come together. (This results in degenerating a split curve at $\partial\Sigma$; we elided this point in the rest of this section.) This is indicated schematically in Figure 14.

Now, suppose we turned the diagram 180°. To avoid re-drawing the figure, we can think of this as sending $T \to 0$ instead of $T \to \infty$. In this case, the two chords in Figure 14 are pushed farther and farther apart; in the limit, the cut goes all the way through to the β-curve, giving a 2-story holomorphic building. Again, this is indicated schematically in Figure 14.

Observe that in both cases, the relevant curves are completely determined, i.e., belong to rigid moduli spaces: there is no "cut" left.

Now, associated to a *set* of Reeb chords $\boldsymbol{\rho}$ is an algebra element $a(\boldsymbol{\rho})$, defined analogously to Equation (3.3); see Exercise 4.8 or [27, Definition 3.23]. Define maps

$$\begin{aligned}
&m_{i+1} : \widehat{CFA}(\mathcal{H}_1) \otimes \mathcal{A}(\mathcal{Z})^{\otimes i} \to \widehat{CFA}(\mathcal{H}_1) \\
&m_{i+1}\big(\mathbf{x}; a(\boldsymbol{\rho}_1), \dots, a(\boldsymbol{\rho}_i)\big) \\
&\quad = \sum_{\mathbf{y}\in\mathfrak{S}(\mathcal{H}_1)} \sum_{\substack{B\in\pi_2(\mathbf{x},\mathbf{y}) \\ \mathrm{ind}(B,\boldsymbol{\rho}_1,\dots,\boldsymbol{\rho}_i)=1}} \big(\#\mathcal{M}^B(\mathbf{x},\mathbf{y};\boldsymbol{\rho}_1,\dots,\boldsymbol{\rho}_i)\big)\mathbf{y}.
\end{aligned}$$

An argument similar to but in some ways easier than the proof of Theorem 4.7 proves:

Theorem 4.15. *For $\mathcal{H}_1$ a provincially-admissible Heegaard diagram and J a generic almost-complex structure, the operations m_{i+1} make $\widehat{CFA}(\mathcal{H}_1)$ into an A_∞-module.*

The argument above is a sketch of the pairing theorem, Theorem 2.17. Specifically, it follows from the sketch above that

$$\widehat{CF}(\mathcal{H}) \simeq \widehat{CFA}(\mathcal{H}_1) \boxtimes \widehat{CFD}(\mathcal{H}_2),$$

where $\boxtimes$ is the model for the tensor product of an A_∞-module with a type D structure described in [27, Section 2.4].

4.5. Exercises

Exercise 4.1. In the setting of Section 4.1, use the Riemann mapping theorem to show that the map $\pi_\Sigma \circ u$ is determined by the position of the branch point (as claimed), and that there are no other elements of $\mathcal{M}(a, d)$.

Exercise 4.2. Suppose that $v : S \to \mathbb{R} \times (\partial\Sigma) \times [0, 1] \times \mathbb{R}$ is a holomorphic curve at east ∞, as discussed in Section 4.2. Show that the restriction of $\pi_{\mathbb{D}} \circ v$ to each component of S is constant.

Exercise 4.3. Prove: If $\mathbf{x}$ is a generator for $\widehat{CFD}(Y)$, where $\partial Y = F(\mathcal{Z})$ then $I(\mathbf{x}) \in \mathcal{A}(\mathcal{Z}, 0) \subset \mathcal{A}(\mathcal{Z})$. (Hint: this is easy.) What is the corresponding statement for the bimodules $\widehat{CFDD}$ associated to arced cobordisms?

Exercise 4.4. The differential on the algebra $\mathcal{A}(T^2, 0)$ associated to the torus is trivial. This means that one of the cases in the proof of Theorem 4.7 does not arise if the boundary is a torus. Which one? Why?

Exercise 4.5. Show that the annular region in Figure 14 is the domain of a holomorphic map $S \to \Sigma \times [0, 1] \times \mathbb{R}$, in two ways:

(1) By adapting the argument from [46, Lemma 9.3].

(2) By using handleslide invariance of Heegaard Floer homology. (After performing the right handleslide on Figure 13, it is easy to compute $\widehat{HF}$.)

Exercise 4.6. Show that when one stretches the neck in Figure 13, as in Example 4.11, the domain in Figure 14 must have a cut passing through the neck.

Exercise 4.7. In Example 4.11 we claimed there is algebraically one length of cut so that the height difference of the two Reeb chords in u_1 agrees with the height difference of the two Reeb chords in u_2. (Since we are working with $\mathbb{F}_2$-coefficients, probably we really meant that there are an odd number of such cut lengths.) Prove this. (Hint: what is the height difference in u_1 when the cut has length 0? When the cut goes all the way to the β-circle?)

Exercise 4.8. Define $a(\boldsymbol{\rho}) \in \mathcal{A}(\mathcal{Z})$ when $\boldsymbol{\rho}$ is a set of chords in $\mathcal{Z}$, no two of which start (respectively end) at points from the same matched pair. (This

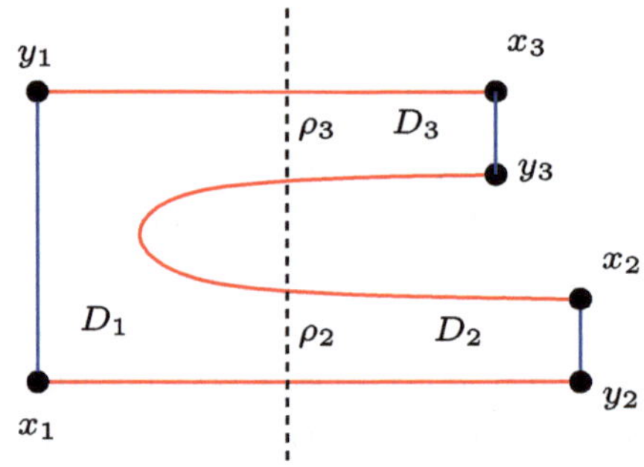

Fig. 15. Degenerating the hexagon. A hexagon in the Heegaard diagram (giving a flow from $\mathbf{x} = \{x_1, x_2, x_3\}$ to $\mathbf{y} = \{y_1, y_2, y_3\}$) is divided into three pieces D_1, D_2, and D_3, grouped as D_1 and $D_2 \cup D_3$. This is [27, Figure 9.3]

is a generalization of Formula (3.3), and should be straightforward. See [27, Definition 3.23] for a solution.)

Exercise 4.9. Figure 15 shows a hexagonal domain connecting $\mathbf{x} = \{x_1, x_2, x_3\}$ to $\mathbf{y} = \{y_1, y_2, y_3\}$. Note that this domain always contributes a term of $\mathbf{y}$ in $\partial(\mathbf{x})$. Consider the result of degenerating this domain along the dashed line, and then deforming the diagonal as in Section 4.4. (In the notation of Section 4.4, consider both the case of sending $T \to \infty$ and the case of sending $T \to 0$.) What happens to the holomorphic representative for this domain in the process? How is this encapsulated algebraically? (See [27, Section 9.6] for a detailed discussion of this example.)

5. Computing with Bordered Floer Homology I: Knot Complements

In this section we will discuss how the torus boundary case of bordered Floer homology can be used to do certain kinds of computations. The main goal is a technique for studying satellite knots, from [27, Chapter 11]. This technique and extensions of it have been used in [12, 24, 25, 55].

We start with a review of knot Floer homology [44, 56], mainly to fix notation (Section 5.1). We then discuss how the knot Floer homology of a knot K in S^3 determines the bordered Floer homology of $S^3 \setminus K$ (Section 5.2). Finally, we turn this around to use our understanding of bordered Floer homology to study the knot Floer homology of satellites (Section 5.3).

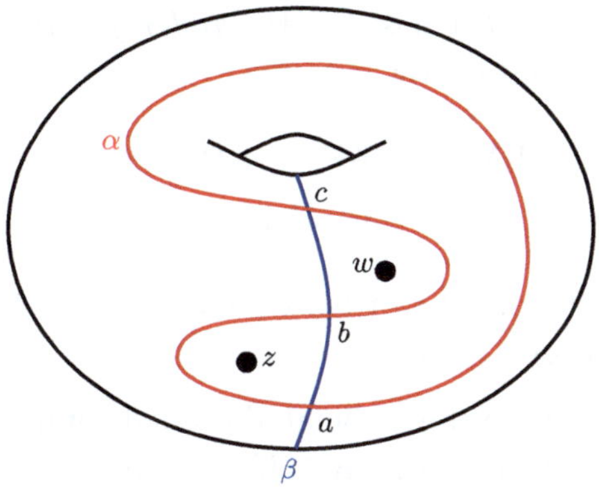

Fig. 16. Doubly pointed Heegaard diagram for the trefoil

5.1. Review of Knot Floer Homology

Let K be a knot in S^3, and let $\mathcal{H} = (\Sigma, \boldsymbol{\alpha}, \boldsymbol{\beta}, z, w)$ be a *doubly pointed Heegaard diagram* for K, in the sense of [44]. (For example, a doubly pointed Heegaard diagram for the trefoil is shown in Figure 16.) Associated to $\mathcal{H}$ are various knot Floer homology groups. The most general of these is $CFK^-(K)$, which is a filtered chain complex over $\mathbb{F}_2[U]$. The complex $CFK^-(K)$ is freely generated (over $\mathbb{F}_2[U]$) by $T_\alpha \cap T_\beta$, the same generators as $\widehat{CF}(\Sigma, \boldsymbol{\alpha}, \boldsymbol{\beta})$. The differential is given by

$$\partial^-(\mathbf{x}) = \sum_{\mathbf{y}} \sum_{\substack{B \in \widetilde{\pi}_2(\mathbf{x},\mathbf{y}) \\ \mu(B)=1}} \#(\mathcal{M}^B(\mathbf{x},\mathbf{y}))U^{n_w(B)} \cdot \mathbf{y}.$$

Here, unlike the discussion above, we allow disks to cross the basepoint z; we have used the notation $\widetilde{\pi}_2(\mathbf{x},\mathbf{y})$ rather than $\pi_2(\mathbf{x},\mathbf{y})$ to indicate this.

The complex $CFK^-(K)$ has an integral grading, called the *Maslov grading*, which is decreased by one by the differential. We will make no particular reference to this additional structure in the present notes; but it will be convenient (for the purposes of taking Euler characteristic, cf. Equations (5.1) and (5.2) below) to have its parity, as encoded in $(-1)^{M(\mathbf{x})}$. This parity is given as the local intersection number of T_α and T_β at $\mathbf{x}$. (As defined, we have specified a function $\mathfrak{S}(\mathcal{H}) \to \{\pm 1\}$ which is well-defined up to overall sign.) Now, the fact that ∂^- respects this parity is equivalent to the statement that if $B \in \widetilde{\pi}_2(\mathbf{x},\mathbf{y})$ has $\mathrm{ind}(B) = 1$, then the local intersection numbers of T_α and T_β at $\mathbf{x}$ and $\mathbf{y}$ are opposite.

The complex $CFK^-(K)$ has an *Alexander filtration* which is uniquely determined up to translation by

$$A(\mathbf{y}) - A(\mathbf{x}) = n_w(B) - n_z(B)$$
$$A(U \cdot \mathbf{y}) = A(\mathbf{y}) - 1$$

where $B \in \pi_2(\mathbf{x}, \mathbf{y})$. In other words, a term of the form $U^{n_w(B)}\mathbf{y}$ in $\partial^-(x)$ has $A(U^{n_w(B)}\mathbf{y}) = A(\mathbf{x}) - n_z(B)$.

Let $gCFK^-(K)$ denote the associated graded complex to $(CFK^-(K), A)$. Explicitly, the differential on $gCFK^-(K)$ is defined in the same way as the differential on $CFK^-(K)$ except that we no longer allow holomorphic curves to cross the z basepoint. Thus, the chain complex $gCFK^-$ splits as a direct sum of complexes, determined by the Alexander grading:

$$gCFK^-(K) = \bigoplus_{s \in \mathbb{Z}} gCFK^-(K, s).$$

Finally, there is the complex $\widehat{CFK}(K)$ obtained from $gCFK^-(K)$ by setting $U = 0$. In other words, $\widehat{CFK}(K)$ is generated over $\mathbb{F}_2$ by $T_\alpha \cap T_\beta$, and the differential counts holomorphic curves which do not cross z or w. Like $gCFK^-$, $\widehat{CFK}$ has a direct sum splitting induced by the Alexander grading.

A key property of knot Floer homology is that its graded Euler characteristic is the Alexander polynomial:

$$\Delta_K(T) = \sum_{s \in \mathbb{Z}} \chi\big(\widehat{CFK}(K, s)\big) T^s; \tag{5.1}$$

and similarly,

$$\Delta_K(T)/(1 - T) = \sum_{s \in \mathbb{Z}} \chi\big(CFK^-(K, s)\big) T^s. \tag{5.2}$$

(Note that the parity of the Maslov grading is used to compute the Euler characteristic. Also, both sides of Formula (5.2) are formal power series.)

The translation indeterminacy in the Alexander grading can then be removed by requiring the graded Euler characteristic of $\widehat{CFK}$ to be the Conway normalized Alexander polynomial (or equivalently $\chi(\widehat{CFK}(K, s)) = \chi(\widehat{CFK}(K, -s))$ for all $s \in \mathbb{Z}$); this normalization can also be used to remove the overall indeterminacy in the parity of the Maslov grading.

There is a numerical invariant for knots derived from knot Floer homology, $\tau(K)$, which will appear in Theorem 5.4 below. This is defined with

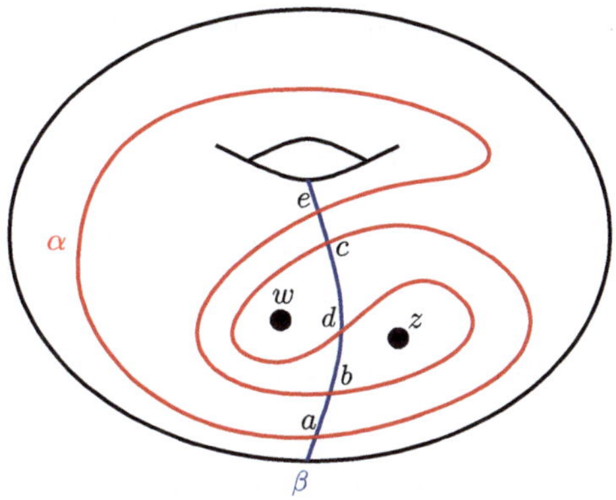

Fig. 17. Doubly pointed Heegaard diagram for the figure eight knot

the help of the following observation. There are U-non-torsion elements in $H_*(gCFK^-(K,s))$, i.e., elements $h \in H_*(gCFK^-(K,s))$ with the property that for all positive integers m, $U^m h$ is homologically non-trivial. We can consider the maximal s for which $H_*(gCFK^-(K,s))$ contains U-non-torsion elements. Multiplying this s by -1 gives the invariant $\tau(K)$.

Example 5.1. Figure 16 shows a doubly-pointed Heegaard diagram for the trefoil knot. The chain complex $CFK^-(\mathcal{H})$ is given by $\mathbb{F}_2[U]\langle a,b,c\rangle$. The differential on $CFK^-(\mathcal{H})$ is given by

$$\partial^-(a) = b \qquad \partial^-(b) = 0 \qquad \partial^-(c) = Ub.$$

The Alexander filtration is given by $A(a) = 1$, $A(b) = 0$, $A(c) = -1$.

The differential on $gCFK^-(\mathcal{H})$ is given by

$$\partial_g^-(a) = 0 \qquad \partial_g^-(b) = 0 \qquad \partial_g^-(c) = Ub.$$

The complex $\widehat{CFK}(\mathcal{H})$ is $\mathbb{F}_2\langle a,b,c\rangle$, with trivial differential.

Another concrete example is furnished by the Figure 8 knot.

Example 5.2. Figure 17 shows a doubly-pointed Heegaard diagram for the figure eight knot. The chain complex $CFK^-(\mathcal{H})$ is given by $\mathbb{F}_2[U]\langle a,b,c,d,e\rangle$. The differential on $CFK^-(\mathcal{H})$ is given by

$$\partial^-(a) = Ub + c \qquad \partial^-(b) = d \qquad \partial^-(c) = Ud \qquad \partial^-(d) = 0$$
$$\partial^-(e) = Ub + c.$$

The Alexander filtration is given by $A(a) = A(d) = A(e) = 0$, $A(b) = 1$, $A(c) = -1$.

The differential on $gCFK^-(\mathcal{H})$ is given by

$$\partial_g^-(a) = Ub \qquad \partial_g^-(b) = 0 \qquad \partial_g^-(c) = Ud \qquad \partial_g^-(d) = 0 \qquad \partial_g^-(e) = Ub.$$

The complex $\widehat{CFK}(\mathcal{H})$ is $\mathbb{F}_2\langle a, b, c, d, e\rangle$, with trivial differential.

We represent the chain complex $CFK^-(\mathcal{H})$ graphically by choosing a basis $\{\xi_i\}$ for $CFK^-(\mathcal{H})$ over $\mathbb{F}_2[U]$—for instance, the standard basis whose elements are points in $T_\alpha \cap T_\beta$—and placing a generator of the form $U^{-x} \cdot \xi_i$ with Alexander depth y on the plane at the position (x, y). Then the differential of a generator at (x, y) can be represented graphically by arrows connecting the point at (x, y) with the coordinates of other generators. These arrows necessarily point (non-strictly) to the left and down.

Up to filtered homotopy equivalence, we can always ensure that the differentials in the chain complex $CFK^-(\mathcal{H})$ change the Alexander grading or the U power, or both; we call a chain complex *reduced* if it has this property. Equivalently, $CFK^-(\mathcal{H})$ is reduced if every arrow changes the x-coordinate or the y-coordinate or both. A reduced complex has two distinct kinds of lowest-order terms: horizontal arrows and vertical arrows. We call the basis $\{\xi_i\}$ *horizontally simplified* (respectively *vertically simplified*) if every element $U^j\xi_i$ is the tail of at most one horizontal (respectively vertical) arrow and the head of at most one horizontal (respectively vertical) arrow. It is reasonably straightforward to verify that a horizontally simplified basis (respectively a vertically simplified basis) always exists; see [27, Proposition 11.52].

Abusing notation, we will say there is a *length ℓ horizontal arrow from ξ_i to ξ_j* if there is a horizontal arrow from ξ_i to $U^\ell\xi_j$.

We can invert U, giving a complex $U^{-1}CFK^-(K) = \mathbb{F}_2[U, U^{-1}] \otimes_{\mathbb{F}_2[U]} CFK^-(K)$. (This complex is also denoted $CFK^\infty(K)$ in the literature.) It still makes sense to talk about horizontal and vertical arrows on $U^{-1}CFK^-(K)$. The homology of $U^{-1}CFK^-(K)$ with respect to the horizontal (respectively vertical) differentials on $U^{-1}CFK^-(K)$ is $\mathbb{F}_2[U, U^{-1}]$. If the basis $\{\xi_i\}$ is horizontally (respectively vertically) simplified then this means there is a single generator η_0 (respectively ξ_0) over $\mathbb{F}_2[U, U^{-1}]$ with no horizontal (vertical) arrows into or out of it (in $U^{-1}CFK^-(K)$).

Example 5.3. Continuing with Example 5.1, we draw the complex $gCFK^-(\mathcal{H})$ as

$$\begin{array}{cccc} & & & a \\ & & & \downarrow \\ & & Ua & b \\ & & \downarrow & \\ & U^2a & Ub & \longleftarrow c \\ & \downarrow & & \\ \cdot^{\cdot^{\cdot}} & U^2b & \longleftarrow Uc & \\ \cdot^{\cdot^{\cdot}} & \cdot^{\cdot^{\cdot}} & & \end{array}$$

In particular:

- This basis is reduced and both horizontally and vertically simplified.
- There is a length 1 horizontal arrow from c to b and a length 1 vertical arrow from a to b.
- The element η_0 is a. The element ξ_0 is c.

Knot Floer homology has been computed extensively. It is determined by the Alexander polynomial for torus knots [48]; it is determined by the Alexander polynomial and the signature for alternating knots [42]; and it has an efficient combinatorial description for knots whose doubly-pointed Heegaard diagram can be drawn on the torus (so that the relevant holomorphic disks are in the torus, rather than some higher symmetric product) [10]. Finally, it admits a purely combinatorial description using grid diagrams [38, 39], which is amenable to computations by computer [2] or via a cube of resolutions [52, 53].

5.2. From $\widehat{CFK}$ to $\widehat{CFD}$: Statement and Example

For convenience, we recall our notation for the torus algebra, from Formula (2.1). It is given by:

$$\mathcal{A}(T^2,0) = \iota_0\bullet \begin{array}{c} \xrightarrow{\rho_1} \\ \xleftarrow{\rho_2} \\ \xrightarrow[\rho_3]{} \end{array} \bullet\iota_1 \;\; /(\rho_2\rho_1 = \rho_3\rho_2 = 0).$$

We have named $\rho_{12} = \rho_1\rho_2$, $\rho_{23} = \rho_2\rho_3$ and $\rho_{123} = \rho_1\rho_2\rho_3$, so $\{\iota_0, \iota_1, \rho_1, \rho_2, \rho_3, \rho_{12}, \rho_{23}, \rho_{123}\}$ is an $\mathbb{F}_2$-basis for $\mathcal{A}(T^2,0)$.

Theorem 5.4 [27, Theorem A.11]. *Let $K \subset S^3$ be a knot and let $CFK^-(K)$ be a reduced model for the knot Floer complex of K. Suppose $CFK^-(K)$ has a basis $\{\xi_i\}$ which is both horizontally and vertically simplified.*

Fix an integer n, and let $Y = S^3 \setminus \mathrm{nbd}(K)$ with framing n. We will describe $\widehat{CFD}(Y)$.

The submodule $\iota_0 \widehat{CFD}(Y)$ has one generator for each basis element ξ_i. The submodule $\iota_1 \widehat{CFD}(Y)$ has basis elements coming from the horizontal and vertical arrows in $CFK^-(K)$. Specifically, for each length ℓ vertical arrow from ξ_i to ξ_j here are ℓ basis elements $\kappa_1^{ij}, \dots, \kappa_\ell^{ij}$ for $\iota_1 \widehat{CFD}(Y)$; and for each length ℓ horizontal arrow from ξ_i to ξ_j there are ℓ basis elements $\lambda_1^{ij}, \dots, \lambda_\ell^{ij}$ for $\iota_1 \widehat{CFD}(Y)$. Finally, there are $m = |2\tau(K) - n|$ more basis elements $\mu_1, \dots, \mu_m$ for $\iota_1 \widehat{CFD}(Y)$.

The differential on $\widehat{CFD}(Y)$ is given as follows. From the vertical arrows we get differentials

$$\xi_i \xrightarrow{\rho_1} \kappa_1^{ij} \xleftarrow{\rho_{23}} \cdots \xleftarrow{\rho_{23}} \kappa_k^{ij} \xleftarrow{\rho_{23}} \kappa_{k+1}^{ij} \xleftarrow{\rho_{23}} \cdots \xleftarrow{\rho_{23}} \kappa_\ell^{ij} \xleftarrow{\rho_{123}} \xi_j.$$

From the horizontal arrows we get differentials

$$\xi_i \xrightarrow{\rho_3} \lambda_1^{ij} \xrightarrow{\rho_{23}} \cdots \xrightarrow{\rho_{23}} \lambda_k^{ij} \xrightarrow{\rho_{23}} \lambda_{k+1}^{ij} \xrightarrow{\rho_{23}} \cdots \xrightarrow{\rho_{23}} \lambda_\ell^{ij} \xrightarrow{\rho_2} \xi_j.$$

Finally, we have the unstable chain:

- *If $n < 2\tau$ the unstable chain has the form*

$$\xi_0 \xrightarrow{\rho_1} \mu_1 \xleftarrow{\rho_{23}} \mu_2 \xleftarrow{\rho_{23}} \cdots \xleftarrow{\rho_{23}} \mu_m \xleftarrow{\rho_3} \eta_0.$$

- *If $n > 2\tau$ the unstable chain has the form*

$$\xi_0 \xrightarrow{\rho_{123}} \mu_1 \xrightarrow{\rho_{23}} \mu_2 \cdots \xrightarrow{\rho_{23}} \mu_m \xrightarrow{\rho_2} \eta_0.$$

- *If $n = 2\tau$ the unstable chain has the form*

$$\xi_0 \xrightarrow{\rho_{12}} \eta_0.$$

It is fairly straightforward to remove the condition that there be a basis which is both horizontally and vertically simplified: one simply works with two bases, one horizontally simplified and one vertically simplified, and keeps track of the transition matrix. See [27, Theorem A.11]. There is also a basis-free version of Theorem 5.4; see [27, Theorem 11.35].

The proof of Theorem 5.4 has two parts. The first part is showing that the theorem holds for large negative surgery coefficients. The argument is somewhat similar to techniques in [11, 44, 51, 56], but is still quite involved. The second part is deducing the result for general surgery coefficients. This is done by changing the framing one step at a time, using the bimodules from Exercise 3.6 (or their type DA analogues).

Example 5.5. Continuing with the trefoil example, recall that the trefoil K has $\tau(K) = -1$. (Compare Exercise 5.1.) The basis $\{a, b, c\}$ is horizontally and vertically simplified. So, $\widehat{CFD}$ of $S^3 \setminus K$ with framing 1, say, is given by

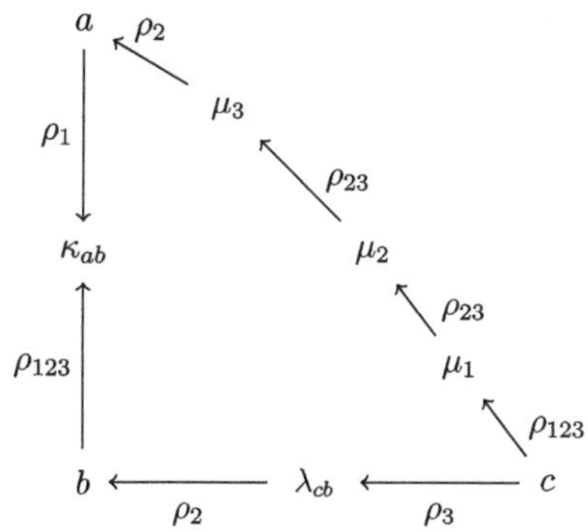

5.3. Studying Satellites

Suppose that $\mathcal{H}_1$ is a bordered Heegaard diagram for $S^3 \setminus \mathrm{nbd}(K)$ with the 0-framing of the boundary. Let $\mathcal{H}_2$ be a bordered Heegaard diagram for $\mathbb{D}^2 \times S^1$ with the ∞-framing. Place an extra basepoint w in $\mathcal{H}_2$, and let $\mathcal{H}_2'$ denote the result. Then $\mathcal{H}_1 \cup_\partial \mathcal{H}_2'$ is a doubly-pointed Heegaard diagram representing a knot L in S^3.

Construction 5.6. Fix a doubly-pointed bordered Heegaard diagram $\mathcal{H} = (\Sigma, \boldsymbol{\alpha}^a, \boldsymbol{\alpha}^c, \boldsymbol{\beta}, z, w)$ for $\mathbb{D}^2 \times S^1$. Consider the knot P in $\mathbb{D}^2 \times S^1$ determined as follows. Connect the basepoints z and w in $\mathcal{H}$ by an arc γ in $\Sigma \setminus (\boldsymbol{\alpha}^a \cup \boldsymbol{\alpha}^c)$ and an arc η in $\Sigma \setminus \boldsymbol{\beta}$. Viewing Σ as $\Sigma \times \{1/2\}$ inside $\Sigma \times [0,1] \subset Y(\mathcal{H})$ (Construction 2.6), let γ' be the result of pushing the interior of γ slightly into $\Sigma \times [0, 1/2)$ and let η' be the result of pushing the interior of η slightly into $\Sigma \times (1/2, 1]$. Then let $P = \gamma' \cup \eta'$. We will say that $\mathcal{H}$ *induces* $(\mathbb{D}^2 \times S^1, P)$.

Lemma 5.7. *With notation as above, suppose that $\mathcal{H}_2'$ induces $(\mathbb{D}^2 \times S^1, P)$. Then L is the satellite knot with companion $K \subset S^3$ and pattern $P \subset \mathbb{D}^2 \times S^1$.*

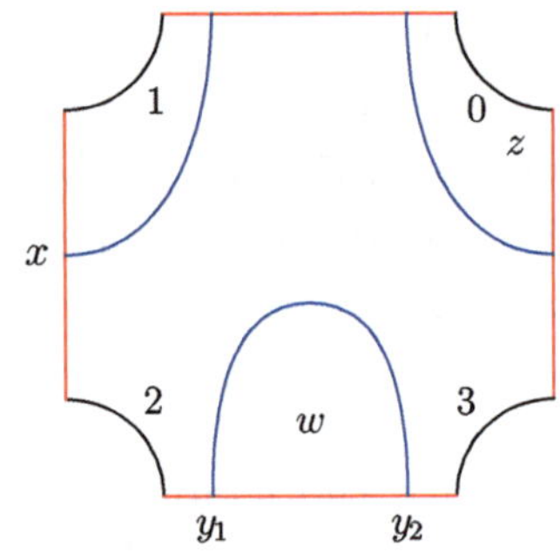

Fig. 18. Heegaard diagram for the $(2,1)$-cabling operation. This is a doubly-pointed Heegaard diagram for the $(2,1)$ cable (of the unknot), thought of as a knot in the solid torus. The basepoint z lies in the region marked with a 0. This picture is adapted from [27, Figure 11.14]

The proof is left as Exercise 5.7.

Example 5.8. Figure 18 shows a doubly-pointed bordered Heegaard diagram inducing the $(2,1)$-cabling operation.

Given a doubly-pointed bordered Heegaard diagram $\mathcal{H}$, let $CFD^-(\mathcal{H},z,w)$ denote $\mathbb{F}_2[U]\otimes_{\mathbb{F}_2}\widehat{CFD}(\mathcal{H})$ with differential given by

$$\partial(\mathbf{x}) = \sum_{\substack{\mathbf{y}\in\mathfrak{S}(\mathcal{H})\\ n\geq 0\\ (\rho_1,\dots,\rho_n)\\ B|\mathrm{ind}(B,\rho_1,\dots,\rho_n)=1}} \big(\#\mathcal{M}^B(\mathbf{x},\mathbf{y};\rho_1,\dots,\rho_n)\big)a(-\rho_1)\cdots a(-\rho_n)U^{n_w(B)}\mathbf{y}.$$

That is, we count curves as before except that we weight the curves which cross w n times by U^n.

Corollary 5.9. *With notation as above,*

$$gCFK^-(L)\cong \mathrm{Mor}\big(\widehat{CFD}(-\mathcal{H}_1), CFD^-(\mathcal{H}_2,z,w)\big).$$

By Theorem 5.4, $\widehat{CFD}(\mathcal{H}_1)$ is determined by $CFK^-(K)$. Thus, if we can compute $CFD^-(\mathcal{H}_2,z,w)$ we obtain a formula for the knot Floer homology $gCFK^-(L)$ in terms of $CFK^-(K)$ (for arbitrary K).

Example 5.10. In [27, Section 11.9] we use these techniques to compute the $(2,-3)$ cable of the left-handed trefoil. However, the computation there uses the type A invariant of the pattern. In the spirit of continuing to avoid

$\widehat{CFA}$, we give a similar computation using the Mor version of the pairing theorem.

Let $\mathcal{H}_2$ denote the doubly-pointed bordered Heegaard diagram shown in Figure 18. The module $CFD^-(\mathcal{H}_2, z, w)$ has generators x, y_1 and y_2 with

$$\iota_1 x = x \qquad \iota_0 y_1 = y_1 \qquad \iota_0 y_2 = y_2.$$

The differentials are given by

$$\begin{aligned} \partial(x) &= U^2 \rho_{23} x \\ \partial(y_1) &= U y_2 + \rho_1 x \\ \partial(y_2) &= U \rho_{123} x. \end{aligned}$$

By Theorem 5.4, the invariant $\widehat{CFD}(Y)$ of the 2-framed left-handed trefoil complement Y is given by

$\widehat{CFD}(Y) =$ 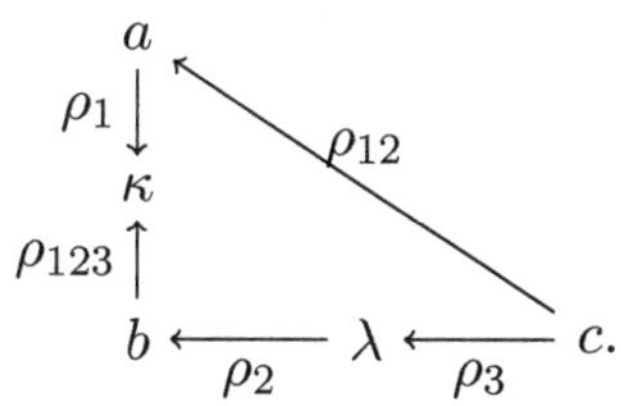

As in Corollary 5.9, $\text{Mor}(\widehat{CFD}(Y), CFD^-(\mathcal{H}_2, z, w))$ is $gCFK^-$ of some cable of the left-handed trefoil. Computing this morphism space, a basis over $\mathbb{F}_2[U]$ is given by:

$a \mapsto y_1$	$a \mapsto \rho_{12} y_1$	$a \mapsto y_2$	$a \mapsto \rho_{12} y_2$
$a \mapsto \rho_1 x$	$a \mapsto \rho_3 x$	$a \mapsto \rho_{123} x$	
$b \mapsto y_1$	$b \mapsto \rho_{12} y_1$	$b \mapsto y_2$	$b \mapsto \rho_{12} y_2$
$b \mapsto \rho_1 x$	$b \mapsto \rho_3 x$	$b \mapsto \rho_{123} x$	
$c \mapsto y_1$	$c \mapsto \rho_{12} y_1$	$c \mapsto y_2$	$c \mapsto \rho_{12} y_2$
$c \mapsto \rho_1 x$	$c \mapsto \rho_3 x$	$c \mapsto \rho_{123} x$	
$\lambda \mapsto x$	$\lambda \mapsto \rho_{23} x$	$\lambda \mapsto \rho_2 y_1$	$\lambda \mapsto \rho_2 y_2$
$\kappa \mapsto x$	$\kappa \mapsto \rho_{23} x$	$\kappa \mapsto \rho_2 y_1$	$\kappa \mapsto \rho_2 y_2.$

(Nobody said this was quick. The complex is smaller if one uses $\widehat{CFA}(\mathcal{H}_2, z, w)$.) The differentials are shown in Figure 19. Cancelling as many

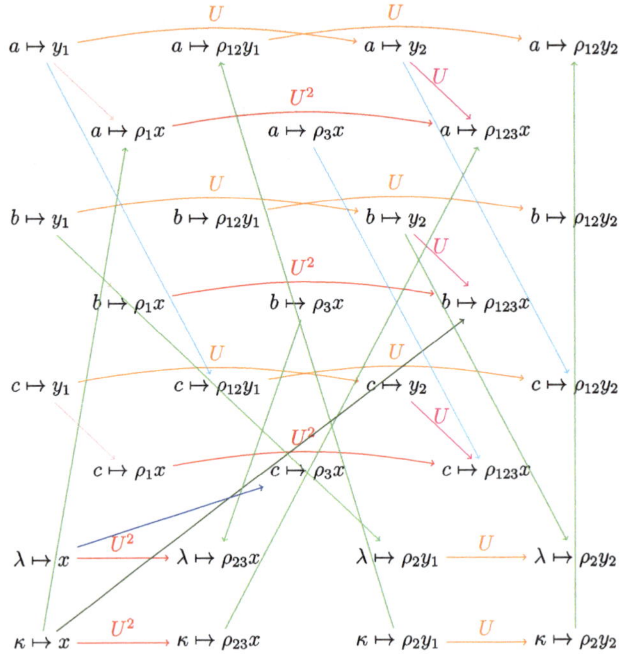

Fig. 19. The complex from Example 5.10

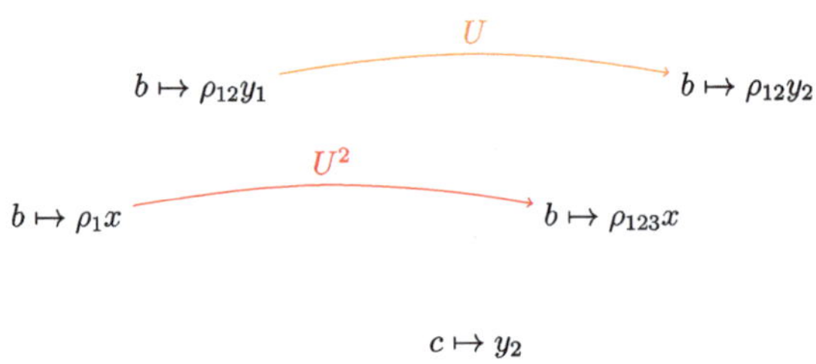

Fig. 20. Result of cancelling differentials in Figure 19

differentials not involving U as possible gives Figure 20. In particular, the homology $gHFK^-(K)$ is given by $\mathbb{F}_2[U] \oplus (\mathbb{F}_2[U]/U^2) \oplus \mathbb{F}_2$; and $\widehat{HFK}(K)$ is given by $\mathbb{F}_2^5$.

In some sense, this strategy works in general:

Lemma 5.11. *Given any pattern P in $\mathbb{D}^2 \times S^1$ there is a doubly-pointed Heegaard diagram inducing P.*

The proof is left as Exercise 5.8.

Corollary 5.12. *Let P be a knot in $\mathbb{D}^2 \times S^1$. Given a knot K in S^3 let K^P denote the satellite of K with pattern P. Then $CFK^-(K)$ determines $gCFK^-(K^P)$ in the following sense: if K_1 and K_2 are knots with $CFK^-(K_1) \cong CFK^-(K_2)$ then $gCFK^-(K_1^P) \cong gCFK^-(K_2^P)$.*

Remark 5.13. The diagram $\mathcal{H}_2'$ specifies more than just a knot in $\mathbb{D}^2 \times S^1$; see Exercise 5.10. Probably the best way to think of $\mathcal{H}_2'$ is as representing a bordered-sutured manifold (in the sense of [64]).

5.4. Exercises

Exercise 5.1. For K the trefoil and the figure eight, compute the $\mathbb{F}_2[U]$ module structure on $H_*(gCFK^-(K))$, using the descriptions of the complexes given in Examples 5.1 and 5.2 respectively. Use this to compute $\tau(K)$ for these knots.

Exercise 5.2. Find a basis for $CFK^-(K)$ when K is the figure eight knot which is both horizontally and vertically simplified.

Exercise 5.3. Let Y be the complement of the unknot in S^3. Compute $\widehat{CFD}(Y)$ in two ways:

(1) Using Theorem 5.4.

(2) Directly from a bordered Heegaard diagram.

(This exercise is courtesy of J. Hom.)

Exercise 5.4. Using Theorem 5.4, write down $\widehat{CFD}$ of the trefoil complement with framings 1 and -2.

Exercise 5.5. Figure 3 gives a bordered Heegaard diagram for the trefoil complement. Compute $\widehat{CFD}$ of that diagram directly, and compare the answer with that given by Theorem 5.4. (This is a fairly challenging computation, after which you are guaranteed to appreciated Theorem 5.4.)

Exercise 5.6. Verify that the modules $\widehat{CFD}(Y)$ given by Theorem 5.4 satisfy $\partial^2 = 0$.

Exercise 5.7. Prove Lemma 5.7.

Exercise 5.8. Prove Lemma 5.11.

Exercise 5.9. Use the bimodules of Exercise 3.6 to show that if Theorem 5.4 holds for surgery coefficient n then it holds for surgery coefficient $n \pm 1$. (This is somewhat messy.)

Exercise 5.10. Find doubly-pointed bordered Heegaard diagrams $\mathcal{H}$, $\mathcal{H}'$ for $\mathbb{D}^2 \times S^1$ so that:

- The singly-pointed Heegaard diagrams obtained from $\mathcal{H}$, $\mathcal{H}'$ by forgetting the w basepoint both specify the same framing for $\mathbb{D}^2 \times S^1$.
- The diagrams $\mathcal{H}$ and $\mathcal{H}'$ represent the same satellite operation in the sense of Construction 5.6.
- The invariants $\mathit{CFD}^-(\mathcal{H}, z, w)$ and $\mathit{CFD}^-(\mathcal{H}', z', w')$ are not homotopy equivalent.

In particular, it is not true that any two diagrams representing the same pattern P are related by a sequence of Heegaard moves in the complement of the basepoints.

Exercise 5.11. We computed $g\mathit{CFK}^-$ of some cable of the trefoil in Example 5.10. Which one?

6. Computing with Bordered Floer Homology II: Factoring Mapping Classes

The goal of this lecture is to discuss an algorithm, coming from bordered Floer homology, for computing the invariant $\widehat{HF}(Y)$ for any closed 3-manifold Y. This is not the first algorithm for computing $\widehat{HF}(Y)$, which is due to Sarkar-Wang [58]; but it is independent of the Sarkar-Wang algorithm and conceptually fairly satisfying.

6.1. Overview of the Algorithm

Fix a closed 3-manifold Y and a Heegaard splitting

$$Y = \mathsf{H}_1 \cup_\psi \mathsf{H}_2$$

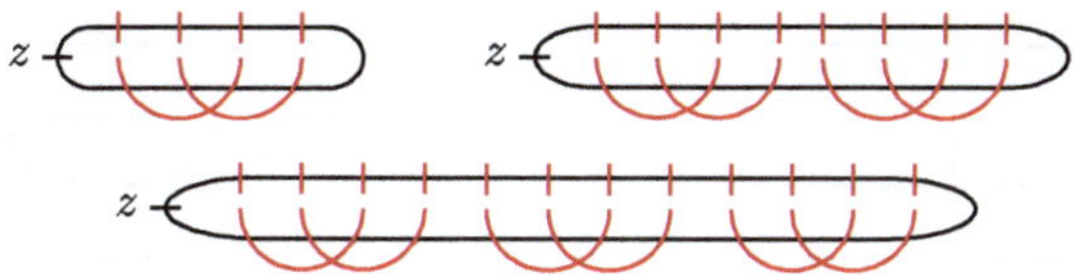

Fig. 21. The pointed matched circles $\mathcal{Z}_k^0$. The cases $k=1$, $k=2$ and $k=3$ are shown

for Y. That is, H_1 and H_2 are handlebodies of some genus k and $\psi : \partial \mathsf{H}_1 \to \partial \mathsf{H}_2$ is an orientation-reversing homeomorphism.

Without loss of generality, we can assume that each H_i is a particular standard bordered handlebody $(\mathsf{H}_k, \phi_0 : F(\mathcal{Z}_k^0) \to \partial \mathsf{H}_k)$. Here, $\mathcal{Z}_k^0$ is a particular pointed matched circle—we will take it to be the k-fold connect sum of the genus 1 pointed matched circle (i.e., the *split matching*); see Figure 21. Then the map ψ is specified by a map $\widetilde{\psi} = \phi_0 \circ \psi \circ \phi_0^{-1} : F(\mathcal{Z}_k^0) \to F(\mathcal{Z}_k^0)$. To specify Y up to homeomorphism we need only specify ψ up to isotopy; so, it is natural to view $\widetilde{\psi}$ as an element of the mapping class group of $F(\mathcal{Z}_k^0)$. Up to isotopy, we can assume that $\widetilde{\psi}$ fixes the preferred disk in $F(\mathcal{Z}_k^0)$, and regard it as an element of the mapping class group of $F^\circ(\mathcal{Z}_k^0)$. (Of course, the lift to the strongly based mapping class group depends on a choice.)

Let $M_{\widetilde{\psi}}$ denote the mapping cylinder of $\widetilde{\psi}$, as in Example 2.13. Then by the relevant pairing theorems, Corollary 2.22 and Theorem 2.23, we have

$$\widehat{CF}(Y) \simeq \mathrm{Mor}\big(\widehat{CFD}(\mathsf{H}_k, \phi_0), \mathrm{Mor}\big(\widehat{CFDD}(-M_{\widetilde{\psi}}), \widehat{CFD}(\mathsf{H}_k, \phi_0)\big)\big).$$

So, we have "reduced" the problem to computing the invariants of (H_k, ϕ_0) and $M_{\widetilde{\psi}}$.

This is not yet useful: there are about as many mapping classes as 3-manifolds. On the other hand, the mapping classes form a group. Suppose that $\psi_1, \dots, \psi_N$ are generators for the mapping class group of $F^\circ(\mathcal{Z}_k^0)$ as a monoid—that is, we include inverses in our list of generators. Then we can write $\widetilde{\psi} = \psi_{i_n} \circ \cdots \circ \psi_{i_1}$ for some sequence of generators $\psi_{i_1}, \dots, \psi_{i_n} \in \{\psi_1, \dots, \psi_N\}$. Repeatedly using Theorem 2.23, we have

$$\widehat{CF}(Y) \simeq \mathrm{Mor}\big(\widehat{CFD}(-\mathsf{H}_k, \phi_0), \mathrm{Mor}\big(\widehat{CFDD}(-M_{\psi_{i_n}}), \mathrm{Mor}(\cdots \\ \dots, \mathrm{Mor}\big(\widehat{CFDD}(-M_{\psi_{i_1}}), \widehat{CFD}(\mathsf{H}_k, \phi_0)\big) \dots)\big)\big).$$

Now we really have reduced the problem: we only need to compute the invariants $\widehat{CFD}(\mathsf{H}_k, \phi_0)$ and $\widehat{CFDD}(M_{\psi_i})$ for our preferred set of generators $\psi_1, \dots, \psi_N$.

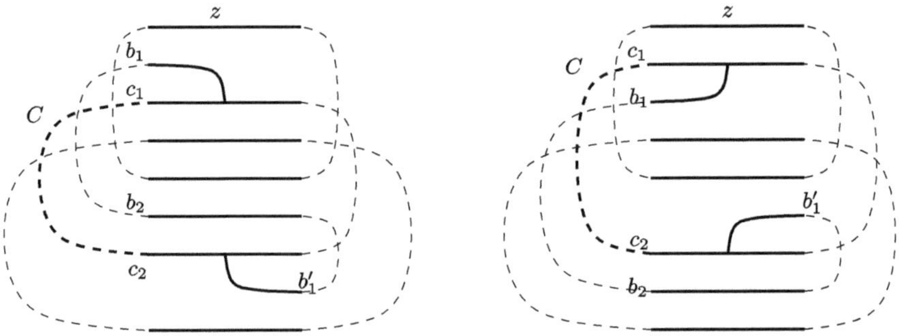

Fig. 22. Arc-slides. Two examples of arc-slides connecting pointed matched circles for genus 2 surfaces. In both cases, the foot b_1 is sliding over the matched pair $C = \{c_1, c_2\}$ (indicated by the darker dotted matching) at c_1. This figure is [31, Figure 2]

6.1.1. Arc-Slides as Generators of the Mapping Class Groupoid. Generalizing the mapping class group to a groupoid leads to a particularly convenient set of generators.

Definition 6.1. The *genus k mapping class groupoid* is the category whose objects are the pointed matched circles representing genus g surfaces, and with $\mathrm{Hom}(\mathcal{Z}_1, \mathcal{Z}_2)$ the set of isotopy classes of strongly-based homeomorphisms $F(\mathcal{Z}_1) \to F(\mathcal{Z}_2)$.

In particular, $\mathrm{Aut}(\mathcal{Z}) = \mathrm{Hom}(\mathcal{Z}, \mathcal{Z})$ is the strongly-based mapping class group.

Definition 6.2. Let $\mathcal{Z}$ be a pointed matched circle, and fix two matched pairs $C = \{c_1, c_2\}$ and $B = \{b_1, b_2\}$ in $\mathcal{Z}$. Suppose moreover that b_1 and c_1 are adjacent, in the sense that there is an arc σ connecting b_1 and c_1 which does not contain the basepoint z or any other point $p_i \in \mathbf{a}$. Then we can form a new pointed matched circle $\mathcal{Z}'$ which agrees everywhere with $\mathcal{Z}$, except that b_1 is replaced by a new distinguished point b'_1, which now is adjacent to c_2 and b'_1 is positioned so that the orientation on the arc from b_1 to c_1 is opposite to the orientation of the arc from b'_1 to c_2. In this case, we say that $\mathcal{Z}'$ and $\mathcal{Z}$ differ by an *arc-slide of* b_1 *over* c_1. (See Figure 22 for two examples.)

In this situation, there is a canonical element in $\mathrm{Hom}(\mathcal{Z}, \mathcal{Z}')$, which we refer to as the *arc-slide diffeomorphism*; see Figure 23.

The diagrams in Figure 22 are shorthand for bordered Heegaard diagrams for the mapping cylinders of the arc-slides. Such a bordered Heegaard diagram for the second arc-slide in Figure 22 is given in Figure 24.

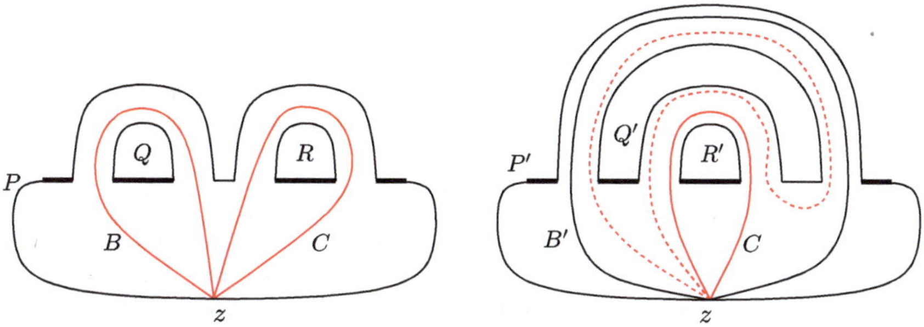

Fig. 23. The local case of an arc-slide diffeomorphism. Left: a pair of pants with boundary components labeled P, Q, and R, and two distinguished curves B and C. Right: another pair of pants with boundary components P', Q', R' and distinguished curves B' and C. The arc-slide diffeomorphism carries B to the dotted curve on the right, the curve labeled C on the left to the curve labeled C on the right, and boundary components P, Q, and R to P', Q' and R' respectively. This diffeomorphism can be extended to a diffeomorphism between surfaces associated to pointed matched circles: in such a surface there are further handles attached along the four dark intervals; however, our diffeomorphism carries the four dark intervals on the left to the four dark intervals on the right and hence extends to a diffeomorphism as stated. (This is only one of several possible configurations of B and C: they could also be nested or linked.) This figure is [31, Figure 3]

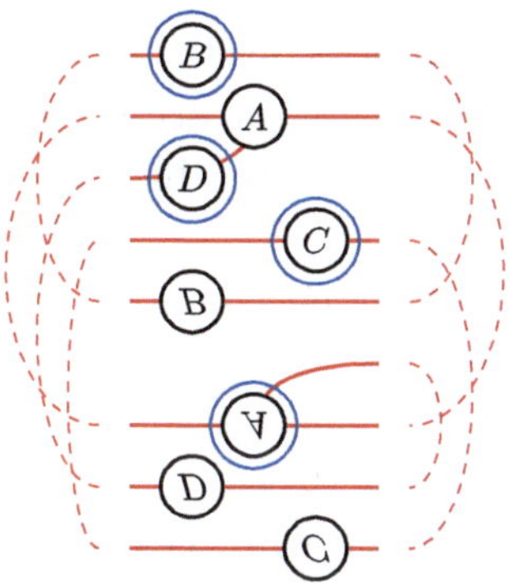

Fig. 24. Heegaard diagram for an arc-slide. This diagram corresponds to the schematic on the right of Figure 22

Lemma 6.3. *The arc-slides generate the mapping class groupoid.*

A proof can be found in [3]. It is perhaps a more familiar fact that the mapping class group is generated by some finite, preferred set of Dehn twists; see for example [13]. Lemma 6.3 can be deduced from this more familiar fact by explicitly factoring that particular collection of Dehn twists into arcslides (see Example 6.4).

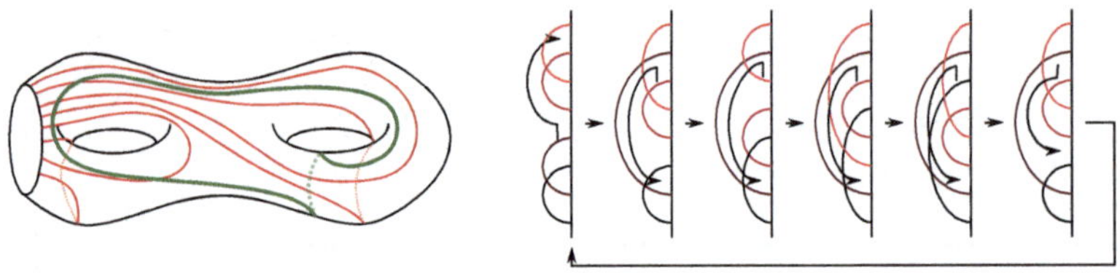

Fig. 25. Factoring a Dehn twist into arc-slides. Left: a genus 2 surface specified by a pointed matched circle, and a curve γ (drawn in thick green) in it. Right: a sequence of arc-slides whose composition is a Dehn twist around γ. This is [31, Figure 7]

Example 6.4. Figure 25 shows a factorization of a (particular) Dehn twist as a product of arc-slides.

So, two steps remain to compute $\widehat{CF}$:

- Compute $\widehat{CFD}(\mathcal{H}_k)$ for some Heegaard diagram $\mathcal{H}_k$ representing the genus k handlebody.
- Compute $\widehat{CFDD}(M_\psi)$ for any arc-slide ψ.

We give these computations in Sections 6.2 and 6.4, respectively. (As a warm-up before computing the invariant of arc-slides we compute the type *DD* module associated to the identity cobordism.)

Remark 6.5. The relations among the arc-slides are also relatively easy to state; see [3].

6.2. The Invariant of a Particular Handlebody

Let $\mathcal{Z}^1$ denote the (unique) pointed matched circle for the torus, and let $\mathcal{Z}^k$ denote the k-fold connect sum of $\mathcal{Z}^1$ with itself, i.e., the genus k *split pointed matched circle.* Label the marked points in $\mathcal{Z}^k$ as $a_1, \dots, a_{4k}$. So, in $\mathcal{Z}^k$ the matched pairs are $\{a_{4i-3}, a_{4i-1}\}$ and $\{a_{4i}, a_{4i-2}\}$.

The 0-*framed solid torus* $\mathsf{H}^1 = (H^1, \phi_0^1)$ is the solid torus with boundary $-F(\mathcal{Z}^1)$ in which the handle determined by $\{a_1, a_3\}$ bounds a disk. Let ϕ_0^1 denote the preferred diffeomorphism $-F(\mathcal{Z}^1) \to \partial H^1$. The 0-*framed handlebody of genus* k $\mathsf{H}^k = (H^k, \phi_0^k)$ is a boundary connect sum of k copies of H^1. Our conventions are illustrated by the bordered Heegaard diagram in Figure 26.

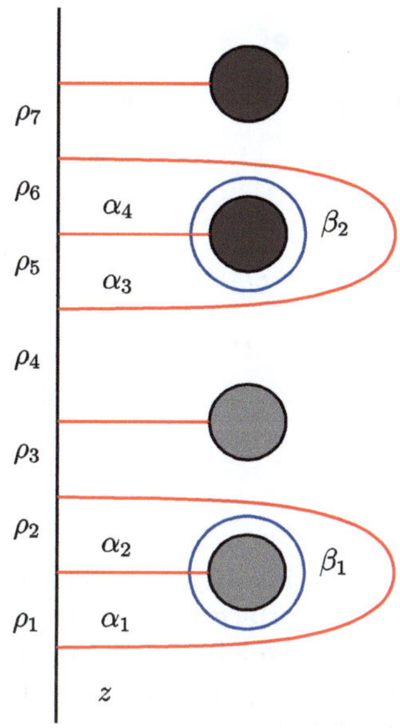

Fig. 26. Heegaard diagram for the 0-framed genus two handlebody. The lighter (respectively darker) shaded pair of circles indicates a handle attached to the diagram. This is [31, Figure 5]

Proposition 6.6. *Let* $\mathbf{s} = \{a_{4i-3}, a_{4i-1}\}_{i=1}^{k}$. *The module* $\widehat{CFD}(\mathsf{H}^k)$ *is generated over the algebra by a single element* $\mathbf{x}$ *with* $I(\mathbf{s})\mathbf{x} = \mathbf{x}$, *and is equipped with the differential determined by*

$$\partial(\mathbf{x}) = \sum_{i=1}^{k} a(\xi_i)\mathbf{x},$$

where ξ_i *is the arc in* $\mathcal{Z}^k$ *connecting* a_{4i-3} *and* a_{4i-1}.

Proof. This is a simple computation from the definitions. Note that the domains of holomorphic curves contributing to the differential on $\widehat{CFD}(\mathsf{H}^k)$ must be connected. It follows that the curves appearing here are simply copies of the curves occurring in the differential on $\widehat{CFD}(\mathsf{H}^1)$. These, in turn, were already studied in Section 3.4. ∎

6.3. The *DD* Identity

Let $\mathbb{I}$ denote the identity arced cobordism of $F(\mathcal{Z})$. As a warm-up to computing the bimodules associated to arc-slides we compute the bimodule $\widehat{CFDD}(\mathbb{I})$. The standard bordered Heegaard diagram $\mathcal{H}(\mathbb{I})$ for the identity

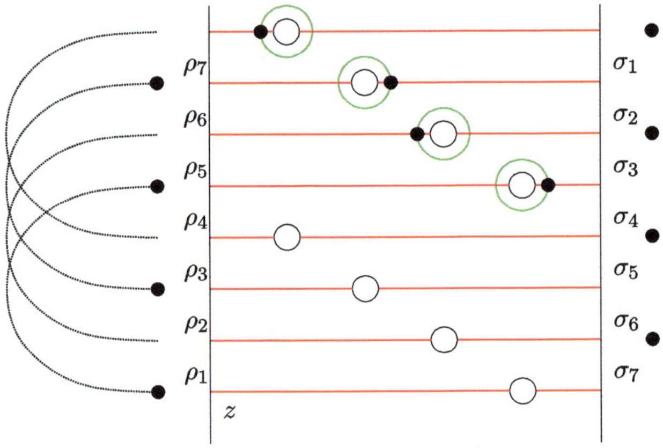

Fig. 27. Heegaard diagram for the identity map. This is a Heegaard diagram for the identity cobordism of the genus two surface with antipodal matching, as indicated by the arcs to the left of the diagram. To the left and the right of the diagram, we have also indicated a pair of complementary idempotents, along with its unique extension into the diagram as a generator for the complex. This figure is [31, Figure 13]

cobordism (for a particular choice of $\mathcal{Z}$) is illustrated in Figure 27. Inspecting the diagram, one has two immediate observations:

(1) Recall that indecomposable idempotents of $\mathcal{A}(\mathcal{Z})$ correspond to subsets of the matched pairs in $\mathcal{Z}$. There is an obvious bijection between matched pairs in $\mathcal{Z}$ and matched pairs in $-\mathcal{Z}$. With respect to this bijection, the generators of $\widehat{CFDD}(\mathbb{I})$ correspond one-to-one with pairs of indecomposable idempotents $I(\mathbf{s}) \otimes I(\mathbf{t}) \in \mathcal{A}(\mathcal{Z}) \otimes \mathcal{A}(-\mathcal{Z})$ with $\mathbf{s} \cap \mathbf{t} = \emptyset$. We call such pairs *complementary idempotents.* (The set of complementary idempotents is also in bijection with the set of idempotents of $\mathcal{A}(\mathcal{Z})$, of course.)

    Given a pair of complementary idempotents $I \otimes I'$ let $\mathbf{x}_{I,I'}$ denote the corresponding generator of $\widehat{CFDD}(\mathbb{I})$.

(2) Any domain in $\mathcal{H}(\mathbb{I})$ has the same multiplicities at the two boundaries of $\mathcal{H}(\mathbb{I})$. Any basic element of $\mathcal{A}(\mathcal{Z})$ has an associated *support* in $H_1(Z \setminus \{z\}, \mathbf{a})$; let $[\xi]$ denote the support of ξ. It follows that if $(\xi \otimes \xi') \otimes \mathbf{x}_{J,J'}$ occurs in $\partial(\mathbf{x}_{I,I'})$ then $[\xi] = [\xi']$ (in the obvious sense).

Formalizing the above, let the *diagonal subalgebra* of $\mathcal{A}(\mathcal{Z}) \otimes \mathcal{A}(-\mathcal{Z})$ denote the subalgebra with basis

$$\big\{(I \cdot \xi \cdot J) \otimes \big(I' \cdot \xi' \cdot J'\big) \mid [\xi] = [\xi'],\ \big(I, I'\big) \text{ complementary},$$
$$\big(J, J'\big) \text{ complementary}\big\}.$$

Proposition 6.7. *The diagonal subalgebra has a* $\mathbb{Z}$*-grading* gr *with the following properties*:

(1) *The grading* gr *respects the differential algebra structure, i.e., for homogeneous elements* a *and* b, $\mathrm{gr}(ab) = \mathrm{gr}(a) + \mathrm{gr}(b)$ *and* $\mathrm{gr}(d(a)) = \mathrm{gr}(a) - 1$.

(2) *The differential on* $\widehat{CFDD}(\mathbb{I})$ *is homogeneous of degree* -1 *with respect to* gr.

(3) *The standard basis elements for the diagonal subalgebra are homogeneous with respect to* gr.

(4) *If* $a \in \mathcal{A}$ *is homogeneous then* $\mathrm{gr}(a) \leq 0$.

(5) *If* $\mathrm{gr}(a) = 0$ *then* a *is an idempotent.*

(6) *If* $\mathrm{gr}(a) = -1$ *then* a *is a linear combination of* chords, *i.e., elements of the form* $a(\rho) \otimes a'(\rho)$ *where* ρ *is a single chord in* $\mathcal{Z}$. *(Here,* $a'(\rho)$ *denotes the element of* $\mathcal{A}(-\mathcal{Z})$ *associated to the chord* ρ.*)*

Sketch of proof. There are at least two ways to go about this proof. One is to show that any element of the diagonal algebra can be factored as a product of chords, and the length of the factorization is unique. (This is the approach taken in [31, Section 3].) Another approach is to observe that there is a *dg* algebra with properties (1) and (2) associated to any type *DD* bimodule (or type *D* module); we call this the *coefficient algebra* [31, Sections 2.3.4 and 2.4.3]. In the case of $\widehat{CFDD}(\mathbb{I})$, the coefficient algebra is exactly the diagonal subalgebra. Verifying the remaining properties above is then a fairly simple computation. (This is the approach taken for arc-slide bimodules in [31, Section 4].) ■

Corollary 6.8. *If* $(a \otimes b) \otimes \mathbf{x}_{J,J'}$ *occurs in* $\partial(\mathbf{x}_{I,I'})$ *then* $a \otimes b$ *is a linear combination of chords* $a(\rho_i) \otimes a(\rho_i')$.

Let $\mathrm{Chord}(\mathcal{Z})$ denote the set of all chords for $\mathcal{Z}$.

Theorem 6.9. *As a bimodule,* $\widehat{CFDD}(\mathbb{I})$ *is given by*

$$\widehat{CFDD}(\mathbb{I}) = \bigoplus_{(I \otimes I') \text{ complementary}} \big(\mathcal{A}(\mathcal{Z}) \cdot I\big) \otimes_{\mathbb{F}_2} \big(\mathcal{A}(-\mathcal{Z}) \cdot I'\big) \otimes \mathbf{x}_{I,I'}.$$

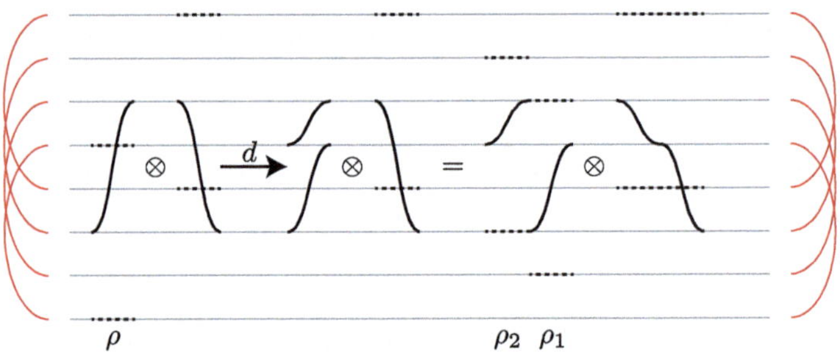

Fig. 28. Illustration of the inductive step in the proof of Theorem 6.9. We want to show the term on the left occurs in ∂ on $\widehat{CFDD}(\mathbb{I})$. The term on the far right occurs in ∂^2, by induction on the length of the chords involved. The only other contribution to ∂^2 which could cancel it is the differential of the term on the left. (The differential of the term on the left also has other terms, not shown)

The differential of $\mathbf{x}_{I,I'}$ *is given by*

$$\partial(\mathbf{x}_{I,I'}) = \sum_{(J,J')} \sum_{\rho \in \mathrm{Chord}(\mathcal{Z})} \left[(I \cdot a(\rho) \cdot J) \otimes (I' \cdot a'(\rho) \cdot J')\right] \otimes \mathbf{x}_{J,J'}.$$

In other word, every term permitted by Corollary 6.8 *to occur in* $\partial(\mathbf{x}_{I,I'})$ *does occur.*

Sketch of proof. All that remains is to show that every term of the form $[(I \cdot a(\rho) \cdot J) \otimes (I' \cdot a'(\rho) \cdot J')] \otimes \mathbf{x}_{J,J'}$ does occur in $\partial \mathbf{x}_{I,I'}$. The argument is by induction on the support to ρ. The base case is when ρ has length 1. In this case, the corresponding domain in $\mathcal{H}(\mathbb{I})$ is a hexagon, so it follows from the Riemann mapping theorem that there is a holomorphic representative.

The rest of the induction argument is illustrated in Figure 28. In words, suppose ρ has length bigger than 1, and suppose there is a position $a \in \mathbf{a}$ so that:

- a lies in the interior of ρ and
- the matched pair containing a is in the idempotent I.

Let ρ_1 be the chord from the start of ρ to the point a and let ρ_2 be the chord from a to the end of ρ. By induction, $\partial^2(\mathbf{x}_{I,I'})$ contains a term of the form $[(I \cdot a(\rho_2)a(\rho_1)J) \otimes (I' \cdot a'(\rho) \cdot J')] \otimes \mathbf{x}_{J,J'}$; this term comes from the sequence

$$\begin{aligned}\mathbf{x}_{I,I'} &\xrightarrow{\partial} \big[\big(I\cdot a(\rho_2)\big)\otimes\big(I'\cdot a'(\rho_2)\big)\big]\otimes\mathbf{x}_{K,K'}\\ &\xrightarrow{\partial} \big[\big(I\cdot a(\rho_2)a(\rho_1)\big)\otimes\big(I'\cdot a'(\rho_2)a'(\rho_1)\big)\big]\otimes\mathbf{x}_{J,J'}\\ &= \big[\big(I\cdot a(\rho_2)a(\rho_1)\big)\otimes\big(I'\cdot a'(\rho)\big)\big]\otimes\mathbf{x}_{J,J'}.\end{aligned}$$

The only term in $\partial^2(\mathbf{x}_{I,I'})$ which could cancel this one is $[(I\cdot\partial a(\rho)\cdot J)\otimes(I\cdot a'(\rho)\cdot J')]\otimes\mathbf{x}_{J,J'}$. Thus, since $\partial^2=0$, the term $[(I\cdot a(\rho)\cdot J)\otimes(I\cdot a'(\rho)\cdot J')]\otimes\mathbf{x}_{J,J'}$ must occur in $\partial(\mathbf{x}_{I,I'})$.

If there is a position a in the interior of ρ occupied in the idempotent I' then a similar argument, with the left and right sides reversed, gives the result. The only other case is that of length three chords in which both of the interior positions are matched to the endpoints. We call such chords *special length* 3 *chords* in [31]. There are various ways to handle this case. A somewhat indirect argument is given in the proof of [31, Theorem 1]. One can also prove the result in this case by a direct computation, as in the proof of [29, Proposition 10.1]. ■

Remark 6.10. The bimodule $\widehat{CFDD}(\mathbb{I})$ exhibits a kind of duality between the algebras $\mathcal{A}(\mathcal{Z})$ and $\mathcal{A}(-\mathcal{Z})$, called *Koszul duality.* See, for instance, [32, Section 8].

6.4. Underslides

To explain the bimodule $\widehat{CFDD}$ associated to an arc-slide we first divide the arc-slides into two classes: underslides and overslides. Specifically, with notation as in Definition 6.2, $Z\setminus C$ has two connected components. One of these components contains the basepoint z; call that component Z_z. Then an arc-slide is an *overslide* if $b_1\in Z_z$, and is an *underslide* if $b_1\notin Z_z$. So, in Figure 22, the example on the left is an overslide while the example on the right is an underslide.

It turns out that the bimodules for underslides are a little simpler, so we will focus on this case, referring the reader to [31, Section 4.5] for the overslide case. So, let $\psi:\mathcal{Z}\to\mathcal{Z}'$ be an underslide and M_ψ the associated mapping cylinder. To describe $\widehat{CFDD}(M_\psi)$ we need two more pieces of terminology:

Definition 6.11. There is an obvious bijection between matched pairs of $\mathcal{Z}$ (i.e., 1-handles of $F(\mathcal{Z})$) and matched pairs of $\mathcal{Z}'$ (i.e., 1-handles of $F(\mathcal{Z})$).

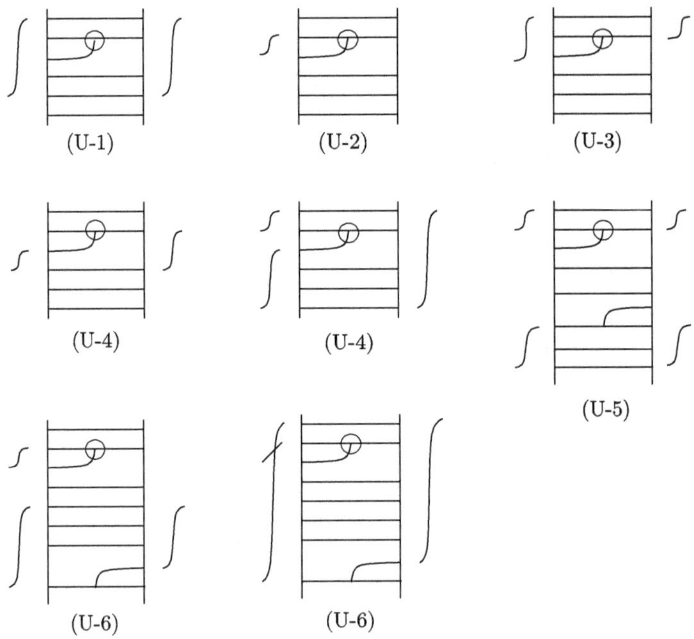

Fig. 29. Near-chords for under-slides

With notation as in Definition 6.2, a pair of indecomposable idempotents $I(\mathbf{s}) \otimes I(\mathbf{s}') \in \mathcal{A}(\mathcal{Z}) \otimes \mathcal{A}(-\mathcal{Z}')$ are called *near-complementary* if either:

- $\mathbf{s}$ is complementary to $\mathbf{s}'$ or
- $\mathbf{s} \cap \mathbf{t}$ consists of the matched pair of the feet of C, while $\mathbf{s} \cup \mathbf{t}$ contains all the matched pairs except for the pair of feet of B.

Definition 6.12. A *near-chord* for the underslide ψ is an algebra element of the form $a(\xi) \otimes a'(\xi')$ where ξ (respectively ξ') is a collection of chords in $\mathcal{Z}$ (respectively $-\mathcal{Z}'$) of one of the forms (U-1)–(U-6) shown in Figure 29.

Let $\mathrm{NChord}(\psi)$ denote the set of near-chords for ψ.

(See [31, Definition 4.17] for a more detailed description of the types (U-1)–(U-6) of near-chords.)

Theorem 6.13. *The bimodule $\widehat{CFDD}(M_\psi)$ has one generator $\mathbf{x}_{I,I'}$ for each near-complementary pair of idempotents $I \otimes I'$; and $\mathbf{x}_{I,I'} = (I \otimes I') \cdot \mathbf{x}_{I,I'}$. (In other words, as a module $\widehat{CFDD}(M_\psi) \cong \bigoplus_{I \otimes I' \text{ near complementary}} (\mathcal{A}(\mathcal{Z}) \cdot I) \otimes$*

$(\mathcal{A}(-\mathcal{Z}')\cdot I')$.) *The differential on* $\widehat{CFDD}(M_\psi)$ *is given by*

$$\begin{aligned}\partial(\mathbf{x}_{I,I'}) &\\ = \sum_{\substack{(J,J')\\ \text{near-complementary}}} \sum_{(\xi,\xi')\in \mathrm{NChord}(\psi)} &\left[(I\cdot a(\xi)\cdot J)\otimes (I'\cdot a'(\xi')\cdot J')\right]\otimes \mathbf{x}_{J,J'}.\end{aligned}$$

Sketch of proof. The proof is similar to, though more involved than, the proof of Theorem 6.9. There is an analogue of the diagonal algebra, called the *near-diagonal algebra*, admitting a $\mathbb{Z}$-grading satisfying analogous properties to Proposition 6.7. In particular, the near-chords are exactly the basic elements in grading -1. So, it only remains to show that every near-chord occurs in the differential. This follows from an inductive argument similar to the proof of Theorem 6.9. For short near-chords—near chords of type (U-2) and minimal-length near-chords of types (U-1) and (U-4)—it follows from the Riemann mapping theorem that the chords occur in the differential. The existence of other near-chords follows by a (somewhat complicated) induction on the support, using only the fact that $\partial^2 = 0$. ■

We do not discuss the case of overslides, which are more complicated than underslides. At the heart of the complication is the fact that, for overslides, the coefficient algebra contains non-idempotent elements in grading 0 (whereas in the underslide case, all non-idempotent elements have negative grading). While in the underslide case, every element in grading -1 appears as a coefficient in the differential, in the overslide case which grading -1 elements appear depends on a choice. Nonetheless, the index zero elements can be used to induce maps between bimodules associated to the various choices, and a somewhat weaker analogue of Theorem 6.13 holds: the overslide bimodule can be computed explicitly after some combinatorial choices are made, and the homotopy type of the answer is independent of those combinatorial choices. The interested reader is referred to [32, Proposition 4.35].

6.5. Exercises

Exercise 6.1. Verify the type *DD* bimodule for the identity cobordism of the torus given in Exercise 3.4 agrees with the answer given by Theorem 6.9.

Exercise 6.2. Verify that the bimodules from Exercise 3.6 agree with the bimodules given by Theorem 6.13. (Note that one can view each of these Dehn twists as an underslide.)

Exercise 6.3. Extend the algorithm above to compute $\widehat{CFD}(Y)$ for any bordered 3-manifold Y.

Acknowledgements. These are notes from a series of lectures by the first author during the CaST conference in Budapest in the summer of 2012. We thank Jennifer Hom for many helpful comments on and corrections to earlier drafts of these notes, and the participants in the CaST summer school for many further corrections. We also thank the Rényi Institute for providing a stimulating environment without which these notes would not have been written, and the Simons Center for providing a peaceful one, without which these notes would not have been revised.

RL was supported by NSF Grant number DMS-0905796 and a Sloan Research Fellowship. PSO was supported by NSF grant number DMS-0804121. DPT was supported by NSF grant number DMS-1008049.

References

[1] D. Auroux, Fukaya categories of symmetric products and bordered Heegaard-Floer homology. J. Gökova Geom. Topol. **4**, 1–54 (2010). arXiv:1001.4323

[2] J.A. Baldwin, W.D. Gillam, Computations of Heegaard-Floer knot homology. J. Knot Theory Ramif. **21**(8), 1250075 (2012). arXiv:math/0610167

[3] A.J. Bene, A chord diagrammatic presentation of the mapping class group of a once bordered surface. Geom. Dedic. **144**, 171–190 (2010). arXiv:0802.2747

[4] J. Bernstein, V. Lunts, *Equivariant Sheaves and Functors*. Lecture Notes in Mathematics, vol. 1578 (Springer, Berlin, 1994)

[5] F. Bourgeois, Y. Eliashberg, H. Hofer, K. Wysocki, E. Zehnder, Compactness results in symplectic field theory. Geom. Topol. **7**, 799–888 (2003). arXiv:math.SG/0308183

[6] V. Colin, P. Ghiggini, K. Honda, The equivalence of Heegaard Floer homology and embedded contact homology via open book decompositions I (2012). arXiv:1208.1074

[7] V. Colin, P. Ghiggini, K. Honda, The equivalence of Heegaard Floer homology and embedded contact homology via open book decompositions II (2012). arXiv:1208.1077

[8] V. Colin, P. Ghiggini, K. Honda, The equivalence of Heegaard Floer homology and embedded contact homology III: from hat to plus (2012). arXiv:1208.1526

[9] P. Ghiggini, Knot Floer homology detects genus-one fibred knots. Am. J. Math. **130**(5), 1151–1169 (2008)

[10] H. Goda, H. Matsuda, T. Morifuji, Knot Floer homology of (1, 1)-knots. Geom. Dedic. **112**, 197–214 (2005)

[11] M. Hedden, On knot Floer homology and cabling. Algebr. Geom. Topol. **5**, 1197–1222 (2005). arXiv:math.GT/0406402

[12] J. Hom, Bordered Heegaard Floer homology and the tau-invariant of cable knots (2012). arXiv:1202.1463

[13] S.P. Humphries, Generators for the mapping class group, in *Topology of Low-Dimensional Manifolds*. Proc. Second Sussex Conf., Chelwood Gate, 1977. Lecture Notes in Math., vol. 722 (Springer, Berlin, 1979), pp. 44–47

[14] S. Jabuka, T.E. Mark, Product formulae for Ozsváth-Szabó 4-manifold invariants. Geom. Topol. **12**(3), 1557–1651 (2008)

[15] A. Juhász, Holomorphic discs and sutured manifolds. Algebr. Geom. Topol. **6**, 1429–1457 (2006). arXiv:math/0601443

[16] A. Juhász, D.P. Thurston, Naturality and mapping class groups in Heegaard Floer homology (2012). arXiv:1210.4996

[17] M. Khovanov, Nilcoxeter algebras categorify the Weyl algebra. Commun. Algebra **29**(11), 5033–5052 (2001). arXiv:math.RT/9906166

[18] M. Khovanov, How to categorify one-half of quantum $\mathfrak{gl}(1 \mid 2)$ (2010). arXiv:1007.3517

[19] Ç. Kutluhan, Y.-J. Lee, C.H. Taubes, HF=HM I: Heegaard Floer homology and Seiberg–Witten Floer homology (2010). arXiv:1007.1979

[20] Ç. Kutluhan, Y.-J. Lee, C.H. Taubes, HF=HM II: Reeb orbits and holomorphic curves for the ech/Heegaard-Floer correspondence (2010). arXiv:1008.1595

[21] Ç. Kutluhan, Y.-J. Lee, C.H. Taubes, HF=HM III: Holomorphic curves and the differential for the ech/Heegaard Floer correspondence (2010). arXiv:1010.3456

[22] Ç. Kutluhan, Y.-J. Lee, C.H. Taubes, HF=HM IV: The Seiberg-Witten Floer homology and ech correspondence (2011). arXiv:1107.2297

[23] Ç. Kutluhan, Y.-J. Lee, C.H. Taubes, HF=HM V: Seiberg-Witten Floer homology and handle additions (2012). arXiv:1204.0115

[24] A.S. Levine, Knot doubling operators and bordered Heegaard Floer homology. J. Topol. **5**(3), 651–712 (2012). arXiv:1008.3349

[25] A.S. Levine, Slicing mixed Bing-Whitehead doubles. J. Topol. **5**(3), 713–726 (2012). arXiv:0912.5222

[26] R. Lipshitz, A cylindrical reformulation of Heegaard Floer homology. Geom. Topol. **10**, 955–1097 (2006). arXiv:math.SG/0502404

[27] R. Lipshitz, P.S. Ozsváth, D.P. Thurston, Bordered Heegaard Floer homology: invariance and pairing (2008). arXiv:0810.0687v4

[28] R. Lipshitz, P.S. Ozsváth, D.P. Thurston, Slicing planar grid diagrams: a gentle introduction to bordered Heegaard Floer homology, in *Proceedings of Gökova Geometry-Topology Conference 2008*, Gökova Geometry/Topology Conference (GGT), Gökova, 2009, pp. 91–119. arXiv:0810.0695

[29] R. Lipshitz, P.S. Ozsváth, D.P. Thurston, Bimodules in bordered Heegaard Floer homology (2010). arXiv:1003.0598v3

[30] R. Lipshitz, P.S. Ozsváth, D.P. Thurston, Bordered Floer homology and the spectral sequence of a branched double cover I (2010). arXiv:1011.0499

[31] R. Lipshitz, P.S. Ozsváth, D.P. Thurston, Computing $\widehat{HF}$ by factoring mapping classes (2010). arXiv:1010.2550v3

[32] R. Lipshitz, P.S. Ozsváth, D.P. Thurston, Heegaard Floer homology as morphism spaces. Quantum Topol. **2**(4), 384–449 (2011). arXiv:1005.1248

[33] R. Lipshitz, P.S. Ozsváth, D.P. Thurston, Tour of bordered Floer theory. Proc. Natl. Acad. Sci. USA **108**(20), 8085–8092 (2011). arXiv:1107.5621

[34] R. Lipshitz, P.S. Ozsváth, D.P. Thurston, A faithful linear-categorical action of the mapping class group of a surface with boundary. J. Eur. Math. Soc. **15**(4), 1279–1307 (2013). arXiv:1012.1032

[35] R. Lipshitz, P.S. Ozsváth, D.P. Thurston, Bordered Floer homology and the spectral sequence of a branched double cover II: the spectral sequences agree, in preparation

[36] R. Lipshitz, P.S. Ozsváth, D.P. Thurston, Computing cobordism maps with bordered Floer homology, in preparation

[37] C. Manolescu, P. Ozsváth, Heegaard Floer homology and integer surgeries on links (2010). arXiv:1011.1317

[38] C. Manolescu, P.S. Ozsváth, Z. Szabó, D.P. Thurston, On combinatorial link Floer homology. Geom. Topol. **11**, 2339–2412 (2007). arXiv:math.GT/0610559

[39] C. Manolescu, P. Ozsváth, S. Sarkar, A combinatorial description of knot Floer homology. Ann. Math. **169**(2), 633–660 (2009). arXiv:math.GT/0607691

[40] C. Manolescu, P. Ozsváth, D. Thurston, Grid diagrams and Heegaard Floer invariants (2009). arXiv:0910.0078

[41] Y. Ni, Heegaard Floer homology and fibred 3-manifolds. Am. J. Math. **131**(4), 1047–1063 (2009)

[42] P.S. Ozsváth, Z. Szabó, Heegaard Floer homology and alternating knots. Geom. Topol. **7**, 225–254 (2003) (electronic)

[43] P.S. Ozsváth, Z. Szabó, Holomorphic disks and genus bounds. Geom. Topol. **8**, 311–334 (2004)

[44] P.S. Ozsváth, Z. Szabó, Holomorphic disks and knot invariants. Adv. Math. **186**(1), 58–116 (2004). arXiv:math.GT/0209056

[45] P.S. Ozsváth, Z. Szabó, Holomorphic disks and three-manifold invariants: properties and applications. Ann. Math. **159**(3), 1159–1245 (2004). arXiv:math.SG/0105202

[46] P.S. Ozsváth, Z. Szabó, Holomorphic disks and topological invariants for closed three-manifolds. Ann. Math. **159**(3), 1027–1158 (2004). arXiv:math.SG/0101206

[47] P.S. Ozsváth, Z. Szabó, Holomorphic triangle invariants and the topology of symplectic four-manifolds. Duke Math. J. **121**(1), 1–34 (2004)

[48] P.S. Ozsváth, Z. Szabó, On knot Floer homology and lens space surgeries. Topology **44**(6), 1281–1300 (2005). arXiv:math/0303017

[49] P.S. Ozsváth, Z. Szabó, On the Heegaard Floer homology of branched double-covers. Adv. Math. **194**(1), 1–33 (2005). arXiv:math.GT/0309170

[50] P.S. Ozsváth, Z. Szabó, Holomorphic triangles and invariants for smooth four-manifolds. Adv. Math. **202**(2), 326–400 (2006). arXiv:math.SG/0110169

[51] P.S. Ozsváth, Z. Szabó, Knot Floer homology and integer surgeries. Algebr. Geom. Topol. **8**(1), 101–153 (2008)

[52] P. Ozsváth, Z. Szabó, A cube of resolutions for knot Floer homology. J. Topol. **2**(4), 865–910 (2009)

[53] P. Ozsváth, A. Stipsicz, Z. Szabó, Floer homology and singular knots. J. Topol. **2**(2), 380–404 (2009)

[54] T. Perutz, Hamiltonian handleslides for Heegaard Floer homology, in *Proceedings of Gökova Geometry-Topology Conference 2007*, Gökova Geometry/Topology Conference (GGT), Gökova, 2008, pp. 15–35

[55] I. Petkova, Cables of thin knots and bordered Heegaard Floer homology (2009). arXiv:0911.2679

[56] J. Rasmussen, Floer homology and knot complements. Ph.D. thesis, Harvard University, Cambridge, MA, 2003. arXiv:math.GT/0306378

[57] L.P. Roberts, Rational blow-downs in Heegaard-Floer homology. Commun. Contemp. Math. **10**(4), 491–522 (2008)

[58] S. Sarkar, J. Wang, An algorithm for computing some Heegaard Floer homologies. Ann. Math. **171**(2), 1213–1236 (2010). arXiv:math/0607777

[59] C.H. Taubes, Embedded contact homology and Seiberg-Witten Floer cohomology I. Geom. Topol. **14**(5), 2497–2581 (2010)

[60] C.H. Taubes, Embedded contact homology and Seiberg-Witten Floer cohomology II. Geom. Topol. **14**(5), 2583–2720 (2010)

[61] C.H. Taubes, Embedded contact homology and Seiberg-Witten Floer cohomology III. Geom. Topol. **14**(5), 2721–2817 (2010)

[62] C.H. Taubes, Embedded contact homology and Seiberg-Witten Floer cohomology IV. Geom. Topol. **14**(5), 2819–2960 (2010)

[63] C.H. Taubes, Embedded contact homology and Seiberg-Witten Floer cohomology V. Geom. Topol. **14**(5), 2961–3000 (2010)

[64] R. Zarev, Bordered Floer homology for sutured manifolds (2009). arXiv:0908.1106

R. Lipshitz (✉)

Department of Mathematics
Columbia University
New York, NY 10027
USA

e-mail: `lipshitz@math.columbia.edu`

D.P. Thurston

Department of Mathematics
Indiana University
Bloomington, IN 47405
USA

e-mail: `dpt@math.berkeley.edu`

P. Ozsváth

Department of Mathematics
Princeton University
Princeton, NJ 08544
USA

e-mail: `petero@math.princeton.edu`

BOLYAI SOCIETY
MATHEMATICAL STUDIES, 26

Contact and Symplectic Topology
pp. 357–388.

Stein Structures: Existence and Flexibility

KAI CIELIEBAK and YAKOV ELIASHBERG

1. The Topology of Stein Manifolds

Throughout this article, (V, J) denotes a smooth manifold (without boundary) of real dimension $2n$ equipped with an *almost complex structure* J, i.e., an endomorphism $J : TV \to TV$ satisfying $J^2 = -\mathrm{id}$. The pair (V, J) is called an *almost complex manifold.* It is called a *complex manifold* if the almost complex structure J is *integrable*, i.e., J is induced by complex coordinates on V. By the theorem of Newlander and Nirenberg [24], a (sufficiently smooth) almost complex structure J is integrable if and only if its Nijenhuis tensor

$$N(X,Y) := [JX, JY] - [X,Y] - J[X, JY] - J[JX,Y], \quad X,Y \in TV,$$

vanishes identically. An integrable almost complex structure is called a *complex structure.* A complex manifold (V, J) is called *Stein* if it admits a proper holomorphic embedding into some $\mathbb{C}^N$. Note that, due to the maximum principle, every Stein manifold is open, i.e., it has no compact components.

By a theorem of Grauert, Bishop and Narasimhan [2, 13, 23], a complex manifold (V, J) is Stein if and only if it admits a smooth function $\phi : V \to \mathbb{R}$ which is

- *exhausting*, i.e., proper and bounded from below, and
- *J-convex* (or strictly plurisubharmonic), i.e., $-dd^{\mathbb{C}}\phi(v, Jv) > 0$ for all $0 \neq v \in TV$, where $d^{\mathbb{C}}\phi := d\phi \circ J$.

F. Bourgeois et al. (eds.), *Contact and Symplectic Topology*,
Bolyai Society Mathematical Studies 26, DOI 10.1007/978-3-319-02036-5_8,

Note that the second condition means that $\omega_\phi := -dd^{\mathbb{C}}\phi$ is a symplectic form compatible with J. Note also that the "only if" follows simply by restricting the i-convex function $\phi(z) = |z|^2$ on $\mathbb{C}^N$ (where i denotes the standard complex structure) to a properly embedded complex submanifold. Here are some examples of Stein manifolds.

(1) $(\mathbb{C}^n, i)$ is Stein, and properly embedded complex submanifolds of Stein manifolds are Stein.

(2) If X is a closed complex submanifold of some projective space $\mathbb{C}P^N$ and $H \subset \mathbb{C}P^N$ is a hyperplane, then $X \setminus H$ is Stein.

(3) All open Riemann surfaces are Stein.

(4) If $\phi : V \to \mathbb{R}$ is J-convex, then so is $f \circ \phi$ for any smooth function $f : \mathbb{R} \to \mathbb{R}$ with $f' > 0$ and $f'' \geq 0$ (such f will be called a *convex increasing function*). Given an exhausting J-convex function $\phi : V \to \mathbb{R}$ and any $c \in \mathbb{R}$, we can pick a diffeomorphism $f : (-\infty, c) \to \mathbb{R}$ with $f' > 0$ and $f'' \geq 0$; then $f \circ \phi$ is an exhausting J-convex function $\{\phi < c\} \to \mathbb{R}$, hence the sublevel set $\{\phi < c\}$ is Stein.

(5) Any strictly convex smooth function $\phi : \mathbb{C}^n \to \mathbb{R}$ is i-convex. As a consequence, using (4), all convex open subsets of $\mathbb{C}^n$ are Stein.

(6) Let $L \subset V$ be a properly embedded *totally real* submanifold, i.e., L has real dimension n and $T_xL \cap J(T_xL) = \{0\}$ for all $x \in L$. Then the squared distance function $\mathrm{dist}_L^2 : V \to \mathbb{R}$ from L with respect to any Hermitian metric on V is J-convex on a neighbourhood of L. As a consequence, L has arbitrarily small Stein tubular neighbourhoods in V (which by (4) can be taken as sublevel sets $\{\mathrm{dist}_L^2 < \varepsilon\}$ if L is compact, but are more difficult to construct if L is noncompact).

Problem 1.1. [1] Prove (1), (2), and the first statements in (4), (5), (6).

Problem 1.2. A quadratic function $\phi(z_1, \dots, z_n) = \sum_{j=1}^n (a_j x_j^2 + b_j y_j^2)$ on $\mathbb{C}^n$ with coordinates $z_j = x_j + iy_j$ is i-convex if and only if $a_j + b_j > 0$ for all $j = 1, \dots, n$. A smooth function $\phi : \mathbb{C} \to \mathbb{R}$ is i-convex iff $\Delta\phi > 0$, i.e., ϕ is strictly subharmonic.

Problem 1.3. For an almost complex manifold (V, J) define $\omega_\phi := -d(d\phi \circ J)$ as in the integrable case. Then $\omega_\phi(\cdot, J\cdot)$ is symmetric for every function $\phi : V \to \mathbb{R}$ if and only if J is integrable.

[1] "Problems" in this survey are meant to be reasonably hard exercises for the reader.

Let us now turn to the following question: *Which smooth manifolds V admit the structure of a Stein manifold?*

Clearly, one necessary condition is the existence of a (not necessarily integrable) almost complex structure on V. This is a topological condition on the tangent bundle of V which can be understood in terms of obstruction theory. For example, the odd Stiefel-Whitney classes of TV must vanish and the even ones must have integral lifts.

A second necessary condition arises from Morse theory. Recall that a smooth function $\phi: V \to \mathbb{R}$ is called *Morse* if all its critical points are nondegenerate, and the *Morse index* $\mathrm{ind}(p)$ of a critical point p is the maximal dimension of a subspace of T_pV on which the Hessian of ϕ is negative definite. The following simple observation, due to Milnor and others, is fundamental for the topology of Stein manifolds.

Lemma 1.4. *The Morse index of each nondegenerate critical point p of a J-convex function $\phi: V \to \mathbb{R}$ satisfies*

$$\mathrm{ind}(p) \leq n = \dim_{\mathbb{C}} V.$$

Proof. [2] Suppose $\mathrm{ind}(p) > n$. Then there exists a complex line $L \subset T_pV$ on which the Hessian of ϕ is negative definite. Pick a small embedded complex curve $C \subset V$ through p in direction L. Then $\phi|_C$ has a local maximum at p, which contradicts the maximum principle because $\Delta(\phi|_C) > 0$. ■

This lemma imposes strong restrictions on the topology of Stein manifolds: Consider a Stein manifold (V, J) with exhausting J-convex function $\phi: V \to \mathbb{R}$. After a C^2-small perturbation (which preserves J-convexity) we may assume that ϕ is Morse. Thus, by Lemma 1.4 and Morse theory, V is obtained from a union of balls by attaching handles $D^k \times D^{2n-k}_\varepsilon$ of indices $k \leq n$. In particular, all homology groups $H_i(V; \mathbb{Z})$ with $i > n$ vanish. Surprisingly, for $n > 2$ these two necessary conditions are also sufficient for the existence of a Stein structure:

Theorem 1.5 ([10]). *A smooth manifold V of real dimension $2n > 4$ admits a Stein structure if and only if it admits an almost complex structure J and an exhausting Morse function ϕ without critical points of index $> n$. More*

[2] "Proofs" in this survey are only sketches of proofs; for details see [7].

precisely, J is homotopic through almost complex structures to a complex structure J' such that ϕ is J'-convex.

The idea of the proof is the following: Pick a sequence $r_0 < r_1 < r_2 < \cdots$ of regular values of ϕ with $r_0 < \min\phi$, $r_i \to \infty$, and such that each interval (r_i, r_{i+1}) contains at most one critical value of ϕ. By Morse theory, each sublevel set $W_i := \{\phi \le r_i\}$ is obtained from W_{i-1} by attaching a finite number of disjoint handles of index $\le n$. Proceeding by induction over i, suppose that on W_{i-1}, J is already integrable and ϕ is J-convex. Then for each $k \le n$ we need to

(i) extend J to a complex structure over a k-handle, and

(ii) extend ϕ to a J-convex function over a k-handle.

The first step is based on h-principles and will be explained in Section 3. The second step requires the construction of certain J-convex model functions on a standard handle and will be explained in Section 2.

2. Constructions of J-Convex Functions

The goal of this section it to construct the J-convex model functions needed for the proof of Theorem 1.5. We begin with some preparations.

J-Convex Hypersurfaces. Consider a smooth hypersurface (of real codimension one) Σ in a complex manifold (V, J). Each tangent space $T_p\Sigma \subset T_pV$, $p \in \Sigma$, contains the unique maximal complex subspace $\xi_p = T_p\Sigma \cap J(T_p\Sigma) \subset T_p\Sigma$. These subspaces form a codimension one distribution $\xi \subset T\Sigma$, the *field of complex tangencies*. Suppose that Σ is cooriented by a transverse vector field ν to Σ in V such that $J\nu$ is tangent to Σ. The hyperplane field ξ can be defined by a Pfaffian equation $\{\alpha = 0\}$, where the sign of the 1-form α is fixed by the condition $\alpha(J\nu) > 0$. The 2-form $\omega_\Sigma := d\alpha|_\xi$, called the *Levi form* of Σ, is then defined uniquely up to multiplication by a positive function. The cooriented hypersurface Σ is called *J-convex* (or strictly Levi pseudoconvex) if $\omega_\Sigma(v, Jv) > 0$ for each nonzero $v \in \xi$.

Problem 2.1. Each regular level set of a J-convex function is J-convex (where we always coorient level sets of a function by its gradient). Conversely, if $\phi : V \to \mathbb{R}$ is a smooth function without critical points all of whose level sets are compact and J-convex, then there exists a convex increasing function $f:\mathbb{R}\to\mathbb{R}$ such that $f\circ\phi$ is J-convex.

Thus, up to composition with a convex increasing function, proper J-convex functions are the same as *J-lc functions* ("lc" stands for "level convex"), i.e., functions that are J-convex near the critical points and have compact J-convex level sets outside a neighbourhood of the critical points.

Problem 2.2. Let $\phi:V\to\mathbb{R}$ be an exhausting J-convex function. Then for every convex increasing function $f:\mathbb{R}\to\mathbb{R}$ with $\lim_{y\to\infty}f'(y)=\infty$ the gradient vector field $\nabla_{f\circ\phi}(f\circ\phi)$ is *complete*, i.e., its flow exists for all time.

Continuous J-Convex Functions. We will need the notion of J-convexity also for continuous functions. To derive this, recall that i-convexity of a function $\phi:U\to\mathbb{R}$ on an open subset $U\subset\mathbb{C}$ is equivalent to $\Delta\phi>0$.

Problem 2.3. A smooth function $\phi:U\to\mathbb{R}$ on an open subset $U\subset\mathbb{C}$ satisfies $\Delta\phi(z)\geq\varepsilon>0$ at $z\in U$ if and only if it satisfies for each sufficiently small $r>0$ the *mean value inequality*

$$\phi(z)+\frac{\varepsilon r^2}{4}\leq\frac{1}{2\pi}\int_0^{2\pi}\phi(z+re^{i\theta})\,d\theta. \tag{1}$$

Since inequality (1) does not involve derivatives of ϕ, we can take it as the definition of i-convexity for a continuous function $\phi:\mathbb{C}\supset U\to\mathbb{R}$, and hence via local coordinates for a continuous function on a complex curve (note however that the value ε depends on the local coordinate). Finally, we call a continuous function $\phi:V\to\mathbb{R}$ on a complex manifold *J-convex* if its restriction to every embedded complex curve $C\subset V$ is J-convex. With this definition, we have

Lemma 2.4. *The maximum* $\max(\phi,\psi)$ *of two continuous J-convex functions is again J-convex.*

Proof. After restriction to complex curves it suffices to consider the case $\phi,\psi:\mathbb{C}\supset U\to\mathbb{R}$. Then the mean value inequalities for ϕ and ψ,

$$\phi(z)+\frac{\varepsilon_\phi r^2}{4}\leq\frac{1}{2\pi}\int_0^{2\pi}\phi(z+re^{i\theta})\,d\theta\leq\frac{1}{2\pi}\int_0^{2\pi}\max(\phi,\psi)(z+re^{i\theta})\,d\theta,$$

$$\psi(z)+\frac{\varepsilon_\psi r^2}{4}\leq\frac{1}{2\pi}\int_0^{2\pi}\psi(z+re^{i\theta})\,d\theta\leq\frac{1}{2\pi}\int_0^{2\pi}\max(\phi,\psi)(z+re^{i\theta})\,d\theta$$

combine to the mean value inequality for $\max(\phi,\psi)$,

$$\max(\phi,\psi)(z)+\frac{\min(\varepsilon_\phi,\varepsilon_\psi)r^2}{4}\le\frac{1}{2\pi}\int_0^{2\pi}\max(\phi,\psi)\big(z+re^{i\theta}\big)d\theta. \quad \blacksquare$$

Smoothing of J-Convex Functions. Continuous J-convex functions are useful for our purposes because of

Proposition 2.5 (Richberg [25]). *Every continuous J-convex function on a complex manifold can be C^0-approximated by smooth J-convex functions.*

Proof. The proof is based on an explicit smoothing procedure for functions on $\mathbb{C}^n$. Fix a smooth nonnegative function $\rho:\mathbb{C}^n\to\mathbb{R}$ with support in the unit ball and $\int_{\mathbb{C}^n}\rho=1$. For $\delta>0$ set $\rho_\delta(x):=\delta^{-2n}\rho(x/\delta)$. For a continuous function $\phi:\mathbb{C}^n\to\mathbb{R}$ define the "mollified" function $\phi_\delta:\mathbb{C}^n\to\mathbb{R}$,

$$\phi_\delta(x):=\int_{\mathbb{C}^n}\phi(x-y)\rho_\delta(y)d^{2n}y=\int_{\mathbb{C}^n}\phi(y)\rho_\delta(x-y)d^{2n}y. \tag{2}$$

The last expression shows that the functions ϕ_δ are smooth for every $\delta>0$, and the first expression shows that $\phi_\delta\to\phi$ as $\delta\to 0$ uniformly on compact subsets. Moreover, if ϕ is i-convex, then the mean value inequality for ϕ yields for all $x,w\in\mathbb{C}$ with $|w|$ sufficiently small

$$\begin{aligned}\phi_\delta(x)+\frac{1}{4}\varepsilon|w|^2&=\int_{\mathbb{C}^n}\Big(\phi(x-y)+\frac{1}{4}\varepsilon|w|^2\Big)\rho_\delta(y)d^{2n}y\\&\le\int_{\mathbb{C}^n}\frac{1}{2\pi}\int_0^{2\pi}\phi\big(x-y+we^{i\theta}\big)d\theta\rho_\delta(y)d^{2n}y\\&=\frac{1}{2\pi}\int_0^{2\pi}\phi_\delta\big(x+we^{i\theta}\big)d\theta,\end{aligned}$$

so ϕ_δ is i-convex. This proves the proposition on $\mathbb{C}^n$. The manifold case follows from this by a patching argument. $\blacksquare$

We will need four corollaries of Proposition 2.5. The first one is just combining it with Lemma 2.4:

Corollary 2.6 (maximum construction for functions). *The maximum $\max(\phi,\psi)$ of two smooth J-convex functions can be C^0-approximated by smooth J-convex functions.* $\square$

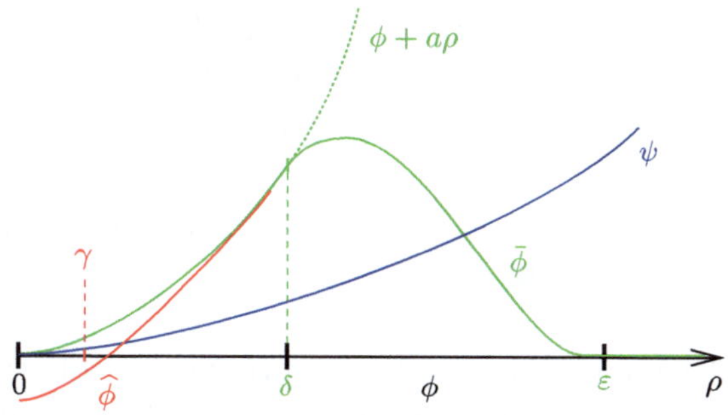

Fig. 1. Construction of the function ϑ[3]

We will denote a smooth approximation of $\max(\phi, \psi)$ by $\operatorname{smooth}\max(\phi, \psi)$. This is a slight abuse of notation because such an approximation is not unique; it is somewhat justified by the fact that the approximation can be chosen smoothly in families.

Corollary 2.7 (interpolation near a totally real submanifold). *Let L be a compact totally real submanifold of a complex manifold (V, J). Let $\phi, \psi : V \to \mathbb{R}$ be two smooth J-convex functions such that $\phi(x) = \psi(x)$ and $d\phi(x) = d\psi(x)$ for all $x \in L$. Then, given any neighborhood U of L, there exists a smooth J-convex function $\vartheta : V \to \mathbb{R}$ which coincides with ϕ outside U and with ψ in a smaller neighborhood of L.*

Proof. For the construction, see Figure 1. Shrink U so that $\rho := \operatorname{dist}_L^2 : U \to \mathbb{R}$ is smooth and J-convex and $U = \{\rho < \varepsilon\}$. Since ϕ and ψ agree to first order along L, we find an $a > 0$ such that $\phi + a\rho > \psi$ on $U \setminus L$. An explicit computation shows that we can find a J-convex function $\bar{\phi} = \phi + f(\rho)$ which agrees with ϕ outside U and with $\phi + a\rho$ on $\{\rho < \delta\}$ for some $\delta < \varepsilon$. Perturb $\bar{\phi}$ inside $\{\rho < \delta\}$ to a J-convex function $\widehat{\phi}$ with $\widehat{\phi} < \psi$ near L. Then the desired function ϑ is given by $\operatorname{smooth}\max(\psi, \widehat{\phi})$ on $\{\rho < \delta\}$, and $\widehat{\phi}$ outside. ∎

Corollary 2.8 (minimum construction for hypersurfaces). *Let Σ, Σ' be two compact J-convex hypersurfaces in a complex manifold $(V = M \times \mathbb{R}, J)$ that are given as graphs of smooth functions $f, g : M \to \mathbb{R}$ and cooriented from below. Then there exists a C^0-close smooth approximation of* $\min(f, g)$ *whose graph Σ'' is J-convex.*

[3]This figure, and all further figures of this Chapter have been taken from our book [7] with the permission of the American Mathematical Society.

Proof. The functions $\phi(x,y) := y - f(x)$ and $\psi(x,y) := y - g(x)$ have J-convex zero sets $\Sigma = \phi^{-1}(0)$ and $\Sigma' = \psi^{-1}(0)$. Note that the zero set of $\max(\phi,\psi) = y - \min(f,g)(x)$ is the graph of the function $\min(f,g)$. Now pick a convex increasing function $h : \mathbb{R} \to \mathbb{R}$ with $h(0) = 0$ such that $h \circ \phi$ and $h \circ \psi$ are J-convex near Σ resp. Σ', and define Σ'' as the zero set of smooth $\max(h \circ \phi, h \circ \psi)$. ■

Corollary 2.9 (from families of hypersurfaces to foliations). *Let $(M \times [0,1], J)$ be a compact complex manifold. Suppose there exists a smooth family of J-convex graphs (cooriented from below) $\Sigma_\lambda = \{y = f_\lambda(x)\}$, $\lambda \in [0,1]$, with $\Sigma_0 = M \times \{0\}$ and $\Sigma_1 = M \times \{1\}$. Then there exists a smooth* foliation *of $M \times [0,1]$ by J-convex graphs $\widetilde{\Sigma}_\lambda = \{y = \widetilde{f}_\lambda(x)\}$ $\lambda \in [0,1]$, with $\widetilde{\Sigma}_0 = M \times \{0\}$ and $\widetilde{\Sigma}_1 = M \times \{1\}$.*

Proof. By a family version of Corollary 2.8, the continuous functions $\bar{f}_\lambda := \min_{\mu \geq \lambda} f_\mu$ can be C^0-approximated by smooth functions $g_\lambda : M \to [0,1]$ whose graphs $\{y = g_\lambda(x)\}$ are J-convex. Since $\bar{f}_\lambda \leq \bar{f}_{\lambda'}$ for $\lambda \leq \lambda'$, this can be done in such a way that $g_\lambda \leq g_{\lambda'}$ for $\lambda \leq \lambda'$. So the graphs of g_λ almost form a foliation, and stretching them slightly in the y-direction yields the desired foliation. ■

Open Question. Does an analogue of Proposition 2.5, or at least of Corollary 2.6, hold for non-integrable J? If this were true, then a lot of the theory in these notes would work in the non-integrable case.

***J*-Convex Model Functions.** Let us fix integers $1 \leq k \leq n$. Consider $\mathbb{C}^n$ with complex coordinates $z_j = x_j + iy_j$, $j = 1, \dots, n$, and set

$$R := \sqrt{\sum_{j=1}^{k} x_j^2}, \qquad r := \sqrt{\sum_{j=k+1}^{n} x_j^2 + \sum_{j=1}^{n} y_j^2}.$$

Fix some $a > 1$ and define the *standard i-convex function*

$$\Psi_{\rm st}(r,R) := ar^2 - R^2.$$

For small $\gamma > 0$, we will use

$$H_\gamma := \{r \leq \gamma,\ R \leq 1 + \gamma\}$$

as a model for a complex k-handle. Its *core disk* is the totally real k-disk $\{r = 0,\ R \leq 1 + \gamma\}$ and it will be attached to the boundary of a Stein domain

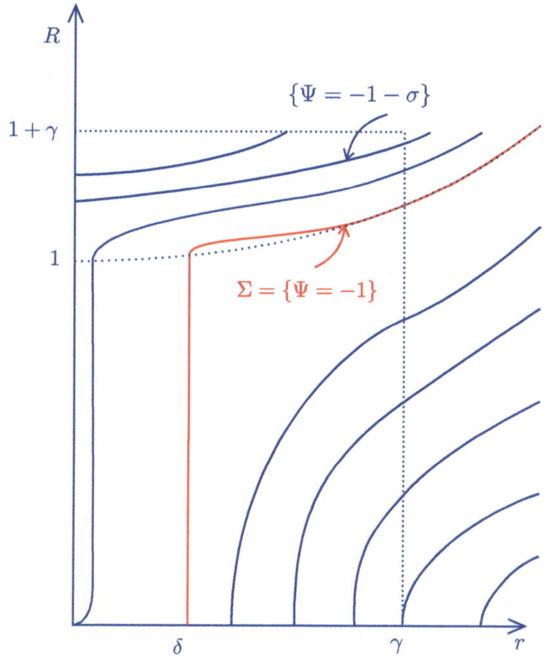

Fig. 2. The function Ψ

along the set $\{r \le \gamma,\ R = 1 + \gamma\}$. The following theorem will allow us to extend a J-convex function over the handle.

Theorem 2.10. *For each $0 < \gamma < 1 < a$ there exists an i-lc function $\Psi(r, R)$ on H_γ with the following properties (see Figure 2):*

(i) $\Psi = \Psi_{\mathrm{st}}$ *near* ∂H_γ;

(ii) Ψ *has a unique index k critical point at the origin;*

(iii) *the level set $\Sigma = \{\Psi = -1\}$* surrounds *the core disk in the sense that* $\{r = 0,\ R \le 1 + \gamma\} \subset \{\Psi < -1\}$.

Proof. Step 1. The first task is the construction of the hypersurface Σ. Let us write Σ as a graph $R = \phi(r)$, which we allow to become vertical at $r = \delta$. One can work out the condition for i-convexity of Σ (cooriented from above), which becomes a rather complicated system of second order differential inequalities for ϕ. However, it turns out that if $\phi > 0$, $\phi' > 0$, and $\phi'' \le 0$, the following simpler condition is *sufficient* for i-convexity:

$$\phi'' + \frac{\phi'^3}{r} - \frac{1}{\phi}(1 + \phi'^2) > 0. \tag{3}$$

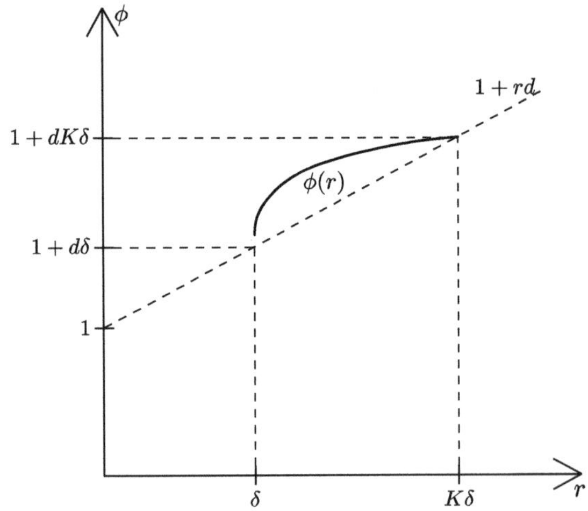

Fig. 3. A solution of Struwe's differential equation

Step 2. To construct solutions of (3), we follow a suggestion by M. Struwe. We will find the function ϕ as a solution of *Struwe's equation*

$$\phi'' + \frac{\phi'^3}{2r} = 0, \tag{4}$$

with $\phi' > 0$ and hence $\phi'' < 0$. Then (3) reduces to

$$\frac{\phi'^3}{2r} - \frac{1}{\phi}(1+\phi'^2) > 0. \tag{5}$$

Now Struwe's equation can be solved explicitly: It is equivalent to

$$\left(\frac{1}{\phi'^2}\right)' = -\frac{2\phi''}{\phi'^3} = \frac{1}{r},$$

thus $1/\phi'^2 = \ln(r/\delta)$ for some constant $\delta > 0$, or equivalently, $\phi'(r) = 1/\sqrt{\ln(r/\delta)}$. By integration, this yields a solution $\phi(r)$ for $r \geq \delta$ which is strictly increasing and concave and satisfies $\phi'(\delta) = +\infty$. Choosing the remaining integration constant appropriately, we find a solution $\phi : [\delta, K\delta] \to \mathbb{R}$ which satisfies (5) and looks as shown in Figure 3. Here $d > 0$ can be chosen arbitrarily and $K\delta$ can be made arbitrarily small.

Step 3. Smoothing the maximum of the function ϕ from Step 2 and the linear function $L(r) = 1 + dr$ yields an i-convex hypersurface which surrounds the core disk and agrees with $\{R = L(r)\}$ for $r \geq K\delta$. To finish the construction of the hypersurface Σ in Theorem 2.10, we still need to interpolate between $L(r)$ and the function $S(r) = \sqrt{1 + ar^2}$ whose graph is the level set $\{\Psi_{\rm st}(r,R) = ar^2 - R^2 = -1\}$. Unfortunately, this cannot be done directly with the maximum construction because the graph of L ceases to define an i-convex hypersurface before it intersects the graph of S. The solution is to interpolate from L to a quadratic function $Q(r) = 1 + br + cr^2/2$ and from there to S. The details are rather involved due to the fact that the simple sufficient condition (3) fails and one needs to invoke the full necessary and sufficient condition to ensure i-convexity during this interpolation.

Step 4. In Step 3 we constructed the level set Σ as a graph $\{R = \phi(r)\}$. To construct the i-lc function $\Psi : H_\gamma \to \mathbb{R}$, in view of Corollary 2.9 it suffices to connect Σ on both sides to level sets of $\Psi_{\rm st}$ by a smooth family of i-convex graphs. Towards larger R this is a simple application of the maximum construction, whereas towards smaller R it requires 1-parametric versions of the constructions in Steps 1–3. This proves Theorem 2.10. ■

3. Existence of Stein Structures

In this section we prove the Existence Theorem 1.5.

Step 1: Extension of complex structures over handles. Consider an almost complex cobordism (W, J) of complex dimension $n \geq 1$ such that J is integrable near $\partial_- W$, and $\partial_- W$ is J-convex when cooriented by an inward pointing vector field. For $k \leq n$ consider an embedding $f : (D^k, \partial D^k) \hookrightarrow (W, \partial_- W)$, where $D^k \subset \mathbb{R}^k \subset \mathbb{C}^n$ is the closed unit disk.

Proposition 3.1. *The almost complex structure J is homotopic rel $\mathcal{O}p(\partial_- W)$ to one which is integrable near $f(D^k)$.*

Proof. After trivializing the relevant bundles, the differential of f defines a map

$$df : \big(D^k, \partial D^k\big) \to (V_{2n,k}, V_{2n-1,k-1}),$$

where $V_{m,\ell}$ is the Stiefel manifold of ℓ-frames in $\mathbb{R}^m$. Let $V^{\mathbb{C}}_{m,\ell} \subset V_{2m,\ell}$ be the Stiefel manifold of complex ℓ-frames in $\mathbb{C}^m$, or equivalently, of totally real ℓ-frames in $\mathbb{R}^{2m}$.

Problem 3.2. For each $n \geq 1$ and $k \leq n$, the map

$$\pi_k\big(V^{\mathbb{C}}_{n,k}, V^{\mathbb{C}}_{n-1,k-1}\big) \to \pi_k(V_{2n,k}, V_{2n-1,k-1})$$

induced by the obvious inclusions is surjective.

Thus there exists a homotopy $F_t : (D^k, \partial D^k) \to (V_{2n,k}, V_{2n-1,k-1})$ from $F_0 = df$ to some $F_1 : (D^k, \partial D^k) \to (V^{\mathbb{C}}_{n,k}, V^{\mathbb{C}}_{n-1,k-1})$. Now a relative version of *Gromov's h-principle for totally real embeddings* [11, 15] yields an isotopy of embeddings $f_t : (D^k, \partial D^k) \hookrightarrow (W, \partial_- W)$ from $f_0 = f$ to a totally real embedding f_1.

By a further isotopy we can achieve that $f_1|_{\partial D^k}$ is real analytic. We complexify $f_1|_{\partial D^k}$ to a holomorphic embedding from a neighbourhood of ∂D^k in $\mathbb{C}^n$ into a slight extension $\widetilde{W}$ of W past $\partial_- W$, and then extend it to an embedding $\widetilde{f}_1 : D^k \times D^{2n-k}_\varepsilon \hookrightarrow \widetilde{W}$ which agrees with f_1 on $D^k = D^k \times 0$ and whose differential is complex linear along D^k. The push-forward $(\widetilde{f}_1)_* i$ of the standard complex structure i on $D^k \times D^{2n-k}_\varepsilon \subset \mathbb{C}^n$ agrees with J on a neighbourhood of $f_1(\partial D^k)$ (since $\widetilde{f}_1$ is holomorphic there) and at points of $f_1(D^k)$. Thus we can extend $(\widetilde{f}_1)_* i$ to an almost complex structure $\widetilde{J}$ on W which coincides with J near $\partial_- W$ and outside a neighbourhood of $f_1(D^k)$ and is integrable near $f_1(D^k)$. An application of the isotopy extension theorem now yields the desired almost complex structure which coincides with J near $\partial_- W$ and is integrable near the original disk $f(D^k)$. ■

By induction over the handles, Proposition 3.1 yields the following special case of the Gromov–Landweber theorem:

Corollary 3.3 (Gromov [14], Landweber [18]). *Let (V, J) be an almost complex manifold of complex dimension $n \geq 1$ which admits an exhausting Morse function $\phi : V \to \mathbb{R}$ without critical points of index $> n$. Then J is homotopic to an integrable complex structure.*

Step 2: Extension of J-convex functions over handles. Consider again (W, J) and $f : (D^k, \partial D^k) \hookrightarrow (W, \partial_- W)$ as in Step 1. After applying Proposition 3.1 we may assume that J is integrable near $\Delta := f(D^k)$. After real analytic approximation and complexification, we may assume that f extends to a holomorphic embedding $F : H_\gamma \hookrightarrow \widetilde{W}$, where H_γ is the standard handle $D^k_{1+\gamma} \times D^{2n-k}_\gamma \subset \mathbb{C}^n$ and $\widetilde{W}$ is a slight extension of W past $\partial_- W$.

Let ϕ be a given J-convex function near $\partial_- W = \{\phi = -1\}$. To finish the proof of Theorem 1.5, we need to extend ϕ to a J-convex function $\widetilde{\phi}$ on a neighbourhood of Δ whose level set $\{\widetilde{\phi} = -1\}$ coincides with $\partial_- W$ outside a neighbourhood of $\partial\Delta$ and surrounds $f(D^k)$ in W as shown in Figure 4.

Equivalently, we need to extend $F^*\phi$ to an i-convex function Ψ on H_γ whose level set $\{\Psi = -1\}$ coincides with $\{F^*\phi = -1\}$ near ∂H_γ and surrounds

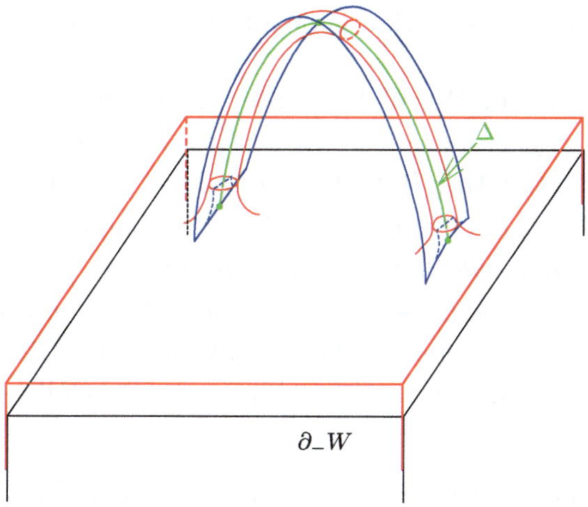

Fig. 4. Surrounding a J-orthogonally attached totally real disk

D^k in H_γ. According to Theorem 2.10 in the previous section, this can be done if we can arrange that $F^*\phi$ equals the standard function $\Psi_{\rm st}(r,R) = ar^2 - R^2$ near ∂D^k.

To analyze the last condition, note that the n-disk D^n meets the level set $\{\Psi_{\rm st} = -1\}$ *i-orthogonally* along ∂D^n in the sense that $i(T_xD^n) \subset T_x\Sigma$ for all $x \in \partial D^n$. Conversely, suppose that D^n is i-orthogonal to the level set $\{F^*\phi = -1\}$ along ∂D^k. Then $F^*\phi$ and $\Psi_{\rm st}$ have the same kernel $T_x\partial D^n \oplus i(T_xD^n)$ at $x \in \partial D^k$. After rescaling we may assume that $F^*\phi$ agrees with $\Psi_{\rm st}$ to first order along ∂D^k, so by Corollary 2.7 we can deform $F^*\phi$ to make it coincide with $\Psi_{\rm st}$ near ∂D^k.

The preceding discussion shows that it suffices to arrange that $F(D^n \cap H_\gamma)$ is J-orthogonal to ∂_-W along $\partial\Delta = f(\partial D^k)$. This can be arranged by appropriate choice of the extension F provided that Δ is J-orthogonal to ∂_-W along $\partial\Delta$. Note that a necessary condition for this is $JT_x\partial\Delta \subset T_x\partial_-W$ for $x \in \partial\Delta$, which means that $\partial\Delta$ is *isotropic* for the contact structure $\xi = T\partial_-W \cap J(T\partial_-W)$ on ∂_-W. Conversely, if this condition holds it is not hard to arrange J-orthogonality. So we have reduced the proof of Theorem 1.5 to

Proposition 3.4. *Consider an almost complex cobordism (W,J) of complex dimension n such that J is integrable near ∂_-W, and ∂_-W is J-convex when cooriented by an inward pointing vector field. If $n > 2$, then any embedding $f : (D^k, \partial D^k) \hookrightarrow (W, \partial_-W)$, $k \le n$, is isotopic to one which is totally real on D^k and isotropic on ∂D^k.*

The remainder of this section is devoted to the proof of this proposition.

The subcritical case. Recall from Step 1 that there exists a homotopy $F_t:(D^k,\partial D^k)\to(V_{2n,k},V_{2n-1,k-1})$ from $F_0=df$ to some $F_1:(D^k,\partial D^k)\to(V^{\mathbb{C}}_{n,k},V^{\mathbb{C}}_{n-1,k-1})$. Restricting it to the boundary provides a homotopy $G_t=F_t|_{\partial D^k}:\partial D^k\to V_{2n-1,k-1}$ from $G_0=df|_{\partial D^k}$ to some $G_1:\partial D^k\to V^{\mathbb{C}}_{n-1,k-1}$. Now *Gromov's h-principle for isotropic immersions* [11, 15] yields a homotopy of immersions $g_t:\partial D^k\to\partial_-W$ from $g_0=f|_{\partial D^k}$ to an isotropic immersion g_1 together with a 2-parameter family of maps $G^s_t:\partial D^k\to V_{2n-1,k-1}$ satisfying $G^0_t=dg_t$, $G^1_t=G_t$, $G^s_0=dg_0$, and $G^s_1:\partial D^k\to V^{\mathbb{C}}_{n-1,k-1}$ for all $s,t\in[0,1]$.

If the g_t can be chosen to be *embeddings* rather than immersions, then the h-principle for totally real embeddings allows us to extend the g_t to embeddings $f_t:D^k\hookrightarrow W$ with f_1 totally real and the proposition follows. In the *subcritical* case $k<n$, this can be achieved simply by a generic perturbation of the g_t (keeping g_1 isotropic).

Remark 3.5. The existence of the 2-parameter family G^s_t is crucial for the application of the h-principle for totally real embeddings. Indeed, we can always connect $g_0=f|_{\partial D^k}$ by embeddings g_t to some isotropic embedding g_1, so if we could extend these g_t to totally real embeddings $D^k\hookrightarrow W$ we would prove Proposition 3.4 also in the case $k=n=2$ where, as we shall see below, it is false in general.

The critical case. In the *critical* case $k=n$, we can still perturb g_1 to a Legendrian embedding, but the g_t need not all be embeddings. To understand the obstruction to this, consider the immersion

$$\Gamma:S^{n-1}\times[0,1]\to\partial_-W\times[0,1],\quad (x,t)\mapsto\big(g_t(x),t\big).$$

After a generic perturbation, we may assume that Γ has finitely many transverse self-intersections and define its *self-intersection index*

$$I_\Gamma:=\sum_p I_\Gamma(p)\in\begin{cases}\mathbb{Z} & \text{if } n \text{ is even,}\\ \mathbb{Z}_2 & \text{if } n \text{ is odd}\end{cases}$$

as the sum over the indices of all self-intersection points p. Here the index $I_\Gamma(p)=\pm1$ is defined by comparing the orientations of the two intersecting branches of Γ to the orientation of $\partial_-W\times[0,1]$. For n even this does not depend on the order of the branches and thus gives a well-defined integer, while

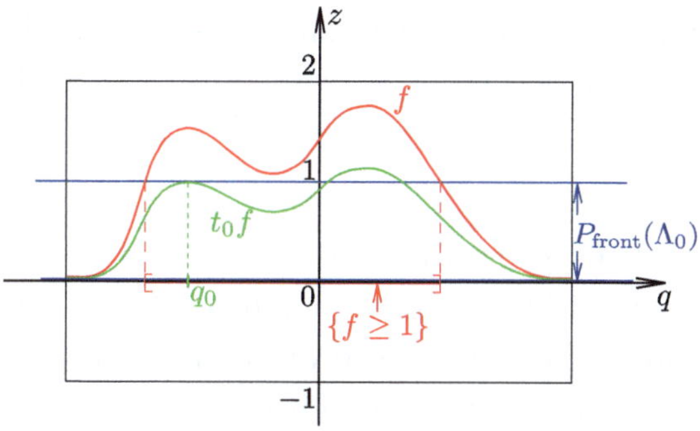

Fig. 5. Stabilization of a Legendrian submanifold

for n odd it is only well-defined mod 2. By a theorem of Whitney [27], for $n > 2$, the regular homotopy g_t can be deformed through regular homotopies fixed at $t = 0, 1$ to an isotopy if and only if $I_\Gamma = 0$.

So if the family g_t satisfies $I_\Gamma = 0$ we are done. If $I_\Gamma \neq 0$ we will connect g_1 to another Legendrian embedding g_2 by a Legendrian regular homotopy g_t, $t \in [1, 2]$, whose self-intersection index equals $-I_\Gamma$. The extended family g_t, $t \in [0, 2]$, then has self-intersection index zero, so applying the previous argument to this family will conclude the proof.

Stabilization of Legendrian submanifolds. Consider a Legendrian submanifold Λ_0 in a contact manifold (M, ξ) of dimension $2n - 1$. Near a point of Λ_0 pick Darboux coordinates $(q_1, p_1, \dots, q_{n-1}, p_{n-1}, z)$ in which $\xi = \ker(dz - \sum_j p_j dq_j)$ and the front projection of Λ_0 is a standard cusp $z^2 = q_1^3$. Deform the two branches of the front to make them parallel over some open ball $B^{n-1} \subset \mathbb{R}^{n-1}$. After rescaling, we may thus assume that the front of Λ_0 has two parallel branches $\{z = 0\}$ and $\{z = 1\}$ over B^{n-1}, see Figure 5.

Pick a non-negative function $f : B^{n-1} \to \mathbb{R}$ with compact support and 1 as a regular value, so $N := \{f \geq 1\} \subset B^{n-1}$ is a compact manifold with boundary. Replacing for each $t \in [0, 1]$ the lower branch $\{z = 0\}$ by the graph $\{z = tf(q)\}$ of the function tf yields the fronts of a path of Legendrian immersions $\Lambda_t \subset M$ connecting Λ_0 to a new Legendrian submanifold Λ_1. Note that Λ_t has a self-intersection for each critical point of tf on level 1.

Problem 3.6. The Legendrian regular homotopy Λ_t, $t \in [0, 1]$, has self-intersection index $(-1)^{(n-1)(n-2)/2} \chi(N)$.

Problem 3.7. For $n > 2$ there exist compact submanifolds $N \subset \mathbb{R}^{n-1}$ of arbitrary Euler characteristic $\chi(N) \in \mathbb{Z}$, while for $n = 2$ the Euler characteristic is always positive.

These two problems show that for $n > 2$ the stabilization construction allows us find a Legendrian regular homotopy Λ_t, $t \in [0,1]$, with arbitrary self-intersection index. In view of the discussion above, this concludes the proof of Proposition 3.4 and hence of Theorem 1.5.

Remark 3.8. The condition $n > 2$ was used twice in the proof of Proposition 3.4: for the application of Whitney's theorem, and to arbitrarily modify the self-intersection index by stabilization.

To illustrate the failure of Theorem 1.5 for $n = 2$, let us analyze for which oriented plane bundles $V \to S^2$ the total space admits a Stein structure. Here V is oriented by minus the orientation of the base followed by that of the fibre. Such bundles are classified by their Euler class $e(V)$, which equals minus the self-intersection number $S \cdot S \in \mathbb{Z}$ of the zero section $S \subset V$.

We can construct each such bundle by attaching a 2-handle to the 4-ball B^4 along a topologically trivial Legendrian knot $\Lambda \subset (S^3, \xi_{\rm st})$. Let $\Delta \subset B^4$ be an embedded 2-disk meeting ∂B^4 transversely along $\partial\Delta = \Lambda$. It fits together with the core disk D of the handle to an embedded 2-sphere $S \subset V$ giving the zero section in V. Recall that the *Thurston-Bennequin invariant* $\mathrm{tb}(\Lambda)$ is defined as the linking number of Λ with a push-off Λ' in the direction of a Reeb vector field on $(S^3, \xi_{\rm st})$.

Problem 3.9. The complex structure on $B^4 \subset \mathbb{C}^2$ extends to a complex structure on V for which the core disk D is totally real (and hence by Theorem 1.5 to a Stein structure on V) if and only if $-e(V) = S \cdot S = \mathrm{tb}(\Lambda) - 1$.

In view of Bennequin's inequality $\mathrm{tb}(\Lambda) \le -1$, this shows that the construction of Theorem 1.5 works to provide a Stein structure on V if and only if $e(V) \ge 2$. A much deeper theorem of Lisca and Matič [19] (proved via Seiberg-Witten theory) asserts that $S \cdot S \le -2$ for every homologically nontrivial embedded 2-sphere S in a Stein surface, hence V admits a Stein structure if and only if $e(V) \ge 2$. For example, the manifold $S^2 \times \mathbb{R}^2$ does not admit any Stein structure.

4. Morse-Smale Theory for J-Convex Functions

Morse-Smale theory deals with the problem of simplification of a Morse function, trying to remove as many critical points as the topology allows. One consequence is the h-cobordism theorem and the proof of the higher-dimensional Poincaré conjecture. In this section we study Morse-Smale theory for J-convex Morse functions, resulting in a Stein version of the h-cobordism theorem.

The h-Cobordism Theorem. Let us begin by recalling the celebrated

Theorem 4.1 (h-cobordism theorem, Smale [26]). *Let W be an h-cobordism, i.e., a compact cobordism such that W and $\partial_\pm W$ are simply connected and $H_*(W, \partial_- W; \mathbb{Z}) = 0$. Suppose that $\dim W \geq 6$. Then W carries a function without critical points and constant on $\partial_\pm W$.*

For the proof, one considers a compact cobordism W with a Morse function $\phi : W \to \mathbb{R}$ having $\partial_\pm W$ as regular level sets and a gradient-like vector field X for ϕ. We will refer to such (W, X, ϕ) as a *Smale cobordism*. It is called *elementary* if $W_p^- \cap W_q^+ = \emptyset$ for all critical points $p \neq q$, where W_p^- and W_p^+ denotes the stable resp. unstable manifold of p with respect to X.

The key geometric ingredients in the proof of the h-cobordism theorem are the following four geometric lemmas about modifications of Smale cobordisms (see [21]). The first three of them are rather simple, while the fourth one is more difficult.

Lemma 4.2 (moving critical levels). *Let (W, X, ϕ_0) be an elementary Smale cobordism. Then there exists a homotopy (W, X, ϕ_t) of elementary Smale cobordisms which arbitrarily changes the ordering of the values of the critical points.*

Lemma 4.3 (moving attaching spheres). *Let (W, X_0, ϕ) be a Smale cobordism and $p \in W$ a critical point whose stable manifold $W_p^-(X_0)$ with respect to X_0 intersects $\partial_- W$ along a sphere $S_0 \subset \partial_- W$. Then given any isotopy $S_t \subset \partial_- W$, $t \in [0,1]$, there exists a homotopy of Smale cobordisms (W, X_t, ϕ) such that the stable manifold $W_p^-(X_t)$ intersects $\partial_- W$ along S_t.*

Lemma 4.4 (creation of critical points). *Let (W, X_0, ϕ_0) be a Smale cobordism without critical points. Then for any $1 \leq k \leq \dim W$ and any $p \in \operatorname{Int} W$*

there exists a Smale homotopy (W, X_t, ϕ_t), $t \in [0,1]$, *fixed outside a neighbourhood of p, which creates a pair of critical points of index* $k-1$ *and* k *connected by a unique trajectory of* X_1 *along which the stable and unstable manifolds intersect transversely.*

Lemma 4.5 (cancellation of critical points). *Suppose that a Smale cobordism* (W, X_0, ϕ_0) *contains exactly two critical points of index* $k-1$ *and* k *which are connected by a unique trajectory of* X *along which the stable and unstable manifolds intersect transversely. Then there exists a Smale homotopy* (W, X_t, ϕ_t), $t \in [0,1]$, *which kills the critical points, so the cobordism* (W, X_1, ϕ_1) *has no critical points.*

Here all the homotopies will be fixed on a neighbourhood of $\partial_\pm W$. The functions ϕ_t in Lemmas 4.4 and 4.5 will be Morse except for one value $t_0 \in (0,1)$ where they have a birth-death type critical point. Here a *birth-death type* critical point of index $k-1$ at t_0 is described by the local model

$$\phi_t(x) = x_1^3 \mp (t-t_0)x_1 - x_2^2 - \cdots - x_k^2 + x_{k+1}^2 + \cdots + x_m^2.$$

Problem 4.6. Prove Lemmas 4.2, 4.3 and 4.4.

Modifications of J-Convex Morse Functions. Let us now state the analogues of the four lemmas for J-convex functions. By a *Stein cobordism* (W, J, ϕ) we will mean a complex cobordism (W, J) with a J-convex Morse function $\phi : W \to \mathbb{R}$ having $\partial_\pm W$ as regular level sets. We will always use the gradient vector field $\nabla_\phi \phi$ of ϕ with respect to the metric $g_\phi = -dd^{\mathbb{C}}\phi(\cdot, J\cdot)$ to obtain a Smale cobordism $(W, \nabla_\phi\phi, \phi)$. Note that in the following four propositions the complex structure J is always fixed.

Proposition 4.7 (moving critical levels). *Let* (W, J, ϕ_0) *be an elementary Stein cobordism. Then there exists a homotopy* (W, J, ϕ_t) *of elementary Stein cobordisms which arbitrarily changes the ordering of the values of the critical points.*

Proposition 4.8 (moving attaching spheres). *Let* (W, J, ϕ_0) *be a Stein cobordism and* $p \in W$ *a critical point whose stable manifold* $W_p^-(\phi_0)$ *with respect to* $\nabla_{\phi_0}\phi_0$ *intersects* $\partial_- W$ *along an isotropic sphere* $S_0 \subset \partial_- W$. *Then given any* isotropic *isotopy* $S_t \subset \partial_- W$, $t \in [0,1]$, *there exists a homotopy of Stein cobordisms* (W, J, ϕ_t) *with fixed critical point p such that the stable manifold* $W_p^-(\phi_t)$ *intersects* $\partial_- W$ *along* S_t.

Proposition 4.9 (creation of critical points). *Let (W, J, ϕ_0) be a Stein cobordism without critical points. Then for any $1 \leq k \leq \dim_{\mathbb{C}} W$ and any $p \in \operatorname{Int} W$ there exists a Stein homotopy (W, J, ϕ_t), $t \in [0,1]$, fixed outside a neighbourhood of p, which creates a pair of critical points of index $k-1$ and k connected by a unique trajectory of $\nabla_{\phi_1}\phi_1$ along which the stable and unstable manifolds intersect transversely.*

Proposition 4.10 (cancellation of critical points). *Suppose that a Stein cobordism (W, J, ϕ_0) contains exactly two critical points of index $k-1$ and k which are connected by a unique trajectory of $\nabla_{\phi_0}\phi_0$ along which the stable and unstable manifolds intersect transversely. Then there exists a Stein homotopy (W, J, ϕ_t), $t \in [0,1]$, which kills the critical points, so the cobordism (W, J, ϕ_1) has no critical points.*

Again, all the homotopies will be fixed on a neighbourhood of $\partial_\pm W$, up to composition of the J-convex functions with some convex increasing function $\mathbb{R} \to \mathbb{R}$. The statements are precise analogues of those in the smooth case, with one notable difference: in Proposition 4.8 we require the isotopy S_t to be *isotropic*. This difference, and the lack of a 1-parametric h-principle for Legendrian embeddings, is largely responsible for all symplectic rigidity phenomena on Stein manifolds. However, in the *subcritical case* $\operatorname{ind}(p) = k < n$ we have an h-principle stating that any smooth isotopy S_t starting at an isotropic embedding S_0 can be C^0-approximated by an isotropic isotopy starting at S_0. With this, the proof of the h-cobordism theorem goes through for J-convex functions and we obtain

Theorem 4.11 (Stein h-cobordism theorem). *Let (W, J, ϕ) be a* subcritical *Stein h-cobordism. Suppose that $\dim_{\mathbb{C}} W \geq 3$. Then W carries a J-convex function without critical points and constant on $\partial_\pm W$.*

Further implications of these results will be discussed in Section 5. The remainder of this section is devoted to the proofs of Propositions 4.7 to 4.10.

Proof of Proposition 4.7. This is an immediate consequence of the J-convex model functions constructed in Section 3: Since the cobordism is elementary, the stable manifolds of the critical points are disjoint embedded disks. For each critical point p, Theorem 2.10 allows us to deform ϕ_0 near W_p^- such that for the new J-lc function the level set containing p is connected to a level set of ϕ_0 slightly above $\partial_- W$. Now we perform this operation for

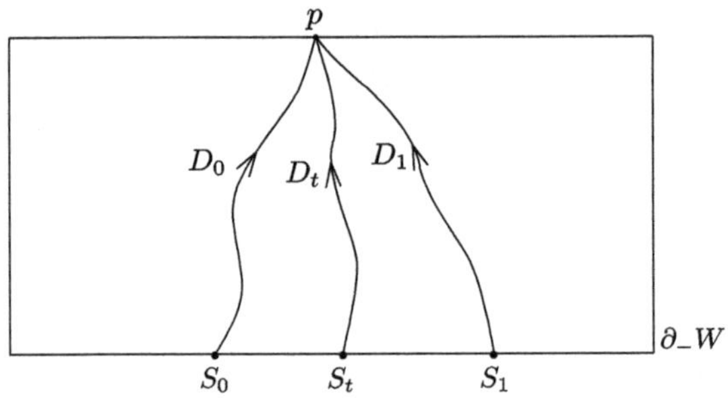

Fig. 6. Moving attaching spheres by isotropic isotopies

each critical point and choose the level sets near $\partial_- W$ to achieve any given ordering. ■

Proof of Proposition 4.8. Let $k := \mathrm{ind}(p) \le n$. We identify level sets of ϕ_0 near $\partial_- W$ via Gray's theorem. Then we construct an isotopy of embedded k-disks $D_t \subset W$ such that $D_0 = W_p^-$, D_t agrees with W_p^- near p, $\partial D_t = S_t$, and D_t intersects all level sets of ϕ below $c := \phi(p)$ transversely in isotropic $(k-1)$-spheres; see Figure 6. The last condition implies that D_t is totally real. If $k < n$ we can further extend D_t to a totally real embedding of $D^k \times D_\varepsilon^{n-k}$ intersecting level sets transversely in isotropic submanifolds, so it suffices to consider the case $k = n$. To conclude the proof, we will construct J-convex functions ϕ_t which agree with ϕ_0 near p and whose gradient $\nabla_{\phi_t}\phi_t$ is tangent to D_t. This is done in two steps.

In the first step we construct J-convex functions ψ_t whose level sets below c are J-orthogonal to D_t. To do this, consider some level set Σ of ϕ_0 intersecting D_t in the isotropic submanifold Λ_t. Let ξ be the induced contact structure on Σ. We deform Σ near Λ_t to a new hypersurface Σ' which agrees with Σ outside a neighbourhood of Λ_t, intersects D_t J-orthogonally in Λ_t, and satisfies $\xi \subset T\Sigma'$ along Λ_t (so we "turn Σ around ξ along Λ_t"); see Figure 7. A careful estimate of the Levi form shows that Σ' can be made J-convex. Deforming all level sets in this way leads to a family of J-convex hypersurfaces, which by Corollary 2.9 can be turned into a foliation and thus into level sets of a J-lc function.

For the second step, consider the J-convex functions ψ_t from the first step whose level sets below c are J-orthogonal to D_t. It is not hard to write down in a local model a J-convex function ϑ_t near D_t which agrees with ψ_t on D_t, whose level sets are J-orthogonal to D_t, and whose gradient $\nabla_{\vartheta_t}\vartheta_t$ is tangent to D_t. Now Corollary 2.7 provides the desired function ϕ_t which

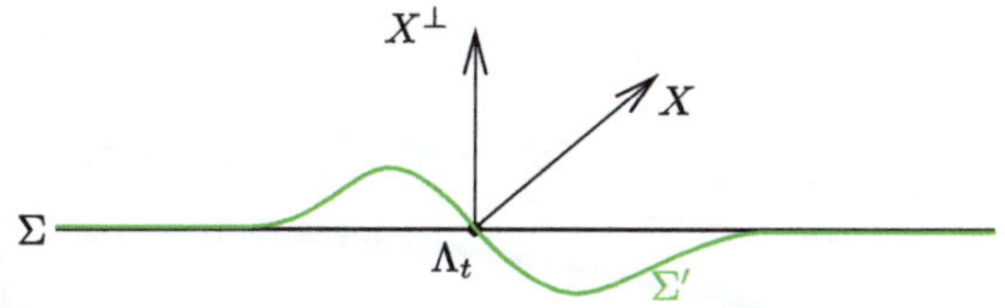

Fig. 7. Turning a J-convex hypersurface along an isotropic submanifold

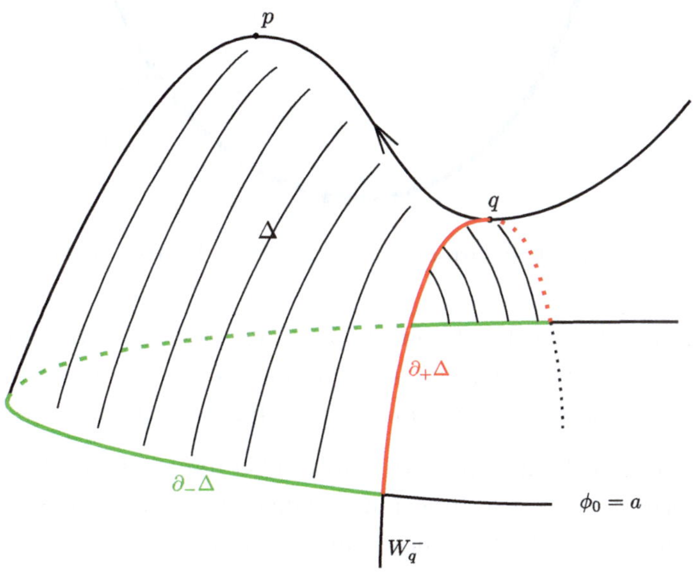

Fig. 8. The half-disk Δ

coincides with ψ_t outside a neighbourhood of D_t and with ϑ_t in a smaller neighborhood of D_t. ■

Proof of Proposition 4.10. Let (W, J, ϕ_0) be a Stein cobordism with exactly two critical points p, q of index k, $k-1$ connected by a unique trajectory of $\nabla_{\phi_0}\phi_0$ along which the stable and unstable manifolds intersect transversely. Set $a := \phi_0|_{\partial_- W}$, $b := \phi_0(q)$ and $c := \phi_0(p)$.

Problem 4.12. In the situation above, suppose that ϕ_0 is quadratic in some holomorphic coordinates near p and q. Then the closure of W_p^- is an embedded k-dimensional half-disk $\Delta \subset W$ with lower boundary $\partial_-\Delta = \Delta \cap \partial_- W$ and upper boundary $\partial_+\Delta = W_q^-$; see Figure 8.

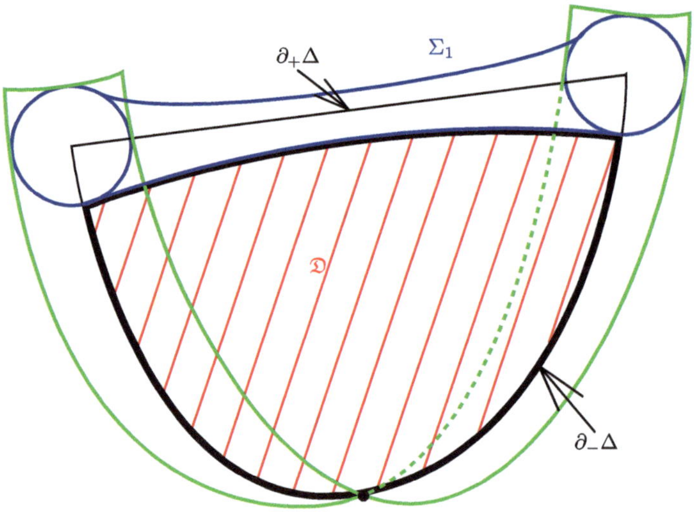

Fig. 9. The first surrounding hypersurface Σ_1 and the disk $\mathfrak{D}$

We will now deform the function ϕ_0 in 4 steps. The first 3 steps modify ϕ_0 outside Δ, without affecting its critical points, to make some level set closely surround Δ; the actual cancellation happens in the last step.

First surrounding. First we apply Theorem 2.10 to the $(k-1)$-disk $\partial_+\Delta$ to deform ϕ_0 to a J-lc function ϕ_1 such that some level set $\Sigma_1 = \{\phi_1 = c_1\}$ closely surrounds $\partial_+\Delta$ as shown in Figure 9.

Second surrounding. Next we apply Theorem 2.10 to the k-disk $\mathfrak{D} := \Delta \cap \{\phi_1 \geq c_1\}$ to deform ϕ_1 to a J-lc function ϕ_2 such that some level set $\Sigma_2 = \{\phi_2 = c_2\}$ closely surrounds Δ as shown in Figure 9. Note that a cross-section of Σ_2 will have a dumbell-like shape as in Figure 10, where $x = (x_1, \dots, x_k)$ and $u = (x_{k+1}, \dots, x_n, y_1, \dots, y_n)$.

Third surrounding. On the other hand, we can construct another hypersurface Σ_3 surrounding Δ as follows: Restrict a very thin model hypersurface Σ provided by Theorem 2.10 to a neighbourhood of the lower half-disk $\{r = 0, R \leq 1, y_k \leq 0\}$ in $\mathbb{C}^n$, implant it onto a neighbourhood of Δ in W, and apply the minimum construction in Corollary 2.8 to this hypersurface and Σ_2. The resulting J-convex hypersurface Σ_3 is shown in Figure 11. The most difficult part is now to connect Σ_3 to Σ_2 by a family of J-convex hypersurfaces. Once this is done, we can apply Corollary 2.9 to deform ϕ_2 to a J-lc function ϕ_3 having Σ_3 as a level set.

The cancellation. Let us extend Δ across $\partial_+\Delta$ to a slightly larger half-disk Δ', still surrounded by Σ_3, so that the critical points p, q lie in

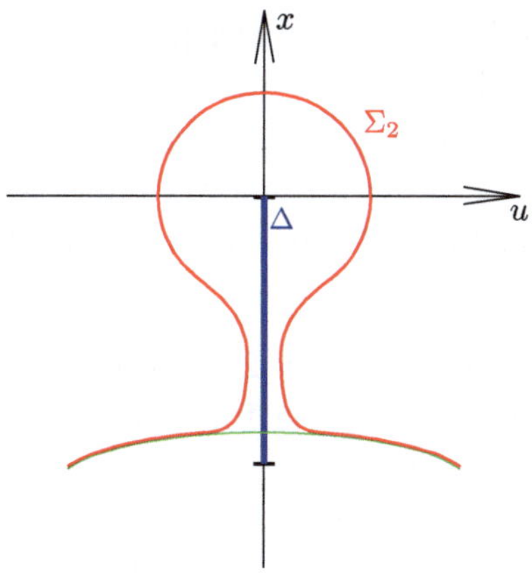

Fig. 10. The dumbell-shaped cross-section of the second surrounding hypersurface Σ_2

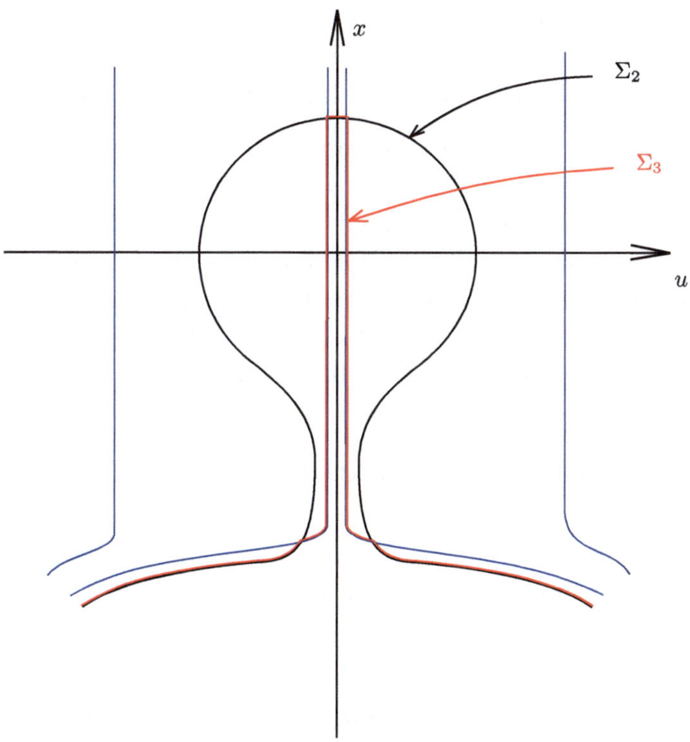

Fig. 11. The third surrounding hypersurface Σ_3

the interior of Δ', and $\nabla_{\phi_3}\phi_3$ is inward pointing along $\partial_-\Delta'$ and outward pointing along $\partial_+\Delta'$. By Lemma 4.5 there exists a family of smooth functions $\beta_t:\Delta'\to\mathbb{R}$, $t\in[3,4]$, fixed near $\partial\Delta'$, such that $\beta_3=\phi_3|_{\Delta'}$ and β_4 has no critical points. Identifying Δ' with the lower half-disk in the standard handle, we can pick a large constant $B>0$ such that the functions $\psi_t:=\beta_t+Br^2$ near Δ' are J-convex for all $t\in[3,4]$.

After an application of Corollary 2.7, we may assume that $\psi_3=\phi_3$ near Δ'. We can choose convex increasing functions $f_t:\mathbb{R}\to\mathbb{R}$ with $f_3=\mathrm{id}$ such that for $t\in[3,4]$ the J-convex function $\phi_t:=\mathrm{smooth}\max(\psi_t,f_t\circ\phi_3)$ agrees with $f_t\circ\phi_3$ in the region outside of Σ_3 and with ψ_t near Δ'. In particular, ϕ_4 has no critical points (for this one needs to check that the maximum constructon does not create new critical points outside Δ'). Hence (W,J,ϕ_{4t}), $t\in[0,1]$, is the desired Stein homotopy and Proposition 4.10 is proved. ■

Proof of Proposition 4.9. The proof is similar to that of Proposition 4.10 but much simpler. Set $a:=\phi_0|_{\partial_-W}$ and $c:=\phi_0(p)$. Pick an isotropic embedded $(k-1)$-sphere S through p in the level set $\phi_0^{-1}(c)$ and let $Z\subset W$ be the totally real cylinder swept out by S under the backward gradient flow of ϕ_0. We identify Z with the cylinder $\{r=0,1/2\le R\le 1\}$ in the standard handle. A slight modification of Theorem 2.10 yields a family of J-convex functions $\phi_t:W\to\mathbb{R}$, $t\in[0,1]$, such that some level set Σ_1 of ϕ_1 surrounds Z in W.

By Lemma 4.4 there exists a family of smooth functions $\beta_t:Z\to\mathbb{R}$, $t\in[1,2]$, fixed near ∂Z, such that $\beta_1=\phi_1|_Z$ and β_2 has exactly two critical points of index $k-1$ and k connected by a unique gradient trajectory along which the stable and unstable manifolds intersect transversely. As above, we can pick a large constant $B>0$ such that the functions $\psi_t:=\beta_t+Br^2$ near Z are J-convex for all $t\in[1,2]$ and set $\phi_t:=\mathrm{smooth}\max(\psi_t,f_t\circ\phi_1)$, $t\in[1,2]$, to obtain the desired family ϕ_{2t}, $t\in[0,1]$. ■

5. Flexibility of Stein Structures

In this section we study the question when two Stein structures on the same manifold can be connected by a Stein homotopy.

Stein Homotopies. Let us first carefully define the notion of a Stein homotopy. Consider first a smooth family (with respect to the C^∞_{loc}-topology) of exhausting functions $\phi_t:V\to\mathbb{R}$, $t\in[0,1]$, on a manifold V. We call it a *simple Morse homotopy* if there exists a family of smooth functions $c_1<c_2<\cdots$

on the interval $[0,1]$ such that for each $t \in [0,1]$, $c_i(t)$ is a regular value of the function ϕ_t and $\bigcup_k \{\phi_t \le c_k(t)\} = V$. Then a *Morse homotopy* is a composition of finitely many simple Morse homotopies, and a *Stein homotopy* is a family of Stein structures (V, J_t, ϕ_t) such that the functions ϕ_t form a Morse homotopy.

The role of the regular levels $c_i(t)$ is to prevent critical points from "escaping to infinity". The following three problems motivate why this is the correct definition. The first one shows that, without this condition, the notion of "homotopy" would become rather trivial:

Problem 5.1. Any two Stein structures (J_0, ϕ_0) and (J_1, ϕ_1) on $\mathbb{C}^n$ can be connected by a smooth family of Stein structures (J_t, ϕ_t) on $\mathbb{C}^n$, allowing critical points to escape to infinity.

The second one shows that the question whether two Stein structures are homotopic does not depend on the chosen J-convex functions:

Problem 5.2. If $\phi_0, \phi_1 : V \to \mathbb{R}$ are two exhausting J-convex functions for the same complex structure J, then (J, ϕ_0) and (J, ϕ_1) can be connected by a Stein homotopy (J, ϕ_t).

The third one makes the question of Stein homotopies accessible to symplectic techniques. Let us call a Stein structure (J, ϕ) *complete* if the gradient vector field $\nabla_\phi \phi$ is complete; by Problem 2.2, any Stein structure can be made complete by composing ϕ with a convex increasing function $f : \mathbb{R} \to \mathbb{R}$.

Problem 5.3. If two complete Stein structures (J_0, ϕ_0) and (J_1, ϕ_1) on a manifold V are Stein homotopic, then the associated symplectic manifolds $(V, -dd^{\mathbb{C}}\phi_0)$ and $(V, -dd^{\mathbb{C}}\phi_1)$ are symplectomorphic.

From now on, when we talk about individual Stein structures (J, ϕ) we will always assume that the function ϕ is Morse, while for Stein homotopies we allow birth-death type singularities.

The 2-Index Theorem. Before studying Stein homotopies, let us first consider the situation in smooth topology. It follows from Problem 5.2 (simply ignoring J-convexity) that any two Morse functions on the same manifold can be connected by a Morse homotopy. In addition, we will need some control over the indices of critical points. This is provided by following immediate consequence of the *two-index theorem* of Hatcher and Wagoner ([16], see also [17]):

Theorem 5.4. *Let $\phi_0, \phi_1 : W \to [0,1]$ be two Morse functions on an m-dimensional cobordism W with $\partial_\pm W$ as regular level sets. For some $k \geq 3$, suppose that ϕ_0, ϕ_1 have no critical points of index $> k$. Then ϕ_0 and ϕ_1 can be connected by a Morse homotopy ϕ_t (all having $\partial_\pm W$ as regular level sets) without critical points of index $> k$.*

We will apply this theorem in the following two cases with $m = 2n$:

- the subcritical case $k + 1 = n \geq 4$;
- the critical case $k = n \geq 3$.

Uniqueness of Subcritical Stein Structures. After these preparations, we can prove our first uniqueness theorem.

Theorem 5.5 (uniqueness of subcritical Stein structures). *Let (J_0, ϕ_0) and (J_1, ϕ_1) be two* subcritical *Stein structures on the same manifold V of complex dimension $n > 3$. If J_0 and J_1 are homotopic as almost complex structures, then (J_0, ϕ_0) and (J_1, ϕ_1) are Stein homotopic.*

Proof. By Theorem 5.4 with $k+1 = n \geq 4$, the functions ϕ_0 and ϕ_1 can be connected a Morse homotopy ϕ_t without critical points of index $\geq n$. We cut the homotopy into a finite number of simple Morse homotopies, and we cut each simple homotopy at the regular levels c_i into countably many compact cobordisms. Let us pick gradient-like vector fields X_t for ϕ_t. After further decomposition of these cobordisms, we may assume that on each cobordism W only one of the following two cases occurs:

(i) all the Smale cobordisms (W, X_t, ϕ_t) are elementary;

(ii) a pair of critical points is created or cancelled.

In the first case, only the levels of the critical points vary and the attaching spheres move by smooth isotopies. By the h-principle for subcritical isotropic embeddings, these isotopies can be C^0-approximated by isotropic isotopies. So we can apply Propositions 4.7 and 4.8 to realize the same moves by J-convex functions. The second case is treated by Propositions 4.9 and 4.10. Applying the four propositions inductively over the simple homotopies, and within each simple homotopy over increasing levels, we hence construct a family of J_0-convex functions (all for the same J_0!) $\psi_t : V \to \mathbb{R}$ such that $\psi_t = \phi_t \circ h_t$ for a smooth family of diffeomorphisms $h_t : V \to V$ with $h_0 = \mathrm{id}$.

Note that $((h_t)_* J_0, \phi_t)$ provides a Stein homotopy from (J_0, ϕ_0) to $(J_2 := (h_1)_* J_0, \phi_1)$. So the theorem is proved if we can connect (J_2, ϕ_1) to (J_1, ϕ_1) by a Stein homotopy (J_t, ϕ_1), $t \in [1,2]$ (with fixed function ϕ_1!). For this, we decompose V into elementary cobordisms containing only one critical level, and we pick a family X_t of gradient-like vector fields for ϕ_1 connecting the gradients with respect to J_1 and J_2. Then for each critical point p on such a cobordism W the attaching spheres with respect to X_t form a smooth isotopy S_t, $t \in [1,2]$, connecting the isotropic spheres S_1 and S_2. Again by the h-principle for subcritical isotropic embeddings, we can make the isotopy S_t isotropic. Now by a 1-parametric version of the Existence Theorem 1.5, we can connect J_1 and J_2 by a smooth family of integrable complex structures J_t on W such that ϕ_1 is J_t-convex for all $t \in [1,2]$. ■

Problem 5.6. Find the major gap in the preceding proof, and consult [7] on how it can be filled.

Exotic Stein Structures. In the critical case, uniqueness fails dramatically. In 2009, McLean [20] constructed infinitely many pairwise non-homotopic Stein structures on $\mathbb{C}^n$ for any $n \geq 4$. Extending McLean's result to $n = 3$ (see [1]) and combining it with the surgery exact sequence from [3], one obtains

Theorem 5.7. *Let (V, J) be an almost complex manifold of real dimension $2n \geq 6$ which admits an exhausting Morse function with finitely many critical points all of which have index $\leq n$. Then V carries infinitely many finite type Stein structures (J_k, ϕ_k), $k \in \mathbb{N}$, such that the J_k are homotopic to J as almost complex structures and (J_k, ϕ_k), (J_ℓ, ϕ_ℓ) are not Stein homotopic for $k \neq \ell$.*

Here a Stein structure (J, ϕ) is said to be of *finite type* if ϕ has only finitely many critical points. The Stein structures (J_k, ϕ_k) are distinguished up to homotopy by showing that the symplectic manifolds $(V, -dd^{\mathbb{C}}\phi_k)$ are pairwise non-symplectomorphic, distinguished by their symplectic homology. Despite this wealth of exotic Stein structures, it has recently turned out that there is still some flexibility in the critical case, which we will describe next.

Murphy's h-Principle for Loose Legendrian Knots. It is well-known that the 1-parametric h-principle fails for Legendrian embeddings. More precisely, a *formal Legendrian isotopy* (f_t, F_t^s) between two Legendrian embeddings $f_0, f_1 : \Lambda \hookrightarrow (M, \xi)$ consists of a smooth isotopy $f_t : \Lambda \hookrightarrow M$, $t \in [0,1]$, together with a 2-parameter family of injective bundle homomorphisms $F_t^s : T\Lambda \to TM$ covering f_t, $s,t \in [0,1]$, such that $F_0^s = df_0$, $F_1^s = df_1$, $F_t^0 = df_t$, and $F_t^1 : T\Lambda \to \xi$ is isotropic for all s,t. By the h-principle for Legendrian immersions, this implies that f_0 and f_1 are connected by a Legendrian regular homotopy. On the other hand, there are many examples of pairs of Legendrian embeddings that are formally Legendrian isotopic but not Legendrian isotopic (see e.g. [5] in dimension 3, and [9] in higher dimensions).

Despite the failure of the h-principle, there are two partial flexibility results for Legendrian knots in dimension 3: Any two formally isotopic Legendrian knots in $(\mathbb{R}^3, \xi_{st})$ become Legendrian isotopic after sufficiently many stabilizations [12], and any two formally isotopic Legendrian knots in the complement of an overtwisted disk are Legendrian isotopic [8]. E. Murphy recently discovered a remarkable class of Legendrian embeddings in dimensions ≥ 5 which satisfy the 1-parametric h-principle:

Theorem 5.8 (Murphy's h-principle for loose Legendrian embeddings [22]). *In contact manifolds (M, ξ) of dimension ≥ 5 there exists a class of* loose *Legendrian embeddings with the following properties*:

(a) *The stabilization construction described in Section* 3 *with $\chi(N) = 0$ turns any Legendrian embedding f_0 into a loose Legendrian embedding f_1 formally isotopic to f_0.*

(b) *Let (f_t, F_t^s), $s,t \in [0,1]$, be a formal Legendrian isotopy connecting two loose Legendrian embeddings $f_0, f_1 : \Lambda \hookrightarrow M$. Then there exists a Legendrian isotopy $\widetilde{f}_t$ connecting $\widetilde{f}_0 = f_0$ and $\widetilde{f}_1 = f_1$ which is C^0-close to f_t and is homotopic to the formal isotopy (f_t, F_t^s) through formal isotopies with fixed endpoints.*

Note that, in contrast to the 3-dimensional case, Legendrian embeddings in dimension ≥ 5 become loose after just *one* stabilization, and the stabilization of a loose Legendrian embedding is Legendrian isotopic to the original one.

Existence and Uniqueness of Flexible Stein Structures. Let us call a Stein manifold (V, J, ϕ) of complex dimension ≥ 3 *flexible* if all attaching spheres on all regular level sets are either subcritical or loose Legendrian. In view of Theorem 5.8(a), we can perform a stabilization in each inductional step of the proof of the Existence Theorem 1.5 to obtain

Theorem 5.9 (existence of flexible Stein structures). *Any smooth manifold V of dimension $2n > 4$ which admits a Stein structure also admits a* flexible *one (in a given homotopy class of almost complex structures).*

Now we can repeat the proof of Theorem 5.5, using Theorem 5.4 in the critical case $k = n \geq 3$ and Theorem 5.8(b) for the Legendrian attaching spheres (always keeping the Stein structures flexible in the process), to obtain

Theorem 5.10 (uniqueness of flexible Stein structures). *Let (J_0, ϕ_0) and (J_1, ϕ_1) be two* flexible *Stein structures on the same manifold V of complex dimension $n > 2$. If J_0 and J_1 are homotopic as almost complex structures, then (J_0, ϕ_0) and (J_1, ϕ_1) are Stein homotopic.*

Remark 5.11. (a) Since subcritical Stein manifolds are flexible, Theorem 5.10 allows us to weaken the hypothesis on the dimension in Theorem 5.5 from $n > 3$ to $n > 2$.

(b) Combining the result in [6] with the surgery exact sequence in [3] implies that flexible Stein manifolds have vanishing symplectic homology.

Applications to Symplectomorphisms and Pseudo-isotopies. Theorem 5.10 has the following consequence for symplectomorphisms of flexible Stein manifolds.

Theorem 5.12. *Let (V, J, ϕ) be a complete* flexible *Stein manifold of complex dimension $n > 2$, and $f : V \to V$ be a diffeomorphism such that f^*J is homotopic to J as almost complex structures. Then there exists diffeotopy (i.e., a smooth family of diffeomorphisms) $f_t : V \to V$, $t \in [0, 1]$, such that $f_0 = f$, and f_1 is a symplectomorphism of (V, ω_ϕ).*

Proof. By Theorem 5.10, there exists a Stein homotopy (J_t, ϕ_t) connecting the flexible Stein structures $(J_0, \phi_0) = (J, \phi)$ and $(J_1, \phi_1) = (f^*J, f^*\phi)$. By Problem 5.3, there exists a diffeotopy $h_t : V \to V$ such that $h_0 = \mathrm{id}$ and $h_t^*\omega_{\phi_t} = \omega_\phi$. In particular, $(f \circ h_1)^*\omega_\phi = h_1^*\omega_{\phi_1} = \omega_\phi$, so $f_t = f \circ h_t$ is the desired diffeotopy. ■

Remark 5.13. Even if (J, ϕ) is of finite type and $f = \mathrm{id}$ outside a compact set, the diffeotopy f_t provided by Theorem 5.12 will in general *not* equal the identity outside a compact set.

For our last application, consider a closed manifold M. A *pseudo-isotopy* of M is a smooth function $\phi: M\times[0,1]\to\mathbb{R}$ without critical points which is constant on $M\times 0$ and $M\times 1$ with $f|_{M\times 0}<f|_{M\times 1}$. We denote by $\mathcal{E}(M)$ the space of pseudo-isotopies equipped with the C^∞-topology. The homotopy group $\pi_0\mathcal{E}(M)$ is trivial if $\dim M\geq 5$ and M is simply connected [4], while in the non-simply connected case for $\dim M\geq 6$ it is often nontrivial [16, 17].

Problem 5.14. Show that $\mathcal{E}(M)$ is homotopy equivalent to the space $\mathcal{P}(M)$ of diffeomorphisms of $M\times[0,1]$ that restrict as the identity to $M\times 0$. (The map $\mathcal{P}(M)\to\mathcal{E}(M)$ assigns to f the pullback $f^*\phi_{\rm st}$ of the function $\phi_{\rm st}(x,t)=t$, and a homotopy inverse is obtained by following trajectories of a gradient-like vector field). This explains the name "pseudo-isotopy" because any isotopy $f_t: M\to M$ with $f_0=\mathrm{id}$ defines an element $f(x,t)=(f_t(x),t)$ in $\mathcal{P}(M)$.

Now consider a topologically trivial Stein cobordism $(M\times[0,1],J,\phi)$ and denote by $\mathcal{E}(M\times[0,1],J)$ the space of J-convex functions $M\times[0,1]\to\mathbb{R}$ without critical points which are constant on $M\times 0$ and $M\times 1$ with $f|_{M\times 0}<f|_{M\times 1}$.

Theorem 5.15. *For any topologically trivial* flexible *Stein cobordism* $(M\times[0,1],J,\phi)$ *of dimension* $2n>4$ *the canonical inclusion* $\mathcal{I}:\mathcal{E}(M\times[0,1],J)\hookrightarrow\mathcal{E}(M)$ *induces a* surjection

$$\mathcal{I}_*:\pi_0\mathcal{E}\big(M\times[0,1],J\big)\to\pi_0\mathcal{E}(M).$$

Proof. Let $\psi\in\mathcal{E}(M)$ be given. By Theorem 5.4 with $k=n\geq 3$, there exists a Morse homotopy $\phi_t: M\times[0,1]\to\mathbb{R}$ without critical points of index $>n$ connecting the J-convex function $\phi_0=\phi$ to $\phi_1=\psi$. Arguing as in the proof of Theorem 5.5, always keeping the Stein structures flexible, we construct a diffeotopy $h_t: M\times[0,1]\to M\times[0,1]$ with $h_0=\mathrm{id}$ such that the functions $\psi_t=\phi_t\circ h_t$ are J-convex for all $t\in[0,1]$. Then the J-convex function $\psi_1=\psi\circ h_1$ is connected to ψ by the path $\psi\circ h_t$ of functions without critical points, so ψ_1 and ψ belong to the same path connected component of $\mathcal{E}(M)$. ■

We conjecture that $\mathcal{I}_*$ in Theorem 5.15 is an isomorphism.

References

[1] M. Abouzaid, P. Seidel, Altering symplectic manifolds by homologous recombination. arXiv:1007.3281

[2] E. Bishop, Mappings of partially analytic spaces. Am. J. Math. **83**, 209–242 (1961)

[3] F. Bourgeois, T. Ekholm, Y. Eliashberg, Effect of Legendrian surgery. arXiv:0911.0026

[4] J. Cerf, La stratification naturelle des espaces de fonctions différentiables réelles et le théorème de la pseudo-isotopie. Publ. Math. IHÉS **39**, 5–173 (1970)

[5] Y. Chekanov, Differential algebra of Legendrian links. Invent. Math. **150**(3), 441–483 (2002)

[6] K. Cieliebak, Handle attaching in symplectic homology and the chord conjecture. J. Eur. Math. Soc. **4**(2), 115–142 (2002)

[7] K. Cieliebak, Y. Eliashberg, *From Stein to Weinstein and Back—Symplectic Geometry of Affine Complex Manifolds*. Colloquium Publications, vol. 59 (Amer. Math. Soc., Providence, 2012)

[8] K. Dymara, Legendrian knots in overtwisted contact structures on S^3. Ann. Glob. Anal. Geom. **19**(3), 293–305 (2001)

[9] T. Ekholm, J. Etnyre, M. Sullivan, Non-isotopic Legendrian submanifolds in $\mathbb{R}^{2n+1}$. J. Differ. Geom. **71**(1), 85–128 (2005)

[10] Y. Eliashberg, Topological characterization of Stein manifolds of dimension > 2. Int. J. Math. **1**(1), 29–46 (1990)

[11] Y. Eliashberg, N. Mishachev, *Introduction to the h-Principle* (Amer. Math. Soc., Providence, 2002)

[12] D. Fuchs, S. Tabachnikov, Invariants of Legendrian and transverse knots in the standard contact space. Topology **36**(5), 1025–1053 (1997)

[13] H. Grauert, On Levi's problem and the imbedding of real-analytic manifolds. Ann. Math. **68**, 460–472 (1958)

[14] M. Gromov, Convex integration of differential relations. I. Izv. Akad. Nauk SSSR, Ser. Mat. **37**, 329–343 (1973)

[15] M. Gromov, *Partial Differential Relations*. Ergebnisse der Mathematik und ihrer Grenzgebiete (3), vol. 9 (Springer, Berlin, 1986)

[16] A. Hatcher, J. Wagoner, *Pseudo-isotopies of Compact Manifolds*. Astérisque, vol. 6 (Soc. Math. de France, Paris, 1973)

[17] K. Igusa, The stability theorem for smooth pseudoisotopies. K-Theory **2**(1–2) (1988)

[18] P. Landweber, Complex structures on open manifolds. Topology **13**, 69–75 (1974)

[19] P. Lisca, G. Matič, Tight contact structures and Seiberg-Witten invariants. Invent. Math. **129**, 509–525 (1997)

[20] M. McLean, Lefschetz fibrations and symplectic homology. Geom. Topol. **13**(4), 1877–1944 (2009)

[21] J. Milnor, *Lectures on the h-Cobordism Theorem* (Princeton Univ. Press, Princeton, 1965). Notes by L. Siebenmann and J. Sondow

[22] E. Murphy, Loose Legendrian embeddings in high dimensional contact manifolds. arXiv:1201.2245

[23] R. Narasimhan, Imbedding of holomorphically complete complex spaces. Am. J. Math. **82**, 917–934 (1960)

[24] A. Newlander, L. Nirenberg, Complex analytic coordinates in almost complex manifolds. Ann. Math. **65**, 391–404 (1957)

[25] R. Richberg, Stetige streng pseudokonvexe Funktionen. Math. Ann. **175**, 251–286 (1968)

[26] S. Smale, On the structure of manifolds. Am. J. Math. **84**, 387–399 (1962)

[27] H. Whitney, The self-intersections of a smooth n-manifold in $2n$-space. Ann. Math. **45**, 220–246 (1944)

K. Cieliebak

Augsburg University
Augsburg
Germany

Y. Eliashberg

Stanford University
Stanford
USA

BOLYAI SOCIETY
MATHEMATICAL STUDIES, 26

Contact and Symplectic Topology
pp. 389–484.

Lecture Notes on Embedded Contact Homology

MICHAEL HUTCHINGS

1. Introduction

We begin by describing an application of ECH to four-dimensional symplectic embedding problems. We will then give an overview of the basic structure of ECH and how it leads to the application.

1.1. Symplectic Embeddings in Four Dimensions

Let (X_0, ω_0) and (X_1, ω_1) be symplectic four-manifolds, possibly with boundary or corners. A *symplectic embedding* of (X_0, ω_0) into (X_1, ω_1) is a smooth embedding $\varphi : X_0 \to X_1$ such that $\varphi^* \omega_1 = \omega_0$. It is interesting to ask when such a symplectic embedding exists.

This is a nontrivial question already for domains in $\mathbb{R}^4$. For example, given $a, b > 0$, define the *ellipsoid*

$$E(a,b) = \left\{ (z_1, z_2) \in \mathbb{C}^2 \;\middle|\; \frac{\pi |z_1|^2}{a} + \frac{\pi |z_2|^2}{b} \le 1 \right\}. \tag{1.1}$$

Here we identify $\mathbb{C}^2 = \mathbb{R}^4$ with coordinates $z_k = x_k + y_k$ for $k = 1, 2$, with the standard symplectic form $\omega = \sum_{k=1}^{2} dx_k dy_k$. In particular, define the *ball*

Partially supported by NSF grant DMS-1105820.

F. Bourgeois et al. (eds.), *Contact and Symplectic Topology*,
Bolyai Society Mathematical Studies 26, DOI 10.1007/978-3-319-02036-5_9,

$B(a) = E(a,a)$. Also, define the *polydisk*

$$P(a,b) = \{(z_1,z_2) \in \mathbb{C}^2 \mid \pi|z_1|^2 \le a,\ \pi|z_2|^2 \le b\}. \tag{1.2}$$

We can now ask, when does one ellipsoid or polydisk symplectically embed into another?

A landmark in the theory of symplectic embeddings is Gromov's non-squeezing theorem from 1985. The four-dimensional case of this theorem asserts that $B(r)$ symplectically embeds into $P(R,\infty)$ if and only if $r \le R$.

The question of when one four-dimensional ellipsoid symplectically embeds into another was answered only in 2010, by McDuff. To state the embedding criterion, let $N(a,b)$ denote the sequence of all nonnegative integer linear combinations of a and b, arranged in nondecreasing order, and indexed starting at 0. For example,

$$N(1,1) = (0,1,1,2,2,2,\ldots) \tag{1.3}$$

and

$$N(1,2) = (0,1,2,2,3,3,4,4,4,5,5,5,\ldots). \tag{1.4}$$

Theorem 1.1 (McDuff [41]). *There is a symplectic embedding* $\operatorname{int}(E(a,b)) \to E(c,d)$ *if and only if* $N(a,b) \le N(c,d)$, *i.e.* $N(a,b)_k \le N(c,d)_k$ *for each* $k \ge 0$.

For example, it is not hard to deduce from Theorem 1.1, together with (1.3) and (1.4), that $\operatorname{int}(E(1,2))$ symplectically embeds into $B(c)$ if and only if $c \ge 2$.

Given more general a,b,c,d, it can be nontrivial to decide whether $N(a,b) \le N(c,d)$. For example, consider the problem of an embedding an ellipsoid into a ball, i.e. the case $c = d$. By scaling, we can encode this problem into a single function $f : [1,\infty) \to [1,\infty)$, where $f(a)$ is defined to be the infimum over c such that $E(1,a)$ symplectically embeds into $B(c) = E(c,c)$.

In general, if there is a symplectic embedding of (X_0,ω_0) into (X_1,ω_1), then necessarily

$$\operatorname{vol}(X_0,\omega_0) \le \operatorname{vol}(X_1,\omega_1), \tag{1.5}$$

where in four dimensions

$$\operatorname{vol}(X,\omega) = \frac{1}{2}\int_X \omega \wedge \omega.$$

In particular, the ellipsoid has volume $\mathrm{vol}(E(a,b)) = ab/2$, cf. Equation (4.12), so it follows from the volume constraint (1.5) that $f(a) \geq \sqrt{a}$.

McDuff-Schlenk computed the function f explicitly and found that the volume constraint is the only constraint if a is sufficiently large, while for smaller a the situation is more interesting. In particular, their calculation implies the following[1]:

Theorem 1.2 (McDuff-Schlenk [43]).

- *On the interval* $[1,(1+\sqrt{5}/2)^4)$, *the function* f *is piecewise linear, given by a "Fibonacci staircase".*
- *The interval* $[(1+\sqrt{5}/2)^4,(17/6)^2]$ *is divided into finitely many intervals, on each of which either* f *is linear or* $f(a)=\sqrt{a}$.
- *On the interval* $[(17/6)^2,\infty)$, *we have* $f(a)=\sqrt{a}$.

Note that Theorems 1.1 and 1.2 were proved by different methods. It is a subtle number-theoretic problem to deduce Theorem 1.2 directly from Theorem 1.1.

1.2. Properties of ECH Capacities

Embedded contact homology can be used to prove the obstruction half of Theorem 1.1, namely the fact that if $\mathrm{int}(E(a,b))$ symplectically embeds into $E(c,d)$ then $N(a,b) \leq N(c,d)$. This follows from the more general theory of "ECH capacities". Here are some of the key properties of ECH capacities; the definition of ECH capacities will be given in Section 1.5.

Theorem 1.3 [24]. *For each symplectic four-manifold* (X,ω) *(not necessarily connected, possibly with boundary or corners), there is a sequence of real numbers*

$$0 = c_0(X,\omega) \leq c_1(X,\omega) \leq \cdots \leq \infty,$$

called ECH capacities, *with the following properties*:

[1] An analogue of Theorem 1.2 for symplectically embedding $\mathrm{int}(E(1,a))$ into $P(c,c)$ was recently worked out in [17]. This is equivalent to symplectically embedding $\mathrm{int}(E(1,a))$ into $E(c,2c)$, by Remark 1.5(b) and Equation (1.10) below.

(Monotonicity) *If* (X_0,ω_0) *symplectically embeds into* (X_1,ω_1), *then*

$$c_k(X_0,\omega_0) \le c_k(X_1,\omega_1) \tag{1.6}$$

for all $k \ge 0$.

(Conformality) *If* r *is a nonzero real number, then*

$$c_k(X,r\omega) = |r| c_k(X,\omega).$$

(Ellipsoid)

$$c_k\big(E(a,b)\big) = N(a,b)_k. \tag{1.7}$$

(Polydisk)

$$c_k\big(P(a,b)\big) = \min\big\{am+bn \mid m,n \in \mathbb{N},\ (m+1)(n+1) \ge k+1\big\}. \tag{1.8}$$

(Disjoint union)

$$c_k\left(\coprod_{i=1}^n (X_i,\omega_i)\right) = \max_{k_1+\cdots+k_n=k} \sum_{i=1}^n c_{k_i}(X_i,\omega_i).$$

(Volume) [11] *If* (X,ω) *is a Liouville domain (see Definition* 1.12) *with all ECH capacities finite (for example a star-shaped domain in* $\mathbb{R}^4$), *then*

$$\lim_{k\to\infty} \frac{c_k(X,\omega)^2}{k} = 4\,\mathrm{vol}(X,\omega). \tag{1.9}$$

In particular, the Monotonicity and Ellipsoid properties immediately imply the obstruction half of Theorem 1.1. Theorem 1.3 does not say anything about the other half of Theorem 1.1, namely the existence of symplectic embeddings.

The Volume property says that for Liouville domains with all ECH capacities finite, the asymptotic behavior of the Monotonicity property (1.6) as $k \to \infty$ recovers the volume constraint (1.5).

Exercise 1.4. Check the volume property (1.9) when (X,ω) is an ellipsoid $E(a,b)$. (See answer in Appendix.)

Remark 1.5. Here is what we know about the sharpness of the ECH obstruction for some other symplectic embedding problems.

(a) ECH capacities give a sharp obstruction to symplectically embedding a disjoint union of balls of possibly different sizes into a ball. This follows by comparison with work of McDuff [40] and Biran [1] from the 1990's which solved this embedding problem. See [25] for details.

(b) It follows from work of Müller that ECH capacities give a sharp obstruction to embedding an ellipsoid into a polydisk, see [25] and [17].

(c) ECH capacities do not give a sharp obstruction to symplectically embedding a polydisk into an ellipsoid. For example, one can check that

$$c_k(P(1,1)) = c_k(E(1,2)) \tag{1.10}$$

for all k, so ECH capacities give no obstruction to symplectically embedding $P(1,1)$ into $E(a,2a)$ when $a>1$. However the Ekeland-Hofer capacities imply that $P(1,1)$ does not symplectically embed into $E(a,2a)$ when $a<3/2$; these capacities are $(1,2,3,\ldots)$ and $(a,2a,2a,3a,4a,4a,\ldots)$ respectively [5, 13]. The Ekeland-Hofer obstruction is sharp, because it follows from (1.1) and (1.2) that $P(1,1)$, as defined, is a subset of $E(3/2,3)$.

(d) We know very little about when one polydisk can be symplectically embedded into another or how good the ECH obstruction to this is.

In Section 4.3 we will compute the ECH capacities of a larger family of examples, namely "toric domains" in $\mathbb{C}^2$.

1.3. Overview of ECH

We now outline the definition of embedded contact homology; details will be given in Section 3.

Let Y be a closed oriented three-manifold. Recall that a *contact form* on Y is a 1-form λ on Y such that $\lambda\wedge d\lambda>0$ everywhere. The contact form λ determines the *contact structure* $\xi=\operatorname{Ker}\lambda$, which is an oriented two-plane field, and the *Reeb vector field* R characterized by $d\lambda(R,\cdot)=0$ and $\lambda(R)=1$.

A *Reeb orbit* is a closed orbit of R, i.e. a map $\gamma:\mathbb{R}/T\mathbb{Z}\to Y$ for some $T>0$, modulo reparametrization, such that $\gamma'(t)=R(\gamma(t))$. A Reeb orbit is either embedded in Y, or an m-fold cover of an embedded Reeb orbit for some integer $m>1$.

We often want to assume that the Reeb orbits are "cut out transversely" in the following sense. Given a Reeb orbit γ as above, the *linearized return map* is a symplectic automorphism P_γ of the symplectic vector space $(\xi_{\gamma(0)}, d\lambda)$, which is defined as the derivative of the time T flow of R. The Reeb orbit γ is called *nondegenerate* if 1 is not an eigenvalue of P_γ. The contact form λ is called nondegenerate if all Reeb orbits are nondegenerate. This holds for generic contact forms.

A nondegenerate Reeb orbit γ is called *elliptic* if the eigenvalues of P_γ are on the unit circle, so that P_γ is conjugate to a rotation. Otherwise γ is *hyperbolic*, meaning that the eigenvalues of P_γ are real. There are two kinds of hyperbolic orbits: *positive hyperbolic* orbits for which the eigenvalues of P_γ are positive, and *negative hyperbolic* orbits for which the eigenvalues of P_γ are negative.

Assume now that λ is nondegenerate, and fix a homology class $\Gamma \in H_1(Y)$. One can then define the *embedded contact homology* $ECH_*(Y,\xi,\Gamma)$ as follows. This is the homology of a chain complex $ECC(Y,\lambda,\Gamma,J)$. The chain complex is freely generated over $\mathbb{Z}/2$ by finite sets of pairs $\alpha = \{(\alpha_i, m_i)\}$ where:

- The α_i are distinct embedded Reeb orbits.
- The m_i are positive integers.
- The total homology class $\sum_i m_i[\alpha_i] = \Gamma$.
- $m_i = 1$ whenever α_i is hyperbolic.

It is a frequently asked question why the last condition is necessary; we will give one answer in Sections 2.6–2.7 and another answer in Section 5.4. Note also that ECH can be defined with integer coefficients, see [30, §9]; however the details of the signs are beyond the scope of these notes, and $\mathbb{Z}/2$ coefficients are sufficient for all the applications we will consider here.

The chain complex differential is defined roughly as follows. We call an almost complex structure J on the "symplectization" $\mathbb{R} \times Y$ *symplectization-admissible* if J is $\mathbb{R}$-invariant, $J(\partial_s) = R$ where s denotes the $\mathbb{R}$ coordinate on $\mathbb{R} \times Y$, and J sends the contact structure ξ to itself, rotating positively with respect to $d\lambda$. These are the standard conditions on J for defining various flavors of contact homology. In the notation for the chain complex, J is a generic symplectization-admissible almost complex structure on $\mathbb{R} \times Y$.

If $\alpha = \{(\alpha_i, m_i)\}$ and $\beta = \{(\beta_j, n_j)\}$ are chain complex generators, then the differential coefficient $\langle \partial\alpha, \beta\rangle \in \mathbb{Z}/2$ is a mod 2 count of J-holomorphic curves C in $\mathbb{R}\times Y$, modulo $\mathbb{R}$ translation and equivalence of currents, satisfying two conditions. The first condition is that, roughly speaking, C converges as a current to $\sum_i m_i\alpha_i$ as $s\to+\infty$, and to $\sum_j n_j\beta_j$ as $s\to-\infty$. The second condition is that C has "ECH index" equal to 1. The definition of the ECH index is the key nontrivial part of the definition of ECH; the original references are [21, 22], and we will spend considerable time explaining this in Section 3. We will see in Proposition 3.7 that our assumption that J is generic implies every ECH index 1 curve is embedded, except possibly for multiple covers of "trivial cylinders" $\mathbb{R}\times\gamma$ where γ is a Reeb orbit; hence the name "embedded contact homology". We will explain in Section 5.3 why ∂ is well-defined. It is shown in [29, §7] that $\partial^2=0$; we will introduce some of what is involved in the proof in Section 5.4.

Let $ECH_*(Y,\lambda,\Gamma,J)$ denote the homology of the chain complex $ECC_*(Y,\lambda,\Gamma,J)$. It turns out that this homology does not depend on the almost complex structure J or on the contact form λ for ξ, and so defines a well-defined $\mathbb{Z}/2$-module $ECH_*(Y,\xi,\Gamma)$. In principle one should be able to prove this by counting holomorphic curves with ECH index zero, but there are unsolved technical problems with this approach which we will describe in Section 5.5. Currently the only way to prove the above invariance is using:

Theorem 1.6 (Taubes [58]). *If Y is connected, then there is a canonical isomorphism of relatively graded modules (with $\mathbb{Z}/2$ or $\mathbb{Z}$ coefficients)*

$$ECH_*(Y,\lambda,\Gamma,J)=\widehat{HM}^{-*}\big(Y,\mathfrak{s}_\xi+\mathrm{PD}(\Gamma)\big). \tag{1.11}$$

Here $\widehat{HM}^*$ denotes the "from" version of Seiberg-Witten Floer cohomology defined by Kronheimer-Mrowka [34], and $\mathfrak{s}_\xi$ denotes a spin-c structure determined by the oriented 2-plane field ξ, see Section 2.8. The relative grading is explained in Section 3.5. Kutluhan-Lee-Taubes [35] and Colin-Ghiggini-Honda [8] also showed that both sides of (1.11) are isomorphic to the Heegaard Floer homology $HF^+(-Y,\mathfrak{s}_\xi+\mathrm{PD}(\Gamma))$ defined in [46]. The upshot is that ECH is a topological invariant of Y, except that one needs to shift Γ if one changes the contact structure.

Remark 1.7. In fact, both Seiberg-Witten Floer cohomology and ECH have absolute gradings by homotopy classes of oriented two-plane fields [22, 34], and Taubes's isomorphism (1.11) respects these absolute gradings [9]. Thus one can write the isomorphism (1.11) as $ECH_{\mathfrak{p}}(Y,\lambda,J)=\widehat{HM}^{\mathfrak{p}}(Y)$ where $\mathfrak{p}$ denotes a homotopy class of oriented two-plane fields on Y.

Although ECH does not depend on the contact form, because it is defined using the contact form it has applications to contact geometry. For example, Theorem 1.6, together with known properties of Seiberg-Witten Floer cohomology, implies the three-dimensional *Weinstein conjecture*: every contact form on a closed connected three-manifold has at least one Reeb orbit. Indeed, Taubes's proof of the Weinstein conjecture in [57] can be regarded as a first step towards proving Theorem 1.6.

The reason that Theorem 1.6 implies the Weinstein conjecture is that if there is no closed orbit, then λ is nondegenerate and

$$ECH_*(Y,\xi,\Gamma)=\begin{cases}\mathbb{Z}/2, & \Gamma=0,\\ 0, & \Gamma\neq 0.\end{cases}$$

Here the $\mathbb{Z}/2$ comes from the empty set of Reeb orbits, which is a legitimate chain complex generator when $\Gamma=0$. However results of Kronheimer-Mrowka [34] imply that if $c_1(\xi)+2\,\mathrm{PD}(\Gamma)\in H^2(Y;\mathbb{Z})$ is torsion (and by a little algebraic topology one can always find a class $\Gamma\in H_1(Y)$ with this property), then $\widehat{HM}^*(Y,\mathfrak{s}_\xi+\Gamma)$ is infinitely generated, which is a contradiction.

Note that although $ECH(Y,\xi,\Gamma)$ is infinitely generated for Γ as above, there might not exist infinitely many embedded Reeb orbits. To give a counterexample, first recall that in any symplectic manifold (M,ω), a *Liouville vector field* is a vector field ρ such that $\mathcal{L}_\rho\omega=\omega$. A hypersurface $Y\subset M$ is of *contact type* if there exists a Liouville vector field ρ transverse to Y defined in a neighborhood of Y. In this case the "Liouville form" $\imath_\rho\omega$ restricts to a contact form on Y, whose Reeb vector field is parallel to the Hamilonian vector field X_H where $H:M\to\mathbb{R}$ is any smooth function having Y as a regular level set. For example, the radial vector field

$$\rho=\frac{1}{2}\sum_{k=1}^{2}\left(x_k\frac{\partial}{\partial x_j}+y_k\frac{\partial}{\partial y_k}\right)$$

is a Liouville vector field defined on all of $\mathbb{R}^4$. It follows that if Y is a hypersurface in $\mathbb{R}^4$ which is "star-shaped", meaning transverse to the radial vector field ρ, then the Liouville form

$$\lambda=\frac{1}{2}\sum_{k=1}^{2}(x_k dy_k-y_k dx_k) \tag{1.12}$$

restricts to a contact form on Y, with Reeb vector field determined as above.

Example 1.8. If $Y = \partial E(a,b)$ is the boundary of an ellipsoid, then it follows from the above discussion that the Liouville form λ in (1.12) restricts to a contact form on Y, whose Reeb vector field is given in polar coordinates by

$$R = \frac{2\pi}{a}\frac{\partial}{\partial\theta_1} + \frac{2\pi}{b}\frac{\partial}{\partial\theta_2}.$$

If a/b is irrational, then there are just two embedded Reeb orbits, which we denote by $\gamma_1 = (z_2 = 0)$ and $\gamma_2 = (z_1 = 0)$. The linearized return map P_{γ_1} is rotation by $2\pi a/b$, and the linearized return map P_{γ_2} is rotation by $2\pi b/a$, so both of these Reeb orbits are elliptic. A generator of the ECH chain complex then has the form $\gamma_1^{m_1}\gamma_2^{m_2}$, where this notation indicates the set consisting of the pair (γ_1, m_1) (if $m_1 \neq 0$) and the pair (γ_2, m_2) (if $m_2 \neq 0$). For grading reasons to be explained in Section 3.7, the differential ∂ is identically zero. Thus $ECH(\partial E(a,b), \lambda, 0)$ has one generator for each pair of nonnegative integers.

By making stronger use of the isomorphism (1.11), one can prove some slight refinements of the Weinstein conjecture. For example, there are always at least two embedded Reeb orbits [10]; and if λ is nondegenerate and Y is not a sphere or a lens space then there at least three embedded Reeb orbits [31]. To put this in perspective, Colin-Honda [6] used linearized contact homology to show that for many contact structures, every contact form has infinitely many embedded Reeb orbits. The only examples of closed contact three-manifolds we know of with only finitely many embedded Reeb orbits are the ellipsoid examples in Example 1.8, and quotients of these on lens spaces, with exactly two embedded Reeb orbits.

Historical Note. The original motivation for the definition of ECH was to find a symplectic model for Seiberg-Witten Floer homology, so that an isomorphism of the form (1.11) would hold[2], analogously to Taubes's Seiberg-Witten = Gromov theorem for closed symplectic four-manifolds. We will explain this motivation in detail in Sections 2–3.

[2]More precisely, we first defined an analogous theory for mapping tori of symplectomorphisms of surfaces, called *periodic Floer homology*, and conjectured that this was isomorphic to Seiberg-Witten Floer homology, see [21, §1.1]. This conjecture was later proved by Lee and Taubes [38]. Initially it was not clear if ECH would also be isomorphic to Seiberg-Witten Floer homology because the geometry of contact manifolds is slightly different than that of mapping tori. However the calculation of the ECH of T^3 then provided nontrivial evidence that this is the case, see [28, §1.1].

1.4. Additional Structure on ECH

The definition of ECH capacities uses four additional structures on ECH, which we now briefly describe.

1. The U map. Assuming that Y is connected, there is a degree -2 map

$$U : ECH_*(Y,\xi,\Gamma) \longrightarrow ECH_{*-2}(Y,\xi,\Gamma). \tag{1.13}$$

This is induced by a chain map which is defined similarly to the differential, except that instead of counting ECH index 1 curves modulo $\mathbb{R}$ translation, it counts ECH index 2 curves that pass through a base point $(0,z) \in \mathbb{R} \times Y$. Since Y is connected, the induced map on homology

$$U : ECH_*(Y,\lambda,\Gamma,J) \longrightarrow ECH_{*-2}(Y,\lambda,\Gamma,J) \tag{1.14}$$

does not depend on the choice of base point z, see Section 3.8 for details. Taubes [59] showed that (1.14) agrees with an analogous U map on Seiberg-Witten Floer cohomology, and in particular it gives a well-defined map (1.13). Thus the U map, like ECH, is in fact a topological invariant of Y.

If Y is disconnected, then there is a different U map for each component of Y. More precisely, suppose that $(Y,\lambda) = \coprod_{i=1}^n (Y_i,\lambda_i)$ with Y_i connected, and let $\Gamma = (\Gamma_1,\ldots,\Gamma_n) \in H_1(Y)$. It follows from the definitions, and the fact that we are using coefficients in a field, that there is a canonical isomorphism

$$ECH(Y,\xi,\Gamma) = \bigotimes_{i=1}^n ECH(Y_i,\xi_i,\Gamma_i).$$

The U map on the left hand side determined by the component Y_i is the tensor product on the right hand side of the U map on $ECH(Y_i,\xi_i,\Gamma_i)$ with the identity on the other factors.

2. The ECH contact invariant. ECH contains a canonical class defined as follows. Observe that for any nondegenerate contact three-manifold (Y,λ), the empty set of Reeb orbits is a generator of the chain complex $ECC(Y,\lambda,0,J)$. It follows from (1.15) below that this chain complex generator is actually a cycle, i.e.

$$\partial\emptyset = 0.$$

(In this equation, the empty set is not the same as zero!) ECH cobordism maps, described below, can be used to show that the homology class of this cycle does not depend on J or λ, and thus represents a well-defined class

$$c(\xi) \in ECH_*(Y,\xi,0),$$

which we call the *ECH contact invariant*. Taubes [59] showed that under the isomorphism (1.11), this agrees with a related contact invariant in Seiberg-Witten Floer cohomology.

Although ECH and the U map on it are topological invariants of the three-manifold Y, the contact invariant can distinguish some contact structures. For example, if ξ is overtwisted then $c(\xi)=0$. This holds because, as shown in the appendix to [63], if ξ is overtwisted then one can find a contact form such that the shortest Reeb orbit γ bounds a unique holomorphic curve (which is a holomorphic plane) in $\mathbb{R}\times Y$; the latter turns out to have ECH index 1, so $\partial\gamma=\emptyset$. On the other hand, it follows using the ECH cobordism maps defined in [26] that $c(\xi)\neq 0$ whenever (Y,ξ) is strongly symplectically fillable; a special case of this is proved in Example 1.10 below.

3. Filtered ECH. There is a refinement of ECH which sees not just the contact *structure* but also the contact *form*. To describe this, recall that if γ is a Reeb orbit, its *symplectic action* is defined by

$$\mathcal{A}(\gamma)=\int_\gamma \lambda.$$

If $\alpha=\{(\alpha_i,m_i)\}$ is an ECH generator, define its symplectic action by

$$\mathcal{A}(\alpha)=\sum_i m_i\mathcal{A}(\alpha_i).$$

It follows from the conditions on the almost complex structure J that the restriction of $d\lambda$ to any J-holomorphic curve in $\mathbb{R}\times Y$ is pointwise nonnegative. Consequently, by Stokes's theorem, the differential decreases[3] the symplectic action, i.e.

$$\langle\partial\alpha,\beta\rangle\neq 0 \Longrightarrow \mathcal{A}(\alpha)\geq\mathcal{A}(\beta). \tag{1.15}$$

Given $L\in\mathbb{R}$, define $ECC^L(Y,\lambda,\Gamma,J)$ to be the span of those generators α with $\mathcal{A}(\alpha)<L$. It follows from (1.15) that this is a subcomplex of $ECC(Y,\lambda,\Gamma,J)$. The homology of this subcomplex is called *filtered ECH*

[3] In fact the inequality on the right side of (1.15) is strict, but we do not need this.

and denoted by $ECH^L(Y,\lambda,\Gamma)$. It is shown in [32, Thm. 1.3] that filtered ECH does not depend on J. There is also a U map (or U maps when Y is disconnected) defined on filtered ECH, which we continue to denote by U.

Unlike the usual ECH, filtered ECH depends heavily on the contact form λ. For example, if $Y=\partial E(a,b)$ with the standard contact form as in Example 1.8, then the symplectic action of a chain complex generator is given by

$$\mathcal{A}\left(\gamma_1^{m_1}\gamma_2^{m_2}\right)=am_1+bm_2. \tag{1.16}$$

Thus the rank of $ECH^L(\partial E(a,b))$ is the number of nonnegative integer linear combinations of a and b that are less than L. Obviously this depends on a and b; but the ellipsoids for different a and b with their contact forms all determine the unique tight contact structure on S^3. There is also a general scaling property: if $r>0$ is a positive constant, then there is a canonical isomorphism

$$ECH^L(Y,\lambda,\Gamma)=ECH^{rL}(Y,r\lambda,\Gamma). \tag{1.17}$$

4. Cobordism maps. We now consider maps on ECH induced by cobordisms. For this purpose there are various kinds of cobordisms that one can consider. To describe these, let (Y_+,λ_+) and (Y_-,λ_-) be closed contact three-manifolds.

A *strong symplectic cobordism* from[4] (Y_+,λ_+) to (Y_-,λ_-) is a compact symplectic four-manifold (X,ω) with boundary

$$\partial X=Y_+-Y_-, \tag{1.18}$$

such that $\omega|_{Y_\pm}=d\lambda_\pm$. Note that the signs in (1.18) are important; here X has an orientation determined by the symplectic structure, while Y_+ and Y_- have orientations determined by the contact structures. In particular, there is a distinction between the positive (or "convex") boundary Y_+ and the negative (or "concave") boundary Y_-.

An *exact symplectic cobordism* is a strong symplectic cobordism as above such that there is a 1-form λ on X with $d\lambda=\omega$ and $\lambda|_{Y_\pm}=\lambda_\pm$.

[4]Our use of the words "from" and "to" in this connection is controversial. In the usual TQFT language, one would say that X is a cobordism from Y_- to Y_+. However cobordism maps on ECH and other kinds of contact homology naturally go from the invariant of Y_+ to the invariant of Y_-. We apologize for any confusion.

A strong (resp. exact) *symplectic filling* of (Y,λ) is a strong (resp. exact) symplectic cobordism from (Y,λ) to the empty set.

For example, if X is a compact star-shaped domain in $\mathbb{R}^4$ with boundary Y, if ω is the standard symplectic form on $\mathbb{R}^4$, and if λ is the Liouville form (1.12), then (X,ω) is an exact symplectic filling of $(Y,\lambda|_Y)$.

Maps on ECH induced by exact symplectic cobordisms were constructed in [32], where they were used to prove the Arnold chord conjecture in three dimensions. More generally, maps on ECH induced by arbitrary strong symplectic cobordisms are constructed in [26].

To set up the theory of ECH capacities, we need a notion in between exact and strong symplectic cobordisms. Define a *weakly exact symplectic cobordism* to be a strong symplectic cobordism as above such that ω is exact (but ω need not have a primitive on X which restricts to the contact forms on the boundary).

Theorem 1.9 ([24, Thm. 2.3]). *Let (X,ω) be a weakly exact symplectic cobordism from (Y_+,λ_+) to (Y_-,λ_-), and assume that the contact forms $\lambda_\pm$ are nondegenerate. Then for each $L>0$ there are maps*

$$\Phi^L(X,\omega): ECH^L(Y_+,\lambda_+,0) \longrightarrow ECH^L(Y_-,\lambda_-,0)$$

with the following properties:

(a) $\phi^L(X,\omega)[\emptyset]=[\emptyset]$.

(b) *If U_+ and U_- are U maps on $ECH^L(Y_\pm,\lambda_\pm,0)$ corresponding to components of $Y_\pm$ that are contained in the same component of X, then*

$$\phi^L(X,\omega)\circ U_+ = U_- \circ \phi^L(X,\omega).$$

Example 1.10. If $Y_-=\emptyset$, i.e. if (X,ω) is a weakly exact symplectic filling of (Y_+,λ_+), then the content of the theorem is that there are maps

$$\Phi^L(X,\omega): ECH^L(Y_+,\lambda_+,0) \longrightarrow \mathbb{Z}/2$$

with $\Phi^L(X,\omega)[\emptyset]=1$. In particular, it follows that $c(\xi_+)\neq 0 \in ECH(Y_+,\xi_+,0)$.

Theorem 1.9 is proved using Seiberg-Witten theory, as we describe in Section 5.5. For now let us see how the above structure can be used to define ECH capacities.

1.5. Definition of ECH Capacities

Before defining ECH capacities of symplectic four-manifolds, we first need another three-dimensional definition.

ECH Spectrum. Let (Y,λ) be a closed contact three-manifold, write $\xi = \mathrm{Ker}(\lambda)$ as usual, and assume that $c(\xi) \neq 0 \in ECH(Y,\xi,0)$. We define a sequence of real numbers

$$0 = c_0(Y,\lambda) < c_1(Y,\lambda) \le c_2(Y,\lambda) \le \cdots \le \infty,$$

called the *ECH spectrum* of (Y,λ), as follows.

Suppose first that λ is nondegenerate and Y is connected. Then $c_k(Y,\lambda)$ is the infimum of L such that there is a class $\eta \in ECH^L(Y,\lambda,0)$ with $U^k\eta = [\emptyset]$. If no such class exists then $c_k(Y,\lambda) = \infty$. In particular, $c_k(Y,\lambda) < \infty$ if and only if $c(\xi)$ is in the image of U^k on $ECH(Y,\xi,0)$.

Example 1.11. Suppose $Y = \partial E(a,b)$ with a/b irrational. Denote the chain complex generators in order of increasing symplectic action by $\zeta_0, \zeta_1, \ldots$. We will see in Section 4.1 that $U\zeta_k = \zeta_{k-1}$ for $k > 0$. It follows from this and (1.16) that

$$c_k\big(\partial E(a,b)\big) = N(a,b)_k. \tag{1.19}$$

Continuing the definition of the ECH spectrum, if $(Y,\lambda) = \coprod_{i=1}^n (Y_i,\lambda_i)$ with Y_i connected and λ_i nondegenerate, let U_i denote the U map corresponding to the ith component. Then $c_k(Y,\lambda)$ is the infimum of L such that there exists a class $\eta \in ECH^L(Y,\lambda,0)$ with

$$U_1^{k_1} \circ \cdots \circ U_n^{k_n}\eta = [\emptyset] \tag{1.20}$$

whenever $k_1 + \cdots + k_n = k$. It follows from some algebra in [24, §5] that

$$c_k\left(\coprod_{i=1}^n (Y_i,\lambda_i)\right) = \max_{k_1+\cdots+k_n=k} \sum_{i=1}^n c_{k_i}(Y_i,\lambda_i). \tag{1.21}$$

Finally, if (Y,λ) is a closed contact three-manifold with λ possibly degenerate, define $c_k(Y,\lambda) = \lim_{n\to\infty} c_k(Y,f_n\lambda)$, where $f_n : Y \to \mathbb{R}^{>0}$ are functions on Y with $f_n\lambda$ nondegenerate and $\lim_{n\to\infty} f_n = 1$ in the C^0 topology. It can be shown using Theorem 1.9 that this is well-defined and still satisfies (1.21). For example, Equation (1.19) also holds when a/b is rational.

ECH Capacities. We are now ready to define ECH capacities.

Definition 1.12. A (four-dimensional) *Liouville domain* is a weakly[5] exact symplectic filling (X, ω) of a contact three-manifold (Y, λ).

Definition 1.13. If (X, ω) is a four-dimensional Liouville domain with boundary (Y, λ), define the *ECH capacities* of (X, ω) by

$$c_k(X, \omega) = c_k(Y, \lambda) \in [0, \infty].$$

To see why this definition makes sense, first note that $c(\xi) \neq 0 \in ECH(Y, \xi, 0)$ by Example 1.10, so $c_k(Y, \lambda)$ is defined. We also need to explain why $c_k(X, \omega)$ does not depend on the choice of contact form λ on Y with $d\lambda = \omega|_Y$. Let λ' be another such contact form. Assume that λ and λ' are nondegenerate (one can handle the degenerate case by taking a limit of nondegenerate forms). Since $d\lambda = d\lambda'$, the Reeb vector fields R and R' for λ and λ' are related by $R' = fR$ where $f : Y \to \mathbb{R}^{>0}$. Let J be an almost complex structure on $\mathbb{R} \times Y$ as needed to define the ECH of λ. Let J' be the almost complex structure on $\mathbb{R} \times Y$ which agrees with J on the contact planes ξ but sends $\partial_s \mapsto R'$. There is then a canonical isomorphism of chain complexes

$$ECC^L(Y, \lambda, 0, J) = ECC^L(Y, \lambda', 0, J') \tag{1.22}$$

which preserves the U maps and the empty set. The reason is that the chain complexes $ECC(Y, \lambda, \Gamma, J)$ and $ECC(Y, \lambda', \Gamma, J')$ have the same generators, and when $\Gamma = 0$ the symplectic actions as defined using λ or λ' agree by Stokes's theorem because $d\lambda = d\lambda'$. Furthermore the J-holomorphic curves in $\mathbb{R} \times Y$ agree with the J'-holomorphic curves after rescaling the $\mathbb{R}$ coordinate on $\mathbb{R} \times Y$ using the function f. And it follows immediately from (1.22) that $c_k(Y, \lambda) = c_k(Y, \lambda')$.

For example, the Ellipsoid property of ECH capacities now follows from (1.19).

Monotonicity for Liouville Domains. We now explain why the Monotonicity property holds when (X_0, ω_0) and (X_1, ω_1) are Liouville domains. By

[5]Our definition of "Liouville domain" is more general than the usual definition, and perhaps should be called a "weak Liouville domain". Ordinarily a "Liouville domain" is an exact symplectic filling.

a limiting argument, one can assume that (X_0, ω_0) symplectically embeds into the interior of (X_1, ω_1). For $i = 0, 1$, let $Y_i = \partial X_i$, and let λ_i be a contact form on Y_i with $\partial\lambda_i = \omega|_{Y_i}$. Then $(X \setminus \varphi(\mathrm{int}(X_0)), \omega_1)$ is a weakly exact symplectic cobordism from (Y_1, λ_1) to (Y_0, λ_0). The Monotonicity property in this case now follows from:

Lemma 1.14. *Let (X, ω) be a weakly exact symplectic cobordism from (Y_+, λ_+) to (Y_-, λ_-). Then*

$$c_k(Y_-, \lambda_-) \le c_k(Y_+, \lambda_+)$$

for each $k \ge 0$.

This lemma follows almost immediately from the fact that c_k is defined solely in terms of the filtration, the U maps, and the contact invariant, and these structures are preserved by the cobordism map. Here are the details.

Proof. By a limiting argument we may assume that the contact forms $\lambda_\pm$ are nondegenerate. Let $U_1^+, \ldots, U_m^+$ denote the U maps on $ECH(Y_+, \lambda_+, 0)$ associated to the components of Y_+, and let $U_1^-, \ldots, U_n^-$ denote the U maps on $ECH(Y_-, \lambda_-, 0)$ associated to the components of Y_-. Let $L > 0$ and suppose that $c_k(Y_+, \lambda) < L$; it is enough to show that $c_k(Y_-, \lambda_-) \le L$. Since $c_k(Y_+, \lambda) < L$, there exists a class $\eta_+ \in ECH^L(Y_+, \lambda_+, 0)$ such that

$$(U_1^+)^{k_1} \cdots (U_m^+)^{k_m} \eta_+ = [\emptyset] \tag{1.23}$$

whenever $k_1 + \cdots + k_m = k$.

Let

$$\eta_- = \Phi^L(X, \omega)\eta_+ \in ECH^L(Y_-, \lambda_-, 0).$$

We claim that

$$(U_1^-)^{k_1} \cdots (U_n^-)^{k_n} \eta_- = [\emptyset] \tag{1.24}$$

whenever $k_1 + \cdots + k_n = k$, so that $c_k(Y_-, \lambda_-) \le L$. To prove this, first note that by Exercise 1.15 below, each component of Y_- is contained in the same component of X as some component of Y_+. Equation (1.24) then follows from Equation (1.23) together with Theorem 1.9. ■

Exercise 1.15. Show that if (X, ω) is a weakly exact symplectic cobordism from (Y_+, λ_+) to (Y_-, λ_-) with $Y_- \ne \emptyset$, then $Y_+ \ne \emptyset$. (See answer in Appendix.)

Non-Liouville Domains. We extend the definition of ECH capacities to symplectic four-manifolds which are not Liouville domains by a simple trick: If (X,ω) is any symplectic four-manifold, define

$$c_k(X,\omega) = \sup\{c_k(X',\omega')\},$$

where the supremum is over Liouville domains (X',ω') that can be symplectically embedded into X. It is a tautology that this new definition of c_k is monotone with respect to symplectic embeddings. And this new definition agrees with the old one when (X,ω) is already a Liouville domain, by the Monotonicity property for the old definition of c_k with respect to symplectic embeddings of Liouville domains.

Properties of ECH Capacities. The remaining properties of ECH capacities in Theorem 1.3 are proved as follows. The Disjoint Union property follows from (1.21). The Conformality property follows from the definitions and the scaling property (1.17) when $r > 0$, and a similar argument[6] when $r < 0$. We will prove the Polydisk property at the end of Section 4.3. The proof of the Volume property is beyond the scope of these notes; it is given in [11], using ingredients from Taubes's proof of the Weinstein conjecture [57].

2. Origins of ECH

One of the main goals of these notes is to explain something about where ECH comes from. The starting point for the definition of ECH is Taubes's "SW=Gr" theorem [55] asserting that the Seiberg-Witten invariants of a symplectic four-manifold agree with a "Gromov invariant" counting holomorphic curves. The basic idea of ECH is that it is a three-dimensional analogue of Taubes's Gromov invariant. So we will now review Taubes's Gromov invariant in such a way as to make the definition of ECH appear as natural as possible. The impatient reader may wish to skip ahead to the definition of ECH in Section 3, and refer back to this section when more motivation is needed.

[6]In particular, there is a canonical isomorphism of chain complexes (with $\mathbb{Z}/2$ coefficients)

$$ECC^L(Y,\lambda,\Gamma,J) = ECC^L(Y,-\lambda,-\Gamma,-J).$$

Note that the resulting isomorphism $ECH(Y,\xi,\Gamma) = ECH(Y,-\xi,-\Gamma)$ corresponds, under Taubes's isomorphism (1.11), to "charge conjugation invariance" of Seiberg-Witten Floer cohomology (with $\mathbb{Z}/2$ coefficients).

2.1. Taubes's "SW=Gr" Theorem

We first briefly recall the statement of Taubes's "SW=Gr" theorem. Let X be a closed connected oriented four-manifold. (All manifolds in these notes are smooth.) Let $b_2^+(X)$ denote the dimension of a maximal positive definite subspace $H_2^+(X;\mathbb{R})$ of $H_2(X;\mathbb{R})$ with respect to the intersection pairing. Let $\mathrm{Spin}^c(X)$ denote the set of spin-c structures[7] on X; this is an affine space over $H^2(X;\mathbb{Z})$. If $b_2^+(X) > 1$, one can define the *Seiberg-Witten invariant*

$$SW(X) : \mathrm{Spin}^c(X) \to \mathbb{Z} \tag{2.1}$$

by counting solutions to the Seiberg-Witten equations, see e.g. [45]. More precisely, the Seiberg-Witten invariant depends on a choice of "homology orientation" of X, namely an orientation of $H_0(X;\mathbb{R}) \oplus H_1(X;\mathbb{R}) \oplus H_2^+(X;\mathbb{R})$. Switching the homology orientation will multiply the Seiberg-Witten invariant by -1. If $b_2^+(X) = 1$, the Seiberg-Witten invariant (2.1) is still defined, but depends on an additional choice of one of two possible "chambers"; one can identify a chamber with an orientation of the line $H_2^+(X;\mathbb{R})$.

While the Seiberg-Witten invariants are very powerful for distinguishing smooth four-manifolds, it is also nearly impossible to compute them directly except in very special cases (although there are axiomatic properties which one can use to compute the invariants for more interesting examples). However, Taubes showed that if X has a symplectic form ω, then the Seiberg-Witten invariants of X are equal to a certain count of holomorphic curves, which are much easier to understand than solutions to the Seiberg-Witten equations. Namely, for each $A \in H_2(X)$, Taubes defines a "Gromov invariant"

$$Gr(X, \omega, A) \in \mathbb{Z},$$

which is a certain count of holomorphic curves in the homology class A, which we will review in Section 2.5 below. Further, the symplectic structure ω determines a distinguished spin-c structure $\mathfrak{s}_\omega$, so that we can identify

$$\begin{aligned} H_2(X) &= \mathrm{Spin}^c(X), \\ A &\leftrightarrow \mathfrak{s}_\omega + \mathrm{PD}(A). \end{aligned} \tag{2.2}$$

[7] A spin-c structure on an oriented n-manifold is a lift of the frame bundle from a principal $SO(n)$ bundle to a principal $\mathrm{Spin}^c(n) = \mathrm{Spin}(n) \times_{\mathbb{Z}/2} U(1)$ bundle. However we will not need this here.

We can now state:

Theorem 2.1 (Taubes). *Let (X,ω) be a closed connected symplectic four-manifold with $b_2^+(X) > 1$. Then X has a homology orientation such that under the identification* (2.2),

$$SW(X) = Gr(X,\omega,\cdot).$$

Remark 2.2. A version of this theorem also holds when $b_2^+(X) = 1$. Here one needs to compute the Seiberg-Witten invariant using the chamber determined by the cohomology class of ω. Also, in this case the definition of the Gromov invariant needs to be modified in the presence of symplectic embedded spheres of square -1, see [39].

2.2. Holomorphic Curves in Symplectic Manifolds

We now briefly review what we will need to know about holomorphic curves in order to define Taubes's Gromov invariant. Proofs of the facts recalled here may be found for example in [42].

Let (X^{2n},ω) be a closed symplectic manifold. An *ω-compatible almost complex structure* is a bundle map $J : TX \to TX$ such that $J^2 = -1$ and $g(v,w) = \langle Jv,w\rangle$ defines a Riemannian metric on X. Given ω, the space of compatible almost complex structures J is contractible. Fix an ω-compatible[8] almost complex structure J.

A *J-holomorphic curve* in (X,ω) is a holomorphic map $u : (\Sigma,j) \to (X,J)$ where (Σ,j) is a compact Riemann surface (i.e. Σ is a compact surface and j is an almost complex structure on Σ), $u : \Sigma \to X$ is a smooth map, and $J \circ du = du \circ j$. The curve u is considered equivalent to $u' : (\Sigma',j') \to (X,J)$ if there exists a holomorphic bijection $\phi : (\Sigma,j) \to (\Sigma',j')$ such that $u' \circ \phi = u$. Thus a J-holomorphic curve is formally an equivalence class of triples (Σ,j,u) satisfying the above conditions.

We call a J-holomorphic curve *irreducible* if its domain is connected.

If $u : (\Sigma,j) \to (X,J)$ is an embedding, then the equivalence class of the J-holomorphic curve u is determined by its image $C = u(\Sigma)$ in X. Indeed,

[8] Taubes's theorem presumably still works if one generalizes from compatible to tame almost complex structures.

an embedded J-holomorphic curve is equivalent to a closed two-dimensional submanifold $C\subset X$ such that $J(TC)=TC$.

More generally, a holomorphic curve $u:\Sigma\to X$ is called *somewhere injective* if there exists $z\in\Sigma$ such that $u^{-1}(u(z))=\{z\}$ and $du_z:T_z\Sigma\to T_{u(z)}X$ is injective. One can show that in this case u is an embedding on the complement of a countable subset of Σ (which is finite in the case of interest where $n=2$), and the equivalence class of u is still determined by its image in X. On the other hand, u is called *multiply covered* if there exists a branched cover $\phi:(\Sigma,j)\to(\Sigma',j')$ of degree $d>1$ and a holomorphic map $u':(\Sigma',j')\to(X,J)$ such that $u=u'\circ\phi$.

It is a basic fact that every irreducible holomorphic curve is either somewhere injective or multiply covered. In particular, every irreducible holomorphic curve is the composition of a somewhere injective holomorphic curve with a branched cover of degree $d\geq 1$. When $d>1$, the holomorphic curve is not determined just by its image in X; it depends also on the degree d, the images of the branch points in X, and the monodromy around the branch points.

Define the *Fredholm index* of a holomorphic curve $u:(\Sigma,j)\to(X,J)$ by

$$\mathrm{ind}(u)=(n-3)\chi(\Sigma)+2\langle c_1(TX),u_*[\Sigma]\rangle. \tag{2.3}$$

Here $c_1(TX)$ denotes the first Chern class of TX, regarded as a complex vector bundle using the almost complex structure J. The isomorphism class of this complex vector bundle depends only on the symplectic structure and not on the compatible almost complex structure.

A transversality argument shows that if J is generic, then for each somewhere injective holomorphic curve u, the moduli space of holomorphic curves near u is a smooth manifold of dimension $\mathrm{ind}(u)$, cut out transversely in a sense to be described below. Unfortunately, this usually does not hold for multiply covered curves. Even if all somewhere injective holomorphic curves are cut out transversely, there can still be multiply covered holomorphic curves u such that $\mathrm{ind}(u)$ is less than the dimension of the moduli space near u, or even negative. This is a major technical problem in defining holomorphic curve counting invariants in general, and it also causes some complications for ECH, as we will see in the proof that $\partial^2=0$ in Section 5.4 and especially in the construction of cobordism maps in Section 5.5.

2.3. Deformations of Holomorphic Curves

We now clarify what it means for a holomorphic curve to be "cut out transversely". To simplify the discussion we restrict attention to immersed curves, which are all we need to consider to define Taubes's Gromov invariant.

Let $u : C \to X$ be an immersed J-holomorphic curve, which by abuse of notation we will usually denote by C. Then C has a well defined normal bundle N_C, which is a complex vector bundle of rank $n-1$ over C. The derivative of the equation for C to be J-holomorphic defines a first-order elliptic differential operator

$$D_C : \Gamma(N_C) \longrightarrow \Gamma\big(T^{0,1}C \otimes N_C\big),$$

which we call the *deformation operator* of C. Here Γ denotes the space of smooth sections.

To explain this operator in more detail, we first recall some general formalism. Suppose $E \to B$ is a smooth vector bundle and $\psi : B \to E$ is a smooth section. Let $x \in B$ be a zero of ψ. Then the derivative of the section ψ at x defines a canonical map

$$\nabla\psi : T_xB \to E_x. \tag{2.4}$$

Namely, the derivative of ψ, regarded as a smooth map $B \to E$, has a differential $d\psi_x : T_xB \to T_{(x,0)}E$, and the map (2.4) is obtained by composing this with the projection $T_{(x,0)}E = T_xB \oplus E_x \to E_x$.

To put holomorphic curves into the above framework, let $\mathcal{B}$ be the infinite dimensional (Frechet) manifold of immersed compact surfaces in X. Given an immersed surface $u : C \to X$, let $N_C = u^*TX/TC$ denote the normal bundle to C, which is a rank $2n-2$ real vector bundle over C, and let $\pi_{N_C} : u^*TX \to N_C$ denote the quotient map. We can identify $T_C\mathcal{B} = \Gamma(N_C)$. There is an infinite dimensional vector bundle $\mathcal{E} \to \mathcal{B}$ whose fiber over C is the space of smooth bundle maps $TC \to N_C$. We define a smooth section $\overline{\partial} : \mathcal{B} \to \mathcal{E}$ by defining $\overline{\partial}(C) : TC \to N_C$ to be the map sending $v \mapsto \pi_{N_C}(Jv)$. Then C is J-holomorphic if and only if $\overline{\partial}(C) = 0$. In this case the derivative of $\overline{\partial}$ defines a map $\Gamma(N_C) \to \Gamma(T^*C \otimes N_C)$. Furthermore, since C is J-holomorphic, the values of this map anticommute with J, so it is actually an operator $\Gamma(N_C) \to \Gamma(T^{0,1}C \otimes N_C)$. This is the deformation operator D_C.

One can write the operator D_C in local coordinates as follows. Let $z = s + it$ be a local coordinate on C, use $id\overline{z}$ to locally trivialize $T^{0,1}C$, and choose a local trivialization of N_C over this coordinate neighborhood. With respect to these coordinates and trivializations, the operator D_C locally has the form

$$D_C = \partial_s + J\partial_t + M(s,t).$$

Here $M(s,t)$ is a real matrix of size $2n-2$ determined by the derivatives of J in the normal directions to C.

We say that C is *regular*, or "cut out transversely", if the operator D_C is surjective. In this case the moduli space of holomorphic curves is a manifold near C, and its tangent space at C is the kernel of D_C.

In the analysis one often needs to extend the operator D_C to suitable Banach space completions of the spaces of smooth sections, for example to extend it to an operator

$$D_C : L^2_1(C, N_C) \longrightarrow L^2(C, T^{0,1}C \otimes N_C). \tag{2.5}$$

Since D_C is elliptic, the extended operator is Fredholm, and its kernel consists of smooth sections. It follows from the Riemann-Roch theorem that the index of this Fredholm operator is the Fredholm index $\operatorname{ind}(C)$ defined in (2.3). This is why the moduli space of holomorphic curves near a regular curve C, under our simplifying assumption that C is immersed, has dimension $\operatorname{ind}(C)$.

2.4. Special Properties in Four Dimensions

In four dimensions, holomorphic curves enjoy three additional special properties which are important for our story. To state the first special property, if p is an isolated intersection point of surfaces S_1 and S_2 in X, let $m_p(S_1 \cap S_2) \in \mathbb{Z}$ denote the intersection multiplicity at p.

Intersection Positivity. *Let C_1 and C_2 be distinct irreducible somewhere injective J-holomorphic curves in a symplectic four-manifold. Then the intersection points of C_1 and C_2 are isolated; and for each $p \in C_1 \cap C_2$, the intersection multiplicity $m_p(C_1 \cap C_2) > 0$. Moreover, $m_p(C_1 \cap C_2) = 1$ if and only if C_1 and C_2 are embedded near p and intersect transversely at p.*

It is easy to see that if C_1 and C_2 are embedded near p and intersect transversely at p, so that $m_p(C_1 \cap C_2) = \pm 1$, then in fact $m_p(C_1 \cap C_2) = +1$, essentially because a complex vector space has a canonical orientation. The hard part of the theorem is to deal with the cases where C_1 and C_2 are not embedded near p or do not intersect transversely at p.

In particular, intersection positivity implies that the homological intersection number

$$[C_1]\cdot[C_2]=\sum_{p\in C_1\cap C_2} m_p(C_1\cap C_2)\geq 0,$$

with equality if and only if C_1 and C_2 are disjoint. Note that the assumption that C_1 and C_2 are distinct is crucial. A single holomorphic curve C can have $[C]\cdot[C]<0$; for example, the exceptional divisor in a blowup is a holomorphic sphere C of square -1. What intersection positivity implies in this case is that the exceptional divisor is the unique holomorphic curve in its homology class.

The second special property of holomorphic curves in four dimensions is the adjunction formula. To state it, define a *singularity* of a somewhere injective J-holomorphic curve C in a symplectic four-manifold X to be a point in X where C is not locally an embedding. A *node* is a singularity given by a transverse self-intersection whose inverse image in the domain of C consists of two points (where C is an immersion). Let $\chi(C)$ denote the Euler characteristic of the domain of C (which may be larger than the Euler characteristic of the image of C in X if there are singularities).

Adjunction Formula. *Let C be a somewhere injective J-holomorphic curve in a symplectic four-manifold (X,ω). Then the singularities of C are isolated, and*

$$\langle c_1(TX),[C]\rangle=\chi(C)+[C]\cdot[C]-2\delta(C) \tag{2.6}$$

where $\delta(C)$ is a count of the singularities of C with positive integer weights. Moreover, a singularity has weight 1 if and only if it is a node.

In particular, we have

$$\chi(C)+[C]\cdot[C]-\langle c_1(TX),[C]\rangle\geq 0, \tag{2.7}$$

with equality if and only if C is embedded.

Exercise 2.3. Prove the adjunction formula in the special case when C is immersed and the only singularities of C are nodes.

The third special property of holomorphic curves in four dimensions is a version of Gromov compactness using currents, which does not require any genus bound. The usual version of Gromov compactness asserts that a sequence of holomorphic curves of fixed genus with an upper bound on the symplectic area has a subsequence which converges in an appropriate sense to a holomorphic curve. In the connection with Seiberg-Witten theory, multiply covered holomorphic curves naturally arise, but the information about the branch points, and hence about the genus of their domains, is not relevant. To keep track of the relevant information, define a *holomorphic current* in X to be a finite set of pairs $\mathcal{C} = \{(C_i, d_i)\}$ where the C_i are distinct irreducible somewhere injective J-holomorphic curves, and the d_i are positive integers.

Gromov Compactness via Currents (Taubes, [54, Prop. 3.3]). *Let (X, ω) be a compact symplectic four-manifold, possibly with boundary, and let J be an ω-compatible almost complex structure. Let $\{\mathcal{C}_n\}_{n\geq 1}$ be a sequence of J-holomorphic currents (possibly with boundary in ∂X) such that $\int_{\mathcal{C}_n} \omega$ has an n-independent upper bound. Then there is a subsequence which converges as a current and as a point set to a J-holomorphic current $\mathcal{C} \subset X$ (possibly with boundary in ∂X).*

Here "convergence as a current" means that if σ is any 2-form then $\lim_{n\to\infty} \int_{\mathcal{C}_n} \sigma = \int_{\mathcal{C}} \sigma$. "Convergence as a point set" means that the corresponding subsets of X converge with respect to the metric on compact sets defined by

$$d(K_1, K_2) = \sup_{x_1 \in K_1} \inf_{x_2 \in K_2} d(x_1, x_2) + \sup_{x_2 \in K_2} \inf_{x_1 \in K_1} d(x_2, x_1).$$

2.5. Taubes's Gromov Invariant

We now have enough background in place to define Taubes's Gromov invariant. While the definition is a bit complicated, we will be able to compute examples in Section 2.6, and this is a useful warmup for the definition of ECH.

What to Count. Let (X^4, ω) be a closed connected symplectic four-manifold, and let $A \in H_2(X)$. We define the Gromov invariant $Gr(X, \omega, A) \in \mathbb{Z}$ as follows. Fix a generic ω-compatible almost complex structure J. The rough idea is to count J-holomorphic currents representing the homology class A in "maximum dimensional moduli spaces".

To explain the latter notion, define an integer

$$I(A) = \langle c_1(TX), A \rangle + A \cdot A. \tag{2.8}$$

In fact one can show that $I(A)$ is always even. The integer $I(A)$ is the closed four-manifold version of the ECH index, a crucial notion which we will introduce in Section 3.4. For now, the significance of the integer $I(A)$ is the following. Let C be a somewhere injective J-holomorphic curve. By (2.3), the Fredholm index of C is given by

$$\operatorname{ind}(C) = -\chi(C) + 2\langle c_1(TX), [C] \rangle. \tag{2.9}$$

It follows from this equation and the adjunction formula (2.6) that

$$\operatorname{ind}(C) = I([C]) - 2\delta(C). \tag{2.10}$$

That is, the maximum possible value of $\operatorname{ind}(C)$ for a somewhere injective holomorphic curve C with homology class $[C] = A$ is $I(A)$, which is attained exactly when C is embedded.

The Gromov invariant $Gr(X, \omega, A) \in \mathbb{Z}$ is now a count of "admissible" holomorphic currents in the homology class A. Here the homology class of a holomorphic current $\mathcal{C} = \{(C_i, d_i)\}$ is defined by

$$[\mathcal{C}] = \sum_i d_i [C_i] \in H_2(X).$$

Furthermore, the current $\mathcal{C}$ is called "admissible" if $d_i = 1$ whenever C_i is a sphere with $[C_i] \cdot [C_i] < 0$.

If $I(A) < 0$, then there are no admissible holomorphic currents in the homology class A as we will show in a moment, and we define $Gr(X, \omega, A) = 0$.

The most important case for our story is when $I(A) = 0$. The admissible holomorphic currents in this case are described by the following lemma.

Lemma 2.4. *Let $\mathcal{C} = \{(C_i, d_i)\}$ be an admissible holomorphic current with homology class $[\mathcal{C}] = A$. Then $I(A) \geq 0$. Moreover, if $I(A) = 0$, then the following hold:*

(a) *The holomorphic curves C_i are embedded and disjoint.*

(b) *$d_i = 1$ unless C_i is a torus with $[C_i] \cdot [C_i] = 0$.*

(c) $\operatorname{ind}(C_i) = I([C_i]) = 0$ *for each i.*

Proof. It follows directly from the definition of I that if $B_1, B_2 \in H_2(X)$ then

$$I(B_1+B_2)=I(B_1)+I(B_2)+2B_1\cdot B_2. \tag{2.11}$$

Applying this to $A=\sum_i d_i[C_i]$ gives

$$I(A)=\sum_i d_i I([C_i])+\sum_i (d_i^2-d_i)[C_i]\cdot[C_i]+\sum_{i\neq j}[C_i]\cdot[C_j]. \tag{2.12}$$

Now the terms on the right hand side are all nonnegative. To see this, first note that $\operatorname{ind}(C_i)\geq 0$, since we are assuming that J is generic so that C_i is regular. So by (2.10) we have $I([C_i])\geq 0$, with equality only if C_i is embedded. In addition, if we combine the inequality $\operatorname{ind}(C_i)\geq 0$ with the adjunction formula (2.7) for C_i, we find that

$$\chi(C_i)+2[C_i]\cdot[C_i]\geq 0 \tag{2.13}$$

with equality only if C_i is embedded. In particular, the only way that $[C_i]\cdot[C_i]$ can be negative is if C_i is an embedded sphere with square -1; and in this case admissibility forces $d_i=1$, so that the corresponding term in (2.12) is zero. Finally, we know by intersection positivity that $[C_i]\cdot[C_j]\geq 0$ with equality if and only if C_i and C_j are disjoint. We conclude that $I(A)\geq 0$, and if $I(A)=0$ then the curves C_i are embedded and disjoint, $\operatorname{ind}(C_i)=I([C_i])=0$, and $d_i>1$ only if C_i is a torus with square zero. (The inequality (2.13) also allows $[C_i]\cdot[C_i]=0$ when C_i is a sphere, but this would require $I([C_i])=2$ and so cannot happen here.) ■

One consequence of this lemma is that when $I(A)=0$, we have a finite set of holomorphic currents to count:

Lemma 2.5. *If $I(A)=0$, then the set of admissible holomorphic currents $\mathcal{C}$ with homology class $[\mathcal{C}]=A$ is finite.*

Proof. Suppose $\{\mathcal{C}_k\}_{k=1,2,\ldots}$ is an infinite sequence of distinct such currents. By Gromov compactness with currents, the sequence converges as a current and a point set to a holomorphic current $\mathcal{C}_\infty$. Convergence as a current implies that $[\mathcal{C}_\infty]=A$. An argument using the Fredholm index which we omit shows that $\mathcal{C}_\infty$ is also admissible. Then by Lemma 2.4, $\mathcal{C}_\infty=\{(C_i,d_i)\}$ where $\operatorname{ind}(C_i)=0$ for each i and $d_i=1$ unless C_i is a torus of square zero. We are assuming that J is generic, so each C_i is isolated in the moduli space of holomorphic curves. If every $d_i=1$, then one can use convergence as a current and a point set to show that possibly after passing to a subsequence, each $\mathcal{C}_k$ has an embedded component such that the sequence of these embedded components converges in the smooth topology to C_i, which is a contradiction. If any $d_i > 1$, one needs an additional lemma from [53] asserting that if J is generic, then the unbranched multiple covers of the tori of square zero are also regular. ■

How to Count. When $I(A) = 0$, we define $Gr(X, \omega, A) \in \mathbb{Z}$ to be the sum, over all admissible holomorphic currents $\mathcal{C} = \{(C_i, d_i)\}$ with homology class $[\mathcal{C}] = A$, of a weight $w(\mathcal{C}) \in \mathbb{Z}$ which we now define. The weight is given by a product of weights associated to the irreducible components,

$$w(\mathcal{C}) = \prod_i w(C_i, d_i).$$

To complete the definition, we need to define the integer $w(C, d)$ when C is an irreducible embedded holomorphic curve with $\text{ind} = 0$, and d is a positive integer (which is 1 unless C is a torus with square 0).

If $d = 1$, then $W(C, 1) = \varepsilon(C) \in \{\pm 1\}$ is defined as follows. Roughly speaking, $\varepsilon(C)$ is the sign of the determinant of the operator D_C, which is the sign of the spectral flow from D_C (extended as in (2.5)) to a complex linear operator. What this means is the following: one can show that there exists a differentiable 1-parameter family of operators $\{D_t\}_{t \in [0,1]}$ between the same spaces such that $D_0 = D_C$; the operator D_1 is complex linear; there are only finitely many t such that D_t is not invertible; and for each such t, the operator D_t has one-dimensional kernel, and the derivative of D_t defines an isomorphism from the kernel of D_t to the cokernel of D_t. Then $\varepsilon(C)$ is simply -1 to the number of such t. One can show that this is well-defined, and we will compute some examples in Section 2.6.

It remains to define the weights $w(C, d)$ when $d > 1$ and C is a torus of square zero. The torus C has three connected unbranched double covers, classified by nonzero elements of $H^1(C; \mathbb{Z}/2)$. By [53], if J is generic then the corresponding doubly covered holomorphic curves are regular. Each of these double covers then has a sign ε defined above. The weight $w(C, d)$ depends only on d, the sign of C, and the number of double covers with each sign. We denote this number by $f_{\pm,k}(d)$, where $\pm$ indicates the sign $\varepsilon(C)$, and $k \in \{0, 1, 2, 3\}$ is the number of double covers whose sign disagrees with that of C. To define the numbers $f_{\pm,k}(d)$, combine them into a generating function

$$f_{\pm,k} = 1 + \sum_{d \geq 1} f_{\pm,k}(d) t^d.$$

Then

$$
\begin{aligned}
f_{+,0} &= \frac{1}{1-t},\\
f_{+,1} &= 1+t,\\
f_{+,2} &= \frac{1+t}{1+t^2},\\
f_{+,3} &= \frac{(1+t)(1-t^2)}{1+t^2},\\
f_{-,k} &= \frac{1}{f_{+,k}}.
\end{aligned}
\tag{2.14}
$$

Where do these generating functions come from? It is shown in [53] that $Gr(X,\omega,A)$ is independent of the choice of J and invariant under deformation of the symplectic form ω; another proof is given in [33]. This invariance requires the generating functions $f_{\pm,k}$ to satisfy certain relations, because of bifurcations of holomorphic curves that can occur as one deforms J or ω. For example, it is possible for a pair of cancelling tori with opposite signs to be created or destroyed, and this forces the relation $f_{+,k}f_{-,k}=1$. We will see another relation in the example in Section 2.6. One still has some leeway in choosing the generating functions to obtain an invariant of symplectic four-manifolds; however the choice above is the one that agrees with Seiberg-Witten theory, for reasons we will explain in Section 2.7.

The Case $I(A)>0$. To define the Gromov invariant $Gr(X,\omega,A)$ when $I(A)\geq 0$, choose $I(A)/2$ generic points $x_1,\ldots,x_{I(A)/2}\in X$. Then $Gr(X,A)$ is a count of admissible holomorphic currents $\mathcal{C}$ in the homology class A that pass through all of the points $x_1,\ldots,x_{I(A)/2}$. We omit the details as this case is less important for motivating the definition of ECH, although it is related to the U map introduced in Section 1.4. The Gromov invariants for classes A with $I(A)>0$ are interesting when $b_2^+(X)=1$. However the "simple type conjecture" for Seiberg-Witten invariants implies that if $b_2^+(X)>1$ and $b_1(X)=0$, then $Gr(X,\omega,A)=0$ for all classes A with $I(A)>0$.

2.6. The Mapping Torus Example

We now compute Taubes's Gromov invariant for an interesting family of examples, namely mapping tori cross S^1, for S^1-invariant homology classes.

This example will indicate what the generators of the ECH chain complex should be.

Mapping Tori. Let (Σ,ω) be a closed connected symplectic two-manifold and let ϕ be a symplectomorphism from (Σ,ω) to itself. The *mapping torus* of ϕ is the three-manifold

$$Y_\phi = [0,1]\times\Sigma/\sim,$$
$$(1,x)\sim\big(0,\phi(x)\big).$$

The three-manifold Y_ϕ fibers over $S^1=\mathbb{R}/\mathbb{Z}$ with fiber Σ, and ω defines a symplectic form on each fiber. We denote the $[0,1]$ coordinate on $[0,1]\times\Sigma$ by t. The vector field ∂_t on $[0,1]\times\Sigma$ descends to a vector field on Y_ϕ, which we also denote by ∂_t. A fixed point of the map ϕ^p determines a periodic orbit of the vector field ∂_t of period p, and conversely a simple periodic orbit of ∂_t of period p determines p fixed points of ϕ^p.

The fiberwise symplectic form ω extends to a closed 2-form on Y_ϕ which annihilates ∂_t, and which we still denote by ω. We then define a symplectic form Ω on $S^1\times Y_\phi$ by

$$\Omega = ds\wedge dt+\omega \tag{2.15}$$

where s denotes the S^1 coordinate.

We will now calculate the Gromov invariant $Gr(S^1\times Y_\phi,\Omega,A)$, where

$$A=\big[S^1\big]\times\Gamma\in H_2\big(S^1\times Y_\phi\big)$$

for some $\Gamma\in H_1(Y_\phi)$. Observe to start that $I(A)=0$, so we just need to count holomorphic currents of the type described in Lemma 2.4.

Almost Complex Structure. Choose a fiberwise ω-compatible almost complex structure J on the fibers of $Y_\phi\to S^1$. That is, for each $t\in S^1=\mathbb{R}/\mathbb{Z}$, choose an almost complex structure J_t on the fiber over t, such that J_t varies smoothly with t. Note that compatibility here just means that J_t rotates positively with respect to the orientation on Σ.

The fiberwise almost complex structure extends to a unique almost complex structure J on $S^1\times Y_\phi$ such that

$$J\partial_s=\partial_t. \tag{2.16}$$

It is an exercise to check that J is Ω-compatible.

Holomorphic Curves. If $\gamma \subset Y_\phi$ is an embedded periodic orbit of ∂_t, then it follows from (2.16) that $S^1 \times \gamma \subset S^1 \times Y$ is an embedded J-holomorphic torus. These are all the holomorphic curves we need to consider, because of the following lemma.

Lemma 2.6. *If $\mathcal{C} = \{(C_i, d_i)\}$ is a J-holomorphic current in $S^1 \times Y_\phi$ with homology class $A = [S^1] \times \Gamma$, then each C_i is a torus $S^1 \times \gamma$ with γ a periodic orbit of ∂_t.*

Proof. We have $\langle A, [\omega]\rangle = 0$, because the class A is S^1-invariant while ω is pulled back via the projection to Y_ϕ. On the other hand, by the construction of J, the restriction of ω to any J-holomorphic curve C is pointwise nonnegative, with equality only where C is tangent to the span of ∂_s and ∂_t (or singular). Thus $\int_{C_i} \omega = 0$ for each i, and then each C_i is everywhere tangent to ∂_s and ∂_t. ■

Transversality and Nondegeneracy. We now determine when the holomorphic tori $S^1 \times \gamma$ are regular.

Let γ be a periodic orbit of period p, and let $x \in \Sigma$ be one of the corresponding fixed points of ϕ^p. The fixed point x of ϕ^p is called *nondegenerate* if the differential $d\phi^p_x : T_x\Sigma \to T_x\Sigma$ does not have 1 as an eigenvalue. In this case, the *Lefschetz sign* is the sign of $\det(1 - d\phi^p_x)$. Also, since the linear map $d\phi^p_x$ is symplectic, we can classify the fixed point x as elliptic, positive hyperbolic, or negative hyperbolic according to the eigenvalues of $d\phi^p_x$, just as we did for Reeb orbits in Section 1.3. In particular, the Lefschetz sign is $+1$ if the fixed point is elliptic or negative hyperbolic, and -1 if the fixed point is positive hyperbolic. We say that the periodic orbit γ is nondegenerate if the fixed point x is nondegenerate. All of the above conditions depend only on γ and not on the choice of corresponding fixed point x.

The following lemma tells us that if all periodic orbits γ are nondegenerate (which will be the case if ϕ is generic), then for any S^1-invariant J, all the J-holomorphic tori that we need to count are regular[9].

Lemma 2.7. *The J-holomorphic torus $C = S^1 \times \gamma$ is regular if and only if the periodic orbit γ is nondegenerate. In this case, the sign $\varepsilon(C)$ agrees with the Lefschetz sign.*

[9]This is very lucky; in other S^1-invariant situations, obtaining transversality for S^1-invariant J may not be possible. See e.g. [15, 16] for examples of this difficulty and ways to deal with it.

Proof. Since the deformation operator

$$D_C : \Gamma(N_C) \longrightarrow \Gamma\left(T^{0,1}C \otimes N_C\right)$$

has index zero, C is regular if and only if $\operatorname{Ker}(D_C) = \{0\}$.

To determine $\operatorname{Ker}(D_C)$, we need to understand the deformation operator D_C more explicitly. To start, identify N_C with the pullback of the normal bundle to γ in Y_ϕ. The latter can be identified with $T^{vert}Y_\phi|_\gamma$, where $T^{vert}Y_\phi$ denotes the vertical tangent bundle of the fiber bundle $Y_\phi \to S^1$. The linearization of the flow ∂_t along γ defines a connection ∇ on the bundle $T^{vert}Y_\phi|_\gamma$.

Exercise 2.8. With the above identifications, if we use $i(ds - idt)$ to trivialize $T^{0,1}C$, then

$$D_C = \partial_s + J\nabla_t.$$

Exercise 2.9. (See answer in Appendix.) Every element of $\operatorname{Ker}(D_C)$ is S^1-invariant, so $\operatorname{Ker}(D_C)$ is identified with the kernel of the operator

$$\nabla_t : \Gamma\left(T^{vert}Y_\phi|_\gamma\right) \longrightarrow \Gamma\left(T^{vert}Y_\phi|_\gamma\right).$$

Exercise 2.10. Let p denote the period of γ and let x be a fixed point of ϕ^p corresponding to γ. Then there is a canonical identification

$$\operatorname{Ker}(\nabla_t) = \operatorname{Ker}\left(1 - d\phi^p_x\right).$$

The above three exercises imply that C is regular if and only if γ is nondegenerate.

To prove that $\varepsilon(C)$ agrees with the Lefschetz sign when γ is nondegenerate, suppose first that γ is elliptic. Then one can choose a basis for $T_x\Sigma$ in which $d\phi^p_x$ is a rotation. It follows that one can choose a trivialization of $T^{vert}Y_\phi|_\gamma$ in which the parallel transport of the connection ∇ between any two points is a rotation. One can now choose J to be the standard almost complex structure in this trivialization. With these choices, the operator D_C is complex linear, so $\varepsilon(C) = 1$. The same will be true for any other choice of J, because one can find a path between any two almost complex structures J, and by the exercises above the operator D_C will never have a nontrivial kernel. On the other hand, the Lefschetz sign is $+1$ in this case because the eigenvalues of $d\phi^p_x$ are complex conjugates of each other.

To prove that $\varepsilon(C)$ agrees with the Lefschetz sign when γ is not elliptic, one deforms the operator D_C in an S^1-invariant fashion to look like the elliptic case and uses the above exercises to show that the spectral flow changes by ± 1 whenever one switches between the elliptic case and the positive hyperbolic case, cf. [48, Lem. 2.6]. ■

How to Count Multiple Covers. Assume now that ϕ is generic so that all periodic orbits γ are nondegenerate. Then by the above lemmas, the Gromov invariant $Gr(S^1 \times Y_\phi, \Omega, [S^1] \times \Gamma)$ counts unions of (possibly multiply covered) periodic orbits of ∂_t in Y_ϕ with total homology class Γ. We now determine the weight with which each union of periodic orbits is counted.

For each embedded torus $C = S^1 \times \gamma$, there is a generating function $f_\gamma(t)$ from (2.14) encoding how its multiple covers are counted; the coefficient of t^d is the number of times we count the current given by the d-fold cover of C.

Lemma 2.11.

$$f_\gamma(t) = \begin{cases} (1-t)^{-1} = 1 + t + t^2 + \cdots, & \gamma \textit{ elliptic}, \\ 1 - t, & \gamma \textit{ positive hyperbolic}, \\ 1 + t, & \gamma \textit{ negative hyperbolic}. \end{cases}$$

Proof. To compute the generating function $f_\gamma(t)$, we need to compute the sign of C (which we have already done in Lemma 2.7) as well as the signs of the three connected double covers of C. Let C_s denote the double cover obtained by doubling in the s direction, let C_t denote the double cover obtained by doubling in the t direction, and let $C_{s,t}$ denote the third connected double cover. We have $\varepsilon(C_s) = \varepsilon(C)$, because one can compute the kernels of the operators D_{C_s} and D_C in the same way. After a change of coordinates, one can similarly show that $\varepsilon(C_{s,t}) = \varepsilon(C)$. Finally $\varepsilon(C_t)$ is the sign corresponding to the double cover of γ, which is positive if γ is elliptic, and negative if γ is positive or negative hyperbolic. So the signs are as shown in the following table:

γ	$\varepsilon(C)$	$\varepsilon(C_s)$	$\varepsilon(C_{s,t})$	$\varepsilon(C_t)$
elliptic	$+1$	$+1$	$+1$	$+1$
positive hyperbolic	-1	-1	-1	-1
negative hyperbolic	$+1$	$+1$	$+1$	-1

The lemma now follows from these sign calculations and (2.14). ■

Conclusion. The above calculation shows the following:

Proposition 2.12. *Let ϕ be a symplectomorphism of a closed connected surface (Σ, ω) such that all periodic orbits of ϕ are nondegenerate. Then $Gr(S^1 \times Y_\phi, \Omega, [S^1] \times \Gamma)$ is a signed count of finite sets of pairs $\{(\gamma_i, d_i)\}$ where:*

(i) *the γ_i are distinct embedded periodic orbits of ϕ_t,*

(ii) *the d_i are positive integers,*

(iii) $\sum_i d_i[\gamma_i] = \Gamma \in H_1(Y)$, *and*

(iv) $d_i = 1$ *whenever γ_i is hyperbolic.*

The sign associated to a set $\{(\gamma_i, d_i)\}$ is -1 to the number of i such that γ_i is positive hyperbolic.

Proof. It follows from Lemma 2.6 that $Gr(S^1 \times Y_\phi, \Omega, [S^1] \times \Gamma)$ is a count, with appropriate weights, of finite sets $\{(\gamma_i, d_i)\}$ satisfying conditions (i)–(iii). The weight associated to a set $\{(\gamma_i, d_i)\}$ is the product over i of the coefficient of t^{d_i} in the generating function $f_{\gamma_i}(t)$. By Lemma 2.11, this weight is zero unless condition (iv) holds, in which case it is ± 1 and given as claimed. ■

2.7. Two Remarks on the Generating Functions

We now attempt to motivate the generating functions (2.14) a bit more, by explaining why they are what they are in the mapping torus example.

1. One could try to define an invariant of the isotopy class of ϕ by counting multiple covers of tori $S^1 \times \gamma$ using other generating functions. For example, suppose we choose generating functions $e(t)$, $h_+(t)$, and $h_-(t)$, and replace the generating functions in Lemma 2.11 by

$$f_\gamma(t) = \begin{cases} e(t), & \gamma \text{ elliptic}, \\ h_+(t), & \gamma \text{ positive hyperbolic}, \\ h_-(t), & \gamma \text{ negative hyperbolic}. \end{cases}$$

These generating functions must satisfy certain relations in order to give an isotopy invariant of ϕ. First, as one isotopes ϕ, it is possible for a bifurcation to occur in which an elliptic orbit cancels a positive hyperbolic orbit of the same period. To obtain invariance under this bifurcation, we must have

$$e(t)h_+(t) = 1. \tag{2.17}$$

Second, a "period-doubling" bifurcation can occur in which an elliptic orbit turns into a negative hyperbolic orbit of the same period and an elliptic orbit of twice the period. For invariance under this bifurcation we need

$$e(t) = h_-(t)e(t^2). \tag{2.18}$$

In fact, any triple of generating functions $e(t)$, $h_+(t)$, and $h_-(t)$ satisfying the relations (2.17) and (2.18) will give rise to an invariant of the isotopy class of ϕ.

The generating functions in Lemma 2.11 are $e(t) = (1-t)^{-1}$ and $h_\pm(t) = 1 \mp t$, which of course satisfy the relations (2.17) and (2.18). If we allowed multiply covered hyperbolic orbits also and counted them with their Lefschetz signs, then the generating functions would be $e(t) = (1-t)^{-1}$, $h_+(t) = 1 - t - t^2 - \cdots$, and $h_-(t) = 1 + t - t^2 + \cdots$, which do not satisfy the above relations. Throwing out all multiple covers and defining $e(t) = h_-(t) = 1 + t$ and $h_+(t) = 1 - t$ would not work either[10].

2. Given that there are different triples of generating functions that satisfy the relations (2.17) and (2.18), why is the triple in Lemma 2.11 the right one for determining the Seiberg-Witten invariant of $S^1 \times Y_\phi$? Here is one answer: Let $[\Sigma] \in H_2(S^1 \times Y_\phi)$ denote the homology class of a fiber of $Y_\phi \to S^1$. One can use Proposition 2.12 and the Lefschetz fixed point theorem to show that for each nonnegative integer d, we have

$$\sum_{\Gamma\cdot[\Sigma]=d} Gr\big(S^1 \times Y_\phi, \Omega, [S^1] \times \Gamma\big) = L\big(\mathrm{Sym}^d \phi\big),$$

where $\mathrm{Sym}^d \phi$ denotes the homeomorphism from the dth symmetric product of Σ to itself determined by ϕ, and L denotes the Lefschetz number. This is what we are supposed to get, because Salamon [49] showed that the

[10] There are of course other triples of generating functions which satisfy the above relations. For example, the Euler characteristic of the mapping torus analogue of symplectic field theory [14] (just using the q variables) is computed by the generating functions

$$\begin{aligned} e(t) &= (1-t)^{-1}\big(1-t^2\big)^{-1}\cdots, \\ h_+(t) &= (1-t)\big(1-t^2\big)\cdots, \\ h_-(t) &= (1-t)^{-1}(1-t)^{-3}\cdots. \end{aligned}$$

Here the omission of even powers of $(1-t)^{-1}$ in $h_-(t)$ corresponds to the omission of "bad" orbits, without which we would not have invariance under period doubling.

corresponding Seiberg-Witten invariant is a signed count of fixed points of a smooth perturbation of $\mathrm{Sym}^d \phi$. (Similar considerations locally in a neighborhood of a holomorphic torus arise in Taubes's work in [55] which originally led to the generating functions.)

2.8. Three Dimensional Seiberg-Witten Theory

We now briefly review two basic ways to use the Seiberg-Witten equations on four-manifolds to define invariants of three-manifolds.

Let Y be a closed oriented connected three-manifold. A *spin-c structure* on Y can be regarded as an equivalence class of oriented two-plane fields (two-dimensional subbundles of TY), where two oriented two-plane fields are considered equivalent if they are homotopic on the complement of a ball in Y. The set of spin-c structures on Y is an affine space over $H^2(Y;\mathbb{Z})$. A spin-c structure $\mathfrak{s}$ has a first Chern class $c_1(\mathfrak{s}) \in H^2(Y;\mathbb{Z})$, and $\mathfrak{s}$ is called "torsion" when $c_1(\mathfrak{s})$ is torsion. A spin-c structure on Y is equivalent to an S^1-invariant spin-c structure on $S^1 \times Y$, or an $\mathbb{R}$-invariant spin-c structure on $\mathbb{R} \times Y$.

The first way to define invariants of Y is to consider the Seiberg-Witten invariants of the four-manifold $S^1 \times Y$ for S^1-invariant spin-c structures. These invariants are the "Seiberg-Witten invariants" of Y, which we denote by $SW(Y,\mathfrak{s}) \in \mathbb{Z}$, and it turns out that they count S^1-invariant solutions to the Seiberg-Witten equations. Since $b_2^+(S^1 \times Y) = b_1(Y)$, these invariants are well-defined[11] when $b_1(Y) > 0$, up to a choice of chamber when $b_1(Y) = 1$. There is also a distinguished "zero" chamber to use when $b_1(Y) = 1$ and $\mathfrak{s}$ is not torsion. Proposition 2.12 computed this invariant when Y is a mapping torus[12]. Indeed, we saw that the invariant counts S^1-invariant holomorphic curves.

In general, however, the Seiberg-Witten invariants of three-manifolds are not very interesting, because it was shown by Meng-Taubes [44] and Turaev [60] that they agree with a kind of Reidemeister torsion of Y.

The second, more interesting way to define invariants of Y, constructed by Kronheimer-Mrowka [34], is to "categorify" the previous invariant by

[11] $S^1 \times Y$ has a canonical homology orientation, so there is no sign ambiguity in the definition.

[12] When $b_1(Y) = 1$, we used the "symplectic" chamber, which disagrees with the "zero" chamber for spin-c structures corresponding to $\Gamma \in H_1(Y_\phi)$ with $\Gamma \cdot [\Sigma] > g(\Sigma) - 1$. If $\Gamma \in H_1(Y)$ corresponds to a torsion spin-s structure then $\Gamma \cdot [\Sigma] = g(\Sigma) - 1$.

defining a chain complex (over $\mathbb{Z}$) whose generators are $\mathbb{R}$-invariant solutions to the Seiberg-Witten equations on $\mathbb{R} \times Y$, and whose differential counts non-$\mathbb{R}$-invariant solutions to the Seiberg-Witten equations on $\mathbb{R} \times Y$ which converge to two different $\mathbb{R}$-invariant solutions as the $\mathbb{R}$-coordinate converges to $\pm\infty$. If the spin-c structure $\mathfrak{s}$ is non-torsion, then the homology of this chan complex is a well-defined invariant $HM_*(Y, \mathfrak{s})$, called "Seiberg-Witten Floer homology" or "monopole Floer homology". This is a relatively $\mathbb{Z}/d$-graded $\mathbb{Z}$-module, where d denotes the divisibility of $c_1(\mathfrak{s})$ in $H^2(Y;\mathbb{Z})$ mod torsion (which turns out to always be an even integer). This means that it splits into d summands, and there is a well-defined grading difference in $\mathbb{Z}/d$ between any two of them, which is additive for the pairwise differences between any three summands. Each summand is finitely generated. There is also a canonical $\mathbb{Z}/2$-grading, with respect to which the Euler characteristic of the Seiberg-Witten Floer homology $HM_*(Y, \mathfrak{s})$ is the Seiberg-Witten invariant $SW(Y, \mathfrak{s})$.

If $\mathfrak{s}$ is torsion, then there is a difficulty in defining Seiberg-Witten Floer homology caused by "reducible" solutions to the Seiberg-Witten equations. There are two ways to resolve this difficulty, which lead to two versions of Seiberg-Witten Floer homology, which are denoted by $\widehat{HM}_*(Y, \mathfrak{s})$ and $\widecheck{HM}_*(Y, \mathfrak{s})$. These are relatively $\mathbb{Z}$-graded; the former is zero in sufficiently positive grading, and the latter is zero in sufficiently negative grading. They fit into an exact triangle

$$\overline{HM}_*(Y, \mathfrak{s}) \to \widecheck{HM}_*(Y, \mathfrak{s}) \to \widehat{HM}_*(Y, \mathfrak{s}) \to \overline{HM}_{*-1}(Y, \mathfrak{s})) \to \cdots$$

where $\overline{HM}_*(Y, \mathfrak{s})$ is a third invariant which is computable in terms of the triple cup product on Y. In particular, $\overline{HM}_*(Y, \mathfrak{s})$ is two-periodic, i.e. $\overline{HM}_*(Y, \mathfrak{s}) = \overline{HM}_{*+2}(Y, \mathfrak{s})$, and nonzero in at least half of the gradings. In conjunction with the above exact triangle, this implies that $\widehat{HM}_*$ (resp. $\widecheck{HM}_*$) is likewise 2-periodic and nontrivial when the grading is sufficiently negative (resp. positive). This fact is the key input from Seiberg-Witten theory to the proof of the Weinstein conjecture, see Section 1.3.

If $\mathfrak{s}$ is not torsion, then both $\widehat{HM}_*(Y, \mathfrak{s})$ and $\widecheck{HM}_*(Y, \mathfrak{s})$ are equal to the invariant $HM_*(Y, \mathfrak{s})$ discussed previously.

2.9. Towards ECH

The original motivation for defining ECH was to find an analogue of Taubes's $SW = Gr$ theorem for a three-manifold. That is, we would like to identify Seiberg-Witten Floer homology with an appropriate analogue of Taubes's

Gromov invariant for a three-manifold Y. The latter should be the homology of a chain complex which is generated by $\mathbb{R}$-invariant holomorphic curves in $\mathbb{R} \times Y$, and whose differential counts non-$\mathbb{R}$-invariant holomorphic curves in $\mathbb{R} \times Y$.

For holomorphic curve counts to make sense, $\mathbb{R} \times Y$ should have a symplectic structure. This is the case for example when Y is the mapping torus of a symplectomorphism ϕ; the symplectic form (2.15) on $S^1 \times Y_\phi$ also makes sense on $\mathbb{R} \times Y_\phi$. The analogue of Taubes's Gromov invariant in this case is the "periodic Floer homology" of ϕ; it is the homology of a chain complex which is generated by the unions of periodic orbits counted in Proposition 2.12, and its differential counts certain holomorphic curves in $\mathbb{R} \times Y$. The definition of PFH is given in [21, 27], and it shown in [38] that PFH agrees with Seiberg-Witten Floer homology.

Which holomorphic curves to count in the PFH differential is a subtle matter which we will explain below. However, since not every three-manifold is a mapping torus, we will instead carry out the analogous construction of ECH for contact three-manifolds[13], which is more general since every oriented three-manifold admits a contact structure. Finding the appropriate definition of the ECH chain complex is not obvious, but Taubes's $SW = Gr$ theorem and the computation of Gr for mapping tori give us a lot of hints.

3. The Definition of ECH

Guided by the discussion in Section 2, we now define the embedded contact homology of a contact three-manifold (Y, λ), using $\mathbb{Z}/2$ coefficients for simplicity.

Assume that λ is nondegenerate and fix $\Gamma \in H_1(Y)$. We wish to define the chain complex $ECC_*(Y, \lambda, \Gamma, J)$, where J is a generic symplectization-admissible almost complex structure on $\mathbb{R} \times Y$, see Section 1.3.

[13]To spell out the analogy here, both mapping tori and contact structures are examples of the more general notion of "stable Hamiltonian structure". A *stable Hamiltonian structure* on an oriented 3-manifold consists of a 1-form λ and a closed 2-form ω such that $\lambda \wedge \omega > 0$ and $d\lambda = f\omega$ with $f : Y \to \mathbb{R}$. These data determine an oriented 2-plane field $\xi = \operatorname{Ker}(\lambda)$ and a "Reeb vector field" R characterized by $\omega(R, \cdot) = 0$ and $\lambda(R) = 1$. For a mapping torus, $\lambda = dt$, $\omega \equiv 0$, $f \equiv 0$, and $R = \partial_t$. For a contact structure, $\omega = d\lambda$, $f \equiv 1$, and R is the usual Reeb vector field. A version of ECH for somewhat more general stable Hamiltonian structures with $f \geq 0$ appears in the work of Kutluhan-Lee-Taubes [36].

Define an *orbit set* in the homology class Γ to be a finite set of pairs $\{(\alpha_i, m_i)\}$ where the α_i are distinct embedded Reeb orbits, the m_i are positive integers, and $\sum_i m_i[\alpha_i] = \Gamma \in H_1(Y)$. Motivated by Proposition 2.12, we define the chain complex to be generated by orbit sets as above such that $m_i = 1$ whenever α_i is hyperbolic. (We also need to study orbit sets not satisfying this last condition in order to develop the theory.) Proposition 2.12 also suggests that there should be a canonical $\mathbb{Z}/2$-grading by the parity of the number of i such that α_i is positive hyperbolic, and we will see in Section 3.5 that this is the case.

The differential should count J-holomorphic currents in $\mathbb{R}\times Y$ by analogy with the Gromov invariant. The three key formulas that entered into the definition of the Gromov invariant were the Fredholm index formula (2.9), the adjunction formula (2.6), and the definition of I in (2.8). To define the ECH differential we need analogues of these three formulas for holomorphic curves in $\mathbb{R}\times Y$, plus one additional ingredient, the "writhe bound". We now explain these.

3.1. Holomorphic Curves and Holomorphic Currents

We consider J-holomorphic curves of the form $u : (\Sigma, j) \to (\mathbb{R}\times Y, J)$ where the domain (Σ, j) is a punctured compact Riemann surface. If γ is a (possibly multiply covered) Reeb orbit, a *positive end* of u at γ is a puncture near which u is asymptotic to $\mathbb{R}\times\gamma$ as $s\to\infty$. This means that a neighborhood of the puncture can be given coordinates $(\sigma,\tau)\in(\mathbb{R}/T\mathbb{Z})\times[0,\infty)$ with $j(\partial_\sigma) = \partial_\tau$ such that $\lim_{\sigma\to\infty}\pi_{\mathbb{R}}(u(\sigma,\tau)) = \infty$ and $\lim_{\sigma\to\infty}\pi_Y(u(s,\cdot)) = \gamma$. A *negative end* is defined analogously with $\sigma\in(-\infty,0]$ and $s\to-\infty$. We assume that all punctures are positive ends or negative ends as above. We mod out by the usual equivalence relation on holomorphic curves, namely composition with biholomorphic maps between domains.

Let $\alpha = \{(\alpha_i, m_i)\}$ and $\beta = \{(\beta_j, n_j)\}$ be orbit sets in the class Γ. Define a *J-holomorphic current* from α to β to be a finite set of pairs $\mathcal{C} = \{(C_k, d_k)\}$ where the C_k are distinct irreducible somewhere injective J-holomorphic curves in $\mathbb{R}\times Y$, the d_k are positive integers, $\mathcal{C}$ is asymptotic to α as a current as the $\mathbb{R}$ coordinate goes to $+\infty$, and $\mathcal{C}$ is asymptotic to β as a current as the $\mathbb{R}$ coordinate goes to $-\infty$. This last condition means that the positive ends of the curves C_k are at covers of the Reeb orbits α_i, the sum over k of d_k times the total covering multiplicity of all ends of C_k at covers of α_i is m_i, and analogously for the negative ends. Let $\mathcal{M}(\alpha,\beta)$ denote the set of J-holomorphic currents from α to β. A holomorphic current $\mathcal{C} = \{(C_k, d_k)\}$ is "somewhere injective" if $d_k = 1$ for each k, in which case it is "embedded" if furthermore each C_k is embedded and the C_k are pairwise disjoint.

Let $H_2(Y, \alpha, \beta)$ denote the set of 2-chains Σ in Y with

$$\partial \Sigma = \sum_i m_i \alpha_i - \sum_j n_j \beta_j,$$

modulo boundaries of 3-chains. Then $H_2(Y, \alpha, \beta)$ is an affine space over $H_2(Y)$, and every J-holomorphic current $\mathcal{C} \in \mathcal{M}(\alpha, \beta)$ defines a class $[\mathcal{C}] \in H_2(Y, \alpha, \beta)$.

3.2. The Fredholm Index in Symplectizations

We now state a symplectization analogue of the index formula (2.3).

Proposition 3.1. *If J is generic, then every somewhere injective J-holomorphic curve C in $\mathbb{R} \times Y$ is regular (i.e. an appropriate deformation operator is surjective), so the moduli space of J-holomorphic curves as above near C is a manifold. Its dimension is the Fredholm index given by Eq.* (3.1) *below.*

If C has k positive ends at Reeb orbits $\gamma_1^+, \ldots, \gamma_k^+$ and l negative ends at Reeb orbits $\gamma_1^-, \ldots, \gamma_l^-$, the *Fredholm index* of C is defined by

$$\operatorname{ind}(C) = -\chi(C) + 2c_\tau(C) + \sum_{i=1}^{k} CZ_\tau(\gamma_i^+) - \sum_{i=1}^{l} CZ_\tau(\gamma_i^-), \tag{3.1}$$

where the terms on the right hand side are defined as follows. First, τ is a trivialization of ξ over the Reeb orbits $\gamma_i^\pm$, which is symplectic with respect to $d\lambda$. Second, $\chi(C)$ denotes the Euler characteristic of the domain of C as usual. Third,

$$c_\tau(C) = c_1(\xi|_C, \tau) \in \mathbb{Z}$$

is the *relative first Chern class* of the complex line bundle $\xi|_C$ with respect to the trivialization τ. To define this, note that the trivialization τ determines a trivialization of $\xi|_C$ over the ends of C, up to homotopy. One chooses a generic section ψ of $\xi|_C$ which on each end is nonvanishing and constant with respect to the trivialization on the ends. One then defines $c_1(\xi|_C, \tau)$ to be the algebraic count of zeroes of ψ.

To say more about what the relative first Chern class depends on, note that $C \in \mathcal{M}(\alpha, \beta)$ for some orbit sets $\alpha = \{(\alpha_i, m_i)\}$ and $\beta = \{(\beta_j, n_j)\}$ in the same homology class. Write $Z = [C] \in H_2(Y, \alpha, \beta)$. Then in fact $c_1(\xi|_C, \tau)$ depends only on α, β, τ, and Z. To see this, let S be a compact oriented surface with boundary, and let $f : S \to [-1,1] \times Y$ be a smooth map, such that $f|_{\partial S}$ consists of positively oriented covers of $\{1\} \times \alpha_i$ with total multiplicity m_i and negatively oriented covers of $\{-1\} \times \beta_j$ with total multiplicity n_j, and the projection of f to Y represents the relative homology class Z. Then $c_1(f^*\xi, \tau) \in \mathbb{Z}$ is defined as before.

Exercise 3.2.

(a) The relative first Chern class $c_1(f^*\xi, \tau)$ above depends only on α, β, τ, and Z, and so can be denoted by $c_\tau(Z)$.

(b) If $Z' \in H_2(Y, \alpha, \beta)$ is another relative homology class, then

$$c_\tau(Z) - c_\tau(Z') = \langle c_1(\xi), Z - Z' \rangle,$$

where on the right hand side, $c_1(\xi) \in H^2(Y; \mathbb{Z})$ denotes the usual first Chern class of the complex line bundle $\xi \to Y$.

Continuing with the explanation of the index formula (3.1), $CZ_\tau(\gamma) \in \mathbb{Z}$ denotes the *Conley-Zehnder index* of γ with respect to the trivialization τ. To define this, pick a parametrization $\gamma : \mathbb{R}/T\mathbb{Z} \to Y$. Let $\{\psi_t\}_{t \in \mathbb{R}}$ denote the one-parameter group of diffeomorphisms of Y given by the flow of R. Then $d\psi_t : T_{\gamma(0)}Y \to T_{\gamma(t)}Y$ induces a symplectic linear map $\phi_t : \xi_{\gamma(0)} \to \xi_{\gamma(t)}$, which using our trivialization τ we can regard as a 2×2 symplectic matrix. In particular, $\phi_0 = 1$, and ϕ_T is the linearized return map (in our trivialization), which does not have 1 as an eigenvalue. We now define $CZ_\tau(\gamma) \in \mathbb{Z}$ to be the Conley-Zehnder index of the family of symplectic matrices $\{\phi_t\}_{t \in [0,T]}$, which is given explicitly as follows. (See e.g. [48, §2.4] for the general definition of the Conley-Zehnder index for paths of symplectic matrices in any dimension.)

If γ is hyperbolic, let $v \in \mathbb{R}^2$ be an eigenvector of ϕ_T; then the family of vectors $\{\phi_t(v)\}_{t \in [0,T]}$ rotates by angle πk for some integer k (which is even in the positive hyperbolic case and odd in the negative hyperbolic case), and

$$CZ_\tau(\gamma) = k.$$

If γ is elliptic, then we can change the trivialization so that each ϕ_t is rotation by angle $2\pi\theta_t \in \mathbb{R}$ where θ_t is a continuous function of $t \in [0,T]$ and $\theta_0 = 0$.

The number $\theta = \theta_T \in \mathbb{R} \setminus \mathbb{Z}$ is called the "rotation angle" of γ with respect to τ, and

$$CZ_\tau(\gamma) = 2\lfloor\theta\rfloor + 1. \tag{3.2}$$

Exercise 3.3. The right hand side of the index formula (3.1) does not depend on τ, even though the individual terms in it do. (See hint in Appendix.)

The proof of Proposition 3.1 consists of a tranversality argument in [12] and an index calculation in [50]. As usual, the somewhere injective assumption is necessary; there is no J for which transversality holds for all multiply covered curves. For example, transversality fails for some branched covers of trivial cylinders, see Exercise 3.14 below.

3.3. The Relative Adjunction Formula

Our next goal is to obtain an analogue of the adjunction formula (2.6) for a somewhere injective holomorphic curve in $\mathbb{R} \times Y$. To do so we need to re-interpret each term in the formula (2.6) in the symplectization context; and there is also a new term arising from the asymptotic behavior of the holomorphic curve.

Relative Adjunction Formula [21, Rmk. 3.2]. *Let $C \in \mathcal{M}(\alpha, \beta)$ be somewhere injective. Then C has only finitely many singularities, and*

$$c_\tau(C) = \chi(C) + Q_\tau(C) + w_\tau(C) - 2\delta(C). \tag{3.3}$$

Here τ is a trivialization of ξ over the Reeb orbits α_i and β_j; the left hand side is the relative first Chern class defined in Section 3.2; $\chi(C)$ is the Euler characteristic of the domain as usual; and $\delta(C) \geq 0$ is an algebraic count of singularities with positive integer weights as in Section 2.4. The term $Q_\tau(C)$ is the "relative intersection pairing", which is a symplectization analogue of the intersection number $[C] \cdot [C]$ in the closed case. The new term $w_\tau(C)$ is the "asymptotic writhe". Let us now explain both of these.

The Relative Intersection Pairing. Given a class $Z \in H_2(Y, \alpha, \beta)$, we want to define the relative intersection pairing $Q_\tau(Z) \in \mathbb{Z}$.

To warm up to this, recall that given a closed oriented 4-manifold X, and given a class $A \in H_2(X)$, to compute $A \cdot A$ one can choose two embedded oriented surfaces $S, S' \subset X$ representing the class A that intersect transversely, and count the intersections of S and S' with signs.

In the symplectization case, we could try to choose two embedded (except at the boundary) oriented surfaces $S, S' \subset [-1,1] \times Y$ representing the class Z such that

$$\partial S = \partial S' = \sum_i m_i \{1\} \times \alpha_i - \sum_j n_j \{-1\} \times \beta_j,$$

and S and S' intersect transversely (except at the boundary), and algebraically count intersections of the interior of S with the interior of S'. However this count of intersections is not a well-defined function of Z, because if one deforms S or S', then intersection points can appear or disappear at the boundary.

To get a well-defined count of intersections, we need to specify something about the boundary behavior. The choice of trivialization τ allows us to do this. We require that the projections to Y of the intersections of S and S' with $(1-\varepsilon, 1] \times Y$ are embeddings, and their images in a transverse slice to α_i are unions of rays which do not intersect and which do not rotate with respect to the trivialization τ as one goes around α_i. Likewise, the projections to Y of the intersections of S and S' with $[-1, -1+\varepsilon) \times Y$ are embeddings, and their images in a transverse slice to β_j are unions of rays which do not intersect and which do not rotate with respect to the trivialization τ as one goes around β_j. If we count the interior intersections of two such surfaces S and S', then we get an integer which depends only on α, β, Z, and τ, and we denote this integer by $Q_\tau(Z)$. For more details see [21, §2.4] and [22, §2.7].

If $\mathcal{C} \in \mathcal{M}(\alpha, \beta)$ is a J-holomorphic current, write $Q_\tau(\mathcal{C}) = Q_\tau([\mathcal{C}])$.

The Asymptotic Writhe. Given a somewhere injective J-holomorphic curve $C \in \mathcal{M}(\alpha, \beta)$, consider the slice $C \cap (\{s\} \times Y)$. If $s \gg 0$, then the slice $C \cap (\{s\} \times Y)$ is an embedded curve which is the union, over i, of a braid ζ_i^+ around the Reeb orbit α_i with m_i strands. This fact, due to Siefring [51], is shown along the way to proving the writhe bound (3.9) below, see Lemma 5.5. This, together with an analogous statement for the negative ends and the fact that the singularities of C are isolated, implies that C has only finitely many singularities. Since the braid ζ_i^+ is embedded for all $s \gg 0$, its isotopy class does not depend on $s \gg 0$.

We can use the trivialization τ to identify the braid ζ_i^+ with a link in $S^1 \times D^2$. The writhe of this link, which we denote by $w_\tau(\zeta_i^+) \in \mathbb{Z}$, is defined by identifying $S^1 \times D^2$ with an annulus cross an interval, projecting ζ_i^+ to the annulus, and counting crossings with signs. We use the sign convention in which counterclockwise rotations in the D^2 direction as one goes counterclockwise around S^1 contribute positively to the writhe; this is opposite the usual convention in knot theory, but makes sense in the present context.

Likewise, the slice $C \cap (\{s\} \times Y)$ for $s \ll 0$ is the union over j of a braid ζ_j^- around the Reeb orbit β_j with n_j strands, and this braid has a writhe $w_\tau(\zeta_j^-) \in \mathbb{Z}$.

We now define the *asymptotic writhe* of C by

$$w_\tau(C) = \sum_i w_\tau(\zeta_i^+) - \sum_j w_\tau(\zeta_j^-).$$

This completes the definition of all of the terms in the relative adjunction formula (3.3).

Exercise 3.4. Show that the two sides of the relative adjunction formula (3.3) change the same way if one changes the trivialization τ. (See hint in Appendix.)

Here is an outline of the proof of the relative adjunction formula (3.3) in the special case where C is immersed and the only singularities of C are nodes. Let N_C denote the normal bundle of C, which can be identified with $\xi|_C$ near the ends of C. We compute $c_1(N_C, \tau)$ in two ways. First, the decomposition $(\mathbb{C} \oplus \xi)|_C = T(\mathbb{R} \times Y)|_C = TC \oplus N_C$ implies that

$$c_\tau(C) = \chi(C) + c_1(N_C, \tau),$$

see [21, Prop. 3.1(a)]. Second, one can count the intersections of C with a nearby surface and compare with the definition of Q_τ to show that

$$c_1(N_C, \tau) = Q_\tau(C) + w_\tau(C) - 2\delta(C),$$

cf. [21, Prop. 3.1(b)].

3.4. The ECH Index

We come now to the key nontrivial part of the definition in ECH, which is to define an analogue of the quantity I in (2.8) for relative homology classes in symplectizations.

Let $C \in \mathcal{M}(\alpha, \beta)$ be somewhere injective. By (3.1), we can write the Fredholm index of C as

$$\operatorname{ind}(C) = -\chi(C) + 2c_\tau(C) + CZ_\tau^{ind}(C),$$

where $CZ_\tau^{ind}(C)$ is shorthand for the Conley-Zehnder term that appears in ind, namely the sum over all positive ends of C at a Reeb orbit γ of $CZ_\tau(\gamma)$ (these Reeb orbits are covers of the Reeb orbits α_i), minus the corresponding sum for the negative ends of C. We know that if J is generic then $\mathcal{M}(\alpha, \beta)$ is a manifold near C of dimension $\operatorname{ind}(C)$. We would like to bound this dimension in terms of the relative homology class $[C]$.

If γ is an embedded Reeb orbit and k is a positive integer, let γ^k denote the k-fold iterate of γ.

Definition 3.5. If $Z \in H_2(Y, \alpha, \beta)$, define the *ECH index*

$$I(\alpha, \beta, Z) = c_\tau(Z) + Q_\tau(Z) + CZ_\tau^I(\alpha, \beta), \tag{3.4}$$

where CZ_τ^I is the Conley-Zehnder term that appears in I, namely

$$CZ_\tau^I(\alpha, \beta) = \sum_i \sum_{k=1}^{m_i} CZ_\tau(\alpha_i^k) - \sum_j \sum_{k=1}^{n_j} CZ_\tau(\beta_j^k). \tag{3.5}$$

If $C \in \mathcal{M}(\alpha, \beta)$, define $I(C) = I(\alpha, \beta, [C])$.

Note that the Conley-Zehnder terms $CZ_\tau^{ind}(C)$ and $CZ_\tau^I(\alpha, \beta)$ are quite different. The former just involves the Conley-Zehnder indices of orbits corresponding to ends of C; while the latter sums up the Conley-Zehnder indices of all iterates of α_i up to multiplicity m_i, minus the Conley-Zehnder indices of all iterates of β_j up to multiplicity n_j. For example, if C has positive ends at α_i^3 and α_i^5 (and no other positive ends at covers of α_i), then the corresponding contribution to $CZ_\tau^{ind}(C)$ is $CZ_\tau(\alpha_i^3) + CZ_\tau(\alpha_i^5)$, while the contribution to $CZ_\tau^I(\alpha, \beta)$ is $\sum_{k=1}^{8} CZ_\tau(\alpha_i^k)$.

Basic Properties of the ECH Index.

(Well Defined) *The ECH index $I(Z)$ does not depend on the choice of trivialization τ.*

(Index Ambiguity Formula) *If $Z' \in H_2(\alpha, \beta)$ is another relative homology class, then*

$$I(Z) - I(Z') = \langle Z - Z', c_1(\xi) + 2\operatorname{PD}(\Gamma) \rangle. \tag{3.6}$$

(Additivity) *If δ is another orbit set in the homology class Γ, and if $W \in H_2(Y, \beta, \delta)$, then $Z + W \in H_2(Y, \alpha, \delta)$ is defined and*

$$I(Z+W) = I(Z) + I(W).$$

(Index Parity) *If α and β are generators of the ECH chain complex (i.e. all hyperbolic orbits have multiplicity 1), then*

$$(-1)^{I(Z)} = \varepsilon(\alpha)\varepsilon(\beta), \tag{3.7}$$

where $\varepsilon(\alpha)$ denotes -1 to the number of positive hyperbolic orbits in α.

Exercise 3.6. Prove the above basic properties. (See [21, §3.3].)

We now have the following analogue of (2.10), which is the key result that gets ECH off the ground.

Index Inequality. *If $C \in \mathcal{M}(\alpha, \beta)$ is somewhere injective, then*

$$\operatorname{ind}(C) \le I(C) - 2\delta(C). \tag{3.8}$$

In particular, $\operatorname{ind}(C) \le I(C)$, with equality only if C is embedded.

The index inequality follows immediately by combining the definition of the ECH index in (3.4), the formula for the Fredholm index in (3.1), the relative adjunction formula (3.3), and the following inequality:

Writhe Bound. *If $C \in \mathcal{M}(\alpha, \beta)$ is somewhere injective, then*

$$w_\tau(C) \le CZ_\tau^I(\alpha, \beta) - CZ_\tau^{ind}(C). \tag{3.9}$$

The proof of the writhe bound will be outlined in Section 5.1.

Holomorphic Curves with Low ECH Index. The index inequality (3.8) is most of what is needed to prove the following analogue of Lemma 2.4. Below, a *trivial cylinder* means a cylinder $\mathbb{R} \times \gamma \subset \mathbb{R} \times Y$ where γ is an embedded Reeb orbit.

Proposition 3.7. *Suppose J is generic. Let α and β be orbit sets and let $\mathcal{C} \in \mathcal{M}(\alpha, \beta)$ be any J-holomorphic current in $\mathbb{R} \times Y$, not necessarily somewhere injective. Then:*

0. $I(\mathcal{C}) \geq 0$, *with equality if and only if* $\mathcal{C}$ *is a union of trivial cylinders with multiplicities.*
1. *If* $I(\mathcal{C}) = 1$, *then* $\mathcal{C} = \mathcal{C}_0 \sqcup C_1$, *where* $I(\mathcal{C}_0) = 0$, *and* C_1 *is embedded and has* $\operatorname{ind}(C_1) = I(C_1) = 1$.
2. *If* $I(\mathcal{C}) = 2$, *and if* α *and* β *are generators of the chain complex* $ECC_*(Y, \lambda, \Gamma, J)$, *then* $\mathcal{C} = \mathcal{C}_0 \sqcup C_2$, *where* $I(\mathcal{C}_0) = 0$, *and* C_2 *is embedded and has* $\operatorname{ind}(C_2) = I(C_2) = 2$.

Proof. Let $\mathcal{C} = \{(C_k, d_k)\}$ be a holomorphic current in $\mathcal{M}(\alpha, \beta)$. We first consider the special case in which $d_k = 1$ whenever C_k is a trivial cylinder.

Since J is $\mathbb{R}$-invariant, any J-holomorphic curve can be translated in the $\mathbb{R}$-direction to make a new J-holomorphic curve. Let C' be the union over k of the union of d_k different translates of C_k. Then C' is somewhere injective, thanks to our simplifying assumption that $d_k = 1$ whenever C_k is a trivial cylinder. So the index inequality applies to C' to give

$$\operatorname{ind}(C') \leq I(C') - 2\delta(C').$$

Now because the Fredholm index ind is additive under taking unions of holomorphic curves, and because the ECH index I depends only on the relative homology class, this gives

$$\sum_k d_k \operatorname{ind}(C_k) \leq I(\mathcal{C}) - 2\delta(C'). \tag{3.10}$$

Since J is generic, we must have $\operatorname{ind}(C_k) \geq 0$, with equality if and only if C_k is a trivial cylinder. Parts (0) and (1) of the Proposition can now be immediately read off from the inequality (3.10).

To prove part (2), we just need to rule out the case where there is one nontrivial C_k with $d_k = 2$. In this case, since α and β are ECH generators, all ends of C_k must be at elliptic Reeb orbits. It then follows from the Fredholm index formula (3.1) that $\operatorname{ind}(C_k)$ is even. Thus $\operatorname{ind}(C_k) \geq 2$, contradicting the inequality (3.10).

To remove the simplifying assumption, one can show that if $\mathcal{C}$ contains no trivial cylinders and if $\mathcal{T}$ is a union of trivial cylinders, then

$$I(\mathcal{C} \cup \mathcal{T}) \geq I(\mathcal{C}) + 2\#(\mathcal{C} \cap \mathcal{T}),$$

compare (2.11). This is proved in [21, Prop. 7.1], and a more general statement bounding the ECH index of any union of holomorphic currents is proved in [22, Thm. 5.1]. Now by intersection positivity, $\#(\mathcal{C}\cap\mathcal{T})\geq 0$, with equality if and only if $\mathcal{C}$ and $\mathcal{T}$ are disjoint. The proposition for $\mathcal{C}\cup\mathcal{T}$ then follows from the proposition for $\mathcal{C}$. ■

3.5. The ECH Differential

We can now define the differential ∂ on the chain complex $ECC_*(Y,\lambda,\Gamma,J)$. If α and β are orbit sets and k is an integer, define

$$\mathcal{M}_k(\alpha,\beta)=\{\mathcal{C}\in\mathcal{M}(\alpha,\beta)\mid I(\mathcal{C})=k\}.$$

If α is a chain complex generator, we define

$$\partial\alpha=\sum_{\beta}\#(\mathcal{M}_1(\alpha,\beta)/\mathbb{R})\beta,$$

where the sum is over chain complex generators β, and '#' denotes the mod 2 count. Here $\mathbb{R}$ acts on $\mathcal{M}_1(\alpha,\beta)$ by translation of the $\mathbb{R}$ coordinate on $\mathbb{R}\times Y$; and by Proposition 3.7 the quotient is a discrete set. We will show in Section 5.3, analogously to Lemma 2.5, that $\mathcal{M}_1(\alpha,\beta)/\mathbb{R}$ is finite so that the count $\#(\mathcal{M}_1(\alpha,\beta)/\mathbb{R})$ is well defined. Next, it follows from the inequality (1.15) and Exercise 3.8 below that for any α, there are only finitely many β with $\mathcal{M}(\alpha,\beta)$ nonempty, so $\partial\alpha$ is well defined.

Exercise 3.8. If λ is a nondegenerate contact form on Y and if L is a real number, then λ has only finitely many Reeb orbits with symplectic action less than L.

The proof that $\partial^2=0$ is much more difficult, and we will give an introduction to this in Section 5.4. Modulo this and the other facts we have not proved, we have now defined $ECH_*(Y,\lambda,\Gamma,J)$, and as reviewed in the introduction this is an invariant $ECH_*(Y,\xi,\Gamma)$.

3.6. The Grading

The chain complex $ECC_*(Y,\lambda,\Gamma,J)$, and hence its homology, is relatively $\mathbb{Z}/d$ graded, where d denotes the divisibility of $c_1(\xi)+2\,\mathrm{PD}(\Gamma)$ in $H^2(Y;\mathbb{Z})$ mod torsion. That is, if α and β are two chain complex generators, we can define their "index difference" $I(\alpha,\beta)$ by choosing an arbitrary $Z\in H_2(Y,\alpha,\beta)$

and setting

$$I(\alpha,\beta) = \big[I(\alpha,\beta,Z)\big] \in \mathbb{Z}/d.$$

This is well defined by the index ambiguity formula (3.6). When the chain complex is nonzero, we can further define an absolute $\mathbb{Z}/d$ grading by picking some generator β and declaring its grading to be zero, so that the grading of any other generator is α is

$$|\alpha| = I(\alpha,\beta).$$

By the Additivity property of the ECH index, the differential decreases this absolute grading by 1.

Remarks 3.9. (1) In particular, if $\Gamma = 0$, then the empty set of Reeb orbits is a generator of the chain complex, which represents a homology class depending only on Y and ξ, see Section 1.4. Thus $ECH_*(Y,\xi,0)$ has a canonical absolute $\mathbb{Z}/d$ grading in which the empty set has grading zero.

(2) It follows from the Index Parity property (3.7) that for every Γ there is a canonical absolute $\mathbb{Z}/2$ grading on $ECH_*(Y,\xi,\Gamma)$ by the parity of the number of positive hyperbolic Reeb orbits.

3.7. Example: The ECH of an Ellipsoid

To illustrate the above definitions, we now compute $ECH_*(Y,\lambda,0,J)$, where Y is the three-dimensional ellipsoid $Y = \partial E(a,b)$ with a/b irrational, and λ is the contact form given by the restriction of the Liouville form (1.12). We already saw in Example 1.8 that the chain complex generators have the form $\gamma_1^{m_1}\gamma_2^{m_2}$ with $m_1, m_2 \geq 0$. Since the Reeb orbits γ_1 and γ_2 are elliptic, it follows from the Index Parity property (3.7) that the grading difference between any two generators is even, so the differential vanishes identically for any J.

The Grading. To finish the computation of the homology, we just need to compute the grading of each generator. We know from Section 3.6 that the chain complex has a canonical $\mathbb{Z}$-grading, where the empty set (corresponding to $m_1 = m_2 = 0$) has grading zero. The grading of $\alpha = \gamma_1^{m_1}\gamma_2^{m_2}$ can then be written as

$$|\alpha| = I(\alpha,\emptyset) = c_\tau(\alpha) + Q_\tau(\alpha) + CZ_\tau^I(\alpha). \tag{3.11}$$

Here $c_\tau(\alpha)$ is shorthand for $c_\tau(Z)$, and $Q_\tau(\alpha)$ is shorthand for $Q_\tau(Z)$, where Z is the unique element of $H_2(Y,\alpha,\emptyset)$; and $CZ_\tau^I(\alpha)$ is shorthand for $CZ_\tau^I(\alpha,\emptyset)$.

To calculate the terms on the right hand side of (3.11), we first need to choose a trivialization τ of ξ over γ_1 and γ_2. Under the identification $T\mathbb{R}^4 = \mathbb{C}\oplus\mathbb{C}$, the restriction of ξ to γ_1 agrees with the second $\mathbb{C}$ summand, and the restriction of ξ to γ_2 agrees with the first $\mathbb{C}$ summand. We use these identifications to define the trivialization τ that we will use.

The calculations in Example 1.8 imply that with respect to this trivialization τ, the rotation angle (see Section 3.2) of γ_1 is a/b, and the rotation angle of γ_2 is b/a. So by the formula (3.2) for the Conley-Zehnder index, we have

$$CZ_\tau^I(\alpha) = \sum_{k=1}^{m_1}(2\lfloor ka/b\rfloor + 1) + \sum_{k=1}^{m_2}(2\lfloor kb/a\rfloor + 1).$$

The remaining terms in (3.11) are given as follows:

Exercise 3.10. $c_\tau(\alpha) = m_1 + m_2$, and $Q_\tau(\alpha) = 2m_1m_2$.

Putting the above together, we get that

$$I(\alpha) = 2\left((m_1+1)(m_2+1) - 1 + \sum_{k=1}^{m_1}\lfloor ka/b\rfloor + \sum_{k=1}^{m_2}\lfloor kb/a\rfloor\right). \tag{3.12}$$

In particular, this is a nonnegative even integer.

How many generators are there of each grading? By Taubes's isomorphism (1.11), together with the calculation of the Seiberg-Witten Floer homology of S^3 in [34], we should get

$$ECH_*(\partial E(a,b),\lambda,0,J) = \begin{cases} \mathbb{Z}/2, & * = 0,2,4,\ldots, \\ 0, & \text{otherwise.} \end{cases} \tag{3.13}$$

Exercise 3.11. Deduce (3.13) from (3.12). That is, show that (3.12) defines a bijection from the set of pairs of nonnegative integers (m_1, m_2) to the set of nonnegative even integers. (See hint in Appendix.)

3.8. The U Map

We now explain some more details of the U map which was introduced in Section 1.4, following [31, §2.5].

Suppose Y is connected, and choose a point $z \in Y$ which is not on any Reeb orbit. Let α and β be generators of the chain complex $ECC_*(Y,\lambda,\Gamma,J)$, and let $\mathcal{C} \in \mathcal{M}_2(\alpha,\beta)$ be a holomorphic current with $(0,z) \in \mathcal{C}$. By Proposition 3.7, we have $\mathcal{C} = \mathcal{C}_0 \sqcup C_2$ where $\mathcal{I}(\mathcal{C}_0) = 0$, and C_2 is embedded and $\operatorname{ind}(C_2) = 2$. Since $\mathcal{C}_0$ is a union of trivial cylinders and z is not on any Reeb orbit, it follows that $(0,z) \in C_2$. Let $N_{(0,z)}C_2$ denote the normal bundle to C_2 at $(0,z)$. There is then a natural map

$$T_{\mathcal{C}}\mathcal{M}_2(\alpha,\beta) \to N_{(0,z)}C_2. \tag{3.14}$$

Transversality arguments as in Proposition 3.1 can be used to show that if J is generic then the map (3.14) is an isomorphism for all holomorphic currents $\mathcal{C}$ as above. In particular, this implies that the set of holomorphic currents $\mathcal{C}$ as above is discrete. For J with this property, we define a chain map

$$U_z : ECC_*(Y,\lambda,\Gamma,J) \longrightarrow ECC_{*-2}(Y,\lambda,\Gamma,J)$$

by

$$U_z\alpha = \sum_\beta \#\{\mathcal{C} \in \mathcal{M}_2(\alpha,\beta) \mid (0,z) \in \mathcal{C}\}\beta,$$

where # denotes the mod 2 count as usual.

A compactness argument similar to the proof that ∂ is defined in Section 5.3 shows that U_z is defined. Likewise, the proof that $\partial^2 = 0$ introduced in Section 5.4 can be modified to show that $\partial U_z = U_z\partial$.

To show that the map (1.14) on ECH induced by U_z does not depend on z, suppose $z' \in Y$ is another point which is not on any Reeb orbit. Since there are only countably many Reeb orbits, we can choose an embedded path η from z to z' which does not intersect any Reeb orbit. Define a map

$$K_\eta : ECC_*(Y,\lambda,\Gamma,J) \longrightarrow ECC_{*-1}(Y,\lambda,\Gamma,J)$$

by

$$K_\eta\alpha = \sum_\beta \#\{(\mathcal{C},y) \in \mathcal{M}_1(\alpha,\beta) \times Y \mid (0,y) \in \mathcal{C}\}\beta.$$

Similarly to the proof that ∂ is well-defined, K_η is well-defined if J is generic. Similarly to the proof that $\partial^2 = 0$, one proves the chain homotopy equation

$$\partial K_\eta + K_\eta\partial = U_z - U_{z'}. \tag{3.15}$$

Remark 3.12. If $z = z'$, then it follows from (3.15) that K_η induces a map on ECH of degree -1. In fact this map depends only on the homology class of the loop η, and thus defines a homomorphism from $H_1(Y)$ to the set of degree -1 maps on $ECH_*(Y,\xi,\Gamma)$. See [28, §12.1] for more about this map and an example where it is nontrivial, and [59] for the proof that it agrees with an analogous map on Seiberg-Witten Floer cohomology.

3.9. Partition Conditions

The definitions of the ECH differential and the U map do not directly specify the topological type of the holomorphic currents to be counted. However it turns out that most of this information is determined indirectly. We now explain how the covering multiplicities of the Reeb orbits at the ends of the nontrivial component of such a holomorphic current are uniquely determined if one knows the trivial cylinder components. (We will further see in Section 5.2 that the genus of the nontrivial part of the holomorphic current is then determined by its relative homology class.)

Let $\alpha=\{(\alpha_i,m_i)\}$ and $\beta=\{(\beta_j,n_j)\}$ be orbit sets, and let $C\in\mathcal{M}(\alpha,\beta)$ be somewhere injective. For each i, the curve C has ends at covers of α_i whose total covering multiplicity is m_i. The multiplicities of these covers are a partition of the positive integer m_i which we denote by $p_i^+(C)$. For example, if C has two positive ends at α_i, and one positive end at the triple cover of α_i, then $m_i=5$ and $p_i^+(C)=(3,1,1)$. Likewise, the covering multiplicities of the negative ends of C at covers of β_j determine a partition of n_j, which we denote by $p_j^-(C)$.

For each embedded Reeb orbit γ and each positive integer m, we will shortly define two partitions of m, the "positive partition" $p_\gamma^+(m)$ and the "negative partition[14]" $p_\gamma^-(m)$. We then have:

Partition Conditions. *Suppose equality holds in the Writhe Bound* (3.9) *for C. (This holds for example if C is the nontrivial component of a holomorphic current that contributes to the ECH differential or the U map.) Then $p_i^+(C)=p_{\alpha_i}^+(m_i)$ and $p_j^-(C)=p_{\beta_j}^-(n_j)$.*

[14] In [21, 22], $p_\gamma^+(m)$ is called the "outgoing partition" and denoted by $p_\gamma^{\mathrm{out}}(m)$, while $p_\gamma^-(m)$ is called the "incoming partition" and denoted by $p_\gamma^{\mathrm{in}}(m)$. It is never too late to change your terminology to make it clearer.

The partitions $p_\gamma^\pm(m)$ are defined as follows. If γ is positive hyperbolic, then

$$p_\gamma^+(m) = p_\gamma^-(m) = (1,\ldots,1).$$

Thus, if equality holds in the writhe bound for C, then C can never have an end at a multiple cover of a positive hyperbolic Reeb orbit. If γ is negative hyperbolic, then

$$p_\gamma^+(m) = p_\gamma^-(m) = \begin{cases} (2,\ldots,2), & m \text{ even}, \\ (2,\ldots,2,1), & m \text{ odd}. \end{cases}$$

Suppose now that γ is elliptic with rotation angle θ with respect to some trivialization τ of $\xi|_\gamma$, see Section 3.2. Then $p_\gamma^\pm(m) = p_\theta^\pm(m)$, where the partitions $p_\theta^\pm(m)$ are defined as follows.

To define $p_\theta^+(m)$, let $\Lambda_\theta^+(m)$ be the maximal concave polygonal path in the plane (i.e. graph of a concave function) with vertices at lattice points which starts at the origin, ends at $(m, \lfloor m\theta \rfloor)$, and lies below the line $y = \theta x$. That is, $\Lambda_\theta^+(m)$ is the non-vertical part of the boundary of the convex hull of the set of lattice points (x,y) with $0 \le x \le m$ and $y \le \theta x$. Then $p_\theta^+(m)$ consists of the horizontal displacements of the segments of $\Lambda_\theta^+(m)$ connecting consecutive lattice points.

The partition $p_\theta^-(m)$ is defined analogously from the path $\Lambda_\theta^-(m)$, which is the minimal convex polygonal path with vertices at lattice points which starts at the origin, ends at $(m, \lceil m\theta \rceil)$, and lies above the line $y = \theta x$. An equivalent definition is $p_\theta^-(m) = p_{-\theta}^+(m)$.

The partition $p_\theta^\pm(m)$ depends only on the class of θ in $\mathbb{R}/\mathbb{Z}$, and so $p_\gamma^\pm(m)$ does not depend on the choice of trivialization τ.

The simplest example, which we will need for the computations in Section 4, is that if $\theta \in (0, 1/m)$, then

$$\begin{aligned} p_\theta^+(m) &= (1,\ldots,1), \\ p_\theta^-(m) &= (m). \end{aligned} \tag{3.16}$$

The partitions are more complicated for other θ, see Figure 1.

If $m > 1$, then $p_\theta^+(m)$ and $p_\theta^-(m)$ are disjoint. (This makes the gluing theory to prove $\partial^2 = 0$ nontrivial, see Section 5.4.) This is a consequence of the following exercise, which may help in understanding the partitions.

<table>
<tr><th></th><th>2</th><th>3</th><th>4</th><th>5</th><th>6</th><th>7</th><th>8</th></tr>
<tr><td>7/8, 1</td><td rowspan="11">2</td><td rowspan="7">3</td><td rowspan="5">4</td><td rowspan="4">5</td><td rowspan="3">6</td><td rowspan="2">7</td><td>8</td></tr>
<tr><td>6/7, 7/8</td><td>7, 1</td></tr>
<tr><td>5/6, 6/7</td><td>6, 1</td><td>6, 2</td></tr>
<tr><td>4/5, 5/6</td><td>5, 1</td><td>5, 2</td><td>5, 3</td></tr>
<tr><td>3/4, 4/5</td><td>4, 1</td><td>4, 2</td><td>4, 3</td><td>4, 4</td></tr>
<tr><td>5/7, 3/4</td><td rowspan="2">3,1</td><td rowspan="2">3,2</td><td rowspan="2">3,3</td><td>7</td><td>7, 1</td></tr>
<tr><td>2/3, 5/7</td><td>3, 3, 1</td><td>3, 3, 2</td></tr>
<tr><td>5/8, 2/3</td><td rowspan="4">2,1</td><td rowspan="4">2,2</td><td rowspan="2">5</td><td rowspan="2">5,1</td><td rowspan="2">5,2</td><td>8</td></tr>
<tr><td>3/5, 5/8</td><td>5, 2, 1</td></tr>
<tr><td>4/7, 3/5</td><td rowspan="2">2,2,1</td><td rowspan="2">2,2,2</td><td>7</td><td>7, 1</td></tr>
<tr><td>1/2, 4/7</td><td>2, 2, 2, 1</td><td>2, 2, 2, 2</td></tr>
<tr><td>3/7, 1/2</td><td rowspan="11">1, 1</td><td rowspan="4">3</td><td rowspan="4">3,1</td><td rowspan="2">5</td><td rowspan="2">5,1</td><td>7</td><td>7, 1</td></tr>
<tr><td>2/5, 3/7</td><td>5, 1, 1</td><td>5, 3</td></tr>
<tr><td>3/8, 2/5</td><td rowspan="2">3,1,1</td><td rowspan="2">3,3</td><td rowspan="2">3,3,1</td><td>8</td></tr>
<tr><td>1/3, 3/8</td><td>3, 3, 1, 1</td></tr>
<tr><td>2/7, 1/3</td><td rowspan="7">1, 1, 1</td><td rowspan="2">4</td><td rowspan="2">4,1</td><td rowspan="2">4,1,1</td><td>7</td><td>7, 1</td></tr>
<tr><td>1/4, 2/7</td><td>4, 1, 1, 1</td><td>4, 4</td></tr>
<tr><td>1/5, 1/4</td><td rowspan="5">1, 1, 1, 1</td><td>5</td><td>5, 1</td><td>5, 1, 1</td><td>5, 1, 1, 1</td></tr>
<tr><td>1/6, 1/5</td><td rowspan="4">1,...,1</td><td>6</td><td>6, 1</td><td>6, 1, 1</td></tr>
<tr><td>1/7, 1/6</td><td rowspan="3">1, ..., 1</td><td>7</td><td>7, 1</td></tr>
<tr><td>1/8, 1/7</td><td rowspan="2">1,...,1</td><td>8</td></tr>
<tr><td>0, 1/8</td><td>1, ..., 1</td></tr>
</table>

Fig. 1. The positive partitions $p_\theta^+(m)$ for $2 \le m \le 8$ and all θ. The left column shows the interval in which θ mod 1 lies, and the top row indicates m. (Borrowed from [21])

Exercise 3.13. (See answer in Appendix.) Write $p_\theta^+(m) = (q_1, \ldots, q_k)$ and $p_\theta^-(m) = (r_1, \ldots, r_l)$.

(a) Show that if (a, b) is an edge vector of the path $\Lambda_\theta^+(m)$, then $b = \lfloor a\theta \rfloor$.

(b) Show that $\sum_{i\in I} \lfloor q_i\theta \rfloor = \lfloor \sum_{i\in I} q_i\theta \rfloor$ for each subset $I \subset \{1, \ldots, k\}$.

(c) Show that there do not exist proper subsets $I \subset \{1, \ldots, k\}$ and $J \subset \{1, \ldots, l\}$ such that $\sum_{i\in I} q_i = \sum_{j\in J} r_j$.

Here is a related combinatorial exercise, some of which is needed for the proofs that ∂ is well-defined and $\partial^2 = 0$ in Section 5.3 and Section 5.4.

Exercise 3.14. (See answer in Appendix.) Fix an irrational number θ and a positive integer m. Suppose γ is an embedded elliptic Reeb orbit with rotation angle θ.

(a) Show that if $u: C \to \mathbb{R} \times \gamma$ is a degree m branched cover, regarded as a holomorphic curve in $\mathbb{R} \times Y$, then the Fredholm index[15] $\operatorname{ind}(u) \geq 0$.

(b) If $(a_1, \ldots, a_k)$ and $(b_1, \ldots, b_l)$ are partitions of m, define $(a_1, \ldots, a_k) \geq (b_1, \ldots, b_l)$ if there is a branched cover u of $\mathbb{R} \times \gamma$ with positive ends at γ^{a_i}, negative ends at γ^{b_j}, and $\operatorname{ind}(u) = 0$. Show that $\geq$ is a partial order on the set of partitions of m.

(c) Show that $p_\theta^-(m) \geq p_\theta^+(m)$.

(d) Show that there does not exist any partition q with $q > p_\theta^-(m)$ or $p_\theta^+(m) > q$.

Remark 3.15. If $\mathcal{C} \in \mathcal{M}(\alpha, \beta)$ contributes to the differential or the U map, and if $\mathcal{C}$ contains trivial cylinders, then additional partition conditions must hold; see [21, Prop. 7.1] and [29, Lem. 7.28] for these conditions.

4. More Examples of ECH

The calculation of the ECH of an ellipsoid in Section 3.7 was fairly simple because we just had to determine the grading of each generator. We now outline some more complicated calculations which require counting holomorphic curves. These are useful for further understanding the machinery, and relevant to the symplectic embedding obstructions described in Section 1.2.

4.1. The U Map on the ECH of an Ellipsoid

We first return to the ellipsoid example from Section 3.7. Recall from (3.13) that $ECH_*(\partial E(a,b), \lambda, 0)$ has one generator of grading $2k$ for each $k = 0, 1, \ldots$; denote this generator by ζ_k. To calculate the ECH capacities of $E(a,b)$ in Section 1.5, we needed:

[15] The Fredholm index of a possibly multiply covered curve $u: C \to \mathbb{R} \times Y$ is defined as in (3.1) with $c_\tau(C)$ replaced by $c_1(u^*\xi, \tau)$.

Proposition 4.1. *For any J, the U map on $ECH_*(\partial E(a,b),\lambda,0,J)$ is given by*

$$U\zeta_k = \zeta_{k-1}, \quad k > 0. \tag{4.1}$$

As mentioned in Example 1.11, this follows from the isomorphism with Seiberg-Witten theory. However it is instructive to try to prove Proposition 4.1 directly in ECH, without using Seiberg-Witten theory.

First of all, we can see directly in this case that the U map does not depend on the almost complex structure J. The idea is that if we generically deform J, then similarly to the compactness part of the proof that $\partial^2 = 0$, see Lemma 5.12, the chain map U_z can change only if at some time there is a broken holomorphic curve containing a level with $I = 1$. But there are no $I = 1$ curves by the Index Parity property (3.7) since all Reeb orbits are elliptic.

We now sketch a direct proof of Proposition 4.1 in the special case when $a = 1-\varepsilon$ and $b = 1+\varepsilon$ where $\varepsilon > 0$ is sufficiently small with respect to k. (One can probably prove the general case similarly with more work.)

If ε is sufficiently small with respect to k, then ζ_k is the kth generator in the sequence

$$1, \gamma_1, \gamma_2, \gamma_1^2, \gamma_1\gamma_2, \gamma_2^2, \gamma_1^3, \gamma_1^2\gamma_2, \gamma_1\gamma_2^2, \gamma_2^3, \ldots$$

(indexed starting at $k = 0$). So to prove Proposition 4.1 in our special case, it is enough to show the following:

Lemma 4.2. *If $a = 1-\varepsilon$ and $b = 1+\varepsilon$, then the U map on $ECH_*(\partial E(a,b), \lambda, 0, J)$ is given by:*

(a) $U(\gamma_1^i\gamma_2^j) = \gamma_1^{i+1}\gamma_2^{j-1}$ *if $j > 0$ and $\varepsilon > 0$ is sufficiently small with respect to $i+j$.*

(b) $U(\gamma_1^i) = \gamma_2^{i-1}$ *if $i > 0$ and $\varepsilon > 0$ is sufficiently small with respect to i.*

Proof. The proof has three steps.

Step 1. We first determine the types of holomorphic curves we need to count.

Let $\mathcal{C}$ be a holomorphic current that contributes to $U_z(\gamma_1^i\gamma_2^j)$ where $i+j > 0$. Write $\mathcal{C} = \mathcal{C}_0 \sqcup C_2$ as in Proposition 3.7. It follows from the partition conditions (3.16) that C_2 has at most one positive end at a cover of γ_1, all positive ends of C_2 at covers of γ_2 have multiplicity 1, all negative ends of C_2 at covers of γ_1 have multiplicity 1, and C_2 has at most one negative end at a cover of γ_2.

Exercise 4.3. Deduce from this and the equation $\mathrm{ind}(C_2)=2$ that if $j=0$, then C_2 is a cylinder if $i>1$, and a plane if $i=1$, assuming that $\varepsilon>0$ is sufficiently small with respect to i. (See answer in Appendix.)

Exercise 4.4. Similarly show that if $j>0$, then C_2 is a cylinder with a positive end at γ_2 and a negative end at γ_1, assuming that $\varepsilon>0$ is sufficiently small with respect to $i+j$. (See answer in Appendix.)

Step 2. We now observe that the transversality conditions needed to define U_z, see Section 3.8, hold automatically for any symplectization-admissible J. This follows from two general facts. First, if C is an immersed irreducible J-holomorphic curve such that

$$2g(C)-2+h_+(C)<\mathrm{ind}(C), \tag{4.2}$$

then C is automatically regular. Here $g(C)$ denotes the genus of the domain of C, and $h_+(C)$ denotes the number of ends of C at positive hyperbolic orbits, including even covers of negative hyperbolic orbits. This and much more general automatic transversality results are proved in [62]. Second, if

$$2g(C)-2+\mathrm{ind}(C)+h_+(C)=0, \tag{4.3}$$

then every nonzero element of the kernel of the deformation operator of C is nonvanishing[16]. If $C=C_2$ where C_2 is one of the holomorphic curves described in Step 1, then C_2 has genus zero, Fredholm index 2, and all ends at elliptic orbits, so both conditions (4.2) and (4.3) hold, and we conclude that C_2 is regular and the map (3.14) has no kernel, which is exactly the transversality needed to define U_z.

Step 3. We now count the holomorphic curves C_2 described in Step 1. To do so, consider the case $a=b=1$. Here the contact form is not nondegenerate, as every point on $Y=S^3$ is on a Reeb orbit. Indeed, the set of embedded

[16]The left side of (4.3) is called the "normal Chern number" by Wendl [61]. Any holomorphic curve u in $\mathbb{R}\times Y$ has normal Chern number ≥ 0, with equality only if the projection of u to Y is an immersion. In favorable cases one can further show that the projection of u to Y is an embedding. One such favorable case is described in [31, Prop. 3.4], which is used to characterize contact three-manifolds in which all Reeb orbits are elliptic.

Reeb orbits can be identified with $\mathbb{C}P^1$, so that the map $S^3 \to \mathbb{C}P^1$ sending a point to the Reeb orbit on which it lies is the Hopf fibration. This is an example of a "Morse-Bott" contact form.

It is explained by Bourgeois [2] how one can understand holomorphic curves for a nondegenerate perturbation of a Morse-Bott contact form in terms of holomorphic curves for the Morse-Bott contact form itself. In the present case, this means that we can understand holomorphic curves for the ellipsoid with $a = 1 - \varepsilon$, $b = 1 + \varepsilon$, in terms of holomorphic curves for the sphere with $a = b = 1$. Specifically, let $p_i \in \mathbb{C}P^1$ denote the point corresponding to the Reeb orbit γ_i for $i = 1, 2$. Choose a Morse function $f : \mathbb{C}P^1 \to \mathbb{R}$ with an index 2 critical point at γ_2 and index 0 critical point at γ_1. Then [2] tells us the following.

First, a holomorphic cylinder for the perturbed contact form with a positive end at γ_2 and a negative end at γ_1 (modulo $\mathbb{R}$ translation) corresponds to a negative gradient flow line of f from p_2 to p_1. If we choose a base point $\overline{z} \in \mathbb{C}P^1 \setminus \{p_1, p_2\}$, then there is exactly one such flow line passing through $\overline{z}$. One can deduce from this that if we choose a base point $z \in Y$ which is not on γ_1 or γ_2, then there is exactly one holomorphic cylinder with a positive end at γ_2 and a negative end at γ_1 passing through $(0, z)$. This proves part (a) of Lemma 4.2.

Second, to prove part (b) of Lemma 4.2, we need to count holomorphic cylinders (or planes when $i = 1$) C for the Morse-Bott contact form with a positive end at γ_1^i, and a negative end at γ_2^{i-1} when $i > 0$, which pass through a base point. To count these, let $\mathcal{L}$ denote the tautological line bundle over $\mathbb{C}P^1$. Let J denote the canonical complex structure on $\mathcal{L}$, and let $Z \subset \mathcal{L}$ denote the zero section.

Exercise 4.5. One can identify $\mathcal{L} \setminus Z \simeq \mathbb{R} \times S^3$ so that J corresponds to a symplectization-admissible almost complex structure. A meromorphic section ψ of $\mathcal{L}$ determines a holomorphic curve in $\mathbb{R} \times S^3$ with positive ends corresponding to the zeroes of ψ, and negative ends corresponding to the poles of ψ. Conversely, a holomorphic curve in $\mathbb{R} \times S^3$ which intersects each fiber of $\mathcal{L} \setminus Z \to \mathbb{C}P^1$, except for the fibers over the Reeb orbits at the positive and negative ends, transversely in a single point, comes from a meromorphic section of $\mathcal{L}$.

If C is a holomorphic curve as in the paragraph preceding the above exercise, then by the definition of linking number in S^3, the curve C has algebraic intersection number 1 with each fiber of $\mathcal{L} \setminus Z$ over $\mathbb{C}P^1 \setminus \{p_1, p_2\}$.

By intersection positivity, C intersects each such fiber transversely in a single point. It follows then from Exercise 4.5 that to compute $U\gamma_1^i$, we need to count meromorphic sections of $\mathcal{L}$ with a zero of order i at p_1, a pole of order $i-1$ at p_2, and no other zeroes or poles, which pass through a base point in $\mathcal{L}\setminus Z$. There is exactly one such meromorphic section, and this completes the proof of Lemma 4.2. ■

4.2. The ECH of T^3

Our next example of ECH is more complicated, but will ultimately be useful in computing many examples of ECH capacities. We consider

$$Y = T^3 = (\mathbb{R}/2\pi\mathbb{Z}) \times (\mathbb{R}/\mathbb{Z})^2.$$

Let θ denote the $\mathbb{R}/2\pi\mathbb{Z}$ coordinate and let x, y denote the two $\mathbb{R}/\mathbb{Z}$ coordinates. We start with the contact form

$$\lambda_1 = \cos\theta\, dx + \sin\theta\, dy. \tag{4.4}$$

Let $\xi_1 = \operatorname{Ker}(\lambda_1)$; we now describe how to compute $ECH_*(T^3, \xi_1, 0)$, following [28].

Perturbing the Contact Form. The Reeb vector field associated to λ_1 is

$$R_1 = \cos\theta \frac{\partial}{\partial x} + \sin\theta \frac{\partial}{\partial y}.$$

If $\tan\theta \in \mathbb{Q} \cup \{\infty\}$, so that the vector $(\cos\theta, \sin\theta)$ is a positive real multiple of a vector (a, b) where a, b are relatively prime integers, then every point on $\{\theta\} \times (\mathbb{R}/\mathbb{Z})^2$ is on an embedded Reeb orbit γ in the homology class $(0, a, b) \in H_1(T^3)$. The symplectic action of the Reeb orbit γ is

$$\mathcal{A}(\gamma) = \sqrt{a^2 + b^2}.$$

In particular, there is a circle $S_{a,b}$ of such Reeb orbits. Thus the contact form λ_1 is not nondegenerate; again it is Morse-Bott.

To compute the ECH of ξ_1, we will perturb λ_1 to a nondegenerate contact form. Given a, b, one can perturb the contact form λ_1 near $S_{a,b}$ so that, modulo longer Reeb orbits, the circle of Reeb orbits $S_{a,b}$ becomes just two embedded Reeb orbits, one elliptic with rotation angle slightly positive, and one positive hyperbolic. We denote these by $e_{a,b}$ and $h_{a,b}$. The orbits $e_{a,b}$ and

$h_{a,b}$ are still in the homology class $(0,a,b)$, and have symplectic action close to $\sqrt{a^2+b^2}$, with the action of $e_{a,b}$ slightly greater than that of $h_{a,b}$. For any given $L>0$, one can perform such a perturbation for all of the finitely many pairs of relatively prime integers (a,b) with $\sqrt{a^2+b^2}<L$, to obtain a contact form λ for which the embedded Reeb orbits with symplectic action less than L are the elliptic orbits $e_{a,b}$ and the hyperbolic orbits $h_{a,b}$ where (a,b) ranges over all pairs of relatively prime integers with $\sqrt{a^2+b^2}<L$.

It is probably not possible to do this for $L=\infty$, i.e. to find a contact form such that the embedded Reeb orbits of all actions are the orbits $e_{a,b}$ and $h_{a,b}$ where (a,b) ranges over all pairs of relatively prime integers. Rather, one can show that to calculate the ECH of ξ_1, we can perturb as above for a given L, compute the filtered ECH in symplectic action less than L, and take the direct limit as $L\to\infty$. In the calculations below, we only consider generators of symplectic action less than L, and we omit L from the notation.

The Generators. A generator of the chain complex $ECC_*(Y,\lambda,0,J)$ now consists of a finite set of Reeb orbits $e_{a,b}$ and $h_{a,b}$ with positive integer multiplicities, where each $h_{a,b}$ has multiplicity 1, and the sum with multiplicities of all the vectors (a,b) is $(0,0)$. To describe this more simply, if (a,b) are relatively prime integers and if m is a positive integer, let $e_{ma,mb}$ denote the elliptic orbit $e_{a,b}$ with multiplicity m; and let $h_{ma,mb}$ denote the hyperbolic orbit $h_{a,b}$, together with the elliptic orbit $e_{a,b}$ with multiplicity $m-1$ when $m>1$. A chain complex generator then consists of a finite set of symbols $e_{a,b}$ and $h_{a,b}$, where each (a,b) is a pair of (not necessarily relatively prime) integers which are not both zero, no pair (a,b) appears more than once, and the sum of the vectors (a,b) that appear is zero. If we arrange the vectors (a,b) head to tail in order of increasing slope, we obtain a convex polygon in the plane. Thus, a generator of the chain complex $ECC_*(Y,\lambda,0,J)$ can be represented as convex polygon Λ in the plane, modulo translation, with vertices at lattice points, with each edge labeled either 'e' or 'h'. The polygon can be a 2-gon (for a generator such as $e_{a,b}e_{-a,-b}$) or a point (for the empty set of Reeb orbits). The symplectic action of the generator is approximately the Euclidean length of the polygon Λ.

The Grading. The two-plane field ξ_1 is trivial; indeed ∂_θ defines a global trivialization τ. Thus $c_1(\xi_1)=0$, and the chain complex $ECC_*(T^3,\lambda,0,J)$ has a canonical $\mathbb{Z}$-grading, in which the empty set has grading zero.

Lemma 4.6. *The canonical $\mathbb{Z}$-grading of a generator Λ is given by*

$$|\Lambda| = 2(\mathcal{L}(\Lambda) - 1) - h(\Lambda), \tag{4.5}$$

where $\mathcal{L}(\Lambda)$ denotes the number of lattice points enclosed by Λ (including lattice points on the edges), and $h(\Lambda)$ denotes the number of edges of Λ that are labeled 'h'.

Proof. As in (3.11), we can write the grading of a generator Λ as

$$|\Lambda| = c_\tau(\Lambda) + Q_\tau(\Lambda) + CZ_\tau^I(\Lambda).$$

Since τ is a global trivialization, $c_\tau(\Lambda) = 0$. We also have $CZ_\tau(e_{a,b}) = 1$ and $CZ_\tau(h_{a,b}) = 0$; consequently,

$$CZ_\tau^I(\Lambda) = m(\Lambda) - h(\Lambda),$$

where $m(\Lambda)$ denotes the total divisibility of all edges of Λ. Finally, it is a somewhat challenging exercise (which can be solved by the argument in [27, Lem. 3.7]) to show that

$$Q_\tau(\Lambda) = 2\,\mathrm{Area}(\Lambda)$$

where $\mathrm{Area}(\Lambda)$ denotes the area enclosed by Λ. Now Pick's formula for the area of a lattice polygon asserts that

$$2\,\mathrm{Area}(\Lambda) = 2\mathcal{L}(\Lambda) - m(\Lambda) - 2.$$

The grading formula (4.5) follows from the above four equations. ■

Combinatorial Formula for the Differential. Define a combinatorial differential

$$\delta : ECC_*(T^3, \lambda, 0, J) \longrightarrow ECC_{*-1}(T^3, \lambda, 0, J)$$

as follows. If Λ is a generator, then $\delta\Lambda$ is the sum over all labeled polygons Λ' that are obtained from Λ by "rounding a corner" and "locally losing one 'h'". Here "rounding a corner" means replacing the polygon Λ by the boundary of the convex hull of the set of enclosed lattice points with one corner removed. "Locally losing one 'h'" means that of the two edges adjacent to the corner that is rounded, at least one must be labeled 'h'; if only one is labeled 'h', then all edges created or shortened by the rounding are labeled 'e'; otherwise exactly one of the edges created or shortened by the rounding is labeled 'h'. All edges not created or shortened by the rounding keep their previous labels. It follows from (4.5) that the combinatorial differential δ decreases the grading by 1, since $\mathcal{L}(\Lambda') = \mathcal{L}(\Lambda) - 1$ and $h(\Lambda') = h(\Lambda) - 1$. A less trivial combinatorial fact, proved in [28, Cor. 3.13], is that $\delta^2 = 0$.

Proposition 4.7 [28, §11.3]. *For every $L > 0$, the perturbed contact form λ and almost complex structure J can be chosen so that up to symplectic action L, the ECH differential ∂ agrees with the combinatorial differential δ.*

We will describe some of the proof of Proposition 4.7 at the end of this subsection.

The homology of the combinatorial differential δ is computed in [28] (there with $\mathbb{Z}$ coefficients), and the conclusion (with $\mathbb{Z}/2$ coefficients) is that

$$ECH_*(T^3, \xi_1, 0) \simeq \begin{cases} (\mathbb{Z}/2)^3, & * \geq 0, \\ 0, & * < 0. \end{cases} \tag{4.6}$$

Exercise 4.8. Prove that the homology of the combinatorial differential δ in degree 0 is isomorphic to $(\mathbb{Z}/2)^3$.

The *U* Map. To compute ECH capacities, we do not need to know the homology (4.6), but rather the following combinatorial formula for the U map. Pick $\theta \in \mathbb{R}/2\pi\mathbb{Z}$ with $\tan\theta$ irrational. Define a combinatorial map

$$U_\theta : ECC_*(T^3, \lambda, 0, J) \longrightarrow ECC_{*-2}(T^3, \lambda, 0, J)$$

as follows. If Λ is a generator, then it has a distinguished corner c_θ such that the oriented line T through c_θ with direction vector $(\cos\theta, \sin\theta)$ intersects Λ only at c_θ, with the rest of Λ lying to the left of T. Then U_θ is the sum over all generators Λ' obtained from Λ by rounding the distinguished corner c_θ and "conserving the h labels". To explain what this last condition means, note that Λ' also has a distinguished corner c'_θ. If the edge of Λ preceding c_θ is labeled 'h', then exactly one of the new or shortened edges of Λ' preceding c'_θ is labeled 'h'; otherwise all new or shortened edges of Λ' preceding c'_θ are labeled 'e'. Likewise for the edge of Λ following c_θ and the new or shortened edges of Λ' following c'_θ. All other edge labels are unchanged.

To connect this with the U map on ECH, let $z = (\theta, x, y) \in T^3$ where $x, y \in \mathbb{R}/\mathbb{Z}$ are arbitrary.

Proposition 4.9 [28, §12.1.4]. *For any $L > 0$, one can choose λ and J as in Proposition 4.7 so that up to symplectic action L, we have $U_z = U_\theta$, modulo terms that decrease the number of 'h' labels.*

In particular, if all edges of Λ are labeled 'e', then $U_z\Lambda$ is the generator Λ' obtained from Λ by rounding the distinguished corner c_θ and keeping all edges labeled 'e'. (If Λ is a point then $U_z\Lambda = 0$.)

ECH Spectrum. We now use the above facts to compute the ECH spectrum of (T^3, λ_1) in terms of a discrete isoperimetric problem.

Proposition 4.10. *The ECH spectrum of (T^3, λ_1) is given by*

$$c_k(T^3, \lambda_1) = \min\{\ell(\Lambda) \mid \mathcal{L}(\Lambda) = k+1\}, \tag{4.7}$$

where the minimum is over closed convex polygonal paths Λ with vertices at lattice points, ℓ denotes the Euclidean length, and $\mathcal{L}(\Lambda)$ denotes the number of lattice points enclosed by Λ, including lattice points on the edges.

Proof. Fix a nonnegative integer k. Let Λ_k be a length-minimizing closed convex polygon with vertices at lattice points subject to the constraint $\mathcal{L}(\Lambda_k) = k+1$. We need to show that $c_k(T^3, \lambda_1) = \ell(\Lambda_k)$.

Fix $z \in T^3$ for use in defining the chain map U_z. Choose $L > \ell(\Lambda_k)$, and let λ and J be a perturbed contact form and almost complex structure supplied by Propositions 4.7 and 4.9. Label all edges of Λ_k by 'e' in order to regard Λ_k as a generator of the chain complex $ECC(T^3, \lambda, 0, J)$. Then Λ_k is a cycle by Proposition 4.7, and $U_z^k \Lambda_k = \emptyset$ by Proposition 4.9. Thus $c_k(T^3, \lambda)$ is less than or equal to the symplectic action of Λ_k, which is approximately $\ell(\Lambda_k)$. It follows from the limiting definition of the ECH spectrum for degenerate contact forms in Section 1.5 that $c_k(T^3, \lambda_1) \le \ell(\Lambda_k)$.

To complete the proof, we now show that $c_k(T^3, \lambda_1) \ge \ell(\Lambda_k)$. It is enough to show that if Λ is any other generator with $\langle U_z^k \Lambda, \emptyset\rangle \ne 0$, then $\ell(\Lambda) \ge \ell(\Lambda_k)$. Since $|\Lambda| = 2k$, it follows from the grading formula (4.5) that

$$\mathcal{L}(\Lambda) = k + 1 + \frac{h(\Lambda)}{2}.$$

We then have

$$\ell(\Lambda) \ge \ell(\Lambda_{k+h(\Lambda)/2}) \ge \ell(\Lambda_k)$$

where the first inequality holds by definition, and the second inequality holds because rounding corners of polygons decreases length[17]. ■

[17] It is a combinatorial exercise to prove that rounding corners of polygons decreases length, see [28, Lem. 2.14].

Computing the Differential. We now indicate a bit of what is involved in the proof of Proposition 4.7; similar arguments prove Proposition 4.9. For the application to ECH capacities, one may skip ahead to Section 4.3.

The easier half of the proof of Proposition 4.7 is to show that λ and J can be chosen so that

$$\langle \partial \Lambda, \Lambda' \rangle \neq 0 \Longrightarrow \langle \delta \Lambda, \Lambda' \rangle \neq 0. \tag{4.8}$$

The following lemma is a first step towards proving (4.8).

Lemma 4.11. *Let $\mathcal{C} \in \mathcal{M}(\Lambda, \Lambda')$ be a holomorphic current that contributes to the differential ∂, and write $\mathcal{C} = \mathcal{C}_0 \sqcup C_1$ as in Proposition* 3.7. *Then C_1 has genus zero, and one of the following three alternatives holds:*

(i) *C_1 is a cylinder with positive end at an embedded elliptic orbit $e_{a,b}$ and negative end at $h_{a,b}$.*

(ii) *C_1 has two positive ends, and the number of positive ends at hyperbolic orbits is one more than the number of negative ends at hyperbolic orbits.*

(iii) *C_1 has three positive ends, all at hyperbolic orbits; and all negative ends of C_1 are at elliptic orbits.*

Proof. Let us first see what the Fredholm index formula (3.1) tells us about C_1. Let g denote the genus of C_1, let e_+ denote the number of positive ends of C_1 at elliptic orbits, let h_+ denote the number of positive ends of C_1 at hyperbolic orbits, and let e_- and h_- denote the number of negative ends of C_1 at elliptic and hyperbolic orbits respectively. Then

$$\chi(C_1) = 2 - 2g - e_+ - h_+ - e_- - h_-$$

and

$$CZ_\tau^{\mathrm{ind}}(C) = e_+ - e_-,$$

so by the Fredholm index formula (3.1) we have

$$\mathrm{ind}(C_1) = 2g - 2 + 2e_+ + h_+ + h_-.$$

Since $\mathrm{ind}(C_1) = 1$, we obtain

$$2g + 2e_+ + h_+ + h_- = 3. \tag{4.9}$$

Since the differential ∂ decreases symplectic action, C_1 has at least one positive end.

Exercise 4.12. Further use the fact that the differential ∂ decreases symplectic action to show that $g=0$. (See answer in Appendix.)

If C_1 has exactly one positive end, then similarly to the solution to Exercise 4.12, this positive end is at an elliptic orbit. By the partition conditions (3.16), this positive end is at an embedded elliptic orbit $e_{a,b}$. Then, similarly to the solution to Exercise 4.12, C_1 has exactly one negative end, which is at $h_{a,b}$, so alternative (i) holds.

If C_1 has more than one positive end, then it follows from Equation (4.9) that alternative (ii) or (iii) holds. ■

Now $\langle \delta\Lambda, \Lambda' \rangle \neq 0$ is only possible in case (ii). So to prove (4.8) we would like to rule out alternatives (i) and (iii). In fact alternative (i) cannot be ruled out; there are two holomorphic cylinders from $e_{a,b}$ to $h_{a,b}$ for each pair of relatively prime integers (a,b). These arise in the Morse-Bott picture from the two flow gradient flow lines of the Morse function on the circle of Reeb orbits $S_{a,b}$ that we used to perturb the Morse-Bott contact form λ_1, similarly to the proof of Proposition 4.1(a). However these cylinders cancel[18] in the ECH differential ∂. Alternative (iii) may occur depending on how exactly one perturbs the Morse-Bott contact form λ_1. However it is shown in [28, §11.3, Step 5] that the perturbation λ and almost complex structure J can be chosen so that alternative (iii) does not happen.

The main remaining step in the proof of (4.8) is to show that λ and J above can be chosen so that if $\langle \partial\Lambda, \Lambda' \rangle \neq 0$, then the polygons Λ and Λ' can be translated so that Λ' is nested inside Λ. The proof uses intersection positivity, see [28, §10.3].

To complete the proof of Proposition 4.7, we need to prove the converse of (4.8), namely that λ and J above can be chosen so that if $\langle \delta\Lambda, \Lambda' \rangle \neq 0$ then $\langle \partial\Lambda, \Lambda' \rangle \neq 0$. One can calculate $\langle \partial\Lambda, \Lambda' \rangle$ by counting appropriate holomorphic curves for the Morse-Bott contact form λ_1. Work of Taubes [56] and Parker [47] determines the latter curves in terms of tropical geometry. Unfortunately it would take us too far afield to explain this story here.

[18]There is also a "twisted" version of ECH in which these cylinders do not cancel in the differential, see [28, §12.1.1].

4.3. ECH Capacities of Convex Toric Domains

We now use the results of Section 4.2 to compute the ECH capacities of a large family of examples. Let Ω be a compact domain in $[0,\infty)^2$ with piecewise smooth boundary. Define a "toric domain" or "Reinhardt domain"

$$X_\Omega = \{(z_1,z_2)\in\mathbb{C}^2 \mid (\pi|z_1|^2,\pi|z_2|^2)\in\Omega\}.$$

For example, if Ω is the triangle with vertices $(0,0)$, $(a,0)$, and $(0,b)$, then X_Ω is the ellipsoid $E(a,b)$. If Ω is the rectangle with vertices $(0,0)$, $(a,0)$, $(0,b)$, and (a,b), then X_Ω is the polydisk $P(a,b)$.

Assume now that Ω is convex and does not touch the axes. We can then state a formula for the ECH capacities of X_Ω, similar to Proposition 4.10. Let $\Omega'\subset\mathbb{R}^2$ be a translate of Ω that contains the origin in its interior. Let $\|\cdot\|$ denote the (not necessarily symmetric) norm on $\mathbb{R}^2$ that has Ω' as its unit ball. Let $\|\cdot\|^*$ denote the dual norm on $(\mathbb{R}^2)^*$, which we identify with $\mathbb{R}^2$ via the Euclidean inner product $\langle\cdot,\cdot\rangle$. That is, if $v\in\mathbb{R}^2$, then

$$\|v\|^* = \max\{\langle v,w\rangle \mid w\in\partial\Omega'\}.$$

If Λ is a polygonal path in $\mathbb{R}^2$, let $\ell_\Omega(\Lambda)$ denote the length of the path Λ as measured using the dual norm $\|\cdot\|^*$, i.e. the sum of the $\|\cdot\|^*$-norms of the edge vectors of Λ.

Exercise 4.13. If Λ is a loop, then $\ell_\Omega(\Lambda)$ does not depend on the choice of translate Ω' of Ω. (See answer in Appendix.)

Theorem 4.14 [24, Thm. 1.11][19]**.** *If Ω is convex and does not intersect the axes, then*

$$c_k(X_\Omega) = \min\{\ell_\Omega(\Lambda) \mid \mathcal{L}(\Lambda) = k+1\}, \tag{4.10}$$

where the minimum is over closed convex polygonal paths Λ with vertices at lattice points, and $\mathcal{L}(\Lambda)$ denotes the number of lattice points enclosed by Λ, including lattice points on the edges.

Remark 4.15. One can weaken and possibly drop the assumption that Ω does not intersect the axes. For example, the formula (4.10) is still correct when Ω is a triangle or rectangle with two sides on the axes, so that X_Ω

[19]The definition of X_Ω in [24] is different, but symplectomorphic to the one given here.

is an ellipsoid or polydisk. This is a consequence of the following exercise, which should help with understanding the combinatorial formula (4.10).

Exercise 4.16.

(a) Suppose that Ω is a convex polygon. Show that the minimum on the right hand side of (4.10) is the same if it is taken over closed convex polygonal paths Λ with arbitrary vertices whose edges are parallel to the edges of Ω.

(b) Use part (a), together with the formulas (1.7) and (1.8) for the ECH capacities of ellipsoids and polydisks, to verify that Equation (4.10) is correct when X_Ω is an ellipsoid or a polydisk.

Proof of Theorem 4.14. We first need to understand a bit about the symplectic geometry of the domains X_Ω. Define coordinates $\mu_1, \mu_2 \in (0,\infty)$ and $\theta_1, \theta_2 \in \mathbb{R}/2\pi\mathbb{Z}$ on $(\mathbb{C}^*)^2$ by writing $z_k = \sqrt{\mu_k/\pi}e^{i\theta_k}$ for $k = 1, 2$. In these coordinates, the standard symplectic form on $\mathbb{C}^2$ restricts to

$$\omega = \frac{1}{2\pi}\sum_{k=1}^{2} d\mu_k\, d\theta_k. \tag{4.11}$$

A useful corollary of this is that

$$\operatorname{vol}(X_\Omega) = \operatorname{area}(\Omega). \tag{4.12}$$

Exercise 4.17. Use (4.11) to show that if Ω_1 and Ω_2 do not intersect the axes, and if Ω_2 can be obtained from Ω_1 by the action of $SL_2\mathbb{Z}$ and translation, then X_{Ω_1} is symplectomorphic to X_{Ω_2}.

Now suppose that Ω has smooth boundary, does not intersect the axes, and is star-shaped with respect to some origin $(\eta_1, \eta_2) \in \operatorname{int}(\Omega)$. This last condition means that each ray starting at (η_1, η_2) intersects $\partial\Omega$ transversely. We claim then that ∂X_Ω is contact type, so that Ω is a Liouville domain. Indeed,

$$\rho = \sum_{k=1}^{2}(\mu_k - \eta_k)\frac{\partial}{\partial\mu_k}$$

is a Liouville vector field transverse to ∂X_Ω, see Section 1.3.

To describe the resulting contact form $\lambda = \imath_\rho\omega$ on ∂X_Ω, suppose further that Ω is strictly convex. Then ∂X_Ω is diffeomorphic to T^3 with coordinates

θ_1, θ_2, ϕ, where θ_1, θ_2 were defined above, and $(\cos\phi, \sin\phi)$ is the unit tangent vector to $\partial\Omega$, oriented counterclockwise. The contact form is now

$$\lambda = \frac{1}{2\pi}\sum_{k=1}^{2}(\mu_k - \eta_k)d\theta_k, \tag{4.13}$$

and the Reeb vector field is

$$R = \frac{2\pi}{\|(\sin\phi, -\cos\phi)\|^*}\left(\sin\phi\frac{\partial}{\partial\theta_1} - \cos\phi\frac{\partial}{\partial\theta_2}\right). \tag{4.14}$$

Here $\|\cdot\|^*$ denotes the dual norm as above, defined using the translate of Ω by $-\eta$. This means that λ has a circle of Reeb orbits for each ϕ for which $(\sin\phi, -\cos\phi)$ is a positive multiple of a vector (a,b) where a,b are relatively prime integers, and the symplectic action of each such Reeb orbit is the dual norm $\|(a,b)\|^*$.

For example, if Ω is a disk of radius 1 centered at η, then the contact form (4.13) agrees with the standard contact form (4.4) on T^3 (via the coordinate change $\theta_1 = 2\pi x$, $\theta_2 = 2\pi y$, $\phi = \theta + \pi/2$), and the norm $\|\cdot\|^*$ is the Euclidean norm. So in this case, Theorem 4.14 follows from Proposition 4.10. In the general case, by the arguments in [28, Prop. 10.15], the calculations in Section 4.2 work just as well for the contact form (4.13), except that symplectic action is computed using the dual norm $\|\cdot\|^*$ instead of the Euclidean norm. This proves Theorem 4.14 whenever the boundary of Ω is smooth and strictly convex. The general case of Theorem 4.14 follows by a limiting argument. ■

The key property of the contact form (4.13) that makes the above calculation work is that the direction of the Reeb vector field (4.14) rotates monotonically with ϕ. It is an interesting open problem to compute the ECH capacities of X_Ω when Ω is star-shaped with respect to some origin but not convex. In this case the direction of the Reeb vector field no longer increases monotonically as one moves along $\partial\Omega$, so the calculations in Section 4.2 do not apply, as there can be more than one circle of Reeb orbits in the same homology class.

Polydisks. We now prove the formula (1.8) for the ECH capacities of a polydisk $P(a,b)$. Let Ω be a rectangle with sides of length a and b parallel to the axes which does not intersect the axes. It follows from Theorem 4.14 and Exercise 4.16(b) that

$$c_k(X_\Omega) = \min\{am + bn \mid m,n \in \mathbb{N}, (m+1)(n+1) \geq k+1\}.$$

So to prove Equation (1.8) for the ECH capacities of a polydisk, it is enough to show that

$$c_k\bigl(P(a,b)\bigr) = c_k(X_\Omega). \tag{4.15}$$

Observe that X_Ω is symplectomorphic to the product of two annuli of area a and b. Also, an annulus can be symplectically embedded into a disk of slightly larger area, and a disk can be symplectically embedded into an annulus of slightly larger area. Consequently, for any $\varepsilon > 0$, there are symplectic embeddings

$$P\bigl((1-\varepsilon)a,(1-\varepsilon)b\bigr) \subset X_\Omega \subset P\bigl((1+\varepsilon)a,(1+\varepsilon)b\bigr).$$

It follows from this and the Monotonicity and Conformality properties of ECH capacities that (4.15) holds. Indeed, *any* symplectic capacity satisfying the Monotonicity and Conformality properties must agree on $P(a,b)$ and X_Ω.

5. Foundations of ECH

We now give an introduction to some of the foundational matters which were skipped over in Section 3. The subsections below introduce foundational issues in the logical order in which they arise in developing the theory, but for the most part can be read in any order.

Below, fix a closed oriented three-manifold Y, a nondegenerate contact form λ on Y, and a generic symplectization-admissible almost complex structure J on $\mathbb{R}\times Y$.

5.1. Proof of the Writhe Bound and the Partition Conditions

We now outline the proof of the writhe bound (3.9) and the partition conditions in Section 3.9. One can prove this one Reeb orbit at a time. That is, let C be a somewhere injective J-holomorphic curve, let γ be an embedded Reeb orbit, and suppose that C has positive ends at covers of γ with multiplicities $a_1,\ldots,a_k$ satisfying $\sum_{i=1}^k a_i = m$. Let N be a tubular neighborhood of γ and let $\zeta = C\cap(\{s\}\times N)$ where $s \gg 0$. Let τ be a trivialization of $\xi|_\gamma$. We then need to prove the following lemma (together with an analogus lemma for the negative ends which will follow by symmetry):

Lemma 5.1. *If $s \gg 0$, then ζ is a braid whose writhe satisfies*

$$w_\tau(\zeta) \leq \sum_{i=1}^{m} CZ_\tau(\gamma^i) - \sum_{i=1}^{k} CZ_\tau(\gamma^{a_i}),$$

with equality only if $(a_1, \ldots, a_k) = p_\gamma^+(m)$.

To sketch the proof of Lemma 5.1, we assume for simplicity that C contains no trivial cylinders, although this assumption is easily dropped. We now need to recall some facts about the asymptotics of holomorphic curves. To set this up, identify $N \simeq (\mathbb{R}/\mathbb{Z}) \times D^2$ so that γ corresponds to $(\mathbb{R}/\mathbb{Z}) \times \{0\}$, and the derivative of the identification along γ sends $\xi|_\gamma$ to $\{0\} \oplus \mathbb{C}$ in agreement with the trivialization τ. It turns out that a nontrivial positive end of C at the d-fold cover of γ can be written using this identification as the image of a map

$$\begin{aligned} u : [s_0, \infty) \times (\mathbb{R}/d\mathbb{Z}) &\longrightarrow \mathbb{R} \times (\mathbb{R}/\mathbb{Z}) \times D^2, \\ (s,t) &\longmapsto \big(s, \pi(t), \eta(s,t)\big) \end{aligned}$$

where $s_0 \gg 0$ and $\pi : \mathbb{R}/d\mathbb{Z} \to \mathbb{R}/\mathbb{Z}$ denotes the projection.

We now want to describe the asymptotics of the function $\eta(s,t)$. Under our tubular neighborhood identification, the almost complex structure J on $\xi|_\gamma$ defines a family of 2×2 matrices J_t with $J_t^2 = -1$ parametrized by $t \in \mathbb{R}/\mathbb{Z}$. Also, the linearized Reeb flow along γ determines a connection $\nabla_t = \partial_t + S_t$ on $\xi|_\gamma$, where $J_t S_t$ is a symmetric matrix for each $t \in \mathbb{R}/\mathbb{Z}$. If d is a positive integer, define the "asymptotic operator" L_d on functions $\mathbb{R}/d\mathbb{Z} \to \mathbb{C}$ by

$$L_d = J_{\pi(t)}(\partial_t + S_{\pi(t)}).$$

Note that the operator L_d is formally self-adjoint, so its eigenvalues are real.

Lemma 5.2 [19]. *Given an end η of a holomorphic curve as above, there exist $c, \kappa > 0$, and a nonzero eigenfunction φ of L_d with eigenvalue $\lambda > 0$, such that*

$$\left|\eta(s,t) - e^{-\lambda s}\varphi(t)\right| < ce^{(-\lambda-\kappa)s}$$

for all $(s,t) \in [s_0, \infty) \times (\mathbb{R}/d\mathbb{Z})$.

To make use of this lemma, we need to know something about the eigenfunctions of L_d with positive eigenvalues.

Lemma 5.3. *Let $\varphi : \mathbb{R}/d\mathbb{Z} \to \mathbb{C}$ be a nonzero eigenfunction of L_d with eigenvalue λ. Then:*

(a) *$\varphi(t) \neq 0$ for all $t \in \mathbb{R}/d\mathbb{Z}$, so φ has a well-defined winding number around 0, which we denote by $\operatorname{wind}_\tau(\varphi)$.*

(b) *If $\lambda > 0$ then $\operatorname{wind}_\tau(\varphi) \leq \lfloor CZ_\tau(\gamma^d)/2 \rfloor$.*

Proof. The eigenfunction φ satisfies the ordinary differential equation

$$\partial_t \varphi = -(S_{\pi(t)} + \lambda J_{\pi(t)})\varphi.$$

Assertion (a) then follows from the uniqueness of solutions to ODE's. Assertion (b) is proved in [18, §3]. ■

Example 5.4. Suppose γ is elliptic with monodromy angle θ with respect to τ. We can then choose the trivialization τ so that

$$\nabla_t = \partial_t - 2\pi i\theta.$$

Suppose J is chosen so that J_t is multiplication by i for each t. Then

$$L_d = i\partial_t + 2\pi\theta.$$

Eigenfunctions of L_d are complex multiples of the functions

$$\varphi_n(t) = e^{2\pi i n t/d}$$

for $n \in \mathbb{Z}$, with eigenvalues

$$\lambda_n = -2\pi n/d + 2\pi\theta \tag{5.1}$$

and winding number

$$\operatorname{wind}_\tau(\varphi_n) = n. \tag{5.2}$$

It follows from (5.1) and (5.2) that $\lambda_n > 0$ if and only if $\operatorname{wind}_\tau(\varphi_n) < d\theta$. This is consistent with Lemma 5.3(b) since by (3.2) we have

$$\lfloor CZ_\tau(\gamma^d)/2 \rfloor = \lfloor d\theta \rfloor.$$

Now return to the slice $\zeta = C \cap (\{s\} \times N)$ where $s \gg 0$. The positive ends of C at covers of γ determine loops $\zeta_1, \ldots, \zeta_k$ in N transverse to the fibers of $N \to \gamma$. We conclude from Lemmas 5.2 and 5.3 that ζ_i has a well-defined winding number around γ with respect to τ, which we denote by $\operatorname{wind}_\tau(\zeta_i)$, and this satisfies

$$\operatorname{wind}_\tau(\zeta_i) \le \lfloor CZ_\tau(\gamma^{a_i})/2 \rfloor. \tag{5.3}$$

To simplify notation, write $\rho_i = \lfloor CZ_\tau(\gamma^{a_i})/2 \rfloor$.

In principle the loops ζ_i might intersect themselves or each other, but we will see below that if s is sufficiently large then they do not, so that their union is a braid ζ. Its writhe is then given by

$$w_\tau(\zeta) = \sum_{i=1}^{k} w_\tau(\zeta_i) + \sum_{i \ne j} \ell_\tau(\zeta_i, \zeta_j). \tag{5.4}$$

Here $\ell_\tau(\zeta_i, \zeta_j)$ is the "linking number" of ζ_i and ζ_j; this is defined like the writhe, except now we count crossings of ζ_i with ζ_j and divide by 2. The terms on the right hand side of (5.4) are bounded as follows:

Lemma 5.5. *If $s \gg 0$, then:*

(a) *Each component ζ_i is embedded and has writhe bounded by*

$$w_\tau(\zeta_i) \le \rho_i(a_i - 1). \tag{5.5}$$

(b) *If $i \ne j$, then ζ_i and ζ_j are disjoint, and*

$$\ell_\tau(\zeta_i, \zeta_j) \le \max(\rho_i a_j, \rho_j a_i).$$

Proof. An analogous result was proved in [21, §6] in an analytically simpler situation. In the present case, parts of the argument require a result of Siefring [51] which generalizes Lemma 5.2 to give "higher order" asymptotics of holomorphic curves. We now outline how this works.

(a) If the integers $\operatorname{wind}_\tau(\zeta_i)$ and a_i are relatively prime, then an elementary argument in [21, Lem. 6.7] related to Lemma 5.3(a) shows that ζ_i is a torus braid, so that

$$w_\tau(\zeta_i) = \operatorname{wind}_\tau(\zeta_i)(a_i - 1). \tag{5.6}$$

The inequality (5.5) now follows from this and the winding bound (5.3). When $\operatorname{wind}_\tau(\zeta_i)$ and a_i have a common factor, one can prove that ζ_i is embedded and satisfies (5.6) using the analysis of Siefring.

(b) Let λ_i and λ_j denote the eigenvalues of the operators L_{a_i} and L_{a_j} associated to the ends ζ_i and ζ_j via Lemma 5.2. If $\lambda_i < \lambda_j$, then it follows from Lemma 5.2 that the braid ζ_j is "inside" the braid ζ_i (assuming as always that we take s sufficiently large), from which it follows that ζ_i and ζ_j do not intersect and

$$\ell_\tau(\zeta_i,\zeta_j) = \operatorname{wind}_\tau(\zeta_i)a_j \le \rho_i a_j. \tag{5.7}$$

The proof that ζ_i and ζ_j do not intersect and satisfy (5.7) when $\lambda_i = \lambda_j$ is more delicate and uses the analysis of Siefring. ∎

Remark 5.6. When ρ_i and a_i have a common factor one can strengthen the inequality (5.5); optimal bounds are given in [52, §3.1]. We do not need this strengthening when γ is elliptic, but it is needed for the proof of the partition conditions when γ is hyperbolic, see [22, Lem. 4.16].

Proof of Lemma 5.1. We will restrict attention to the most interesting case where γ is elliptic with monodromy angle θ. (See [22, §4] for the proof when γ is hyperbolic.) By Equation (3.2) we have $\rho_i = \lfloor a_i\theta\rfloor$. So by Equation (5.4) and Lemma 5.5, it is enough to show that

$$\begin{aligned}\sum_{i=1}^{k}\lfloor a_i\theta\rfloor(a_i-1) &+ \sum_{i\neq j}\max\big(\lfloor a_i\theta\rfloor a_j, \lfloor a_j\theta\rfloor a_i\big)\\ &\le \sum_{i=1}^{m}\big(2\lfloor i\theta\rfloor+1\big) - \sum_{i=1}^{k}\big(2\lfloor a_i\theta\rfloor+1\big),\end{aligned}$$

with equality only if $(a_1,\ldots,a_k) = p_\theta^+(m)$. We can write the above inequality a bit more simply as

$$\sum_{i,j=1}^{n}\max\big(\lfloor a_i\theta\rfloor a_j, \lfloor a_j\theta\rfloor a_i\big) \le 2\sum_{i=1}^{m}\lfloor i\theta\rfloor - \sum_{i=1}^{k}\lfloor a_i\theta\rfloor + m - k. \tag{5.8}$$

To prove (5.8), following [22], order the numbers $a_1,\ldots,a_k$ so that

$$\frac{\lfloor a_1\theta\rfloor}{a_1} \ge \frac{\lfloor a_2\theta\rfloor}{a_2} \ge \cdots \ge \frac{\lfloor a_k\theta\rfloor}{a_k}.$$

Let Λ be the path in the plane starting at $(0,0)$ with edge vectors $(a_1,\lfloor a_1\theta\rfloor)$, $\ldots,(a_k,\lfloor a_k\theta\rfloor)$ in that order. Let P denote the region bounded by the path Λ, the horizontal line from $(0,0)$ to $(m,0)$, and the vertical line from $(m,0)$ to $(m,\sum_{i=1}^{k}\lfloor a_i\theta\rfloor)$. Let L denote the number of lattice points contained in P (including the boundary), let A denote the area of P, and let B denote the number of lattice points on the boundary of P.

By dividing P into rectangles and triangles, we find that the left hand side of (5.8) is twice the area of P, i.e.

$$(5.9)\qquad 2A = \sum_{i,j=1}^{n} \max\big(\lfloor a_i\theta\rfloor a_j, \lfloor a_j\theta\rfloor a_i\big).$$

Counting by vertical strips, we find that the number of lattice points enclosed by P is bounded by

$$(5.10)\qquad L \leq m+1+\sum_{i=1}^{m}\lfloor i\theta\rfloor,$$

with equality if and only if the image of the path Λ agrees with the image of the path $\Lambda_\theta^+(m)$ defined in Section 3.9. In addition, the number of boundary lattice points satisfies

$$(5.11)\qquad B \geq m+k+\sum_{i=1}^{k}\lfloor a_i\theta\rfloor,$$

with equality if and only if none of the edge vectors of the path Λ is divisible in $\mathbb{Z}^2$. Now Pick's formula for the area of a lattice polygon asserts that

$$(5.12)\qquad 2A = 2L - B - 2.$$

Combining (5.9), (5.10), (5.11), and (5.12), we conclude that the inequality (5.8) holds, with equality if and only if $\Lambda = \Lambda_\theta^+(m)$. ■

Remark 5.7. Counts of lattice points in polygons have now arisen in two, seemingly independent, ways in our story: first in the above proof of the writhe bound and the partition conditions, and second in the calculation of the ECH of T^3 and the ECH capacities of toric domains in Section 4.2 and Section 4.3. We do not know if there is a deep explanation for this.

5.2. Topological Complexity of Holomorphic Curves

The definitions of the ECH differential and the U map do not directly specify the genus of the (nontrivial component of the) holomorphic currents to be counted. However this is determined indirectly by the relative homology class of the holomorphic current if one knows the trivial cylinder components, as we now explain. This fact is useful for understanding the holomorphic currents (see e.g. [31, §4.5] and [37, Appendix] for applications), and also in the compactness argument in Section 5.3 below.

Let $\alpha = \{(\alpha_i, m_i)\}$ and $\beta = \{(\beta_j, n_j)\}$ be orbit sets in the homology class Γ, and let $Z \in H_2(Y, \alpha, \beta)$. Define[20]

$$J_0(\alpha, \beta, Z) = -c_\tau(Z) + Q_\tau(Z) + CZ_\tau^{J_0}(\alpha, \beta), \tag{5.13}$$

where

$$CZ_\tau^{J_0}(\alpha, \beta) = \sum_i \sum_{k=1}^{m_i - 1} CZ_\tau(\alpha_i^k) - \sum_j \sum_{k=1}^{n_j - 1} CZ_\tau(\beta_j^k). \tag{5.14}$$

The definition of J_0 is very similar to the definition of the ECH index in (3.4) and (3.5); however the sign of the relative first Chern class term is switched, and the Conley-Zehnder term is slightly different. More importantly, while I bounds the *Fredholm index* of holomorphic curves, J_0 bounds the "*topological complexity*" of holomorphic curves.

To give a precise statement in the case that we will need to consider, let $\mathcal{C} \in \mathcal{M}(\alpha, \beta)$ be a holomorphic current. Suppose that $\mathcal{C} = \mathcal{C}_0 \sqcup C$ where $\mathcal{C}_0$ is a union of trivial cylinders with multiplicities, and C is somewhere injective. Let n_i^+ denote "the number of positive ends of $\mathcal{C}$ at covers of α_i^+", namely the number of positive ends of C at α_i^+, plus 1 if $\mathcal{C}_0$ includes the trivial cylinder $\mathbb{R} \times \alpha_i^+$ with some multiplicity. Likewise, let n_j^- denote "the number of negative ends of $\mathcal{C}$ at covers of β_j^-", namely the number of negative ends of C at β_j^-, plus 1 if $\mathcal{C}_0$ includes the trivial cylinder $\mathbb{R} \times \beta_j^-$ with some multiplicity. Write $J_0(\mathcal{C}) = J_0(\alpha, \beta, [\mathcal{C}])$.

Proposition 5.8. *Let $\alpha = \{(\alpha_i, m_i)\}$ and $\beta = \{(\beta_j, n_j)\}$ be generators of the ECH chain complex, and let $\mathcal{C} = \mathcal{C}_0 \sqcup C \in \mathcal{M}(\alpha, \beta)$ be a holomorphic current as above. Then*

$$-\chi(C) + \sum_i (n_i^+ - 1) + \sum_j (n_j^- - 1) \le J_0(\mathcal{C}). \tag{5.15}$$

If $\mathcal{C}$ is counted by the ECH differential or the U map, then equality holds in (5.15).

For example, it follows from (5.15) that $J_0(\mathcal{C}) \ge -1$, with equality only if C is a plane with positive end at a cover of some Reeb orbit γ, and $\mathcal{C}_0$

[20] It is perhaps not optimal to denote this number by J_0 since J usually denotes an almost complex structure. However the idea is that J_0 is similar to I and so should be denoted by a nearby letter.

does not contain any trivial cylinders over γ. Proposition 5.8 is proved in [31, Lem. 3.5], using more general results from [22, §6].

Exercise 5.9. (See answer in Appendix.) Use the relative adjunction formula (3.3) and the partition conditions in Section 3.9 to prove the following special case of Proposition 5.8: If $C \in \mathcal{M}(\alpha, \beta)$ is an embedded holomorphic curve which is counted by the ECH differential or the U map, then

$$-\chi(C) + \sum_i (n_i^+ - 1) + \sum_j (n_j^- - 1) = J_0(C).$$

5.3. Proof that ∂ Is Well Defined

Assume now that the almost complex structure J on $\mathbb{R} \times Y$ is generic. To complete the proof in Section 3.5 that the ECH differential ∂ is well-defined, we need to prove the following:

Lemma 5.10. *If α and β are orbit sets, then $\mathcal{M}_1(\alpha, \beta)/\mathbb{R}$ is finite.*

To prove this we want to assume that $\mathcal{M}_1(\alpha, \beta)$ is infinite and apply a compactness argument to obtain a contradiction. A relevant version of Gromov compactness was proved by Bourgeois-Eliashberg-Hofer-Wysocki-Zehnder [4]. To describe this result, say that a holomorphic curve u is "nontrivial" if it is not a union of trivial cylinders; branched covers of trivial cylinders with a positive number of branched points are considered nontrivial. If u_+ and u_- are nontrivial holomorphic curves, define "gluing data" between u_+ and u_- to consist of the following:

- A bijection between the negative ends of u^+ and the positive ends of u^- such that ends paired up under the bijection are at the same (possibly multiply covered) Reeb orbit γ.
- When γ above is multiply covered with degree m, then the negative end of u^+ and the positive end of u^- each determine a degree m cover of the underlying embedded Reeb orbit, and the gluing data includes an isomorphism of these covering spaces (there are m possible choices for this).

Finally, define a *broken holomorphic curve* to be a finite sequence $(u^0, \ldots, u^k)$, where each u_i is a nontrivial holomorphic curve in $\mathbb{R} \times Y$ modulo $\mathbb{R}$ translation, called a "level", together with gluing data as above between u^{i-1} and

u^i for each $i = 1, \ldots, k$. It is shown in [4] that if $\{u_\nu\}_{\nu \geq 0}$ is a sequence of holomorphic curves with fixed genus between the same sets of Reeb orbits, then there is a subsequence which converges in an appropriate sense to a broken holomorphic curve.

Unfortunately we cannot directly apply the above result to a sequence of holomorphic currents in $\mathcal{M}_1(\alpha, \beta)/\mathbb{R}$, because we do not have an a priori bound on the genus of their nontrivial components. One can obtain a bound on the genus of a holomorphic curve from Proposition 5.8, but this bound depends on the relative homology class of the holomorphic curve. To get control over the relative homology classes of holomorphic currents in $\mathcal{M}_1(\alpha, \beta)/\mathbb{R}$, we will first use a second version of Gromov compactness which we now state.

If α and β are orbit sets, define a *broken holomorphic current* from α to β to be a finite sequence of nontrivial J-holomorphic currents $(\mathcal{C}^0, \ldots, \mathcal{C}^k)$ in $\mathbb{R} \times Y$, modulo $\mathbb{R}$ translation, for some $k \geq 0$ such that there are orbit sets $\alpha = \gamma^0, \gamma^1, \ldots, \gamma^{k+1} = \beta$ with $\mathcal{C}^i \in \mathcal{M}(\gamma^i, \gamma^{i+1})/\mathbb{R}$ for $i = 0, \ldots, k$. Here "nontrivial" means not a union of trivial cylinders with multiplicities. Let $\overline{\mathcal{M}(\alpha, \beta)/\mathbb{R}}$ denote the set of broken holomorphic currents from α to β.

We say that a sequence of holomorphic currents $\{\mathcal{C}_\nu\}_{\nu \geq 0}$ in $\mathcal{M}(\alpha, \beta)/\mathbb{R}$ *converges* to the broken holomorphic current $(\mathcal{C}^0, \ldots, \mathcal{C}^k)$ if for each $i = 0, \ldots, k$ there are representatives $\mathcal{C}_\nu^i \in \mathcal{M}(\alpha, \beta)$ of the equivalence classes $\mathcal{C}_\nu \in \mathcal{M}(\alpha, \beta)/\mathbb{R}$ such that the sequence $\{\mathcal{C}_\nu^i\}_{\nu \geq 0}$ converges as a current and as a point set on compact sets to $\mathcal{C}^i$, see Section 2.4.

Lemma 5.11.

(a) *Any sequence $\{\mathcal{C}_\nu\}_{\nu \geq 0}$ of holomorphic currents in $\mathcal{M}(\alpha, \beta)/\mathbb{R}$ has a subsequence which converges to a broken holomorphic current $(\mathcal{C}^0, \ldots, \mathcal{C}^k) \in \overline{\mathcal{M}(\alpha, \beta)/\mathbb{R}}$.*

(b) *If the sequence $\{\mathcal{C}_\nu\}_{\nu \geq 0}$ converges to $(\mathcal{C}^0, \ldots, \mathcal{C}^k)$, then*

$$\sum_{i=0}^{k} [\mathcal{C}^i] = [\mathcal{C}_\nu] \in H_2(Y, \alpha, \beta)$$

for all ν sufficiently large.

Proof. (a) The proof has three steps.

Step 1. For each ν, suppose that $\mathcal{C}_\nu^* \in \mathcal{M}(\alpha, \beta)$ is a representative of the equivalence class $\mathcal{C}_\nu \in \mathcal{M}(\alpha, \beta)/\mathbb{R}$. We claim that $\{\mathcal{C}_\nu^*\}_{\nu \geq 0}$ has a subsequence which converges as a current and a point set on compact sets to some holomorphic current $\hat{\mathcal{C}}$ in $\mathbb{R}\times Y$.

To prove the claim, let $a<b$. We apply Gromov compactness via currents, see Section 2.4, to the sequence of intersections $\mathcal{C}_\nu^*\cap([a,b]\times Y)$. To see why this is applicable, note that $[a,b]\times Y$ is equipped with the symplectic form $\omega=d(e^s\lambda)$ where s denotes the $\mathbb{R}$ coordinate, and J is ω-compatible. Assume that $\mathcal{C}_\nu^*$ is transverse to $\{a\}\times Y$ and $\{b\}\times Y$, which we can arrange by perturbing a and b. Then by Stokes's theorem,

$$\int_{\mathcal{C}_\nu^*\cap((-\infty,a]\times Y)} e^a d\lambda+\int_{\mathcal{C}_\nu^*\cap([a,b]\times Y)}\omega+\int_{\mathcal{C}_\nu^*\cap([b,\infty)\times Y)} e^b d\lambda=e^b\mathcal{A}(\alpha)-e^a\mathcal{A}(\beta).$$

The conditions on J imply that $d\lambda$ is everywhere nonnegative on $\mathcal{C}_\nu^*$. Thus we obtain the a priori bound

$$\int_{\mathcal{C}_\nu^*\cap([a,b]\times Y)}\omega\le e^b\mathcal{A}(\alpha).$$

Gromov compactness via currents now implies that we can pass to a subsequence so that the sequence $\{\mathcal{C}_\nu^*\cap([a,b]\times Y)\}_{\nu\ge 0}$ converges as a current and as a point set to some holomorphic current in $[a,b]\times Y$. By diagonalizing, we can pass to a subsequence so that the sequence $\{\mathcal{C}_\nu^*\}_{\nu\ge 0}$ converges as a current and as a point set on compact sets to some holomorphic current $\hat{\mathcal{C}}$ in $\mathbb{R}\times Y$.

Steps 2 and 3 are a fairly standard argument which we will just outline. See e.g. [21, Lem. 9.8] for details in a similar situation.

Step 2. By applying Step 1 to translates of $\hat{\mathcal{C}}$, one shows that $\hat{\mathcal{C}}\in\mathcal{M}(\gamma^+,\gamma^-)$, where γ^+ and γ^- are orbit sets with $\mathcal{A}(\alpha)\ge\mathcal{A}(\gamma^+)\ge\mathcal{A}(\gamma^-)\ge\mathcal{A}(\beta)$.

Step 3. One can now choose representatives $\mathcal{C}_\nu^*\in\mathcal{M}(\alpha,\beta)$ of the equivalence classes $\mathcal{C}_\nu$ so that the intersection of each $\mathcal{C}_\nu^*$ with $\{0\}\times Y$ contains a point with distance at least ε from all Reeb orbits of action less than or equal to $\mathcal{A}(\alpha)$. One then applies Steps 1 and 2 to this sequence $\mathcal{C}_\nu^*$. The limiting current $\hat{\mathcal{C}}$ must be nontrivial. If $\gamma^+=\alpha$ and $\gamma^-=\beta$, then we are done. Otherwise one applies the same argument to different choices of $\mathcal{C}_\nu^*$ to find the other holomorphic currents $\mathcal{C}^i$ in the limiting broken holomorphic current.

(b) If this fails, then one uses arguments from the proof of part (a) to pass to a further subsequence which converges to a broken holomorphic current including $\mathcal{C}^0,\ldots,\mathcal{C}^k$ together with at least one additional level. But this is impossible by symplectic action considerations. ■

We can now complete the proof that the differential ∂ is well-defined.

Proof of Lemma 5.10. Suppose to get a contradiction that there is an infinite sequence $\{\mathcal{C}_\nu\}_{\nu\geq 0}$ of distinct elements of $\mathcal{M}_1(\alpha,\beta)/\mathbb{R}$.

For each ν, by Proposition 3.7 we can write $\mathcal{C}_\nu = \mathcal{C}_{\nu,0} \sqcup C_{\nu,1}$, where $\mathcal{C}_{\nu,0}$ is a union of trivial cylinders with multiplicities, and $C_{\nu,1}$ is somewhere injective with $I(C_{\nu,1}) = \operatorname{ind}(C_{\nu,1}) = 1$. Since there are only finitely many possibilities for the trivial part $\mathcal{C}_{\nu,0}$, we can pass to a subsequence so that $\mathcal{C}_{\nu,0}$ is the same for all ν. There are then orbit sets α' and β' which do not depend on ν such that $C_{\nu,1} \in \mathcal{M}_1(\alpha',\beta')$ for each ν.

By Lemma 5.11, we can pass to a further subsequence such that the holomorphic curves $C_{\nu,1}$ all represent the same relative homology class $Z \in H_2(Y,\alpha',\beta')$.

By Proposition 5.8, there is a ν-independent upper bound on the genus of $C_{\nu,1}$ in terms of $J_0(\alpha,\beta,Z)$. Thus we can pass to a further subsequence so that the holomorphic curves $C_{\nu,1}$ all have the same genus.

Now we can apply the compactness result of [4] to pass to a further subsequence so that the sequence of holomorphic curves $\{C_{\nu,1}\}_{\nu\geq 0}$ converges in the sense of [4] to a broken holomorphic curve $(u^0,\ldots,u^k)$.

By the Additivity property of the ECH index, see Section 3.4, we have $\sum_{i=0}^k I(u^i) = 1$. By Proposition 3.7, one of the curves u_i has $I = 1$, and the rest of the curves u^i have $I = 0$ and are unions of branched covers of trivial cylinders.

We will now be a bit sketchy for the rest of the proof. By a similar additivity property of the Fredholm index which follows from (3.1), we also have $\sum_{i=0}^k \operatorname{ind}(u^i) = 1$. It then follows from Exercise 3.14 that in fact there is no level u^i with $I(u^i) = 0$. Hence the limiting broken holomorphic curve is a single holomorphic curve u^0, which is somewhere injective and has $ind(u^0) = 1$. Since J is generic, u^0 is an isolated point in the moduli space of holomorphic curves modulo translation. But this contradicts the fact that u^0 is the limit of the sequence of distinct curves $\{C_{\nu,1}\}_{\nu\geq 0}$. ■

5.4. Proof that $\partial^2 = 0$

The proof that $\partial^2 = 0$ is much more subtle than the proof that ∂ is defined, for reasons which we now explain.

Fix a generic J. Let α_+ and α_- be generators of the chain complex $ECC_*(Y,\lambda,\Gamma,J)$. We would like to show that the coefficient $\langle\partial^2\alpha_+,\alpha_-\rangle = 0$. To do so, consider the moduli space of $I=2$ holomorphic currents $\mathcal{M}_2(\alpha_+,\alpha_-)/\mathbb{R}$.

Lemma 5.12. *Any sequence $\{\mathcal{C}_\nu\}_{\nu\geq 0}$ of holomorphic currents in $\mathcal{M}_2(\alpha_+,\alpha_-)/\mathbb{R}$ has a subsequence which converges either to an element of $\mathcal{M}_2(\alpha_+,\alpha_-)/\mathbb{R}$, or to a broken holomorphic current $(\mathcal{C}^+,\mathcal{C}^-) \in \overline{\mathcal{M}_2(\alpha_+,\alpha_-)/\mathbb{R}}$ with $I(\mathcal{C}^+)=I(\mathcal{C}^-)=1$.*

Proof. By Lemma 5.11, there is a subsequence which converges to a broken holomorphic current $(\mathcal{C}^0,\ldots,\mathcal{C}^k)$, where by definition each $\mathcal{C}^i$ is nontrivial. By the Additivity property of the ECH index, $\sum_{i=0}^k I(\mathcal{C}^i)=2$. By Proposition 3.7, $I(\mathcal{C}^i)\geq 1$ for each i. The lemma follows from these two facts. ∎

The usual strategy now would be to add one point to each end of $\mathcal{M}_2(\alpha_+,\alpha_-)/\mathbb{R}$ to form a compact one-manifold with boundary, whose boundary points correspond to ends converging to broken holomorphic currents as above. In the present situation this is not quite correct; in fact we do not even know a priori that the moduli space $\mathcal{M}_2(\alpha_+,\alpha_-)/\mathbb{R}$ has only finitely many components[21]. Instead, one can *truncate* the moduli space $\mathcal{M}_2(\alpha_+,\alpha_-)$, i.e. remove holomorphic currents which are "close to breaking" in an appropriate sense, to obtain a compact one-manifold with boundary $\mathcal{M}_2'(\alpha_+,\alpha_-)/\mathbb{R}$. The boundary is equipped with a natural map

$$\partial\left(\frac{\mathcal{M}_2'(\alpha_+,\alpha_-)}{\mathbb{R}}\right) \longrightarrow \bigsqcup_{\alpha_0} \frac{\mathcal{M}_1(\alpha_+,\alpha_0)}{\mathbb{R}} \times \frac{\mathcal{M}_1(\alpha_0,\alpha_-)}{\mathbb{R}} \tag{5.16}$$

which sends a boundary point to the broken holomorphic current that it is "close to breaking into". The details of this truncation procedure are explained in [29, §1.3].

To complete the proof that $\langle\partial^2\alpha_+,\alpha_-\rangle = 0$, we want to show that $\langle\partial^2\alpha_+,\alpha_-\rangle$ counts boundary points of $\mathcal{M}_2'(\alpha_+,\alpha_-)/\mathbb{R}$. For this purpose let α_0 be an orbit set and let $(\mathcal{C}^+,\mathcal{C}^-)\in(\mathcal{M}_1(\alpha_+,\alpha_0)/\mathbb{R})\times(\mathcal{M}_1(\alpha_0,\alpha_-)/\mathbb{R})$. We then want to show the following:

[21]The compactness result of [4] does not imply that $\mathcal{M}_2(\alpha_+,\alpha_-)/\mathbb{R}$ has only finitely many components, because of the failure of transversality of branched covers of trivial cylinders that can arise as levels in limits of sequences of ind $=2$ holomorphic curves.

(1) If α_0 is a generator of the chain complex $ECC_*(Y,\lambda,\Gamma,J)$, then $(\mathcal{C}^+,\mathcal{C}^-)$ has 1 (mod 2) inverse image under the map (5.16).

(2) If α_0 is not a generator of the chain complex $ECC_*(Y,\lambda,\Gamma,J)$, i.e. if α_0 includes a hyperbolic Reeb orbit with multiplicity greater than one, then $(\mathcal{C}^+,\mathcal{C}^-)$ has 0 (mod 2) inverse images under the map (5.16).

The standard picture from symplectic field theory is that if (u^+,u^-) is a broken holomorphic curve such that u^+ and u^- are regular and have $\operatorname{ind}(u^+)=\operatorname{ind}(u^-)=1$, then for each choice of gluing data between u^+ and u^-, see Section 5.3, one can "glue" u^+ and u^- to obtain a unique end of the moduli space of index 2 holomorphic curves.

To describe the proof of (1) and (2) above, let us restrict attention to the case where α_0 consists of a single pair (γ,m) where γ is an embedded Reeb orbit and $m\geq 1$. Write $\mathcal{C}^\pm=\mathcal{C}_0^\pm\sqcup C_1^\pm$ where $\mathcal{C}_0^\pm$ is a union of trivial cylinders with multiplicities and $C_1^\pm$ is somewhere injective with $\operatorname{ind}(C_1^\pm)=I(C_1^\pm)=1$. To further simplify the discussion, let us also assume that there are no trivial cylinders involved, i.e. $\mathcal{C}_0^\pm=\emptyset$.

Gluing in the Hyperbolic Case. Suppose first that γ is positive hyperbolic. In this case, the partition conditions from Section 3.9 tell us that C_1^+ has m negative ends at γ, and C_1^- has m positive ends at γ. It follows that there are $m!$ choices of gluing data between C_1^+ and C_1^-, see Section 5.3. Hence SFT gluing implies that C_1^+ and C_1^- can be glued to obtain $m!$ different ends of the moduli space of index 2 curves. The number of gluings $m!$ is odd (namely 1) when $m=1$ and even when $m>1$, which is exactly what we want in order to prove (1) and (2) above.

Suppose next that γ is negative hyperbolic. Let $k=\lfloor m/2\rfloor$. Then by the partition conditions in Section 3.9, the curve C_1^+ (resp. C_1^-) has k negative (resp. positive) ends at the double cover of γ, together with one negative (resp. positive) end at γ when m is odd. It follows that there are $2^k k!$ choices of gluing data between C_1^+ and C_1^-. Again, this is odd (namely 1) when $m=1$ and even when $m>1$, as desired.

Although we are using $\mathbb{Z}/2$ coefficients here, we remark that in the proof that $\partial^2=0$ with $\mathbb{Z}$ coefficients, work of Bourgeois-Mohnke [3] implies that in the above cases when $m>1$, half of the gluings have one sign and half of the gluings have the other sign, so that the signed count of gluings is still zero.

Gluing in the Elliptic Case. Suppose now that γ is elliptic. If $m=1$ then there is one gluing as usual. But if $m>1$, then it follows from Exercise 3.13(c) that $p_\gamma^+(m)$ and $p_\gamma^-(m)$ are disjoint, so the covering multiplicities of the negative ends of C_1^+ at covers of γ are disjoint from the covering multiplicities of the positive ends of C_1^- at covers of γ. Hence, there does not exist *any* gluing data between C_1^+ and C_1^-. So how can we glue them?

It helps to think backwards from the process of breaking. If a sequence of holomorphic currents in $\mathcal{M}_2(\alpha_+,\alpha_-)/\mathbb{R}$ converges to the broken holomorphic current (C_1^+,C_1^-), then as in the proof of Lemma 5.10, we can pass to a subsequence which converges in the sense of [4] to a broken holomorphic curve $(u^0,\ldots,u^k)$, with $\sum_{i=0}^k \operatorname{ind}(u^i)=\sum_{i=0}^k I(u^i)=2$. Since $\sum_{i=0}^k \operatorname{ind}(u^i)=2$, Exercise 3.14 implies that $u^0=C_1^+$, $u^k=C_1^-$, and each u^i with $0<i<k$ is a union of branched covers of trivial cylinders.

To reverse this process, let u^+ and u^- be any irreducible somewhere injective holomorphic curves with $\operatorname{ind}=1$, but not necessarily with $I=1$. Suppose that u^+ has negative ends at covers of the embedded elliptic orbit γ of multiplicities $a_1,\ldots,a_k$ with $\sum_{i=1}^k a_i=m$ and no other negative ends, and u^- has positive ends at covers of γ of multiplicities $b_1,\ldots,b_l$ with $\sum_{j=1}^l b_j=m$ and no other positive ends. We can try to glue u^+ and u^- to an $\operatorname{ind}=2$ curve as follows. First, try to find an $\operatorname{ind}=0$ branched cover u^0 of $\mathbb{R}\times\gamma$ of degree m with positive ends at covers of γ with multiplicities $a_1,\ldots,a_k$ and negative ends at covers of γ with multiplicities $b_1,\ldots,b_l$; see Exercise 3.14 for a discussion of when such a branched cover exists. Second, try to glue u^+, u^0, and u^- to a holomorphic curve. There is an obstruction to gluing here because u^0 is not regular. However one can also vary u^0. The obstructions to gluing for various u^0 comprise a section of an "obstruction bundle" over the moduli space of all branched covers u^0. The (signed) number of ways to glue is then the (signed) number of zeroes of this section of the obstruction bundle. See [30, §1] for an introduction to this analysis.

This signed count of gluings is denoted by $\#G(u^+,u^-)$ and computed in [29, Thm. 1.13]. The result is that $\#G(u^+,u^-)=\pm c_\gamma(a_1,\ldots,a_k|b_1,\ldots,b_l)$, where $c_\gamma(a_1,\ldots,a_k|b_1,\ldots,b_l)$ is a nonnegative integer which depends only on (the monodromy angle of) γ and the multiplicities a_i and b_j. It turns out that $c_\gamma(a_1,\ldots,a_k|b_1,\ldots,b_l)=1$ if (and only if) $(a_1,\ldots,a_k)=p_\gamma^-(m)$ and $(b_1,\ldots,b_l)=p_\gamma^+(m)$, see [29, Ex. 1.29]. Thus, up to signs, the number of gluings is 1 in the case needed to show that $\partial^2=0$ (and in no other case).

5.5. Cobordism Maps

We now discuss what is involved in the construction of cobordism maps on ECH, as introduced in Section 1.4.

Holomorphic Curves in Exact Symplectic Cobordisms. We begin with the nicest kind of cobordism. Let (Y_+,λ_+) and (Y_-,λ_-) be nondegenerate contact three-manifolds, and let (X,ω) be an exact symplectic cobordism from (Y_+,λ_+) to (Y_-,λ_-). In this situation, one can define for each $L\in\mathbb{R}$ a cobordism map

$$\Phi^L(X,\omega): ECH^L(Y_+,\lambda_+) \longrightarrow ECH^L(Y_-,\lambda_-) \tag{5.17}$$

satisfying various axioms [32, Thm. 1.9]. Here

$$ECH^L(Y,\lambda) = \bigoplus_{\Gamma\in H_1(Y)} ECH^L(Y,\lambda,\Gamma).$$

The first step in the construction of the map (5.17) is to "complete" the cobordism as follows. Let λ be a primitive of ω on X with $\lambda|_{Y_\pm}=\lambda_\pm$. If $\varepsilon>0$ is sufficiently small, then there is a neighborhood N_+ of Y_+ in X, identified with $(-\varepsilon,0]\times Y_+$, such that $\lambda=e^s\lambda_+$ where s denotes the $(-\varepsilon,0]$ coordinate. The neighborhood identification is the one for which $\partial/\partial s$ corresponds to the unique vector field ρ on X with $\imath_\rho\omega=\lambda$. Likewise there is a neighborhood N_- of Y_- in X, identified with $[0,\varepsilon)\times Y_-$, on which $\lambda=e^s\lambda_-$. We now define the "symplectization completion"

$$\overline{X} = \big((-\infty,0]\times Y_-\big)\cup_{Y_-} X \cup_{Y_+} \big([0,\infty)\times Y_+\big),$$

glued using the above neighborhood identifications.

Call an almost complex structure J on $\overline{X}$ "cobordism-admissible" if it agrees with a symplectization-admissible almost complex structure J_+ for λ_+ on $[0,\infty)\times Y_+$, if it agrees with a symplectization-admissible almost complex structure J_- for λ_- on $(-\infty,0]\times Y_-$, and if it is ω-compatible on X.

Given a cobordism-admissible almost complex structure J, one can consider J-holomorphic curves in $\overline{X}$ with positive ends at Reeb orbits in Y_+ and negative ends at Reeb orbits in Y_-, by a straightforward modification of the definition in the symplectization case in Section 3.2. If J is generic, and if C is a somewhere injective holomorphic curve as above, then the moduli space of holomorphic curves near C is a manifold of dimension $\operatorname{ind}(C)$, where $\operatorname{ind}(C)$ is defined as in (3.1), except that in the relative first Chern class term, the complex line bundle ξ is replaced by $\det(TX)$.

Likewise, if $\alpha_\pm$ are orbit sets for $\lambda_\pm$, then there is a corresponding moduli space $\mathcal{M}(\alpha_+,\alpha_-)$ of J-holomorphic currents in $\overline{X}$. One can define the ECH index I of a holomorphic current in $\overline{X}$ as in (3.4), again replacing ξ by

$\det(TX)$ in the first Chern class term. The index inequality (3.8) then holds for somewhere injective holomorphic curves C in $\overline{X}$, by the same proof as in the symplectization case, see [22, §4]. As in Section 3.5, let $\mathcal{M}_k(\alpha_+,\alpha_-)$ denote the set of holomorphic currents $\mathcal{C}\in\mathcal{M}(\alpha_+,\alpha_-)$ with ECH index $I(\mathcal{C})=k$.

We have the following important generalization of (1.15): If there exists $\mathcal{C}\in\mathcal{M}(\alpha_+,\alpha_-)$, then

$$\mathcal{A}(\alpha_+)\geq\mathcal{A}(\alpha_-). \tag{5.18}$$

The reason is that by Stokes's theorem,

$$\mathcal{A}(\alpha_+)-\mathcal{A}(\alpha_-)=\int_{\mathcal{C}\cap([0,\infty)\times Y_+)}d\lambda_+ +\int_{\mathcal{C}\cap X}\omega+\int_{\mathcal{C}\cap((-\infty,0]\times Y_-)}d\lambda_-, \tag{5.19}$$

and the conditions on J imply that each integrand is pointwise nonnegative on $\mathcal{C}$.

The Trouble with Multiple Covers. One would now like to define a chain map

$$\phi: ECC(Y_+,\lambda_+,J_+)\longrightarrow ECC(Y_-,\lambda_-,J_-)$$

by declaring that if $\alpha_\pm$ are ECH generators for $\lambda_\pm$, then $\langle\phi\alpha_+,\alpha_-\rangle$ is the mod 2 count of $I=0$ holomorphic currents in $\mathcal{M}_0(\alpha_+,\alpha_-)$. The inequality (5.18) implies that only finitely many α_- could arise in $\phi\alpha_+$, and moreover we would get a map on the filtered chain complexes

$$\phi^L: ECC^L(Y_+,\lambda_+,J_+)\longrightarrow ECC^L(Y_-,\lambda_-,J_-)$$

for each $L>0$.

Unfortunately this does not work. The problem is that $\mathcal{M}_0(\alpha_+,\alpha_-)$ need not be finite, even if J is generic. The compactness argument from Section 5.3 does not carry over here, because the key Proposition 3.7 can fail in cobordisms. In particular, multiply covered holomorphic currents may have negative ECH index. We do know from [22, Thm. 5.1] that the ECH index of a d-fold cover of a somewhere injective irreducible curve C satisfies

$$I(dC)\geq dI(C)+\left(\frac{d^2-d}{2}\right)\big(2g(C)-2+\operatorname{ind}(C)+h(C)\big), \tag{5.20}$$

where $g(C)$ denotes the genus of C, and $h(C)$ denotes the number of ends of C at (positive or negative) hyperbolic orbits[22]. If J is generic then the index inequality implies that $I(C) \geq 0$; but $I(dC) < 0$ is still possible when $2g(C) - 2 + \operatorname{ind}(C) + h(C) < 0$.

To correctly define the coefficient $\langle \phi\alpha_+, \alpha_- \rangle$, one needs to take into account the entire "compactification" of $\mathcal{M}_0(\alpha_+, \alpha_-)$, namely the set of all broken holomorphic currents from α_+ to α_- with total ECH index 0. This moduli space may have many components of various dimensions, and each may make some contribution to the coefficient $\langle \phi\alpha_+, \alpha_- \rangle$. In fact, there is a simple example in which the coefficient $\langle \phi\alpha_+, \alpha_- \rangle$ must be nonzero, but there does not exist *any* $I = 0$ holomorphic current from α_+ to α_-; rather, the contribution to $\langle \phi\alpha_+, \alpha_- \rangle$ comes from a broken holomorphic current with two levels, one of which is an $I = -1$ double cover. The example is the cobordism where $X = [0,1] \times Y$ which one obtains in trying to prove that ECH is unchanged under a period-doubling bifurcation. Even more interestingly, the orbit set in between the two levels is not a generator of the ECH chain complex, because it includes a doubly covered negative hyperbolic orbit.

Because of the above complications, it is a highly nontrivial, and currently unsolved problem, to define a chain map directly from the compactified moduli space of $I = 0$ holomorphic currents.

Seiberg-Witten Theory to the Rescue. The definition of the cobordism map (5.17) in [32] instead counts solutions to the Seiberg-Witten equations, perturbed as in the proof of the isomorphism (1.11). The cobordism maps satisfy a "Holomorphic Curves axiom" which says among other things that for any cobordism-admissible J, the cobordism map is induced by a (non-canonical) chain map ϕ such that the coefficient $\langle \phi\alpha_+, \alpha_- \rangle \neq 0$ only if there exists a broken J-holomorphic current from α_+ to α_-. In particular, the coefficient $\langle \phi\alpha_+, \alpha_- \rangle \neq 0$ only if (5.18) holds, which is why the cobordism map preserves the symplectic action filtration.

The Weakly Exact Case. If (X, ω) is only a weakly exact symplectic cobordism from (Y_+, λ_+) to (Y_-, λ_-), see Section 1.4, then using Seiberg-Witten theory as above, one still gets a cobordism map

$$\Phi^L(X, \omega) : ECH^L(Y_+, \lambda_+, 0) \longrightarrow ECH^L(Y_-, \lambda_-, 0)$$

[22]Note that the magic number $2g(C) - 2 + \operatorname{ind}(C) + h(C)$ in (5.20) is similar to the normal Chern number in (4.3).

which satisfies the Holomorphic Curves axiom. The reason why this map preserves the symplectic action filtrations is that a modification of the calculation in (5.19) shows that in the weakly exact case, if there exists a holomorphic current $\mathcal{C} \in \mathcal{M}(\alpha_+, \alpha_-)$, and if moreover $[\alpha_\pm] = 0 \in H_1(Y_\pm)$, then the inequality (5.18) still holds, see [24, Thm. 2.3]. It is this inequality which ultimately leads to all of the symplectic embedding obstructions coming from ECH capacities.

6. Comparison of ECH with SFT

To conclude, we now outline how ECH compares to the symplectic field theory (SFT) of Eliashberg-Givental-Hofer [14]. Although both theories are defined using the same ingredients, namely Reeb orbits and holomorphic curves, their features are quite different.

Dimensions. ECH is only defined for three-dimensional contact manifolds (and in some cases stable Hamiltonian structures) and certain four-dimensional symplectic cobordisms between them. SFT is defined in all dimensions. It is an interesting question whether there exists an analogue of ECH in higher dimensions, and what that would mean.

Multiply Covered Reeb Orbits. In an ECH generator, we only care about the total multiplicity of each Reeb orbit. One can think of an ECH generator as a "Reeb current". In an SFT generator, one keeps track of individual covering multiplicities of Reeb orbits. For example, if γ_1 is an elliptic Reeb orbit, and if γ_k denotes the k-fold multiple cover of γ_1, then γ_1^2 and γ_2 are distinct SFT generators which correspond to the same ECH generator $\{(\gamma_1, 2)\}$. Likewise, the SFT generators γ_1^3, $\gamma_2\gamma_1$ and γ_3 all correspond to the ECH generator $\{(\gamma_1, 3)\}$.

Holomorphic Curves. The full version of SFT counts all Fredholm index 1 holomorphic curves (after suitable perturbation to make the moduli spaces transverse). Other versions of SFT just count genus 0 Fredholm index 1 curves (rational SFT), or genus 0 Fredholm index 1 curves with one positive end (the contact homology algebra).

ECH counts holomorphic currents with ECH index 1, without explicitly specifying their genus (although the genus is more or less determined indirectly by the theory as explained in Section 5.2). These also have Fredholm index 1, although the way we are selecting a subset of the Fredholm index 1 curves to count (by setting the ECH index equal to 1) is very different from the way this is done in SFT (by setting the genus to 0, etc.).

Grading. SFT is relatively graded by the Fredholm index. ECH is relatively graded by the ECH index, and has an absolute grading by homotopy classes of oriented 2-plane fields.

Topological Invariance. ECH depends only on the three-manifold, if one uses the absolute grading, as explained in Remark 1.7. SFT depends heavily on the contact structure; for example, the basic versions are trivial for overtwisted contact structures. On the other hand, ECH does contain the contact invariant (the homology class of the empty set of Reeb orbits) which can distinguish some contact structures, as explained in Section 1.4. The ECH contact invariant is analogous to the unit in the contact homology algebra.

Disallowed Reeb Orbits. In ECH, hyperbolic orbits cannot have multiplicity greater than 1. In SFT, "bad" Reeb orbits are thrown out; in the three-dimensional case, a bad Reeb orbit is an even cover of a negative hyperbolic orbit. The reasons for discarding bad orbits in SFT are similar to the reasons for disallowing multiply covered hyperbolic orbits in ECH, see Section 2.7 and Section 5.4.

Keeping Track of Topological Complexity. In SFT, there is a formal variable $\hbar$ which keeps track of the topological complexity of holomorphic curves; whenever one counts a curve with genus g and p positive ends, one multiplies by $\hbar^{p+g-1}$. In ECH, topological complexity is measured by the number J_0 defined in Section 5.2. There is also a variant of J_0, denoted by J_+, which is closer to the exponent of $\hbar$, see [22, §6] and [37, Appendix].

U Maps. ECH has a U map counting holomorphic curves passing through a base point, and also operations determined by elements of H_1 of the three-manifold, counting holomorphic curves intersecting a 1-cycle, see Section 3.8. There are analogous structures on SFT, which can be more interesting for higher dimensional contact manifolds with lots of homology.

Algebra Structure. SFT has some algebra structure (for example the contact homology algebra is an algebra). ECH does not. There is a natural way to "multiply" two ECH generators, by adding the multiplicities of all Reeb orbits in the two generators, but the differential and grading are not well behaved with respect to this "multiplication".

Legendrian Knots. SFT defines invariants of Legendrian knots by counting holomorphic curves with boundary in $\mathbb{R}$ cross the Legendrian knot. No analogous construction in ECH is known, although one can define invariants of Legendrian knots using sutured ECH, see [7, §7.3].

Technical Difficulties with Multiply Covered Holomorphic Curves. Both SFT and ECH have serious technical difficulties arising from multiply covered holomorphic curves of negative Fredholm index or ECH index. In SFT, it is expected that the polyfold theory of Hofer-Wysocki-Zehnder [20] will resolve these difficulties. In ECH, we could manage these difficulties to prove that $\partial^2 = 0$ using holomorphic curves as outlined in Section 5.4. Defining cobordism maps on ECH is harder, and it is not clear whether polyfolds will help, but fortunately one can define ECH cobordism maps using Seiberg-Witten theory, as described in Section 5.5.

Field Theory Structure. SFT can recover Gromov-Witten invariants of closed symplectic manifolds by cutting them into pieces along contact-type hypersurfaces. ECH can similarly recover Taubes's Gromov invariant of closed symplectic four-manifolds [26].

Symplectic Capacities. ECH can be used to define symplectic capacities. Other kinds of contact homology or SFT can also be used to define symplectic capacities, and this is an interesting topic for further research. For example, one can define an analogue of ECH capacities using linearized contact homology, and these turn out to agree with the Ekeland-Hofer capacities, at least for four-dimensional ellipsoids and polydisks, see Remark 1.5.

Appendix: Answers and Hints to Selected Exercises

1.4. We need to show that

$$\lim_{k\to\infty} \frac{N(a,b)_k^2}{k} = 2ab. \tag{A.1}$$

Given nonnegative integers m and n, let $T(m,n)$ denote the triangle in the plane bounded by the x and y axes and the line L through (m,n) with slope $-b/a$. Then $N(a,b)_k = am+bn$ where $T(m,n)$ encloses $k+1$ lattice points (including the edges). When k is large, the number of lattice points enclosed by $T(m,n)$ is the area of the triangle, plus an $O(k^{1/2})$ error. The line L intersects the axes at the points $(a^{-1}N(a,b)_k, 0)$ and $(0, b^{-1}N(a,b)_k)$, so its area is

$$\operatorname{area}\big(T(m,n)\big) = \frac{N(a,b)_k^2}{2ab}.$$

Thus

$$k = \frac{N(a,b)_k^2}{2ab} + O\big(k^{1/2}\big).$$

This implies (A.1).

1.15. It is enough to show that

$$2\operatorname{vol}(X,\omega) = \operatorname{vol}(Y_+,\lambda_+) - \operatorname{vol}(Y_-,\lambda_-), \tag{A.2}$$

where $\operatorname{vol}(Y,\lambda) = \int_Y \lambda\wedge d\lambda$. To prove (A.2), let λ be a primitive of ω on X. Then by Stokes's theorem,

$$2\operatorname{vol}(X,\omega) = \int_{Y_+} \lambda\wedge d\lambda_+ - \int_{Y_-} \lambda\wedge d\lambda_-.$$

Since $d\lambda = d\lambda_\pm$ on $Y_\pm$, by Stokes's theorem again we have

$$\int_{Y_\pm} \lambda\wedge d\lambda_\pm = \operatorname{vol}(Y_\pm,\lambda_\pm).$$

2.9. We use an infinitesimal analogue of the proof of Lemma 2.6. Let $\psi\in \operatorname{Ker}(D_C)$. Let $\varepsilon>0$ be small and let C' be the image of the map $C\to S^1\times Y_\phi$ sending $z\mapsto \exp_z(\varepsilon\psi(z))$. Then

$$\int_{C'}\omega = \varepsilon^2\int_C \omega(\partial_s\psi,\nabla_t\psi)ds\,dt + O\big(\varepsilon^3\big).$$

Since C' is homologous to C, we have $\int_{C'}\omega = 0$, so

$$\int_C \omega(\partial_s\psi,\nabla_t\psi)ds\,dt = 0. \tag{A.3}$$

On the other hand, since $\psi\in\operatorname{Ker}(D_C)$, we have $\nabla_t\psi = J\partial_s\psi$, so the integrand above is

$$\omega(\partial_s\psi,\nabla_t\psi) = \|\partial_s\psi\|^2, \tag{A.4}$$

where $\|\cdot\|$ denotes the metric on $T^{\text{vert}}Y|_\gamma$ determined by ω and J. It follows from (A.3) and (A.4) that $\partial_s\psi \equiv 0$.

3.3. Given a Reeb orbit γ, the set of homotopy classes of trivializations of $\xi|_\tau$ is an affine space over $\mathbb{Z}$. For an appropriate sign convention, shifting the trivialization over $\gamma_i^\pm$ by 1 shifts c_1 by ∓ 1 and shifts $CZ_\tau(\gamma_i^\pm)$ by 2.

3.3. For an appropriate sign convention, shifting the trivialization τ over α_i by 1 shifts c_1 by $-m_i$, shifts Q_τ by m_i^2, and shifts w_τ by $-m_i(m_i-1)$.

3.11. Let T be the triangle in the plane which is bounded by the coordinate axes together with the line through (m_1,m_2) with slope $-a/b$, cf. the answer to Exercise 1.4. Then $\frac{1}{2}I(\alpha)$ can be interpreted as the number of lattice points in the triangle T (including the boundary) minus 1.

3.13. (a) Since the path $\Lambda_\theta^+(m)$ starts at the origin and stays below the line $y=\theta x$, the initial edge has slope less than θ. Since the path is the graph of a concave function, every subsequent edge also has slope less than θ. Thus $b \le \lfloor a\theta \rfloor$. If $b < \lfloor a\theta \rfloor$ then there is a lattice point which is above the path $\Lambda_\theta^+(m)$ but below the line $y=\theta x$, contradicting the definition of $\Lambda_\theta^+(m)$.

(b) Since the total vertical displacement of the path $\Lambda_\theta^+(m)$ is $\lfloor m\theta \rfloor$, it follows from part (a) that

$$\sum_{i=1}^k \lfloor q_i\theta \rfloor = \left\lfloor \sum_{i=1}^k q_i\theta \right\rfloor.$$

Since $\lfloor x \rfloor + \lfloor y \rfloor \le \lfloor x+y \rfloor$ for any real numbers x,y, we have

$$\sum_{i\in I} \lfloor q_i\theta \rfloor \le \left\lfloor \sum_{i\in I} q_i\theta \right\rfloor,$$

$$\sum_{i\in\{1,\ldots,k\}\setminus I} \lfloor q_i\theta \rfloor \le \left\lfloor \sum_{i\in\{1,\ldots,k\}\setminus I} q_i\theta \right\rfloor.$$

Adding the above two inequalities and comparing with the previous equation, we see that both inequalities must be equalities.

(c) Suppose that such proper subsets I,J exist. Let $m_1 = \sum_{i\in I} q_i = \sum_{j\in J} r_j$ and let $m_2 = m - m_1$. By part (b) applied to the subsets I, $\{1,\ldots,k\}\setminus I$, and $\{1,\ldots,k\}$, we have

$$\lfloor m_1\theta \rfloor + \lfloor m_2\theta \rfloor = \lfloor m\theta \rfloor.$$

By the analogue of part (b) for $p_\theta^-(m)$, we have

$$\lceil m_1\theta\rceil + \lceil m_2\theta\rceil = \lceil m\theta\rceil.$$

Subtracting the above two equations gives $2=1$.

3.14. (a) Without loss of generality C is connected. Let $a_1,\ldots,a_k$ denote the covering multiplicities of the positive ends of u, and let $b_1,\ldots,b_l$ denote the covering multiplicities of the negative ends of u. Let g denote the genus of C. By the Fredholm index formula (3.1) and the Conley-Zehnder index formula (3.2), we have

$$\begin{aligned}\operatorname{ind}(u) &= 2g-2+k+l+\sum_{i=1}^k(2\lfloor a_i\theta\rfloor+1)-\sum_{j=1}^l(2\lfloor b_j\theta\rfloor+1)\\ &= 2\left(g-1+\sum_{i=1}^k\lceil a_i\theta\rceil-\sum_{j=1}^l\lfloor b_j\theta\rfloor\right)\\ &\ge 2(g-1+\lceil m\theta\rceil-\lfloor m\theta\rfloor).\end{aligned}$$

Since $\lceil m\theta\rceil-\lfloor m\theta\rfloor=1$, it follows that $\operatorname{ind}(u)\ge 0$.

(b) We need to check: (i) if $p\ge q$ and $q\ge r$ then $p\ge r$, and (ii) if $p\ge q$ and $q\ge p$ then $p=q$.

Suppose u_1 is a branched cover with positive ends corresponding to p and negative ends corresponding to q, and u_2 is a branched cover with positive ends corresponding to q and negative ends corresponding to r. Gluing these together gives a branched cover $u_1\#u_2$ (defined up to sliding the branched points around) with positive ends corresponding to p and negative ends corresponding to r. It follows immediately from the index formula (3.1) that $\operatorname{ind}(u_1\#u_2)=\operatorname{ind}(u_1)+\operatorname{ind}(u_2)$. So if $\operatorname{ind}(u_1)=\operatorname{ind}(u_2)=0$, then $\operatorname{ind}(u_1\#u_2)=0$ also, and this proves (i). Now suppose further that $p=r$. Then $q=r$, because otherwise $u_1\#u_2$ has at least two branch points, so its domain has $\chi\le -2$, so $\operatorname{ind}(u_1\#u_2)\ge 2$, a contradiction. This proves (ii).

(c) Let u be a connected genus 0 branched cover with positive ends corresponding to $p_\theta^-(m)$ and negative ends corresponding to $p_\theta^+(m)$. Write $p_\theta^-(m)=(a_1,\ldots,a_k)$ and $p_\theta^+(m)=(b_1,\ldots,b_l)$. By the calculation in part (a), we have

$$\operatorname{ind}(u)=2\left(\sum_{i=1}^k\lceil a_i\theta\rceil-\sum_{j=1}^l\lfloor b_j\theta\rfloor-1\right).$$

By Exercise 3.13(b) we have $\sum_{i=1}^{k}\lceil a_i\theta\rceil = \lceil m\theta\rceil$, and by symmetry $\sum_{j=1}^{l}\lfloor b_j\theta\rfloor = \lfloor m\theta\rfloor$. Hence $\operatorname{ind}(u) = 0$.

(d) Suppose there exists a partition q with $p_\theta^+(m) > q$. Write $p_\theta^+(m) = (a_1,\ldots,a_k)$ and $q = (b_1,\ldots,b_l)$. By Exercise 3.13(b) we have $\sum_{i=1}^{k}\lfloor a_i\theta\rfloor = \lfloor m\theta\rfloor$. By the calculation in part (a) above we have $\sum_{i=1}^{k}\lceil a_i\theta\rceil = \lceil m\theta\rceil$. These two equations imply that $k=1$. Thus the path $\Lambda_\theta^+(m)$ is just the line segment from $(0,0)$ to $(m,\lfloor m\theta\rfloor)$.

Now the calculation in part (a) above also implies that $\sum_{j=1}^{l}\lfloor b_j\theta\rfloor = \lfloor m\theta\rfloor$. But this is impossible. To see why, order the numbers b_j so that $\lfloor b_j\theta\rfloor/b_j \geq \lfloor b_{j+1}\theta\rfloor/b_{j+1}$. Let Λ' be the path in the plane that starts at $(0,0)$ and whose edge vectors are the segments $(b_j,\lfloor b_j\theta\rfloor)$ in order of increasing j. Since $(b_1,\ldots,b_l) \neq (m)$ and since there are no lattice points above the path $\Lambda_\theta^+(m)$ and below the line $y=\theta x$, it follows that the path Λ' is below the path $\Lambda_\theta^+(m)$, with the two paths intersecting only at $(0,0)$. Hence the right endpoint of Λ' is below the right endpoint of $\Lambda_\theta^+(m)$, which means that $\sum_j\lfloor b_j\theta\rfloor < \lfloor m\theta\rfloor$.

By symmetry, there also does not exist a partition q with $q > p_\theta^-(m)$.

4.3. By Exercise 3.10 we have $c_\tau(C_2) = 1$. Since $\operatorname{ind}(C_2) = 2$, it follows from (3.1) that

$$\chi(C_2) = CZ_\tau^{\mathrm{ind}}(C_2).$$

If ε is sufficiently small with respect to i, then $CZ_\tau(\gamma_1^i) = 2i-1$ when $i>0$, and $CZ_\tau(\gamma_2^{i-1}) = 2i-1$ when $i>1$. It follows that $CZ_\tau^{\mathrm{ind}}(C_2) = 0$ when $i>1$, and $CZ_\tau^{\mathrm{ind}}(C_2) = 1$ when $i=1$.

4.4. Without loss of generality, $\mathcal{C}_0 = \emptyset$. We then compute that

$$CZ_\tau^{\mathrm{ind}}(C_2) = \begin{cases} i+j-1, & i>0, j>1,\\ i+1, & i>0, j=1,\\ j, & i=0, j>1,\\ 2, & i=0, j=1.\end{cases}$$

On the other hand, letting g denote the genus of C_2, we have

$$\chi(C_2) = \begin{cases} -2g-i-j-1, & i>0, j>1,\\ -2g-i-1, & i>0, j=1,\\ -2g-j, & i=0, j>1,\\ -2g, & i=0, j=1.\end{cases}$$

Since $c_\tau(C_2)=0$ (by Exercise 3.10) and $\operatorname{ind}(C_2)=2$, it follows from (3.1) that

$$\chi(C_2)=CZ_\tau^{\mathrm{ind}}(C_2)-2.$$

Combining the above three equations, we find that if $i>0$ or $j>1$ then $g<0$, which is a contradiction. Thus $i=0$ and $j=1$, and combining the above three equations again we find that $g=0$.

4.12. Otherwise $g=1$. Then Equation (4.9) (together with the fact that C_1 has at least one positive end) implies that C_1 has exactly one positive end at some hyperbolic orbit $h_{a,b}$, and all negative ends of C_1 are elliptic. Let $(a_1,b_1),\ldots,(a_k,b_k)$ denote the vectors corresponding to the negative ends. The action of $h_{a,b}$ is slightly less than $\sqrt{a^2+b^2}$, and the sum of the symplectic actions of the negative ends is slightly greater $\sum_{i=1}^k\sqrt{a_i^2+b_i^2}$. Since the differential decreases symplectic action,

$$\sum_{i=1}^k\sqrt{a_i^2+b_i^2}<\sqrt{a^2+b^2}.$$

But this contradicts the triangle inequality, since $\sum_{i=1}^k(a_i,b_i)=(a,b)$, since $h_{a,b}$ is homologous in T^3 to $\sum_{i=1}^k e_{a_i,b_i}$.

4.13. Let Λ be any polygonal path with edge vectors $v_1,\ldots,v_k$. Then

$$\ell_\Omega(\Lambda)=\sum_{i=1}^k\langle v_i,w_i\rangle$$

where $w_i\in\partial\Omega'$ is a point at which an outward normal vector to Ω' is a positive multiple of v_i. (When w_i is a corner of $\partial\Omega'$, "an outward normal vector" means a vector whose direction is between the directions of the limits of the normal vectors on either side of w_i.) If we replace Ω' by its translate by some vector η, then the above formula is replaced by

$$\ell_\Omega(\Lambda)=\sum_{i=1}^k\langle v_i,w_i+\eta\rangle.$$

If Λ is a loop, then the two formulas for $\ell_\Omega(\Lambda)$ agree since $\sum_i v_i=0$.

5.9. By the relative adjunction formula (3.3), and since equality holds in the writhe bound (3.9), we have

$$-\chi(C)=-c_\tau(C)+Q_\tau(C)+CZ_\tau^I(C)-CZ_\tau^{\mathrm{ind}}(C).$$

So by the definition of J_0 in (5.13) and (5.14), we need to show that

$$\sum_i (n_i^+ - 1) + \sum_j (n_j^- - 1) = CZ_\tau^{J_0}(C) - CZ_\tau^I(C) + CZ_\tau^{\mathrm{ind}}(C).$$

This equation can be proved one Reeb orbit at a time. Namely, it is enough to show that for each i, if C has positive ends at covers of α_i with multiplicities $q_1, \ldots, q_{n_i^+}$ where $\sum_{k=1}^{n_i^+} q_k = m_i$, then

$$n_i^+ - 1 = -CZ_\tau(\alpha_i^{m_i}) + \sum_{k=1}^{n_i^+} CZ_\tau(\alpha^{q_k}), \tag{A.5}$$

and an analogous equation for each Reeb orbit β_j.

To prove (A.5), first note that if α_i is hyperbolic, then $n_i^+ = m_i = 1$ and the equation is trivial. Suppose now that α_i is elliptic with rotation angle θ with respect to τ. Then (A.5) becomes

$$0 = -2\lfloor m_i\theta \rfloor + \sum_{k=1}^{n_i^+} 2\lfloor q_k\theta \rfloor.$$

This last equation holds by the partition conditions and Exercise 3.13(b).

References

[1] P. Biran, From symplectic packing to algebraic geometry and back, in *European Congress of Mathematicians, vol. II.* Progress in Math, vol. 202 (Birkhäuser, Basel, 2000), pp. 507–524

[2] F. Bourgeois, A Morse-Bott approach to contact homology, in *Symplectic and Contact Topology: Interactions and Perspectives*, Toronto, 2001. Fields Inst. Commun., vol. 35 (AMS, Providence, 2003), pp. 55–77

[3] F. Bourgeois, K. Mohnke, Coherent orientations in symplectic field theory. Math. Z. **248**, 123–146 (2004)

[4] F. Bourgeois, Y. Eliashberg, H. Hofer, K. Wysocki, E. Zehnder, Compactness results in symplectic field theory. Geom. Topol. **7**, 799–888 (2003)

[5] K. Cieliebak, H. Hofer, J. Latschev, F. Schlenk, Quantitative symplectic geometry, in *Dynamics, Ergodic Theory, and Geometry.* MSRI Publ, vol. 54 (Cambridge University Press, Cambridge, 2007), pp. 1–44

[6] V. Colin, K. Honda, Reeb vector fields and open book decompositions. J. Eur. Math. Soc. **15**, 443–507 (2013)

[7] V. Colin, P. Ghiggini, K. Honda, M. Hutchings, Sutures and contact homology I. Geom. Topol. **15**, 1749–1842 (2011)

[8] V. Colin, P. Ghiggini, K. Honda, HF = ECH via open book decompositions: a summary. arXiv:1103.1290

[9] D. Cristofaro-Gardiner, The absolute gradings in embedded contact homology and Seiberg-Witten Floer cohomology. Algebr. Geom. Topol. **13**, 2239–2260 (2013)

[10] D. Cristofaro-Gardiner, M. Hutchings, From one Reeb orbit to two. arXiv:1202.4839

[11] D. Cristofaro-Gardiner, V. Gripp, M. Hutchings, The asymptotics of ECH capacities. arXiv:1210.2167

[12] D. Dragnev, Fredholm theory and transversality for noncompact pseudoholomorphic maps in symplectizations. Commun. Pure Appl. Math. **57**, 726–763 (2004)

[13] I. Ekeland, H. Hofer, Symplectic topology and Hamiltonian dynamics. II. Math. Z. **203**, 553–567 (1990)

[14] Y. Eliashberg, A. Givental, H. Hofer, Introduction to symplectic field theory. Geom. Funct. Anal. (2000), Special Volume, Part II, 560–673

[15] O. Fabert, Contact homology of Hamiltonian mapping tori. Comment. Math. Helv. **85**, 203–241 (2010)

[16] D. Farris, The embedded contact homology of nontrivial circle bundles over Riemann surfaces. UC Berkeley PhD thesis, 2011

[17] D. Frenkel, D. Müller, Symplectic embeddings of 4-dimensional ellipsoids into cubes. arXiv:1210.2266

[18] H. Hofer, K. Wysocki, E. Zehnder, Properties of pseudo-holomorphic curves in symplectisations. II. Embedding controls and algebraic invariants. Geom. Funct. Anal. **5**, 270–328 (1995)

[19] H. Hofer, K. Wysocki, E. Zehnder, Properties of pseudoholomorphic curves in symplectisations. I. Asymptotics. Ann. Inst. Henri Poincaré, Anal. Non Linéaire **13**, 337–379 (1996)

[20] H. Hofer, K. Wysocki, E. Zehnder, A general Fredholm theory. I. A splicing-based differential geometry. J. Eur. Math. Soc. **9**, 841–876 (2007)

[21] M. Hutchings, An index inequality for embedded pseudoholomorphic curves in symplectizations. J. Eur. Math. Soc. **4**, 313–361 (2002)

[22] M. Hutchings, The embedded contact homology index revisited, in *New Perspectives and Challenges in Symplectic Field Theory*. CRM Proc. Lecture Notes, vol. 49 (Amer. Math. Soc., Providence, 2009), pp. 263–297

[23] M. Hutchings, Embedded contact homology and its applications, in *Proceedings of the 2010 ICM*, vol. II (2010), pp. 1022–1041

[24] M. Hutchings, Quantitative embedded contact homology. J. Differ. Geom. **88**, 231–266 (2011)

[25] M. Hutchings, Recent progress on symplectic embedding problems in four dimensions. Proc. Natl. Acad. Sci. USA **108**, 8093–8099 (2011)

[26] M. Hutchings, Embedded contact homology as a (symplectic) field theory, in preparation

[27] M. Hutchings, M. Sullivan, The periodic Floer homology of a Dehn twist. Algebr. Geom. Topol. **5**, 301–354 (2005)

[28] M. Hutchings, M. Sullivan, Rounding corners of polygons and the embedded contact homology of T^3. Geom. Topol. **10**, 169–266 (2006)

[29] M. Hutchings, C.H. Taubes, Gluing pseudoholomorphic curves along branched covered cylinders I. J. Symplectic Geom. **5**, 43–137 (2007)

[30] M. Hutchings, C.H. Taubes, Gluing pseudoholomorphic curves along branched covered cylinders II. J. Symplectic Geom. **7**, 29–133 (2009)

[31] M. Hutchings, C.H. Taubes, The Weinstein conjecture for stable Hamiltonian structures. Geom. Topol. **13**, 901–941 (2009)

[32] M. Hutchings, C.H. Taubes, Proof of the Arnold chord conjecture in three dimensions II. Geom. Topol. **17**, 2601–2688 (2013)

[33] E. Ionel, T. Parker, The Gromov invariants of Ruan-Tian and Taubes. Math. Res. Lett. **4**, 521–532 (1997)

[34] P.B. Kronheimer, T.S. Mrowka, *Monopoles and Three-Manifolds* (Cambridge University Press, Cambridge, 2008)

[35] C. Kutluhan, Y.-J. Lee, C. Taubes, $HF = HM$ I: Heegaard Floer homology and Seiberg-Witten Floer homology. arXiv:1007.1979

[36] C. Kutluhan, Y.-H. Lee, C. Taubes, $HF = HM$ II: Reeb orbits and holomorphic curves for the ech/Heegaard-Floer correspondence. arXiv:1008.1595

[37] J. Latschev, C. Wendl, Algebraic torsion in contact manifolds. Geom. Funct. Anal. **21**, 1144–1195 (2011), corrected version of appendix at arXiv:1009.3262

[38] Y.-J. Lee, C. Taubes, Periodic Floer homology and Seiberg-Witten Floer cohomology. J. Symplectic Geom. **10**, 81–164 (2012)

[39] T.-J. Li, A.-K. Liu, The equivalence between SW and Gr in the case where $b^+ = 1$. Int. Math. Res. Not. **7**, 335–345 (1999)

[40] D. McDuff, From symplectic deformation to isotopy, in *Topics in Symplectic 4-Manifolds*, Irvine, CA, 1996. First Int. Press Lect. Ser. I (International Press, Cambridge, 1996), pp. 85–99

[41] D. McDuff, The Hofer conjecture on embedding symplectic ellipsoids. J. Differ. Geom. **88**, 519–532 (2011)

[42] D. McDuff, D. Salamon, *J-Holomorphic Curves and Symplectic Topology.* AMS Colloquium Publications, vol. 52

[43] D. McDuff, F. Schlenk, The embedding capacity of 4-dimensional symplectic ellipsoids. Ann. Math. **175**, 1191–1282 (2012)

[44] G. Meng, C.H. Taubes, $\underline{SW}$ = Milnor torsion. Math. Res. Lett. **3**, 661–674 (1996)

[45] J. Morgan, *The Seiberg-Witten Equations and Applications to the Topology of Smooth Four-Manifolds.* Mathematical Notes, vol. 44 (Princeton Univ. Press, Princeton, 1996)

[46] P. Ozsváth, Z. Szabó, Holomorphic disks and topological invariants for closed three-manifolds. Ann. Math. **159**, 1027–1158 (2004)

[47] B. Parker, Holomorphic curves in Lagrangian torus fibrations. Stanford University PhD thesis, 2005

[48] D. Salamon, Lectures on Floer homology, in *Symplectic Geometry and Topology*, Park City, UT, 1997. IAS/Park City Math. Ser., vol. 7 (AMS, Providence, 1999)

[49] D. Salamon, Seiberg-Witten invariants of mapping tori, symplectic fixed points, and Lefschetz numbers. Turk. J. Math. **23**, 117–143 (1999). 6th Gökova Geometry-Topology Conference

[50] M. Schwarz, Cohomology operations from S^1-cobordisms in Floer homology. PhD thesis, ETH Zürich, 1995

[51] R. Siefring, Relative asymptotic behavior of pseudoholomorphic half-cylinders. Commun. Pure Appl. Math. **61**, 1631–1684 (2008)

[52] R. Siefring, Intersection theory of punctured pseudoholomorphic curves. Geom. Topol. **15**, 2351–2457 (2011)

[53] C.H. Taubes, Counting pseudo-holomorphic submanifolds in dimension four. J. Differ. Geom. **44**, 818–893 (1996)

[54] C.H. Taubes, The structure of pseudoholomorphic subvarieties for a degenerate almost complex structure and symplectic form on $S^1 \times B^3$. Geom. Topol. **2**, 221–332 (1998)

[55] C.H. Taubes, *Seiberg-Witten and Gromov Invariants for Symplectic 4-Manifolds*. First International Press Lecture Series, vol. 2 (International Press, Cambridge, 2000)

[56] C.H. Taubes, Pseudolomorphic punctured spheres in $\mathbb{R} \times (S^1 \times S^2)$: properties and existence. Geom. Topol. **10**, 785–928 (2006)

[57] C.H. Taubes, The Seiberg-Witten equations and the Weinstein conjecture. Geom. Topol. **11**, 2117–2202 (2007)

[58] C.H. Taubes, Embedded contact homology and Seiberg-Witten Floer homology I. Geom. Topol. **14**, 2497–2581 (2010)

[59] C.H. Taubes, Embedded contact homology and Seiberg-Witten Floer homology V. Geom. Topol. **14**, 2961–3000 (2010)

[60] V. Turaev, A combinatorial formulation for the Seiberg-Witten invariants of 3-manifolds. Math. Res. Lett. **5**, 583–598 (1998)

[61] C. Wendl, Compactness for embedded pseudoholomorphic curves in 3-manifolds. J. Eur. Math. Soc. **12**, 313–342 (2010)

[62] C. Wendl, Automatic transversality and orbifolds of punctured holomorphic curves in dimension four. Comment. Math. Helv. **85**, 347–407 (2010)

[63] M.-L. Yau, Vanishing of the contact homology of overtwisted contact 3-manifolds. Bull. Inst. Math. Acad. Sin. **1**, 211–229 (2006), with an appendix by Y. Eliashberg

M. Hutchings

University of California
Berkeley
USA

BOLYAI SOCIETY
MATHEMATICAL STUDIES, 26

Contact and Symplectic Topology
pp. 485–530.

A Topological Introduction to Knot Contact Homology

LENHARD NG

1. Introduction

This article is intended to serve as a general introduction to the subject of knot contact homology. There are two related sides to the theory: a geometric side devoted to the contact geometry of conormal bundles and explicit calculation of holomorphic curves, and an algebraic, combinatorial side emphasizing ties to knot theory and topology. We will focus on the latter side and only treat the former side lightly. The present notes grew out of lectures given at the Contact and Symplectic Topology Summer School in Budapest in July 2012.

The strategy of studying the smooth topology of a smooth manifold via the symplectic topology of its cotangent bundle is an idea that was advocated by V.I. Arnold and has been extensively studied in symplectic geometry in recent years. It is well-known that if M is smooth then T^*M carries a natural symplectic structure, with symplectic form $\omega = -d\lambda_{\mathrm{can}}$, where $\lambda_{\mathrm{can}} \in \Omega^1(T^*M)$ is the Liouville form; the idea then is to analyze T^*M as a symplectic manifold to recover topological data about M.

In recent years this strategy has been executed quite successfully by examining Gromov-type moduli spaces of holomorphic curves on T^*M. For instance, one can show that the symplectic structure on T^*M recovers homotopic information about M, as shown in various guises by Viterbo [41], Salamon–Weber [40], and Abbondandolo–Schwarz [1], who each prove some

F. Bourgeois et al. (eds.), *Contact and Symplectic Topology*,
Bolyai Society Mathematical Studies 26, DOI 10.1007/978-3-319-02036-5_10,

version of the following result (where technical restrictions have been omitted for simplicity):

Theorem 1.1 ([1, 40, 41]). *The Hamiltonian Floer homology of T^*M is isomorphic to the singular homology of the free loop space of M.*

Subsequent work has related certain additional Floer-theoretic constructions on T^*M to the Chas–Sullivan loop product and string topology; see for example [2, 9].

In a slightly different direction, M. Abouzaid has used holomorphic curves to show that the symplectic structure on T^*M can contain more than topological information about M:

Theorem 1.2 ([3]). *If Σ is an exotic $(4k+1)$-sphere that does not bound a parallelizable manifold, then $T^*\Sigma$ is not symplectomorphic to T^*S^{4k+1}.*

At the time of this writing, it is still possible that the smooth type of a closed smooth manifold M (up to diffeomorphism) is determined by the symplectic type of T^*M (up to symplectomorphism), which would be a very strong endorsement of Arnold's idea. (See however [26] for counterexamples when M is not closed.) For a nice discussion of this and related problems, see [39].

In this survey article, we discuss a relative version of Arnold's strategy. The setting is as follows. Let $K \subset M$ be an embedded submanifold (or an immersed submanifold with transverse self-intersections). Then one can construct the *conormal bundle* of K:

$$L_K = \{(q,p) \mid q \in K,\ \langle p, v\rangle = 0 \text{ for all } v \in T_qK\} \subset T^*M.$$

It is a standard exercise to check that L_K is a Lagrangian submanifold of T^*M.

One can work in one dimension lower by considering the *cosphere* (unit cotangent) *bundle* ST^*M of unit covectors in T^*M with respect to some metric; then ST^*M is a contact manifold with contact form $\alpha = \lambda_{\text{can}}$, and it can be shown that the contact structure on ST^*M is independent of the metric. The *unit conormal bundle* of K,

$$\Lambda_K = L_K \cap ST^*M \subset ST^*M,$$

is then a Legendrian submanifold of ST^*M, with $\alpha|_{\Lambda_K} = 0$. See Figure 1.

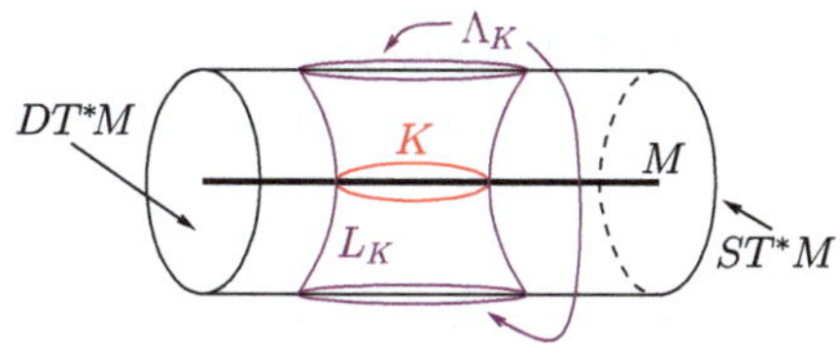

Fig. 1. A schematic depiction of cotangent and conormal bundles. Only the disk bundle portion DT^*M of T^*M is shown, with boundary ST^*M. Note that both L_K and the zero section M are Lagrangian in T^*M, and their intersection is K

By construction, if K changes by smooth isotopy in M, then Λ_K changes by Legendrian isotopy (isotopy within the class of Legendrian submanifolds) in ST^*M. One can then ask what the Legendrian isotopy type of Λ_K remembers about the smooth isotopy type of K; see Question 1.3 below.

For the remainder of the section and article, we restrict our focus by assuming that $M = \mathbb{R}^3$ and $K \subset \mathbb{R}^3$ is a knot or link. In this case, ST^*M is contactomorphic to the 1-jet space $J^1(S^2) = T^*S^2 \times \mathbb{R}$ equipped with the contact form $dz - \lambda_{\text{can}}$, where z is the coordinate on $\mathbb{R}$ and λ_{can} is the Liouville form on S^2, via the diffeomorphism $ST^*\mathbb{R}^3 \to J^1(S^2)$ sending (q,p) to $((p, q - \langle q,p\rangle p), \langle q,p\rangle)$ where $\langle\cdot,\cdot\rangle$ is the standard metric on $\mathbb{R}^3$.

In the 5-manifold $ST^*\mathbb{R}^3$, the unit conormal bundle Λ_K is topologically a 2-torus (or a disjoint union of tori if K has multiple components). This can for instance be seen in the dual picture in TR^3, where the unit normal bundle can be viewed as the boundary of a tubular neighborhood of K. The topological type of $\Lambda_K \cong T^2 \subset S^2 \times \mathbb{R}^3$ contains no information: if K_1, K_2 are arbitrary knots, then Λ_{K_1} and Λ_{K_2} are smoothly isotopic. (Choose a one-parameter family of possibly singular knots K_t joining K_1 to K_2, and perturb Λ_{K_t} slightly when K_t is singular to eliminate double points.)

However, there is no reason for Λ_{K_1} and Λ_{K_2} to be Legendrian isotopic. This suggests the following question.

Question 1.3. How much of the topology of $K \subset \mathbb{R}^3$ is encoded in the Legendrian structure of $\Lambda_K \subset ST^*\mathbb{R}^3$? If Λ_{K_1} and Λ_{K_2} are Legendrian isotopic, are K_1 and K_2 necessarily smoothly isotopic knots?

At the present, the answer to the second part of this question is unknown but could possibly be "yes". The answer is known to be "yes" if either knot is the unknot; see below.

In order to tackle Question 1.3, it is useful to have invariants of Legendrian submanifolds under Legendrian isotopy. One particularly powerful invariant is Legendrian contact homology, which is a Floer-theoretic count of holomorphic curves associated to a Legendrian submanifold and is discussed in more detail in Section 2.

Definition 1.4. Let $K \subset \mathbb{R}^3$ be a knot or link. The *knot contact homology* of K, written $HC_*(K)$, is the Legendrian contact homology of Λ_K.

Knot contact homology is the homology of a differential graded algebra associated to a knot, the *knot DGA* $(\mathcal{A}, \partial)$. By the general invariance result for Legendrian contact homology, the knot DGA and knot contact homology are topological invariants of knots and links.

This article is a discussion of knot contact homology and its properties. Despite the fact that the original definition of knot contact homology involves holomorphic curves, there is a purely combinatorial formulation of knot contact homology. The article [15], which does most of the heavy lifting for the results presented here, derives this combinatorial formula and can be viewed as the first reasonably involved computation of Legendrian contact homology in high dimensions.

Viewed from a purely knot theoretic perspective, knot contact homology is a reasonably strong knot invariant. For instance, it detects the unknot (see Corollaries 4.10 and 5.10): if K is a knot such that $HC_*(K) \cong HC_*(O)$ where O is the unknot, then $K = O$. This implies in particular that the answer to Question 1.3 is yes if one of the knots is unknotted. It is currently an open question whether knot contact homology is a complete knot invariant.

Connections between knot contact homology and other knot invariants are gradually beginning to appear. It is known that $HC_*(K)$ determines the Alexander polynomial (Theorem 3.18). A portion of the homology also has a natural topological interpretation, via an object called the *cord algebra* that is closely related to string topology. In addition, one can use $HC_*(K)$ to define a three-variable knot invariant, the *augmentation polynomial*, which is closely related to the A-polynomial and conjecturally determines a specialization of the HOMFLY-PT polynomial. Very recently, a connection between knot contact homology and string theory has been discovered, and this suggests that the augmentation polynomial may in fact determine many known knot invariants, including the HOMFLY-PT polynomial and certain knot homologies, and may also be determined by a recursion relation for colored HOMFLY-PT polynomials.

Knot contact homology also produces a strong invariant of transverse knots, which are knots that are transverse to the standard contact structure

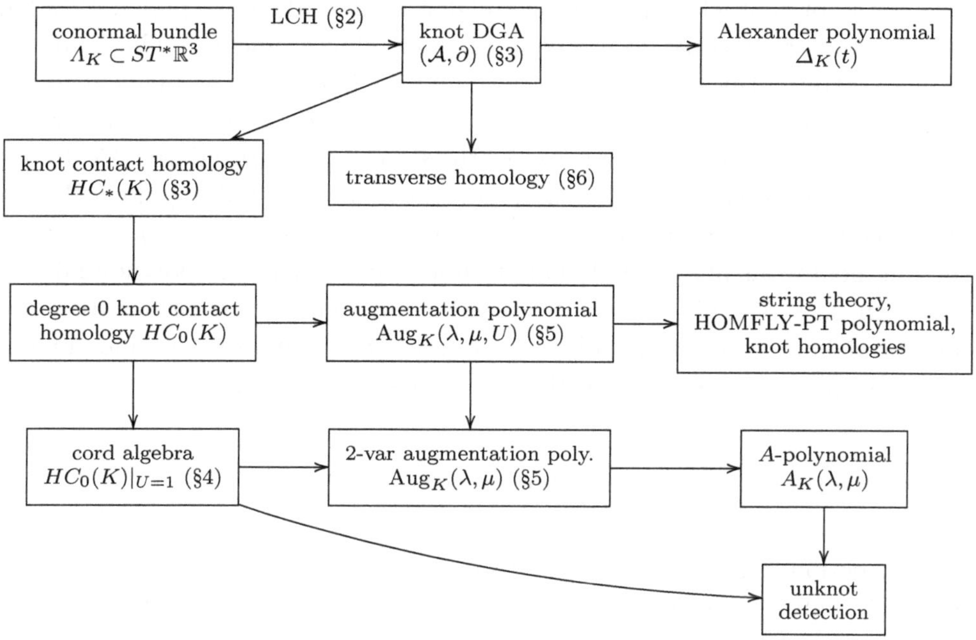

Fig. 2. The knot invariants and interconnections described in this article

on $\mathbb{R}^3$. For a transverse knot, the knot contact homology of the underlying topological knot contains an additional filtered structure, *transverse homology*, which is invariant under transverse isotopy. This has been shown to be an effective transverse invariant (Theorem 6.9), one of two that are currently known (the other comes from Heegaard Floer theory).

In the rest of the article, we expand on the properties of knot contact homology mentioned above; see Figure 2 for a schematic chart. In Section 2, we review the general definition of Legendrian contact homology. We apply this to knots and conormal bundles in Section 3 to give a combinatorial definition of knot contact homology and present a few of its properties. In Section 4, we discuss the cord algebra, which gives a topological interpretation of knot contact homology in degree 0. Section 5 defines the augmentation polynomial and relates it to other knot invariants; this includes a speculative discussion of the relation to string theory. In Section 6, we present transverse homology and consider its effectiveness as an invariant of transverse knots. Some technical details (a definition of the "fully noncommutative" version of knot contact homology, and a comparison of the conventions used in this article to conventions in the literature) are included in the Appendix.

As this is a survey article, many details will be omitted in favor of what we hope is an accessible exposition of the subject. (For more introductory material on knot contact homology, the reader is referred to two papers [12, 32]; note however that these do not contain recent developments.) There are exercises scattered through the text as a concrete, hands-on complement to the main discussion. There is not much new mathematical content in this article beyond what has already appeared in the literature, particularly [15, 16] on the geometric side and [30, 31, 33, 34] on the combinatorial/topological side. One exception is a representation-theoretic interpretation of some factors of the augmentation polynomial that do not appear in the A-polynomial; see Theorem 5.11. We have also introduced a number of conventions for combinatorial knot contact homology in this article that are new and, in the author's opinion, more natural than previous conventions.

2. Legendrian Contact Homology

In this section, we give a cursory introduction to Legendrian contact homology and augmentations, essentially the minimum necessary to motivate the construction of knot contact homology in Section 3. The reader interested in further details is referred to the various references given in this section.

Legendrian contact homology (LCH), introduced by Eliashberg and Hofer in [17], is an invariant of Legendrian submanifolds in suitable contact manifolds. This invariant is defined by counting certain holomorphic curves in the symplectization of the contact manifold, and is a part of the (much larger) Symplectic Field Theory package of Eliashberg, Givental, and Hofer [18]. LCH is the homology of a differential graded algebra (DGA) that we now describe, and in some sense the DGA (up to an appropriate equivalence relation), rather than the homology, is the "true" invariant of the Legendrian submanifold.

In this section, we will work exclusively in a contact manifold of the form $V = J^1(M) = T^*M \times \mathbb{R}$ with the standard contact form $\alpha = dz - \lambda_{\mathrm{can}}$. LCH can be defined for much more general contact manifolds, but the proof of invariance in general has not been fully carried out, and even the definition is more complicated than the one given below when the contact manifold has closed Reeb orbits. Note that for $V = J^1(M)$, the Reeb vector field R_α is $\partial/\partial z$ and thus $J^1(M)$ has no closed Reeb orbits.

Let $\Lambda \subset V$ be a Legendrian submanifold. We assume for simplicity that Λ has trivial Maslov class (e.g., for Legendrian knots in $\mathbb{R}^3 = J^1(\mathbb{R})$, this means that Λ has rotation number 0), and that Λ has finitely many Reeb chords, integral curves for the Reeb field R_α with endpoints on Λ. We label the Reeb chords formally as $a_1, \dots, a_n$. Finally, let R denote (here and throughout the article) the coefficient ring $R = \mathbb{Z}[H_2(V, \Lambda)]$, the group ring of the relative homology group $H_2(V, \Lambda)$.

Definition 2.1. The LCH *differential graded algebra* associated to Λ is $(\mathcal{A}, \partial)$, defined as follows:

1. **Algebra**: $\mathcal{A} = R\langle a_1, \dots, a_n\rangle$ is the free noncommutative unital algebra over R generated by $a_1, \dots, a_n$. As an R-module, $\mathcal{A}$ is generated by all words $a_{i_1} \cdots a_{i_k}$ for $k \geq 0$ (where $k = 0$ gives the empty word 1).
2. **Grading**: Define $|a_i| = CZ(a_i) - 1$, where CZ denotes Conley–Zehnder index (see [14] for the definition in this context) and $|r| = 0$ for $r \in R$. Extend the grading to all of $\mathcal{A}$ in the usual way: $|xy| = |x| + |y|$.
3. **Differential**: Define $\partial(r) = 0$ for $r \in R$ and
$$\partial(a_i) = \sum_{\dim \mathcal{M}(a_i; a_{j_1}, \dots, a_{j_k})/\mathbb{R} = 0} \; \sum_{\Delta \in \mathcal{M}/\mathbb{R}} (\mathrm{sgn}(\Delta)) e^{[\Delta]} a_{j_1} \cdots a_{j_k}$$
where $\mathcal{M}(a_i; a_{j_1}, \dots, a_{j_k})$ is the moduli space defined below, $\mathrm{sgn}(\Delta)$ is an orientation sign associated to Δ, and $[\Delta]$ is the homology class[1] of Δ in $H_2(V, \Lambda)$.
   Extend the differential to all of $\mathcal{A}$ via the signed Leibniz rule: $\partial(xy) = (\partial x)y + (-1)^{|x|} x(\partial y)$.

The key to Definition 2.1 is the moduli space $\mathcal{M}(a_i; a_{j_1}, \dots, a_{j_k})$. To define this, let J be a (suitably generic) almost complex structure on the symplectization $(\mathbb{R} \times V, d(e^t \alpha))$ of V (where α is the contact form on V and t is the $\mathbb{R}$ coordinate) that is compatible with the symplectization in the following sense: J is $\mathbb{R}$-invariant, $J(\partial/\partial t) = R_\alpha$, and J maps $\xi = \ker \alpha$ to itself. With respect to this almost complex structure, $\mathbb{R} \times a_i$ is a holomorphic strip for any Reeb chord a_i of Λ.

Let $D_k^2 = D^2 \setminus \{p^+, p_1^-, \dots, p_k^-\}$ be a closed disk with $k+1$ punctures on its boundary, labeled $p^+, p_1^-, \dots, p_k^-$ in counterclockwise order around ∂D^2. For (not necessarily distinct) Reeb chords a_i and $a_{j_1}, \dots, a_{j_k}$ for some $k \geq 0$, let $\mathcal{M}(a_i; a_{j_1}, \dots, a_{j_k})$ be the moduli space of J-holomorphic maps
$$\Delta : \left(D_k^2, \partial D_k^2\right) \to (\mathbb{R} \times V, \mathbb{R} \times \Lambda)$$

[1] To define this homology class, we assume that "capping half-disks" have been chosen in V for each Reeb chord a_i, with boundary given by a_i along with a path in Λ joining the endpoints of a_i. Some additional care must be taken if Λ has multiple components.

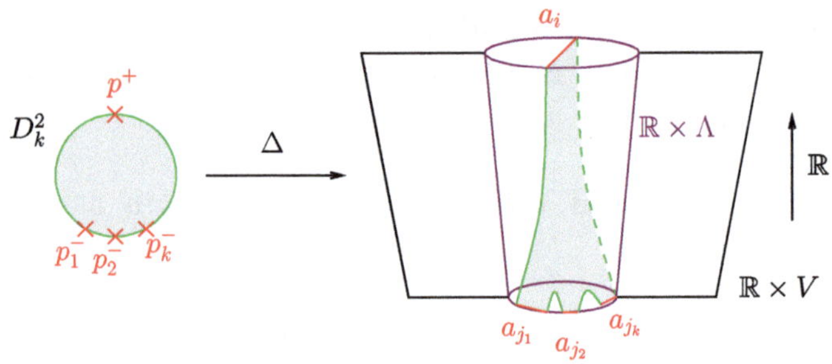

Fig. 3. A holomorphic disk $\Delta : (D_k^2, \partial D_k^2) \to (\mathbb{R} \times V, \mathbb{R} \times \Lambda)$ contributing to $\mathcal{M}(a_i; a_{j_1,\ldots,a_{j_k}})$ and the differential $\partial(a_i)$

up to domain reparametrization, such that:

- near p^+, Δ is asymptotic to a neighborhood of the Reeb strip $\mathbb{R} \times a_i$ near $t = +\infty$;
- near p_l^- for $1 \le l \le k$, Δ is asymptotic to a neighborhood of $\mathbb{R} \times a_{j_l}$ near $t = -\infty$.

See Figure 3.

When everything is suitably generic, $\mathcal{M}(a_i; a_{j_1}, \ldots, a_{j_k})$ is a manifold of dimension $|a_i| - \sum_l |a_{j_l}|$. The moduli space also has an $\mathbb{R}$ action given by translation in the $\mathbb{R}$ direction, and the differential $\partial(a_i)$ counts moduli spaces $\mathcal{M}(a_i; a_{j_1}, \ldots, a_{j_k})$ that are rigid after quotienting by this $\mathbb{R}$ action.

Remark 2.2. If $H_2(V, \Lambda) \cong H_2(V) \oplus H_1(\Lambda)$, as is true in the case that we will consider, one can "improve" the DGA $(\mathcal{A}, \partial)$ to a DGA that we might call the *fully noncommutative DGA* $(\widetilde{\mathcal{A}}, \partial)$, defined as follows. For simplicity, assume that Λ is connected; there is a similar but slightly more involved construction otherwise. The algebra $\widetilde{\mathcal{A}}$ is the tensor algebra over the group ring $\mathbb{Z}[H_2(V)]$, generated by the Reeb chords $a_1, \ldots, a_n$ along with elements of $\pi_1(\Lambda)$, with no relations except for the ones inherited from $\pi_1(\Lambda)$. Thus $\widetilde{\mathcal{A}}$ is generated as a $\mathbb{Z}[H_2(V)]$-module by words of the form

$$\gamma_0 a_{i_1} \gamma_1 a_{i_2} \gamma_2 \cdots \gamma_{k-1} a_{i_k} \gamma_k$$

where $a_{i_1}, \ldots, a_{i_k}$ are Reeb chords of Λ, $\gamma_0, \ldots, \gamma_k \in \pi_1(\Lambda)$, and $k \ge 0$. Note that $\mathcal{A}$ is a quotient of $\widetilde{\mathcal{A}}$: just abelianize $\pi_1(\Lambda)$ to $H_1(\Lambda)$, and allow Reeb chords a_i to commute with homology classes $\gamma \in H_1(\Lambda)$.

To define the differential, let Δ be a disk in $\mathcal{M}(a_i; a_{j_1}, \ldots, a_{j_k})$. The projection map $\pi : H_2(V, \Lambda) \to H_2(V)$ gives a class $\pi([\Delta]) \in H_2(V)$. The boundary of the image of Δ consists of an ordered collection of $k+1$ paths in Λ joining endpoints of Reeb chords. By fixing paths in Λ joining each Reeb chord endpoint to a fixed point on Λ, one can close these $k+1$ paths into $k+1$ loops in Λ. Let $\gamma_0(\Delta), \ldots, \gamma_k(\Delta)$ denote the homotopy classes of these loops in $\pi_1(\Lambda)$, where the loops are ordered in the order that they appear in the image of ∂D^2, traversed counterclockwise. Finally, define $\partial(\gamma) = 0$ for $\gamma \in \pi_1(\Lambda)$ and

$$\partial(a_i) = \sum_{\dim \mathcal{M}(a_i; a_{j_1}, \ldots, a_{j_k})/\mathbb{R} = 0} \; \sum_{\Delta \in \mathcal{M}/\mathbb{R}} (\operatorname{sgn}(\Delta)) e^{\pi([\Delta])} \times \gamma_0(\Delta) a_{j_1} \gamma_1(\Delta) \cdots a_{j_k} \gamma_k(\Delta),$$

and extend the differential to $\widetilde{\mathcal{A}}$ by the Leibniz rule.

Note that the quotient that sends $\widetilde{\mathcal{A}}$ to $\mathcal{A}$ also sends the differential on $\widetilde{\mathcal{A}}$ to the differential on $\mathcal{A}$. The fully noncommutative DGA $(\widetilde{\mathcal{A}}, \partial)$ satisfies the same properties as $(\mathcal{A}, \partial)$ (Theorem 2.3 below), with a suitable alteration of the definition of stable tame isomorphism. For the majority of this article, we will stick to the usual LCH DGA $(\mathcal{A}, \partial)$, which is enough for most purposes, because it simplifies notation; see however the discussion after Theorem 4.8, as well as the Appendix.

We now state some fundamental properties of the LCH DGA $(\mathcal{A}, \partial)$. These began with the work of Eliashberg–Hofer [17]; Chekanov [7] wrote down the precise statement and gave a combinatorial proof for the case $V = \mathbb{R}^3$ (see also [19]). The formulation given here is due to, and proven by, Ekholm–Etnyre–Sullivan [14].

Theorem 2.3 ([7, 14, 17]). *Given suitable genericity assumptions*:

1. ∂ *decreases degree by* 1;
2. $\partial^2 = 0$;
3. *up to stable tame isomorphism,* $(\mathcal{A}, \partial)$ *is independent of all choices (of contact form for the contact structure on* V*, and of* J*), and is an invariant of* Λ *up to Legendrian isotopy*;
4. *up to isomorphism,* $H_*(\mathcal{A}, \partial) =: HC_*(V, \Lambda)$ *is also an invariant of* Λ *up to Legendrian isotopy.*

Here "stable tame isomorphism" is an equivalence relation between DGAs defined in Definition 2.4 below, which is a special case of quasi-isomorphism;

thus item 3 in Theorem 2.3 directly implies item 4. The homology $HC_*(V,\Lambda)$ is called the *Legendrian contact homology* of Λ.

Definition 2.4 ([7], see also [19]).

1. Let $\mathcal{A} = R\langle a_1,\dots,a_n\rangle$. An *elementary automorphism* of $\mathcal{A}$ is an algebra map $\phi : \mathcal{A} \to \mathcal{A}$ of the form: for some i, $\phi(a_j) = a_j$ for all $j \neq i$, and $\phi(a_i) = a_i + v$ for some $v \in R\langle a_1,\dots,a_{i-1},a_{i+1},\dots,a_n\rangle$.
2. A *tame automorphism* of $\mathcal{A}$ is a composition of elementary automorphisms.
3. DGAs $(\mathcal{A} = R\langle a_1,\dots,a_n\rangle, \partial)$ and $(\mathcal{A}' = R\langle a_1',\dots,a_n'\rangle, \partial')$ are *tamely isomorphic* if there is an algebra isomorphism $\psi = \phi_2 \circ \phi_1$ such that $\phi_1 : \mathcal{A} \to \mathcal{A}$ is a tame automorphism and $\phi_2 : \mathcal{A} \to \mathcal{A}'$ is given by $\phi_2(a_i) = a_i'$ for all i, and ψ intertwines the differentials: $\psi \circ \partial = \partial' \circ \psi$.
4. A *stabilization* of $(\mathcal{A} = R\langle a_1,\dots,a_n\rangle, \partial)$ is $(S(\mathcal{A}), \partial)$, where $S(\mathcal{A}) = R\langle a_1,\dots,a_n,e_1,e_2\rangle$ with grading inherited from $\mathcal{A}$ along with $|e_1| = |e_2| + 1$, and ∂ is induced on $S(\mathcal{A})$ by ∂ on $\mathcal{A}$ along with $\partial(e_1) = e_2$, $\partial(e_2) = 0$.
5. DGAs $(\mathcal{A},\partial)$ and $(\mathcal{A}',\partial')$ are *stable tame isomorphic* if they are tamely isomorphic after stabilizing each of them some (possibly different) number of times.

Exercise 2.5.

1. Prove that $H(S(\mathcal{A}),\partial) \cong H(\mathcal{A},\partial)$ and thus stable tame isomorphism implies quasi-isomorphism.
2. Prove that if $(\mathcal{A},\partial)$ is a DGA with a generator a satisfying $|a| = 1$ and $\partial(a) = 1$, then $H(\mathcal{A},\partial) = 0$. Conclude that quasi-isomorphism does not necessarily imply stable tame isomorphism.
3. If all generators of $\mathcal{A}$ are in degree ≥ 0, and S is a unital ring, show that there is a one-to-one correspondence between augmentations of $(\mathcal{A},\partial)$ to S (see Definition 2.6 below) and ring homomorphisms $H_0(\mathcal{A},\partial) \to S$. Find an example to show that this is not true in general without the degree condition.
4. Find the stable tame isomorphism in Example 3.13 below.

We conclude this section by introducing the notion of an augmentation, which is an important algebraic tool for studying DGAs.

Definition 2.6. Let $(\mathcal{A}, \partial)$ be a DGA over R, and let S be a unital ring. An *augmentation* of $(\mathcal{A}, \partial)$ to S is a graded ring homomorphism

$$\epsilon : \mathcal{A} \to S$$

sending ∂ to 0; that is, $\epsilon \circ \partial = 0$, $\epsilon(1) = 1$, and $\epsilon(a) = 0$ unless $|a| = 0$.

Note that augmentations use the multiplicative structure on the DGA $(\mathcal{A}, \partial)$. An augmentation allows one to construct a linearized version of the homology of $(\mathcal{A}, \partial)$.

Exercise 2.7. Let $(\mathcal{A}, \partial)$ be the LCH DGA for a Legendrian Λ, and let ϵ an augmentation of $(\mathcal{A}, \partial)$ to S.

1. Write $\mathcal{A} = R\langle a_1, \ldots, a_n\rangle$. The augmentation ϵ induces an augmentation $\epsilon_S : S\langle a_1, \ldots, a_n\rangle \to S$ that acts as the identity on S and as ϵ on the a_i's. Prove that $(\ker \epsilon_S)/(\ker \epsilon_S)^2$ is a finitely generated, graded S-module.
2. Prove that ∂ descends to a map here: then

$$HC_*^{\text{lin}}(\Lambda, \epsilon) := H_*\big((\ker \epsilon)/(\ker \epsilon)^2, \partial\big)$$

is a graded S-module, the *linearized Legendrian contact homology* of Λ with respect to the augmentation ϵ.

Remark 2.8. Here is a less concise, but possibly more illuminating, description of linearized contact homology. We can define a differential ∂_S on $\mathcal{A}_S := S\langle a_1, \ldots, a_n\rangle$ by composing ∂ by the map $R \to S$ induced by ϵ (this map fixes all a_i's). Define an S-algebra automorphism $\phi_\epsilon : \mathcal{A}_S \to \mathcal{A}_S$ by $\phi_\epsilon(a_i) = a_i + \epsilon(a_i)$ for all i and $\phi_\epsilon(s) = s$ for all $s \in S$. Then the map

$$\partial_{S,\epsilon} := \phi_\epsilon \circ \partial_S \circ \phi_\epsilon^{-1}$$

is a differential on $\mathcal{A}_S$. Furthermore, if we define $\mathcal{A}_S^+$ to be the subalgebra of $\mathcal{A}_S$ generated by $a_1, \ldots, a_n$, so that $\mathcal{A}_S \cong S \oplus \mathcal{A}_S^+$ as S-modules, then $\partial_{S,\epsilon}$ restricts to a map from $\mathcal{A}_S^+$ to itself, and so it induces a differential from $\mathcal{A}_S^+/(\mathcal{A}_S^+)^2$ to itself. The homology of the complex $(\mathcal{A}_S^+/(\mathcal{A}_S^+)^2, \partial_{S,\epsilon})$ is the linearized contact homology of Λ with respect to ϵ.

Remark 2.9. Let $\Lambda \subset V$ have LCH DGA $(\mathcal{A}, \partial)$, and write $R = \mathbb{Z}[H_2(V, \Lambda)]$ as usual. Any augmentation ϵ of $(\mathcal{A}, \partial)$ to a ring S induces a map $\epsilon|_R : R \to S$, since $R \subset \mathcal{A}$. This motivates the following definition: define the *augmentation* *variety* of Λ to S to be

$$\mathrm{Aug}(\Lambda,S)=\{\varphi:R\to S\,|\,\varphi=\epsilon|_R \text{ for some augmentation } \epsilon \text{ from } (\mathcal{A},\partial) \text{ to } S\}$$
$$\subset \mathrm{Hom}(R,S).$$

It follows from Theorem 2.3 that $\mathrm{Aug}(\Lambda,S)$ is an invariant of Λ under Legendrian isotopy.

In the simplest case, when $V=\mathbb{R}^3$ and Λ is a Legendrian knot, one can consider the augmentation variety

$$\mathrm{Aug}(\Lambda,S)\subset \mathrm{Hom}\big(\mathbb{Z}[\mathbb{Z}],S\big)\cong S^{\times}$$

where $S^{\times}$ is the multiplicative group of units in S. It can then be shown (by upcoming work of Caitlin Leverson) that $\mathrm{Aug}(\Lambda,S)$ is either $\{-1\}$ if Λ has a (graded) ruling, or $\emptyset$ otherwise; the augmentation variety contains fairly minimal information about Λ. However, in the main case of interest in this article, where $V=J^1(S^2)$ and $\Lambda=\Lambda_K$, the augmentation variety contains a great deal of information about Λ_K. See Section 5.

Remark 2.10. A geometric motivation for augmentations comes from exact Lagrangian fillings. Here is a somewhat imprecise description. Suppose that the contact manifold V is a convex end of an open exact symplectic manifold (W,ω); for instance, W could be the symplectization of V, or an exact symplectic filling of V. Let $L\subset W$ be an oriented exact Lagrangian submanifold whose boundary is the Legendrian $\Lambda\subset V$. Then L induces an augmentation ϵ of the LCH DGA of Λ, to the ring $S=\mathbb{Z}[H_2(W,L)]$, which restricts on the coefficient ring to the usual map $\mathbb{Z}[H_2(V,\Lambda)]\to\mathbb{Z}[H_2(W,L)]$. This augmentation is defined as follows: $\epsilon(a_i)$ is the sum of all rigid holomorphic disks in W with boundary on L and positive boundary puncture limiting to the Reeb chord a_i of Λ, where each holomorphic disk contributes its homology class in $H_2(W,L)$. The fact that ϵ is an augmentation is established by an argument similar to the proof that $\partial^2=0$ in Theorem 2.3 above, which involves two-story holomorphic buildings.

3. Knot Contact Homology

In this section, we present a combinatorial calculation of knot contact homology, which is Legendrian contact homology in the particular case where the contact manifold is $ST^*\mathbb{R}^3\cong J^1(S^2)$ and the Legendrian submanifold is the unit conormal bundle Λ_K to some link $K\subset\mathbb{R}^3$. The version of knot contact homology we give here is a theory over the coefficient ring $\mathbb{Z}[\lambda^{\pm1}, \mu^{\pm1}, U^{\pm1}]$, and has appeared in the literature in several places and guises,[2] up to various changes of variables (see the Appendix). Our presentation corresponds to what is called the "infinity" version of transverse homology in [16, 34], and is the most general (as of now) version of knot contact homology for topological knots and links. Setting $U=1$, one obtains an invariant called "framed knot contact homology" in [33] and simply "knot contact homology" in [15]. If we set $U=\lambda=1$ and $\mu=-1$, we obtain the original version of knot contact homology from [30, 31].

For simplicity, we assume that $K \subset \mathbb{R}^3$ is an oriented knot; see Remark 3.2 below for the case of a multi-component link. The unit conormal bundle $\Lambda_K \subset J^1(S^2)$ is a Legendrian T^2. As discussed in the previous section, the LCH DGA of Λ_K is a topological link invariant. The coefficient ring for this DGA is

$$R = \mathbb{Z}\left[H_2\left(J^1\left(S^2\right), \Lambda_K\right)\right] \cong \mathbb{Z}\left[\lambda^{\pm1}, \mu^{\pm1}, U^{\pm1}\right],$$

where λ, μ correspond to the longitude and meridian generators of $H_1(\Lambda_K)$ and U corresponds to the generator of $H_2(J^1(S^2)) = H_2(S^2)$. Note that the choice of λ, μ relies on a choice of (orientation and) framing for K; we choose the Seifert framing for definiteness.

Definition 3.1. $K \subset \mathbb{R}^3$ knot. The *knot DGA* of K is the LCH differential graded algebra of $\Lambda_K \subset J^1(S^2)$, an algebra over the ring $R = \mathbb{Z}[\lambda^{\pm1}, \mu^{\pm1}, U^{\pm1}]$. The homology of this DGA is the *knot contact homology* of K, $HC_*(K) = HC_*(ST^*\mathbb{R}^3, \Lambda_K)$.

Remark 3.2. If K is an oriented r-component link, one can similarly define the "knot DGA", now an algebra over

$$\mathbb{Z}\left[H_2\left(J^1\left(S^2\right), \Lambda_K\right)\right] \cong \mathbb{Z}\left[\lambda_1^{\pm1}, \dots, \lambda_r^{\pm1}, \mu_1^{\pm1}, \dots, \mu_r^{\pm1}, U^{\pm1}\right].$$

Here, as in the knot case, we choose the 0-framing on each link component to fix the above isomorphism. The combinatorial description for the DGA in the link case is a bit more involved than for the knot case; see the Appendix for details.

We now return to the case where K is a knot. It follows directly from Theorem 2.3 that knot contact homology $HC_*(K)$ is an invariant up to R-

[2]The profusion of terms and specializations is an unfortunate byproduct of the way that the subject evolved over a decade.

algebra isomorphism, as is the knot DGA up to stable tame isomorphism. What we describe next is a combinatorial form for the knot DGA, given a braid presentation of K; this follows the papers [16, 34], which build on previous work [15, 30, 31, 33]. The fact that the combinatorial DGA agrees with the holomorphic-curve DGA described in Section 2 is a rather intricate calculation and the subject of [15].

Let B_n be the braid group on n strands. Define $\mathcal{A}_n$ to be the free noncommutative unital algebra over $\mathbb{Z}$ generated by $n(n-1)$ generators a_{ij} with $1 \le i,j \le n$ and $i \ne j$. We consider the following representation of B_n as a group of automorphisms of $\mathcal{A}_n$, which was first introduced (in a slightly different form) in [29].

Definition 3.3. The braid homomorphism $\phi : B_n \to \operatorname{Aut} \mathcal{A}_n$ is the map defined on generators σ_k ($1 \le k \le n-1$) of B_n by:

$$\phi_{\sigma_k} : \begin{cases} a_{ij} \mapsto a_{ij}, & i,j \ne k, k+1 \\ a_{k+1,i} \mapsto a_{ki}, & i \ne k, k+1 \\ a_{i,k+1} \mapsto a_{ik}, & i \ne k, k+1 \\ a_{k,k+1} \mapsto -a_{k+1,k} & \\ a_{k+1,k} \mapsto -a_{k,k+1} & \\ a_{ki} \mapsto a_{k+1,i} - a_{k+1,k}a_{ki}, & i \ne k, k+1 \\ a_{ik} \mapsto a_{i,k+1} - a_{ik}a_{k,k+1}, & i \ne k, k+1. \end{cases}$$

This extends to a map on B_n (see the following exercise).

Exercise 3.4.

1. Check that ϕ_{σ_k} is invertible.
2. Check that ϕ respects the braid relations: $\phi_{\sigma_k}\phi_{\sigma_{k+1}}\phi_{\sigma_k} = \phi_{\sigma_{k+1}}\phi_{\sigma_k}\phi_{\sigma_{k+1}}$ and $\phi_{\sigma_i}\phi_{\sigma_j} = \phi_{\sigma_j}\phi_{\sigma_i}$ for $|i-j| \ge 2$.
3. For the braid $B = (\sigma_1 \cdots \sigma_{n-1})^m \in B_n$ for $m \ge 1$, calculate ϕ_B. (The answer is quite simple.)

Remark 3.5. As a special case of Exercise 3.4(3), when B is a full twist $(\sigma_1 \cdots \sigma_{n-1})^n$, ϕ_B is the identity map; thus $\phi : B_n \to \operatorname{Aut} \mathcal{A}_n$ is not a faithful representation. However, one can create a faithful representation of B_n from ϕ, as follows. Embed B_n into B_{n+1} by adding an extra (noninteracting) strand to any braid in B_n; then the composition

$$B_n \hookrightarrow B_{n+1} \stackrel{\phi}{\to} \operatorname{Aut} \mathcal{A}_{n+1}$$

is a faithful representation of B_n as a group of algebra automorphisms of $\mathcal{A}_{n+1}$. See [31].

Before we proceed with the combinatorial definition of the knot DGA, we present a possibly illustrative reinterpretation of ϕ that begins by viewing B_n as the mapping class group of $D^2 \setminus \{p_1, \dots, p_n\}$; this will be useful in Section 4. To this end, let $p_1, \dots, p_n$ be a collection of n points in D^2, which we arrange in order in a horizontal line.

Definition 3.6. An *arc* is a continuous path $\gamma : [0,1] \to D^2$ such that $\gamma^{-1}(\{p_1, \dots, p_n\}) = \{0,1\}$; that is, the path begins at some p_i, ends at some p_j (possibly the same point), and otherwise does not pass through any of the p's. We consider arcs up to endpoint-fixing homotopy through arcs: two arcs are identified if, except at their endpoints, they are homotopic in $D^2 \setminus \{p_1, \dots, p_n\}$. Let $\widetilde{\mathcal{A}}$ denote the tensor algebra over $\mathbb{Z}$ generated by arcs, modulo the (two-sided ideal generated by the) relations:

1. 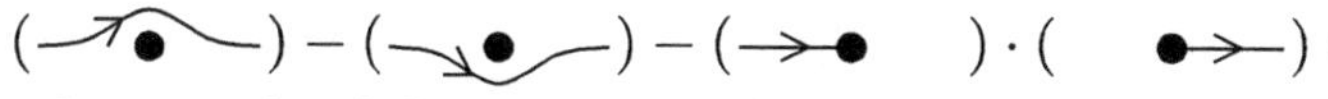 $= 0$,
   where each of these dots indicates the same point p_i;
2. any contractible arc with both endpoints at some p_i is equal to 0.

Remark 3.7. There is a notion of a framed arc that generalizes Definition 3.6, and a corresponding version of $\widetilde{\mathcal{A}}$ in which 0 is replaced by $1 - \mu$. Framed arcs are used to relate knot contact homology to the cord algebra (see Section 4), but we omit their definition here in the interest of simplicity. See [33] for more details.

One can now relate the homomorphism ϕ with the algebra $\widetilde{\mathcal{A}}$ generated by arcs.

Theorem 3.8 ([31]).

1. *For $i \neq j$, let γ_{ij} denote the arc depicted below (left diagram for $i < j$, right for $i > j$):*

   Then the map sending a_{ij} to γ_{ij} for $i < j$ and $-\gamma_{ij}$ for $i > j$ induces an algebra isomorphism $\Phi : \mathcal{A}_n \xrightarrow{\cong} \widetilde{\mathcal{A}}$.

2. *For any* $B \in B_n$ *and any* i, j, *we have*

$$\Phi(\phi_B(a_{ij})) = B \cdot \Phi(a_{ij}),$$

where B *acts on* $\widetilde{\mathcal{A}}$ *by the mapping class group action: if* a *is an arc, then* $B \cdot a$ *is the arc obtained by applying to* a *the diffeomorphism of* $D^2 \setminus \{p_1, \ldots, p_n\}$ *given by* B.

As an illustration of Theorem 3.8(2), the braid $B = \sigma_k$ sends the arc γ_{ki} for $i > k+1$ to

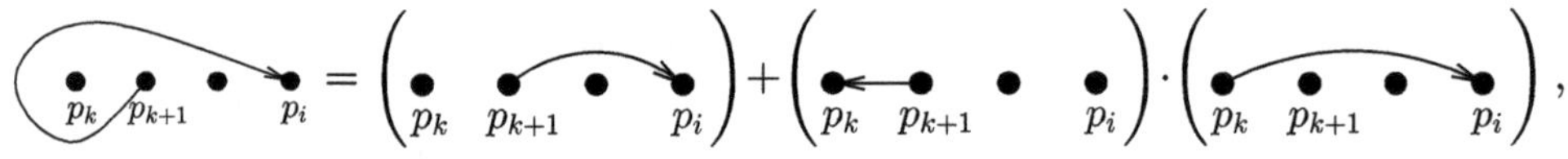

where the equality is in $\widetilde{\mathcal{A}}$ and uses the skein relation in Definition 3.6; the right hand side is the image under Φ of $a_{k+1,i} - a_{k+1,k}a_{ki} = \phi_{\sigma_k}(a_{ki})$.

We now proceed with the definition of the knot DGA. We will need two $n \times n$ matrices $\boldsymbol{\Phi}^L_B, \boldsymbol{\Phi}^R_B$ that arise from the representation ϕ (or, more precisely, its extension as described in Remark 3.5).

Definition 3.9 ([30]). Let $B \in B_n \hookrightarrow B_{n+1}$, and label the additional strand in B_{n+1} by $*$. Define $\boldsymbol{\Phi}^L_B, \boldsymbol{\Phi}^R_B \in \mathrm{Mat}_{n\times n}(\mathcal{A}_n)$ by:

$$\phi_B(a_{i*}) = \sum_{j=1}^{n} \left(\boldsymbol{\Phi}^L_B\right)_{ij} a_{j*}$$

$$\phi_B(a_{*i}) = \sum_{i=1}^{n} a_{*j} \left(\boldsymbol{\Phi}^R_B\right)_{ji}$$

for $1 \le i \le n$.

Exercise 3.10.

1. For $B = \sigma_1^3 \in B_3$, use arcs and Theorem 3.8 to check that

$$\phi_B(a_{13}) = -2a_{21}a_{13} + a_{21}a_{12}a_{21}a_{13} + a_{23} - a_{21}a_{12}a_{23}.$$

2. Now view $B = \sigma_1^3$ as living in B_2. Verify:

$$\boldsymbol{\Phi}^L_B = \begin{pmatrix} -2a_{21} + a_{21}a_{12}a_{21} & 1 - a_{21}a_{12} \\ 1 - a_{12}a_{21} & a_{12} \end{pmatrix}$$

$$\boldsymbol{\Phi}_B^R = \begin{pmatrix} -2a_{12} + a_{12}a_{21}a_{12} & 1 - a_{12}a_{21} \\ 1 - a_{21}a_{12} & a_{21} \end{pmatrix}.$$

3. For general B, $\boldsymbol{\Phi}_B^L$ and $\boldsymbol{\Phi}_B^R$ can be thought of as "square roots" of ϕ_B, in the following sense. Let $\mathbf{A}$ and $\phi_B(\mathbf{A})$ be the $n \times n$ matrices defined in Definition 3.11 below; roughly speaking, $\mathbf{A}$ is the matrix of the a_{ij}'s and $\phi_B(\mathbf{A})$ is the matrix of the $\phi_B(a_{ij})$'s. Then we have

$$\phi_B(\mathbf{A}) = \boldsymbol{\Phi}_B^L \cdot \mathbf{A} \cdot \boldsymbol{\Phi}_B^R; \tag{1}$$

see [33, 34] for the proof. Verify (1) for $B = \sigma_1^3$.

Definition 3.11 ([15, 34][3]). Let K be a knot given by the closure of a braid $B \in B_n$. The *(combinatorial) knot DGA* for K is the differential graded algebra $(\mathcal{A}, \partial)$ over $R = \mathbb{Z}[\lambda^{\pm 1}, \mu^{\pm 1}, U^{\pm 1}]$ given as follows.

1. Generators: $\mathcal{A} = R\langle a_{ij}, b_{ij}, c_{ij}, d_{ij}, e_{ij}, f_{ij}\rangle$ with generators
   - a_{ij}, where $1 \le i, j \le n$ and $i \ne j$, of degree 0 ($n(n-1)$ of these)
   - b_{ij}, where $1 \le i, j \le n$ and $i \ne j$, of degree 1 ($n(n-1)$ of these)
   - c_{ij} and d_{ij}, where $1 \le i, j \le n$, of degree 1 (n^2 of each)
   - e_{ij} and f_{ij}, where $1 \le i, j \le n$, of degree 2 (n^2 of each).
2. Differential: assemble the generators into $n \times n$ matrices $\mathbf{A}, \hat{\mathbf{A}}, \mathbf{B}, \hat{\mathbf{B}}, \mathbf{C}, \mathbf{D}, \mathbf{E}, \mathbf{F}$, defined as follows. For $1 \le i, j \le n$, the ij entry of the matrices $\mathbf{C}, \mathbf{D}, \mathbf{E}, \mathbf{F}$ is $c_{ij}, d_{ij}, e_{ij}, f_{ij}$, respectively. The other matrices $\mathbf{A}, \hat{\mathbf{A}}, \mathbf{B}, \hat{\mathbf{B}}$ are given by:

$$\mathbf{A}_{ij} = \begin{cases} a_{ij} & i < j \\ -\mu a_{ij} & i > j \\ 1 - \mu & i = j \end{cases} \qquad \mathbf{B}_{ij} = \begin{cases} b_{ij} & i < j \\ -\mu b_{ij} & i > j \\ 0 & i = j \end{cases}$$

$$(\hat{\mathbf{A}})_{ij} = \begin{cases} U a_{ij} & i < j \\ -\mu a_{ij} & i > j \\ U - \mu & i = j \end{cases} \qquad (\hat{\mathbf{B}})_{ij} = \begin{cases} U b_{ij} & i < j \\ -\mu b_{ij} & i > j \\ 0 & i = j. \end{cases}$$

Also define a matrix $\boldsymbol{\Lambda}$ as the diagonal matrix

$$\boldsymbol{\Lambda} = \operatorname{diag}\left(\lambda \mu^w U^{-(w-n+1)/2}, 1, \ldots, 1\right),$$

[3]See the Appendix for differences in convention between our definition and the ones from [34] and [15].

where w is the writhe of B (the sum of the exponents in the braid word).

The differential is given in matrix form by:

$$\begin{aligned}
\partial(\mathbf{A}) &= 0 \\
\partial(\mathbf{B}) &= \mathbf{A} - \boldsymbol{\Lambda} \cdot \phi_B(\mathbf{A}) \cdot \boldsymbol{\Lambda}^{-1} \\
\partial(\mathbf{C}) &= \hat{\mathbf{A}} - \boldsymbol{\Lambda} \cdot \boldsymbol{\Phi}^L_B \cdot \mathbf{A} \\
\partial(\mathbf{D}) &= \mathbf{A} - \hat{\mathbf{A}} \cdot \boldsymbol{\Phi}^R_B \cdot \boldsymbol{\Lambda}^{-1} \\
\partial(\mathbf{E}) &= \hat{\mathbf{B}} - \mathbf{C} - \boldsymbol{\Lambda} \cdot \boldsymbol{\Phi}^L_B \cdot \mathbf{D} \\
\partial(\mathbf{F}) &= \mathbf{B} - \mathbf{D} - \mathbf{C} \cdot \boldsymbol{\Phi}^R_B \cdot \boldsymbol{\Lambda}^{-1}.
\end{aligned}$$

Here $\partial(\mathbf{A})$ is the matrix whose ij entry is $\partial(\mathbf{A}_{ij})$, $\phi_B(\mathbf{A})$ is the matrix whose ij entry is $\phi_B(\mathbf{A}_{ij})$, and similarly for $\partial(\mathbf{B})$, etc. (For $U = 1$ as in the setting of [33], we can omit the hats.)

The homology of $(\mathcal{A}, \partial)$ is the (combinatorial) knot contact homology $HC_*(K)$.

Remark 3.12. Combinatorial knot DGAs and related invariants are readily calculable by computer. There are a number of *Mathematica* packages to this end available at http://www.math.duke.edu/~ng/math/programs.html.

Example 3.13. For the unknot, the knot DGA is the algebra over $\mathbb{Z}[\lambda^{\pm 1}, \mu^{\pm 1}, U^{\pm 1}]$ generated by four generators, c, d in degree 1 and e, f in degree 2, with differential:

$$\begin{aligned}
\partial c &= U - \lambda - \mu + \lambda\mu \\
\partial d &= 1 - \mu - \lambda^{-1} U + \lambda^{-1}\mu \\
\partial e &= -c - \lambda d \\
\partial f &= -d - \lambda^{-1} c.
\end{aligned}$$

Up to stable tame isomorphism, this is the same as the DGA generated by c and e with differential $\partial c = U - \lambda - \mu + \lambda\mu$, $\partial e = 0$. See Exercise 2.5(4).

The main result of [15] is that the combinatorial knot DGA of K, described above, agrees with the LCH DGA of Λ_K, after one changes Λ_K by Legendrian isotopy in $J^1(S^2)$ in a particular way and makes other choices that do not affect LCH. The proof of this result is far outside the scope of this article, but we will try to indicate the strategy; see also [12] for a nice summary with a bit more detail.

Theorem 3.14 ([15, 16]). *The combinatorial knot DGA of K in the sense of Definition* 3.11 *is the LCH DGA of Λ_K in the sense of Definition* 3.1.

Idea of proof. Braid K around an unknot U. Then Λ_K is contained in a neighborhood of $\Lambda_U \cong T^2$, and so we can view

$$\Lambda_K \subset J^1(T^2) \subset J^1(S^2)$$

by the Legendrian neighborhood theorem. Reeb chords for Λ_K split into two categories: "small" chords lying in $J^1(T^2)$, corresponding to the a_{ij}'s and b_{ij}'s, and "big" chords that lie outside of $J^1(T^2)$, corresponding to the $c_{ij}, d_{ij}, e_{ij}, f_{ij}$ generators (which themselves correspond to four Reeb chords for Λ_U). Holomorphic disks similarly split into small disks lying in $J^1(T^2)$, and big disks that lie outside of $J^1(T^2)$. The small disks produce the subalgebra of the knot DGA generated by the a_{ij}'s and b_{ij}'s. The big disks produce the rest of the differential, and can be computed in the limit degeneration when K approaches U. These disk counts use gradient flow trees in the manner of [11]. ■

It follows from Theorem 3.14 that the combinatorial knot DGA, up to stable tame isomorphism, is a knot invariant, as is its homology $HC_*(K)$. Alternatively, one can prove this directly without counting holomorphic curves, just by using algebraic properties of the representation ϕ and the matrices $\boldsymbol{\Phi}^L_B, \boldsymbol{\Phi}^R_B$.

Theorem 3.15 ([33] for $U=1$, [34] in general). *For the combinatorial knot DGA*:

1. $\partial^2 = 0$ (*see Exercise* 3.16);
2. $(\mathcal{A}, \partial)$ *is a knot invariant: up to stable tame isomorphism, it is invariant under Markov moves.*

Exercise 3.16.

1. Use (1) from Exercise 3.10 to prove that $\partial^2 = 0$ for the combinatorial knot DGA.

2. Show that the two-sided ideal in $\mathcal{A}$ generated by the entries of any two of the three matrices $\mathbf{A} - \boldsymbol{\Lambda} \cdot \phi_B(\mathbf{A}) \cdot \boldsymbol{\Lambda}^{-1}$, $\hat{\mathbf{A}} - \boldsymbol{\Lambda} \cdot \boldsymbol{\Phi}^L_B \cdot \mathbf{A}$, $\mathbf{A} - \hat{\mathbf{A}} \cdot \boldsymbol{\Phi}^R_B \cdot \boldsymbol{\Lambda}^{-1}$ contains the entries of the third. (Note that these three matrices are the matrices of differentials $\partial(\mathbf{B})$, $\partial(\mathbf{C})$, $\partial(\mathbf{D})$ in the knot DGA.) This fact will appear later; see Remark 4.2.

It is natural to ask how effective the knot DGA is as a knot invariant. In order to answer this, one needs to find practical ways of distinguishing between stable tame isomorphism classes of DGAs. One way, outlined in the following exercise, is by linearizing, as in Exercise 2.7; another, which we will employ and discuss extensively later, is by considering the space of augmentations, as in Remark 2.9.

Exercise 3.17.

1. Show that the knot DGA has an augmentation to $\mathbb{Z}[\lambda^{\pm 1}]$ that sends μ, U to 1, and another augmentation to $\mathbb{Z}[\mu^{\pm 1}]$ that sends λ, U to 1. (In general there are many more augmentations, but these are "canonical" in some sense.) Hint: this is easiest to do using the cord algebra (see Section 4) rather than the knot DGA directly.
2. Consider the right-handed trefoil K, expressed as the closure of $\sigma_1^3 \in B_2$. If we further compose the second augmentation from the previous part with the map $\mathbb{Z}[\mu^{\pm 1}] \to \mathbb{Z}$ that sends μ to -1, then we obtain an augmentation of the knot DGA of K to $\mathbb{Z}$. This is explicitly given as the map $\epsilon : \mathcal{A} \to \mathbb{Z}$ with $\epsilon(\lambda) = 1$, $\epsilon(\mu) = -1$, $\epsilon(U) = 1$, $\epsilon(a_{12}) = \epsilon(a_{21}) = -2$.
   For this augmentation, show that the linearized contact homology (see Exercise 2.7) $HC^{\mathrm{lin}}_*(\Lambda_K, \epsilon)$ is given as follows:
$$HC^{\mathrm{lin}}_* \cong \begin{cases} \mathbb{Z}_3 & * = 0 \\ \mathbb{Z} \oplus (\mathbb{Z}_3)^3 & * = 1 \\ \mathbb{Z} & * = 2 \\ 0 & \text{otherwise.} \end{cases}$$
3. By contrast, check that for the unknot (whose DGA is given at the end of Example 3.13), there is a unique augmentation to $\mathbb{Z}$ with $\epsilon(\lambda) = 1$, $\epsilon(\mu) = -1$, $\epsilon(U) = 1$, with respect to which $HC^{\mathrm{lin}}_0 \cong 0$, $HC^{\mathrm{lin}}_1 \cong \mathbb{Z}$, $HC^{\mathrm{lin}}_2 \cong \mathbb{Z}$. It can be shown (see [7]) that the collection of all linearized homologies over all possible augmentations is an invariant of the stable tame isomorphism class of a DGA. Thus the knot DGAs for the unknot and right-handed trefoil are not stable tame isomorphic.

We close this section by discussing some properties of the knot DGA, which are proved using the combinatorial formulation from Definition 3.11.

Theorem 3.18 ([33]).

1. *Knot contact homology encodes the Alexander polynomial: there is a canonical augmentation of the knot DGA* $(\mathcal{A}, \partial)$ *to* $\mathbb{Z}[\mu^{\pm 1}]$ *(see Exercise* 3.17)*, with respect to which the linearized contact homology* $HC_*^{\text{lin}}(K)$*, as a module over* $\mathbb{Z}[\mu^{\pm 1}]$*, is such that* $HC_1^{\text{lin}}(K)$ *determines the Alexander module of* K *(see* [33] *for the precise statement).*

2. *Knot contact homology detects mirrors and mutants: counting augmentations to* $\mathbb{Z}_3$ *shows that the knot DGAs for the right-handed and left-handed trefoils and the Kinoshita–Terasaka and Conway mutants are all distinct.*

Remark 3.19. Since the knot DGA $(\mathcal{A}, \partial)$ is supported in nonnegative degree, augmentations to $\mathbb{Z}_3$ (or arbitrary rings) are the same as ring homomorphisms from $HC_0(K)$ to $\mathbb{Z}_3$; see Exercise 2.5. Thus the number of such augmentations is a knot invariant. Counting augmentations to finite fields is easy to do by computer.

Remark 3.20. It is not known if there are nonisotopic knots K_1, K_2 whose knot contact homologies are the same. Thus at present it is conceivable that any of the following are *complete* knot invariants, in decreasing order of strength of the invariant (except possibly for the last two items, which do not determine each other in any obvious way):

- the Legendrian isotopy class of $\Lambda_K \subset ST^*\mathbb{R}^3$;
- the knot DGA $(\mathcal{A}, \partial)$ up to stable tame isomorphism;
- degree 0 knot contact homology $HC_0(K)$ over $R = \mathbb{Z}[\lambda^{\pm 1}, \mu^{\pm 1}, U^{\pm 1}]$;
- the cord algebra (see Section 4);
- the augmentation polynomial $\text{Aug}_K(\lambda, \mu, U)$ (see Section 5).

Even if these are not complete invariants, they are rather strong. For instance, physics arguments suggest that the augmentation polynomial may be at least as strong as the HOMFLY-PT polynomial and possibly some knot homologies; see Section 5.

4. Cord Algebra

In the previous section, we introduced the (combinatorial) knot DGA. The fact that the knot DGA is a topological invariant can be shown in two ways: computation of holomorphic disks and an appeal to the general theory of Legendrian contact homology as in Section 2 [15], or combinatorial verification of invariance under the Markov moves [34]. The first approach is natural but difficult, while the second is technically easier but somewhat opaque from a topological viewpoint, a bit like the usual proofs that the Jones polynomial is a knot invariant.

In this section, we present a direct topological interpretation for a significant part (though not the entirety) of knot contact homology, namely the degree 0 homology $HC_0(K)$ with $U=1$, in terms of a construction called the "cord algebra". Our aim is to give some topological intuition for what knot contact homology measures as a knot invariant. It is currently an open problem to extend this interpretation to all of knot contact homology.

We begin with the observation that $HC_*(K)$ is supported in degree $*\geq 0$, and that for $*=0$ it can be written fairly explicitly:

Theorem 4.1. *Let* $R=\mathbb{Z}[\lambda^{\pm 1},\mu^{\pm 1},U^{\pm 1}]$. *Then*

$$HC_0(K)\cong(\mathcal{A}_n\otimes R)/\big(\textit{entries of } \mathbf{A}-\boldsymbol{\Lambda}\cdot\phi_B(\mathbf{A})\cdot\boldsymbol{\Lambda}^{-1},\ \hat{\mathbf{A}}-\boldsymbol{\Lambda}\cdot\boldsymbol{\Phi}^L_B\cdot\mathbf{A},\ \mathbf{A}-\hat{\mathbf{A}}\cdot\boldsymbol{\Phi}^R_B\cdot\boldsymbol{\Lambda}^{-1}\big).$$

Proof. Since the knot DGA $(\mathcal{A},\partial)$ is supported in degree ≥ 0, all degree 0 elements of $\mathcal{A}$, i.e., elements of $\mathcal{A}_n\otimes R$, are cycles. The ideal of $\mathcal{A}_n\otimes R$ consisting of boundaries is precisely the ideal generated by the entries of the three matrices. ∎

Remark 4.2. In fact, one can drop any single one of the matrices $\mathbf{A}-\boldsymbol{\Lambda}\cdot\phi_B(\mathbf{A})\cdot\boldsymbol{\Lambda}^{-1}$, $\hat{\mathbf{A}}-\boldsymbol{\Lambda}\cdot\boldsymbol{\Phi}^L_B\cdot\mathbf{A}$, $\mathbf{A}-\hat{\mathbf{A}}\cdot\boldsymbol{\Phi}^R_B\cdot\boldsymbol{\Lambda}^{-1}$ in the statement of Theorem 4.1. See Exercise 3.16(2).

Remark 4.3. It does not appear to be an easy task to find an analogue of Theorem 4.1 for $HC_*(K)$ with $*\geq 1$, in part because not all elements of $\mathcal{A}$ of the appropriate degree are cycles.

Although the expression for $HC_0(K)$ from Theorem 4.1 is computable in examples, it has a particularly nice interpretation if we set $U=1$, as we will do for the rest of this section. With $U = 1$, the coefficient ring for the knot DGA becomes $R_0 = \mathbb{Z}[\lambda^{\pm 1}, \mu^{\pm 1}]$, and we can express $HC_0(K)|_{U=1}$ as an algebra over R_0 generated by "cords".

Definition 4.4 ([31, 33]).

1. Let $(K, *) \subset S^3$ be an oriented knot with a basepoint. A *cord* of $(K, *)$ is a continuous path $\gamma : [0,1] \to S^3$ with $\gamma^{-1}(K) = \{0, 1\}$ and $\gamma^{-1}(\{*\}) = \emptyset$.
2. Define $\mathcal{A}_K$ to be the tensor algebra over R_0 freely generated by homotopy classes of cords (note: the endpoints of the cord can move along the knot, as long as they avoid the basepoint $*$).
3. The *cord algebra* of K is the algebra $\mathcal{A}_K$ modulo the relations:

(a) 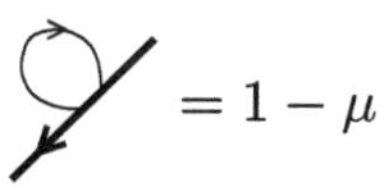

(b) 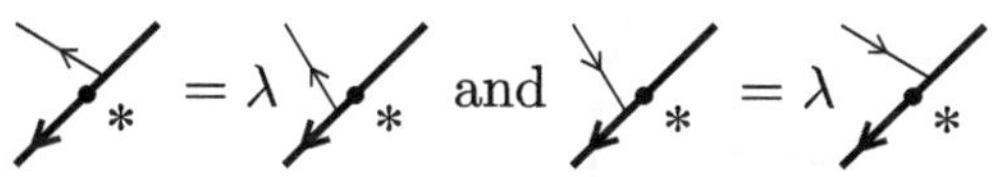

(c) 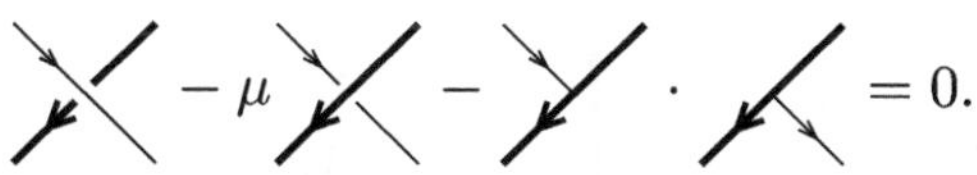

The "skein relations" in Definition 4.4 are understood to be depictions of relations in $\mathbb{R}^3$, and not just relations as planar diagrams. For instance, relation (c) is equivalent to:

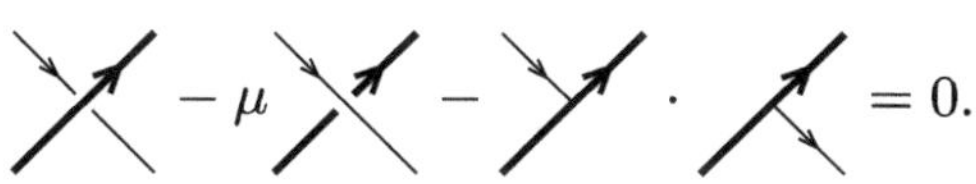

It is then evident that the cord algebra is a topological knot invariant.

Exercise 4.5. One can heuristically think of cords as corresponding to Reeb chords of Λ_K. More precisely:

1. Let $K \subset \mathbb{R}^3$ be a smooth knot. A *binormal chord* of K is an oriented (nontrivial) line segment with endpoints on K that is orthogonal to K at both endpoints. Show that binormal chords are exactly the same as Reeb chords of Λ_K.
2. For generic K, all binormal chords are cords in the sense of Definition 4.4. Show that any element of the cord algebra of K can be ex-

pressed in terms of just binormal chords, i.e., in terms of Reeb chords of Λ_K.

3. Prove that the cord algebra of a m-bridge knot has a presentation with (at most) $m(m-1)$ generators. (It is currently unknown whether this also holds for HC_0 if we do not set $U=1$.)

4. Prove that the cord algebra of the torus knot $T(m,n)$ has a presentation with at most $\min(m,n)-1$ generators, as indeed does $HC_0(T(m,n))$ without setting $U=1$. (For this last statement, see Exercise 3.4(3).)

Exercise 4.6. Here we calculate the cord algebra in two simple examples.

1. Prove that the cord algebra of the unknot is $R_0/((\lambda-1)(\mu-1))$.
2. Next consider the right-handed trefoil K, shown below with five cords labeled:

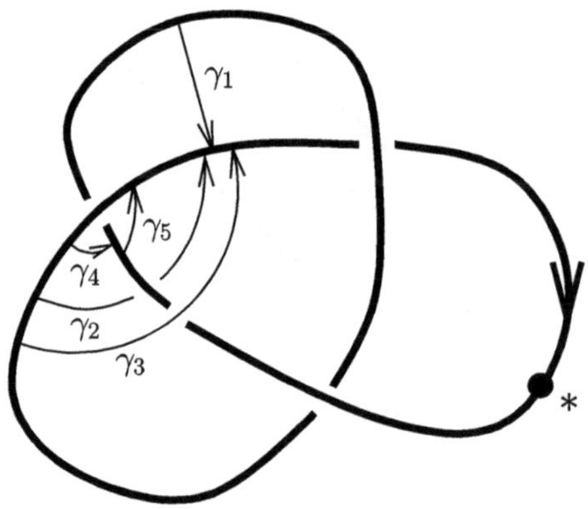

In the cord algebra of K, denote γ_1 by x. Show that $\gamma_2=\gamma_5=x$, $\gamma_4=\lambda x$, and $\gamma_3=1-\mu$. Conclude the relation

$$\lambda x^2 - x + \mu - \mu^2 = 0.$$

3. Use the skein relations in another way to derive another relation in the cord algebra of K:

$$\lambda x^2 + \lambda\mu x + \mu - 1 = 0.$$

4. Prove that the cord algebra of K is generated by x.
5. It can be shown that the above two relations generate all relations: the cord algebra of the right-handed trefoil is

$$R_0[x]\,/\,\big(\lambda x^2 - x + \mu - \mu^2,\ \lambda x^2 + \lambda\mu x + \mu - 1\big).$$

Suppose that there is a ring homomorphism from the cord algebra of K to $\mathbb{C}$, mapping λ to λ_0 and μ to μ_0. Show that

$$(\lambda_0 - 1)(\mu_0 - 1)(\lambda_0 \mu_0^3 + 1) = 0.$$

The left hand side is the two-variable augmentation polynomial for the right-handed trefoil (see Section 5 and Example 5.8).

We now present the relation between the cord algebra and knot contact homology.

Theorem 4.7 ([31, 33]). *The cord algebra of K is isomorphic as an R_0-algebra to $HC_0(K)|_{U=1}$.*

Idea of proof. Let K be the closure of a braid $B \in B_n$, and embed B in S^3 with braid axis L. A page of the resulting open book decomposition of S^3 is D^2 with $\partial D^2 = L$, and D^2 intersects B in n points $p_1, \dots, p_n$. Any arc in $D^2 \subset S^3$ in the sense of Definition 3.6 is a cord of K. Under this identification, skein relations (c) and (a) from Definition 4.4 become relations 1 and 2 from Definition 3.6 (at least when $\mu = 1$; for general μ, one needs to use a variant of Definition 3.6 involving framed cords, cf. Remark 3.7).

Any cord of K is homotopic to a cord lying in the D^2 slice of S^3. It then follows from Theorem 3.8 that there is a surjective R_0-algebra map from $\mathcal{A}_n \otimes R_0$ to the cord algebra. Thus the cord algebra is the quotient of $\mathcal{A}_n \otimes R_0$ by relations that arise from considering homotopies between arcs in D^2 given by one-parameter families of cords that do not lie in the D^2 slice. If this family avoids intersecting L, we obtain the relations given by the entries of $\partial(\mathbf{B}) = \mathbf{A} - \boldsymbol{\Lambda} \cdot \phi_B(\mathbf{A}) \cdot \boldsymbol{\Lambda}^{-1}$. Considering families that pass through L once gives the entries of $\partial(\mathbf{C}) = \hat{\mathbf{A}} - \boldsymbol{\Lambda} \cdot \boldsymbol{\Phi}^L_B \cdot \mathbf{A}$ and $\partial(\mathbf{D}) = \mathbf{A} - \hat{\mathbf{A}} \cdot \boldsymbol{\Phi}^R_B \cdot \boldsymbol{\Lambda}^{-1}$ as relations in the cord algebra. ■

For various purposes, it is useful to reformulate the cord algebra of a knot K in terms of homotopy-group information. In particular, this gives a proof that knot contact homology detects the unknot (Corollary 4.10); in Section 5, we will also use this to relate the augmentation polynomial to the A-polynomial. Here we give a brief description of this perspective and refer the reader to [33] for more details.

We can view cords of K as elements of the knot group $\pi_1(S^3 \setminus K)$ by pushing the endpoints slightly off of K and joining them via a curve parallel to K. One can then present the cord algebra entirely in terms of the knot group π and the peripheral subgroup $\hat{\pi} = \pi_1(\partial(\text{nbd}(K))) \cong \mathbb{Z}^2$. Write l, m for the longitude, meridian generators of $\hat{\pi}$.

Theorem 4.8 ([33]). *The cord algebra of K is isomorphic to the tensor algebra over R_0 freely generated by elements of $\pi_1(S^3 \setminus K)$ (denoted with brackets), quotiented by the relations:*

1. $[e] = 1 - \mu$, *where e is the identity element;*
2. $[\gamma l] = [l\gamma] = \lambda[\gamma]$ *and* $[\gamma m] = [m\gamma] = \mu[\gamma]$ *for* $\gamma \in \pi_1(S^3 \setminus K)$;
3. $[\gamma_1\gamma_2] - [\gamma_1 m \gamma_2] - [\gamma_1][\gamma_2] = 0$ *for any* $\gamma_1, \gamma_2 \in \pi_1(S^3 \setminus K)$.

If $(\mathcal{A}, \partial)$ is the knot DGA of K, then Theorem 4.8 (along with Theorem 4.7) gives an expression for $HC_0(K)|_{U=1} = H_0(\mathcal{A}|_{U=1}, \partial)$ as an R_0-algebra. One can readily "improve" this result to give an analogous expression for the degree 0 homology of the fully noncommutative knot DGA $(\widetilde{\mathcal{A}}, \partial)$ of K (see Remark 2.2 and the Appendix), which we write as

$$\widetilde{HC}_0(K)|_{U=1} = H_0(\widetilde{\mathcal{A}}|_{U=1}, \partial);$$

note that this is a $\mathbb{Z}$-algebra rather than a R_0-algebra, but contains R_0 as a subalgebra. Details are contained in joint work in progress with K. Cieliebak, T. Ekholm, and J. Latschev, which is also the reference for Theorem 4.9 and Corollary 4.10 below.

Theorem 4.9. *Write $\pi = \pi_1(S^3 \setminus K)$ and $\hat{\pi} = \pi_1(\partial(\text{nbd}(K))) = \langle m, l\rangle$. There is an injective ring homomorphism*

$$\widetilde{HC}_0(K)|_{U=1} \hookrightarrow \mathbb{Z}[\pi_1(S^3 \setminus K)]$$

under which $\widetilde{HC}_0(K)|_{U=1}$ maps isomorphically to the subring of $\mathbb{Z}[\pi]$ generated by $\hat{\pi}$ and elements of the form $\gamma - m\gamma$ for $\gamma \in \pi$. This map sends λ to l and μ to m.

Idea of proof. The homomorphism is induced by the map sending λ to l, μ to m, and $[\gamma]$ to $\gamma - m\gamma$ for $\gamma \in \pi$. ∎

Corollary 4.10. *Knot contact homology, in its fully noncommutative form, detects the unknot.*

Idea of proof. Use the Loop Theorem and consider the action of multiplication by λ on the cord algebra. ■

For a proof that ordinary (not fully noncommutative) knot contact homology detects the unknot, see the next section.

5. Augmentation Polynomial

In this section, we describe how knot contact homology can be used to produce a three-variable knot invariant, the augmentation polynomial. We then discuss the relation of a two-variable version of the augmentation polynomial to the A-polynomial, and of the full augmentation polynomial to the HOMFLY-PT polynomial and to mirror symmetry and physics.

The starting point is the space of augmentations from the knot DGA $(\mathcal{A}, \partial)$ to $\mathbb{C}$, as in Remark 2.9.

Definition 5.1 ([33, 34]). Let $(\mathcal{A}, \partial)$ be the knot DGA of a knot K, with the usual coefficient ring $\mathbb{Z}[\lambda^{\pm 1}, \mu^{\pm 1}, U^{\pm 1}]$. The *augmentation variety* of K is

$$V_K = \{(\epsilon(\lambda), \epsilon(\mu), \epsilon(U)) \,|\, \epsilon \text{ an augmentation from } (\mathcal{A}, \partial) \text{ to } \mathbb{C}\} \subset (\mathbb{C}^*)^3.$$

When the maximal-dimension part of the Zariski closure of V_K is a codimension 1 subvariety of $(\mathbb{C}^*)^3$, this variety is the vanishing set of a reduced polynomial[4] $\mathrm{Aug}_K(\lambda, \mu, U)$, the *augmentation polynomial*[5] of K.

Remark 5.2. The augmentation polynomial is well-defined only up to units in $\mathbb{C}[\lambda^{\pm 1}, \mu^{\pm 1}, U^{\pm 1}]$. However, because the differential on the knot DGA involves only integer coefficients, we can choose $\mathrm{Aug}_K(\lambda, \mu, U)$ to have integer coefficients with overall gcd equal to 1. We can further stipulate that $\mathrm{Aug}_K(\lambda, \mu, U)$ contains no negative powers of λ, μ, U, and that it is divisible by none of λ, μ, U. The result is an augmentation polynomial $\mathrm{Aug}_K(\lambda, \mu, U) \in \mathbb{Z}[\lambda, \mu, U]$, well-defined up to an overall $\pm$ sign.

Conjecture 5.3. *The condition about the Zariski closure in Definition* 5.1 *holds for all knots K; the augmentation polynomial is always defined.*

[4] I.e., no repeated factors.

[5] Caution: the polynomial described here differs from the augmentation polynomial from [34] by a change of variables $\mu \mapsto -1/\mu$. See the Appendix.

A fair number of augmentation polynomials for knots have been computed and are available at http://www.math.duke.edu/~ng/math/programs.html; see also Exercise 5.5 below. We note in passing some symmetries of the augmentation polynomial:

Theorem 5.4. *Let K be a knot and $m(K)$ its mirror. Then*

$$\mathrm{Aug}_K(\lambda, \mu, U) \doteq \mathrm{Aug}_K(\lambda^{-1}U, \mu^{-1}U, U)$$

and

$$\mathrm{Aug}_{m(K)}(\lambda, \mu, U) \doteq \mathrm{Aug}_K(\lambda U^{-1}, \mu^{-1}, U^{-1}),$$

where $\doteq$ denotes equality up to units in $\mathbb{Z}[\lambda^{\pm 1}, \mu^{\pm 1}, U^{\pm 1}]$.

The first equation in Theorem 5.4 follows from [34, Propositions 4.2, 4.3], while the second can be proved using the results from [34, §4].

Exercise 5.5. Here are a couple of computations of augmentation polynomials.

1. Show that the augmentation polynomial for the unknot is

$$\mathrm{Aug}_O(\lambda, \mu, U) = U - \lambda - \mu + \lambda\mu.$$

2. The cord algebra $HC_0|_{U=1}$ for the right-handed trefoil was computed in Exercise 4.6. It can be checked directly from the definition of the knot DGA that the full degree 0 knot contact homology is

$$HC_0(\text{RH trefoil}) \cong R[a_{12}]/\big(Ua_{12}^2 - \mu U a_{12} + \lambda\mu^3(1-\mu),\\ Ua_{12}^2 + \lambda\mu^2 a_{12} + \lambda\mu^2(\mu - U)\big).$$

Use resultants to deduce the augmentation polynomial:

$$\mathrm{Aug}_{\text{RH trefoil}}(\lambda, \mu, U) = (U^3 - \mu U^2) + (-U^3 + \mu U^2 - 2\mu^2 U + 2\mu^2 U^2\\ + \mu^3 U - \mu^4 U)\lambda + (-\mu^3 + \mu^4)\lambda^2.$$

From Theorem 5.4, we can then also deduce the polynomial for the left-handed trefoil:

$$\mathrm{Aug}_{\text{LH trefoil}}(\lambda, \mu, U) = (\mu^3 U^2 - \mu^4 U) + (U^2 - \mu U^2 - 2\mu^2 U + 2\mu^2 U^2\\ - \mu^3 U + \mu^4)\lambda + (-U^2 + \mu U^2)\lambda^2.$$

We next turn to the two-variable augmentation polynomial.

Definition 5.6 ([33]). If the $U = 1$ slice of the augmentation variety, $V_K \cap \{U = 1\} \subset (\mathbb{C}^*)^2$, is such that the maximal-dimensional part of its Zariski closure is a (co)dimension 1 subvariety of $(\mathbb{C}^*)^2$, then this subvariety is the vanishing set of a reduced polynomial $\mathrm{Aug}_K(\lambda, \mu)$, the *two-variable augmentation polynomial* of K. As in Remark 5.2, $\mathrm{Aug}_K(\lambda, \mu)$ can be chosen to lie in $\mathbb{Z}[\lambda, \mu]$.

Conjecture 5.7. *The two-variable augmentation polynomial* $\mathrm{Aug}_K(\lambda, \mu)$ *is always defined, and the two augmentation polynomials are related in the obvious way:*

$$\mathrm{Aug}_K(\lambda, \mu) = \mathrm{Aug}_K(\lambda, \mu, U = 1).$$

The two-variable augmentation polynomial has a number of interesting factors. For instance, it follows from Exercise 3.17 that

$$(\lambda - 1)(\mu - 1) \mid \mathrm{Aug}_K(\lambda, \mu)$$

for all knots K.

Example 5.8. For the unknot and trefoils, the two-variable augmentation polynomials are

$$\begin{aligned} \mathrm{Aug}_O(\lambda, \mu) &= (\lambda - 1)(\mu - 1) \\ \mathrm{Aug}_{\text{RH trefoil}}(\lambda, \mu) &= (\lambda - 1)(\mu - 1)(\lambda\mu^3 + 1) \\ \mathrm{Aug}_{\text{LH trefoil}}(\lambda, \mu) &= (\lambda - 1)(\mu - 1)(\lambda + \mu^3). \end{aligned}$$

The polynomial for the right-handed trefoil follows from Exercise 4.6, while the polynomial for the left-handed trefoil follows from the behavior of the polynomial (and knot contact homology generally) under mirroring, cf. Theorem 5.4.

The observant reader may notice that the two-variable augmentation polynomials for the unknot and trefoils are essentially the same as another knot polynomial, the A-polynomial. Recall that the A-polynomial is defined as follows. Given an $SL_2\mathbb{C}$ representation of the knot group

$$\rho : \pi_1(S^3 \setminus K) \to SL_2\mathbb{C},$$

simultaneously diagonalize $\rho(l)$, $\rho(m)$ to get $\rho(l) = \begin{pmatrix} \lambda & * \\ 0 & \lambda^{-1} \end{pmatrix}$, $\rho(m) = \begin{pmatrix} \mu & * \\ 0 & \mu^{-1} \end{pmatrix}$. The (maximal-dimensional part of the Zariski closure of the) collection of (λ, μ) over all $SL_2\mathbb{C}$ representations is the zero set of the *A-polynomial* of K, $A_K(\lambda, \mu)$.

Theorem 5.9 ([33]). *$(\mu^2 - 1)A_K(\lambda, \mu)$ divides* $\mathrm{Aug}_K(\lambda, \mu^2)$.

We outline the proof of Theorem 5.9 in Exercise 5.12 below.

Corollary 5.10. *The cord algebra detects the unknot.*

Proof. By a result of Dunfield and Garoufalidis [10], based on gauge-theoretic work of Kronheimer and Mrowka [27], the A-polynomial detects the unknot. It follows that when K is knotted, either $\mathrm{Aug}_K(\lambda, \mu)$ is not defined (if the augmentation variety is 2-dimensional), or $\mathrm{Aug}_K(\lambda, \mu^2)$ has a factor besides $(\lambda - 1)(\mu - 1)$. In either case, the augmentation variety for K is distinct from the variety for the unknot, which is $\{\lambda = 1\} \cup \{\mu = 1\}$ (see Example 5.8). ■

Note that the statement of unknot detection in Corollary 5.10 differs from, and is slightly stronger than, the statement from Corollary 4.10, because of the issue of commutativity. However, the proof of Corollary 4.10 uses only the Loop Theorem, rather than the deep Kronheimer–Mrowka result that leads to Corollary 5.10.

To expand on Theorem 5.9, it is sometimes, but not always, the case that

$$\mathrm{Aug}_K(\lambda, \mu^2) = (\mu^2 - 1)A_K(\lambda, \mu).$$

In general, the left hand side can contain factors that do not appear in the right hand side. For example,

$$A_{T(3,4)}(\lambda, \mu) = (\lambda - 1)(\lambda\mu^{12} + 1)(\lambda\mu^{12} - 1)$$
$$\mathrm{Aug}_{T(3,4)}(\lambda, \mu) = (\lambda - 1)(\mu - 1)(\lambda\mu^6 + 1)(\lambda\mu^6 - 1)(\lambda\mu^8 - 1),$$

and the last factor in $\mathrm{Aug}_{T(3,4)}$ has no corresponding factor in $A_{T(3,4)}$.

An explanation for (at least some of the) extra factors in the augmentation polynomial is given by the following result, which shows that representations of the knot group besides SU_2 representations can contribute to the augmentation polynomial.

Theorem 5.11. *Suppose that* $\rho : \pi_1(S^3 \setminus K) \to GL_m\mathbb{C}$ *is a representation of the knot group of* K *for some* $m \geq 2$, *such that* ρ *sends the meridian and longitude to the diagonal matrices*

$$\rho(m) = \text{diag}(\mu_0, 1, 1, \ldots, 1)$$
$$\rho(l) = \text{diag}(\lambda_0, *, *, \ldots, *)$$

where the asterisks indicate arbitrary complex numbers. Then there is an augmentation of the knot DGA of K *sending* (λ, μ, U) *to* $(\lambda_0, \mu_0, 1)$.

This result, which has not previously appeared in the literature, is proven in the following exercise, and also implies Theorem 5.9.

Exercise 5.12. Here we give a proof of Theorems 5.9 and 5.11.

1. Suppose $\rho : \pi_1(S^3 \setminus K) \to GL_m\mathbb{C}$ is a representation as in Theorem 5.11. Define a $\mathbb{C}$-valued map ϵ by
   - $\epsilon(\mu) = \mu_0$;
   - $\epsilon(\lambda) = \lambda_0$;
   - $\epsilon([\gamma]) = (1 - \mu_0)(\rho(\gamma))_{11}$, where M_{11} is the $(1, 1)$ entry of a matrix M, for all $\gamma \in \pi_1(S^3 \setminus K)$.

   Show that ϵ extends to an augmentation of the cord algebra of K, where we use the description of the cord algebra from Theorem 4.8. Deduce Theorem 5.11.

2. If ρ is an SU_2 representation of $\pi_1(S^3 \setminus K)$ with $\rho(m) = \left(\begin{smallmatrix} \mu & 0 \\ 0 & \mu^{-1} \end{smallmatrix}\right)$ and $\rho(l) = \left(\begin{smallmatrix} \lambda & 0 \\ 0 & \lambda^{-1} \end{smallmatrix}\right)$, then show that

   $$\widetilde{\rho}(\gamma) = \mu^{\text{lk}(K,\gamma)}\rho(\gamma)$$

   for $\gamma \in \pi_1(S^3 \setminus K)$ defines a $GL_2(\mathbb{C})$ representation satisfying the condition of Theorem 5.11 with $\mu_0 = \mu^2$ and $\lambda_0 = \lambda$. (Here $\text{lk}(K, \gamma)$ is the linking number of K with γ, i.e., the image of γ in $H_1(S^3 \setminus K) \cong \mathbb{Z}$.) Deduce Theorem 5.9.

3. For $K = T(3, 4)$ and $\lambda_0 = \mu_0^{-8}$ with arbitrary $\mu_0 \in \mathbb{C}^*$, find a $GL_3(\mathbb{C})$ representation of $\pi_1(S^3 \setminus K) \cong \langle x, y \,|\, x^3 = y^4 \rangle$ satisfying the condition of Theorem 5.11. (Note that in this presentation, $m = xy^{-1}$ and $l = x^3 m^{-12}$.) This shows that $\lambda\mu^8 - 1$ is a factor of $\text{Aug}_{T(3,4)}(\lambda, \mu)$; as discussed above, this factor does not appear in the A-polynomial of $T(3, 4)$.

We now turn to some recent developments linking the augmentation polynomial to physics. Our discussion is very sketchy and imprecise; see [4, 5] for more details. Recently the (three-variable) augmentation polynomial has appeared in various string theory papers [4, 20], in the context of studying topological strings for SU_N Chern–Simons theory on S^3. A very sketchy description of the idea, whose origins in the physics literature include [23, 36], is as follows.

Start with a knot $K \subset S^3$, with conormal bundle $L_K \subset T^*S^3$. (Note that this differs slightly from our usual setting of $K \subset \mathbb{R}^3$, though not in a substantial way, either topologically or contact-geometrically.) Collapse the zero section of T^*S^3 to a point, resulting in a conifold singularity; we can then resolve the singularity to a $\mathbb{CP}^1$ to obtain the "resolved conifold" given as the total space of the bundle

$$\mathcal{O}(-1) \oplus \mathcal{O}(-1) \to \mathbb{CP}^1.$$

(In physics language, this conifold transition is motivated by placing N branes on the zero section of T^*S^3 and taking the $N \to \infty$ limit.) One would like to follow L_K through this conifold transition to obtain a special Lagrangian $\tilde{L}_K \subset \mathcal{O}(-1) \oplus \mathcal{O}(-1)$. In [4], Aganagic and Vafa propose a generalized SYZ conjecture by which $\tilde{L}_K$ produces a mirror Calabi–Yau of $\mathcal{O}(-1) \oplus \mathcal{O}(-1)$ given by a variety of the form

$$uv = \boldsymbol{A}_K(e^x, e^p, Q)$$

where $(u, v, x, p) \subset \mathbb{C}^4$, Q is a parameter measuring the complexified Kähler class of $\mathbb{CP}^1$, and $\boldsymbol{A}_K$ is a three-variable polynomial that Aganagic and Vafa [4] refers to as the "Q-deformed A-polynomial".[6]

Surprisingly, we can make the following conjecture, for which there is strong circumstantial evidence [5]:

Conjecture 5.13 ([4, 5]). *The three-variable augmentation polynomial and the Q-deformed A-polynomial agree for all K:*

$$\boldsymbol{A}_K(e^x, e^p, Q) = \mathrm{Aug}_K(\lambda = e^x, \mu = e^p, U = Q).$$

Although Conjecture 5.13 has yet to be rigorously proven, it would have significant implications for the augmentation polynomial. By physical ar-

[6]In a related vein, Fuji, Gukov, and Sulkowski [20] have proposed a four-variable "super-A-polynomial" that specializes to the Q-deformed A-polynomial.

guments (see in particular [24] and [4]), $\boldsymbol{A}_K$ satisfies a number of interesting properties. In particular, $\boldsymbol{A}_K$ encodes a large amount of information about the knot K, possibly including the HOMFLY-PT polynomial as well as Khovanov–Rozansky HOMFLY-PT homology [25] and other knot homologies (or some portion thereof). The knot homologies appear in studying Nekrasov deformation of topological strings and refined Chern–Simons theory [24].

Thus, assuming Conjecture 5.13, one can make purely mathematical predictions about the augmentation polynomial. One such prediction begins with the observation (whose proof we omit here) that for any knot K,

$$\mathrm{Aug}_K(\lambda = 0, \mu = U, U) = 0$$

for all U. It appears that the first-order behavior of the augmentation variety near the curve $\{(0, U, U)\} \subset (\mathbb{C}^*)^3$ determines a certain specialization of the HOMFLY-PT polynomial:

Conjecture 5.14. *Let K be any knot in S^3. Let $f(U)$ be the polynomial such that near $(\lambda, \mu, U) = (0, U, U)$, the zeroes of the augmentation polynomial Aug_K satisfy*

$$\mu = U + f(U)\lambda + O(\lambda^2)$$

($f(U)$ can be explicitly written in terms of the λ^1 and λ^0 coefficients of Aug_K). Then

$$\frac{f(U)}{U-1} = P_K(U^{-1/2}, 1),$$

where $P_K(a, q)$ is the HOMFLY-PT polynomial of K (sometimes written as $P_K(a, z = q - q^{-1})$).

Conjecture 5.14 has been checked for all knots where the augmentation polynomial is currently known, including many where the Q-deformed A-polynomial has not been computed.

Exercise 5.15. Verify Conjecture 5.14 for the unknot and the right-handed and left-handed trefoils, using the augmentation polynomials computed in Exercise 5.5. Note that the HOMFLY-PT polynomials for the unknot and the RH trefoil are 1 and $-a^{-4} + a^{-2}q^{-2} + a^{-2}q^2$, respectively.

In a different direction, the physics discussion of $\boldsymbol{A}_K$ in [4] also predicts that the augmentation polynomial is determined by the recurrence relation for the colored HOMFLY-PT polynomials:

Conjecture 5.16. *Let $\{P_{K;n}(a,q)\}$ denote the colored HOMFLY-PT polynomials of K, colored by the n-th symmetric power of the fundamental representation. Define operations L, M by $L(P_{K;n}(a,q)) = P_{K;n+1}(a,q)$ and $M(P_{K;n}(a,q)) = q^n P_{K;n}(a,q)$. These polynomials satisfy a minimal recurrence relation of the form*

$$\widehat{A}_K(a,q,M,L)P_{K;n}(a,q) = 0,$$

where $\widehat{A}_K$ is a polynomial in noncommuting variables L, M and commuting parameters a,q; see [22]. Then sending $q \to 1$ and applying an appropriate change of variables sends $\widehat{A}_K(a,q,M,L)$ to the augmentation polynomial $\mathrm{Aug}_K(\lambda,\mu,U)$.

The precise change of variables depends on the conventions used for $P_{K;n}(a,q)$. In the conventions of [20] (where their x,y are our M,L), a more exact statement is that $\mathrm{Aug}_K(\lambda,\mu,U)$ and

$$\widehat{A}_K\left(a = U, q = 1, M = \mu^{-1}, L = \frac{\mu-1}{\mu-U}\lambda\right)$$

agree up to trivial factors.

Conjecture 5.16 is a direct analogue of the AJ conjecture [21] (quantum volume conjecture, in the physics literature) relating colored Jones polynomials to the A-polynomial, with colored HOMFLY-PT replacing colored Jones, and the augmentation polynomial replacing the A-polynomial. See also [20] for an extended discussion of this topic.

6. Transverse Homology

In this section, we discuss a concrete application of knot contact homology to contact topology, and in particular to transverse knots. Here one obtains additional filtrations on the knot DGA that produce effective invariants of transverse knots. So far our construction of knot contact homology begins with a smooth knot in $\mathbb{R}^3$; we now explore what happens if the knot is assumed to be transverse to a contact structure on $\mathbb{R}^3$ (note that this is independent of the canonical contact structure on $ST^*\mathbb{R}^3$!).

Definition 6.1. Let $\xi = \ker(\alpha = dz + r^2 d\theta)$ be the standard contact structure on $\mathbb{R}^3$. An oriented knot $T \subset \mathbb{R}^3$ is *transverse* if $\alpha > 0$ along T.

One usually studies transverse knots up to *transverse isotopy*: isotopy through transverse knots. There is a standard transverse unknot in $\mathbb{R}^3$ given by the unit circle in the xy plane. By work of Bennequin [6], any braid produces a transverse knot by gluing the closure of the braid into a neighborhood of the standard unknot. Conversely, all transverse knots are obtained in this way, up to transverse isotopy: the map from braids to transverse knots is surjective. The following theorem precisely characterizes failure of injectivity.

Theorem 6.2 (Transverse Markov Theorem [37, 42]). *Two braids produce transverse knots that are transversely isotopic if and only if they are related by:*

- *conjugation in the braid groups*
- *positive Markov stabilization and destabilization:* $(B \in B_n) \longleftrightarrow (B\sigma_n \in B_{n+1})$.

Transverse knots have two "classical" invariants of transverse knots:

- underlying topological knot type
- self-linking number (for a braid, $sl = w - n$).

It is of considerable interest to find other, "effective" transverse invariants, which can distinguish between transverse knots with the same classical invariants. One such invariant is the transverse invariant in knot Floer homology [28, 38]. This (more precisely, one version of it) associates, to a transverse knot T of topological type K, an element $\widehat{\theta}(T) \in \widehat{HFK}(m(K))$. The HFK invariant has been shown to be effective at distinguishing transverse knots; see e.g. [35].

The purpose of this section is to discuss how one can refine knot contact homology to produce another effective transverse invariant. Geometrically, the idea is as follows (see [16] for details). Given a transverse knot $T \subset (\mathbb{R}^3, \xi)$, one constructs the conormal bundle $\Lambda_T \subset ST^*R^3$ as usual. Now the cooriented contact plane field ξ on $\mathbb{R}^3$ also has a conormal lift $\widetilde{\xi} \subset ST^*\mathbb{R}^3$: concretely, this is the section of $ST^*\mathbb{R}^3$ given by $\alpha/|\alpha|$ where α is the contact form. Since T is transverse to ξ, $\Lambda_T \cap \widetilde{\xi} = \emptyset$.

One can choose an almost complex structure on the symplectization $\mathbb{R} \times ST^*\mathbb{R}^3$ (and change the metric on $\mathbb{R}^3$ that determines $ST^*\mathbb{R}^3$) so that $\mathbb{R} \times \widetilde{\xi}$ is holomorphic. Given a holomorphic disk with boundary on $\mathbb{R} \times \Lambda_T$ as in the LCH of Λ_T, one can then count intersections with $\mathbb{R} \times \widetilde{\xi}$, and all of these intersections are positive. Thus we can filter the LCH differential of Λ_T:

$$\partial(a_i) = \sum_{\dim \mathcal{M}(a_i;a_{j_1},\ldots,a_{j_k})/\mathbb{R}=0} \quad \sum_{\Delta\in\mathcal{M}/\mathbb{R}} (\mathrm{sgn}) U^{\#(\Delta\cap(\mathbb{R}\times\widetilde{\xi}))} e^{[\partial\Delta]} a_{j_1}\cdots a_{j_k}.$$

Here $[\partial\Delta]$ is the homology class of $\partial\Delta$ in $H_1(\Lambda_T)$ and $\#(\Delta\cap(\mathbb{R}\times\widetilde{\xi}))$ is always nonnegative. This gives a filtered version for the knot DGA for T, which is now a DGA over $R_0[U]$ (recall that $R_0 = \mathbb{Z}[\lambda^{\pm 1}, \mu^{\pm 1}]$).

Definition 6.3. The *transverse DGA* $(\mathcal{A}^-, \partial^-)$ associated to a transverse knot $T \subset \mathbb{R}^3$ is the resulting DGA over $R_0[U]$.

The minus signs in the notation $(\mathcal{A}^-, \partial^-)$ are by analogy with Heegaard Floer homology.

When the transverse knot T is the closure of a braid B, there is a straightforward combinatorial description for the transverse DGA:

Definition 6.4. Let B be a braid. The *combinatorial transverse DGA* for B is the DGA over $R_0[U]$ with the same generators and differential as in Definition 3.11, but with $\boldsymbol{\Lambda} = \mathrm{diag}(\lambda\mu^w, 1, \ldots, 1)$ rather than $\mathrm{diag}(\lambda\mu^w U^{-(w-n+1)/2}, 1, \ldots, 1)$.

With this new definition of $\boldsymbol{\Lambda}$, the differential in Definition 3.11 contains only nonnegative powers of U, and we indeed obtain a DGA over $R_0[U]$ (versus $R_0[U^{\pm 1}]$ in Definition 3.11).

Theorem 6.5 ([16]). *The transverse DGA and the combinatorial transverse DGA agree.*

We now have the following invariance result.

Theorem 6.6 ([16, 34]). *Given a braid B, the DGA $(\mathcal{A}^-, \partial^-)$ over $R_0[U]$, up to stable tame isomorphism, is an invariant of the transverse knot corresponding to B.*

Theorem 6.6 follows from the general theory of Legendrian contact homology (and a few details that we omit here). Alternatively, one can prove directly that the combinatorial transverse DGA is a transverse invariant by checking invariance under braid conjugation and positive braid stabilization, and invoking the Transverse Markov Theorem; this approach is carried out in [34]. In any case, the homology of $(\mathcal{A}^-,\partial^-)$ is also a transverse invariant and is called *transverse homology.*

Remark 6.7. In fact, a transverse knot gives *two* filtrations on the knot DGA, given by U and another parameter V; what we have presented is the specialization $V=1$. One can extend this to a DGA over $R_0[U,V]$ that, like $(\mathcal{A}^-,\partial^-)$, has a combinatorial description. The generators of the DGA are the usual ones from Definition 3.11, while the differential is given by:

$$\begin{aligned}
\partial(\mathbf{A}) &= 0\\
\partial(\mathbf{B}) &= \mathbf{A} - \boldsymbol{\Lambda}\cdot\phi_B(\mathbf{A})\cdot\boldsymbol{\Lambda}^{-1}\\
\partial(\mathbf{C}) &= \hat{\mathbf{A}} - \boldsymbol{\Lambda}\cdot\boldsymbol{\Phi}^L_B\cdot\check{\mathbf{A}}\\
\partial(\mathbf{D}) &= \check{\mathbf{A}} - \hat{\mathbf{A}}\cdot\boldsymbol{\Phi}^R_B\cdot\boldsymbol{\Lambda}^{-1}\\
\partial(\mathbf{E}) &= \hat{\mathbf{B}} - \mathbf{C} - \boldsymbol{\Lambda}\cdot\boldsymbol{\Phi}^L_B\cdot\mathbf{D}\\
\partial(\mathbf{F}) &= \check{\mathbf{B}} - \mathbf{D} - \mathbf{C}\cdot\boldsymbol{\Phi}^R_B\cdot\boldsymbol{\Lambda}^{-1}.
\end{aligned}$$

Here $\boldsymbol{\Lambda}=\operatorname{diag}(\lambda\mu^w,1,\dots,1)$; $\mathbf{A},\hat{\mathbf{A}},\mathbf{B},\hat{\mathbf{B}},\mathbf{C},\mathbf{D},\mathbf{E},\mathbf{F}$ are as in Definition 3.11; and $\check{\mathbf{A}},\check{\mathbf{B}}$ are defined by:

$$(\check{\mathbf{A}})_{ij} = \begin{cases} a_{ij} & i<j\\ -\mu V a_{ij} & i>j\\ 1-\mu V & i=j \end{cases} \qquad (\check{\mathbf{B}})_{ij} = \begin{cases} b_{ij} & i<j\\ -\mu V b_{ij} & i>j\\ 0 & i=j. \end{cases}$$

Geometrically, the powers of V count intersections with the "negative" lift of ξ to $ST^*\mathbb{R}^3$, given by $-\alpha/|\alpha|$. The full DGA over $R_0[U,V]$ has some nice formal properties, such as its behavior under transverse stabilization, but for known applications it suffices to set $V=1$ and thus ignore V.

We now return to the transverse DGA $(\mathcal{A}^-,\partial^-)$ over $R_0[U]$. In a manner familiar from Heegaard Floer theory, one can obtain several other flavors of transverse homology from $(\mathcal{A}^-,\partial^-)$. Two particularly interesting ones are:

- The "hat version": $(\widehat{\mathcal{A}},\widehat{\partial})$, a DGA over $R_0=\mathbb{Z}[\lambda^{\pm 1},\mu^{\pm 1}]$, by setting $U=0$. This is a transverse invariant.
- The "infinity version": $(\mathcal{A},\partial)$, the usual knot DGA over $R=R_0[U^{\pm 1}]$, by tensoring $(\mathcal{A}^-,\partial^-)$ with $R_0[U^{\pm 1}]$ and replacing λ by $\lambda U^{-(w-n+1)/2}$. This is an invariant of the underlying topological knot, as usual.

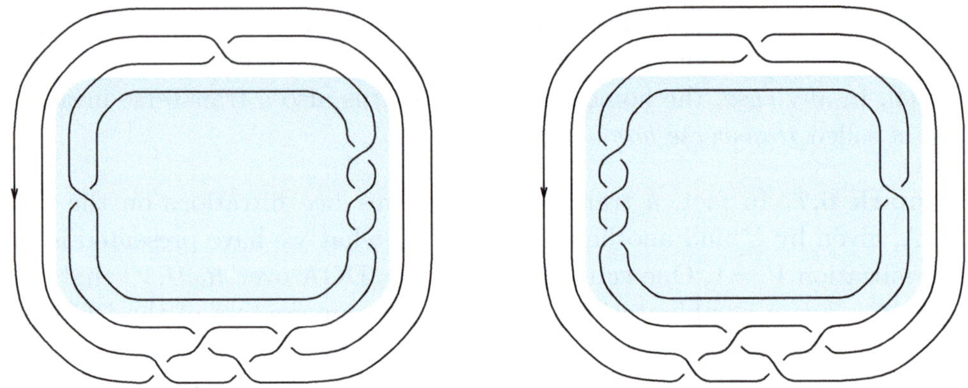

Fig. 4. Two braids B_1, B_2 whose closure is the knot $m(7_6)$. To see that they produce the same knot, note that their closures are related by a negative flype (the shaded regions)

Remark 6.8. Independent of the fact that the infinity version is the usual knot DGA, we can see geometrically that the infinity version is a topological knot invariant, as follows. If we disregard positivity of intersection, then powers of U in the differential ∂ merely encode homological data about the holomorphic disk Δ; a bit of thought shows that $\#(\Delta \cap (\mathbb{R} \times \widetilde{\xi}))$ is equal to the class of Δ in $H_2(S^2) \cong \mathbb{Z}$. Thus this indeed reduces to the usual LCH DGA of Λ_K.

We now have the following result.

Theorem 6.9 ([16, 34]). *The hat version of the transverse DGA, $(\widehat{\mathcal{A}}, \widehat{\partial})$, is an effective invariant of transverse knots.*

As one example, consider the transverse knots given by the closures of the braids B_1, B_2 given in Figure 4, both of which are of topological type $m(7_6)$ and have self-linking number -1. For each braid, one can count the number of augmentations of $(\widehat{\mathcal{A}}, \widehat{\partial})$ to $\mathbb{Z}_3$; this augmentation number is a transverse invariant. A computer calculation shows that the augmentation number is 0 for B_1 and 5 for B_2. It follows that the transverse knots corresponding to B_1 and B_2 are not transversely isotopic.

One can heuristically gauge the relative effectiveness of various transverse invariants by using the Legendrian knot atlas [8], which provides a conjecturally complete list of all Legendrian knots representing topological knots of arc index ≤ 9. The atlas proposes 13 knots with arc index ≤ 9 that have at least two transverse representatives with the same self-linking number. Of these 13:

- 6 ($m(7_2)$, $m(10_{132})$, $m(10_{140})$, $m(10_{145})$, $m(10_{161})$, $12n_{591}$) have transverse representatives that can be distinguished by both the HFK invariant and by transverse homology;
- 4 ($m(7_6)$, 9_{44}, 9_{48}, 10_{136}) can be distinguished by transverse homology but not the HFK invariant;
- 3 ($m(9_{45})$, 10_{128}, 10_{160}) cannot yet be distinguished by either HFK or transverse homology.

Of these last 3, preliminary joint work with Dylan Thurston suggests that $m(9_{45})$ and 10_{128} can be distinguished by naturality in conjunction with the HFK invariant, but the third cannot.[7] It is conceivable that some or all of these last 3 can be distinguished by transverse homology, but they are related by an operation known as "transverse mirroring" that is relatively difficult to detect by transverse homology.

It appears that the two known effective transverse invariants, the transverse HFK invariant and transverse homology, are functionally independent, but it would be very interesting to know if there is some connection between them.

Acknowledgements. I am grateful to the organizers and participants of the 2012 Contact and Symplectic Topology Summer School in Budapest, and particularly Chris Cornwell and Tobias Ekholm, for a great deal of helpful feedback, and to Dan Rutherford for catching a typo in an earlier draft. This work was partially supported by NSF CAREER grant DMS-0846346.

Appendix: Conventions and the Fully Noncommutative DGA

In the literature on knot contact homology, a number of mutually inconsistent conventions are used. The conventions that we have adopted in this article are unfortunately different again from the existing ones, but we would like to

[7]The transverse representatives of $m(7_6)$, 9_{44}, 9_{48}, 10_{136}, and 10_{160} cannot be distinguished by the transverse HFK invariant, with or without naturality, because $\widehat{HFK} = 0$ and HFK^- has rank 1 in the relevant bidegree.

advocate these new conventions as combining the best qualities of previous ones while avoiding some disadvantages that have become apparent in the interim.

First we describe how to extend the definition of knot contact homology from Section 3 in two directions: first, by allowing for multi-component links, and second, by extending to the fully noncommutative DGA (see Remark 2.2), in which homology classes do not commute with Reeb chords. The result is a "stronger" formulation of (combinatorial) knot contact homology than usually appears in the literature. After this, we will discuss how this definition compares to previous conventions.

If K is a link given by the closure of a braid $B \in B_n$, we can define a slightly more complicated version of the braid homomorphism ϕ_B from Section 3 as follows. Let $\tilde{\mathcal{A}}_n$ denote the tensor algebra over $\mathbb{Z}$ freely generated by a_{ij}, $1 \le i \ne j \le n$, and by $\tilde{\mu}_i^{\pm 1}$, $1 \le i \le n$. (Here the $\tilde{\mu}_i$'s do not commute with the a_{ij}'s, or indeed with each other, and the only nontrivial relations are $\tilde{\mu}_i \cdot \tilde{\mu}_i^{-1} = \tilde{\mu}_i^{-1} \cdot \tilde{\mu}_i = 1$.) For $1 \le k \le n-1$, define $\phi_{\sigma_k} : \tilde{\mathcal{A}}_n \to \tilde{\mathcal{A}}_n$ by:

$$\phi_{\sigma_k} : \begin{cases} a_{ij} \mapsto a_{ij}, & i,j \ne k,k+1 \\ a_{k+1,i} \mapsto a_{ki}, & i \ne k,k+1 \\ a_{i,k+1} \mapsto a_{ik}, & i \ne k,k+1 \\ a_{k,k+1} \mapsto -a_{k+1,k} & \\ a_{k+1,k} \mapsto -\tilde{\mu}_k a_{k,k+1} \tilde{\mu}_{k+1}^{-1} & \\ a_{ki} \mapsto a_{k+1,i} - a_{k+1,k} a_{ki}, & i \ne k,k+1 \\ a_{ik} \mapsto a_{i,k+1} - a_{ik} a_{k,k+1}, & i < k \\ a_{ik} \mapsto a_{i,k+1} - a_{ik} \tilde{\mu}_k a_{k,k+1} \tilde{\mu}_{k+1}^{-1}, & i > k+1 \\ \tilde{\mu}_i^{\pm 1} \mapsto \tilde{\mu}_i^{\pm 1}, & i \ne k,k+1 \\ \tilde{\mu}_k^{\pm 1} \mapsto \tilde{\mu}_{k+1}^{\pm 1} & \\ \tilde{\mu}_{k+1}^{\pm 1} \mapsto \tilde{\mu}_k^{\pm 1}. & \end{cases}$$

This extends to a group homomorphism $\phi : B_n \to \operatorname{Aut} \tilde{\mathcal{A}}_n$ and thus defines a map $\phi_B \in \operatorname{Aut} \tilde{\mathcal{A}}_n$.

Suppose that K has r components, and number the components of K $1, \dots, r$. For $i = 1, \dots, n$, define $\alpha(i) \in \{1, \dots, r\}$ to be the number of the component containing strand i of the braid B whose closure is K. If we now define $\mathcal{A}_n$ to be the tensor algebra over $\mathbb{Z}$ freely generated by the a_{ij}'s and by variables $\mu_1^{\pm 1}, \dots, \mu_r^{\pm 1}$, then it is easy to check that ϕ_B descends to an algebra automorphism of $\mathcal{A}_n$ by setting $\tilde{\mu}_i = \mu_{\alpha(i)}$ for all $1 \le i \le n$. We can define $\Phi^L_B, \Phi^R_B \in \mathrm{Mat}_{n\times n}(\mathcal{A}_n)$ as in Definition 3.9, with the important caveat that the extra strand $*$ is treated as strand 0 rather than strand $n+1$; for multi-component links, this makes a difference because of the form of the definition of ϕ_{σ_k} above.

Define $\mathcal{A}$ to be the tensor algebra over $\mathbb{Z}[U^{\pm 1}]$ freely generated by $\mu_1^{\pm 1}, \ldots, \mu_r^{\pm 1}$ along with the generators $a_{ij}, b_{ij}, c_{ij}, d_{ij}, e_{ij}, f_{ij}$ as in Definition 3.11. Assemble $n \times n$ matrices $\mathbf{A}, \hat{\mathbf{A}}, \mathbf{B}, \hat{\mathbf{B}}, \mathbf{C}, \mathbf{D}, \mathbf{E}, \mathbf{F}$, where $\mathbf{C}, \mathbf{D}, \mathbf{E}, \mathbf{F}$ are as in Definition 3.11, while

$$\mathbf{A}_{ij} = \begin{cases} a_{ij} & i<j \\ -a_{ij}\mu_{\alpha(j)} & i>j \\ 1-\mu_{\alpha(i)} & i=j \end{cases} \qquad \mathbf{B}_{ij} = \begin{cases} b_{ij} & i<j \\ -b_{ij}\mu_{\alpha(j)} & i>j \\ 0 & i=j \end{cases}$$

$$(\hat{\mathbf{A}})_{ij} = \begin{cases} Ua_{ij} & i<j \\ -a_{ij}\mu_{\alpha(j)} & i>j \\ U-\mu_{\alpha(i)} & i=j \end{cases} \qquad (\hat{\mathbf{B}})_{ij} = \begin{cases} Ub_{ij} & i<j \\ -b_{ij}\mu_{\alpha(j)} & i>j \\ 0 & i=j. \end{cases}$$

Also define a matrix $\boldsymbol{\Lambda}$ as follows: choose one strand of B belonging to each component of the closure K, and call the resulting r strands *leading*; then define

$$(\boldsymbol{\Lambda})_{ij} = \begin{cases} \lambda_{\alpha(i)}\mu_{\alpha(i)}^{w(\alpha(i))}U^{-(w(\alpha(i))-n(\alpha(i))+1)/2} & i=j \text{ and strand } i \text{ leading} \\ 1 & i=j \text{ and strand } i \text{ not leading} \\ 0 & i \neq j, \end{cases}$$

where $n(\alpha)$ is the number of strands belonging to component α and $w(\alpha)$ is the writhe of component α viewed as an $n(\alpha)$-strand braid (with the other components deleted).

With this notation, one can now define the differential ∂ on $\mathcal{A}$ exactly as in Definition 3.11. The resulting DGA has the same properties as in Theorem 3.15: $\partial^2 = 0$ and $(\mathcal{A}, \partial)$ is an isotopy invariant of the link K viewed as an oriented link with numbered components, up to stable tame isomorphisms that act as the identity on U and on each of $\lambda_1, \ldots, \lambda_r, \mu_1, \ldots, \mu_r$.

Note that the definition of the DGA given above is for the topological knot/link invariant as discussed in Sections 2 and 3. This corresponds to "infinity transverse homology" from [34] (also mentioned in [16]). There is an analogous definition of transverse homology as in Section 6 or [16, 34] but we omit its definition here.

We now compare our definition to the two previous conventions for the knot DGA: the convention from [15, 16] and the convention from [33, 34]. Note that all versions of the DGA from these references first quotient so that homology classes $\lambda_1, \ldots, \lambda_r, \mu_1, \ldots, \mu_r$ commute with all Reeb chords. Also, the versions from [16] and [34] involve an additional variable V, but we set $V = 1$ for this discussion; as explained in [34, §4], this does not lose any information. Finally, the conventions from [15] and [33] agree with the conventions from [16] and [34], respectively, after setting $U = V = 1$.

We claim that we can then obtain the knot DGAs in the conventions of [16] and [34] from the knot DGA presented in this article, up to isomorphism, as follows:

- for [16], replace $\lambda_\alpha \mapsto -\lambda_\alpha$ and $\mu_\alpha \mapsto -\mu_\alpha$ for each component α;
- for [34], which only considers the single-component case, keep λ as is, and replace $\mu \mapsto -\mu^{-1}$.

We check the claim for the convention of [16]; the claim for [34] then follows from [34, §3.4].[8] Note that negating each λ_α and μ_α causes our definitions to line up precisely with the definitions from [16], except for $\boldsymbol{\Lambda}$ (denoted in [16] by -$\boldsymbol{\lambda}$), which differs by the presence or absence of a power of U, along with some signs. But the power of U is merely a notational/framing issue (cf. [34]), while the sign discrepancy disappears because it can be checked that the products of the diagonal entries of $\boldsymbol{\Lambda}$ and $-\boldsymbol{\lambda}$ corresponding to any particular link component are exactly equal including sign, whence the DGAs given by the two conventions are isomorphic by the argument of [34, Proposition 3.1].

Remark A.1. Except for the $\mu \mapsto \mu^{-1}$ issue, all differences between conventions consist just of negating some subset of $\{\lambda, \mu\}$. This is explained by the fact that the signs in the differential in Legendrian contact homology depend on a choice of spin structure on the Legendrian submanifold Λ; see [13] for full discussion. In our setting, if K is a knot, $\Lambda_K \cong T^2$ has four spin structures, and changing from one spin structure to another sends $\lambda \mapsto \pm\lambda$ and $\mu \mapsto \pm\mu$. Thus the different choices of signs arise from different choices of spin structure on Λ_K.

[8]Note that [34], building on work from [33], uses an unusual convention for braids, so that a *positive* generator σ_k of the braid group is given topologically as a *negative* crossing in the usual knot theory sense. This has the effect of mirroring all topological knots and explains the μ^{-1} difference in conventions.

A summary of the relations between conventions in different articles is as follows:

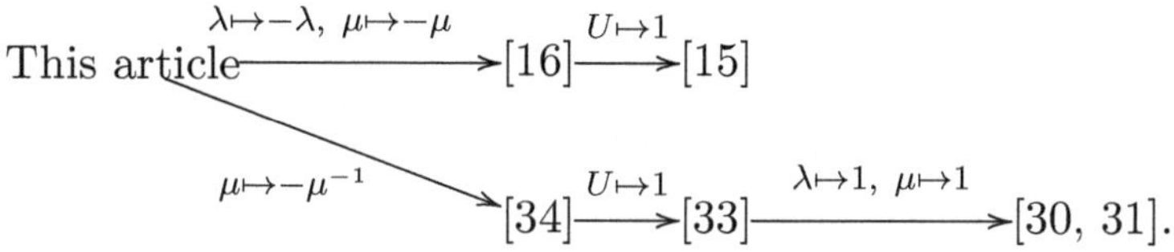

We close by noting that our current choice of conventions allows for some cleaner results than the conventions from [16] or [34]. In particular, our signs are more natural than the signs from either [16] or [34] when we consider the relation to representations of the knot group as in Section 5. For instance, our two-variable augmentation polynomials are divisible by $(\lambda-1)(\mu-1)$ as opposed to $(\lambda+1)(\mu+1)$ in [16] or $(\lambda-1)(\mu+1)$ in [33, 34], and $A_K(\lambda,\mu)$ divides $\mathrm{Aug}_K(\lambda,\mu^2)$ in our convention rather than $\mathrm{Aug}_K(-\lambda,-\mu^2)$ or $\mathrm{Aug}_K(\lambda,-\mu^2)$ in the other two.

There is another technical reason for preferring our signs or those from [16] to the ones from [34], or more precisely to the extrapolation of [34] to the link case. In either of the first two cases but not the third, we have the following statement, which we leave as an exercise.

Proposition A.2. *Let K be a link given by the closure of braid B, and let $K' \subset K$ be a sublink given by the closure of a subbraid $B' \subset B$ obtained by erasing some strands of B. Then the DGA for K' is a quotient of the DGA for K, given by setting all Reeb chords a_{ij}, b_{ij}, etc. to 0 unless strands i and j both belong to B'.*

This result is used in [5] and is a special case of a general result relating the Legendrian contact homology of a multi-component Legendrian to the LCH of some subset of components.

References

[1] A. Abbondandolo, M. Schwarz, On the Floer homology of cotangent bundles. Commun. Pure Appl. Math. **59**(2), 254–316 (2006). arXiv:math/0408280

[2] A. Abbondandolo, M. Schwarz, Floer homology of cotangent bundles and the loop product. Geom. Topol. **14**(3), 1569–1722 (2010). arXiv:0810.1995

[3] M. Abouzaid, Framed bordism and Lagrangian embeddings of exotic spheres. Ann. Math. **175**(1), 71–185 (2012). arXiv:0812.4781

[4] M. Aganagic, C. Vafa, Large N duality, mirror symmetry, and a Q-deformed A-polynomial. arXiv:1204.4709

[5] M. Aganagic, T. Ekholm, L. Ng, C. Vafa, Topological strings, D-model, and knot contact homology. arXiv:1304.5778

[6] D. Bennequin, Entrelacements et équations de Pfaff, in *Third Schnepfenried Geometry Conference, vol. 1*, Schnepfenried, 1982. Astérisque, vol. 107 (Soc. Math. France, Paris, 1983), pp. 87–161

[7] Y. Chekanov, Differential algebra of Legendrian links. Invent. Math. **150**(3), 441–483 (2002)

[8] W. Chongchitmate, L. Ng, An atlas for Legendrian knots. Exp. Math. **22**(1), 26–37 (2013). arXiv:1010.3997

[9] K. Cieliebak, J. Latschev, The role of string topology in symplectic field theory, in *New Perspectives and Challenges in Symplectic Field Theory*. CRM Proc. Lecture Notes, vol. 49 (Amer. Math. Soc., Providence, 2009), pp. 113–146. arXiv:0706.3284

[10] N.M. Dunfield, S. Garoufalidis, Non-triviality of the A-polynomial for knots in S^3. Algebr. Geom. Topol. **4**, 1145–1153 (2004) (electronic). arXiv:math/0405353

[11] T. Ekholm, Morse flow trees and Legendrian contact homology in 1-jet spaces. Geom. Topol. **11**, 1083–1224 (2007). arXiv:math/0509386

[12] T. Ekholm, J.B. Etnyre, Invariants of knots, embeddings and immersions via contact geometry, in *Geometry and Topology of Manifolds*. Fields Inst. Commun., vol. 47 (Amer. Math. Soc., Providence, 2005), pp. 77–96. arXiv:math/0412517

[13] T. Ekholm, J. Etnyre, M. Sullivan, Orientations in Legendrian contact homology and exact Lagrangian immersions. Int. J. Math. **16**(5), 453–532 (2005). arXiv:math/0408411

[14] T. Ekholm, J. Etnyre, M. Sullivan, Legendrian contact homology in $P \times \mathbb{R}$. Trans. Am. Math. Soc. **359**(7), 3301–3335 (2007) (electronic). arXiv:math/0505451

[15] T. Ekholm, J. Etnyre, L. Ng, M. Sullivan, Knot contact homology. Geom. Topol. **17**, 975–1112 (2013). arXiv:1109.1542

[16] T. Ekholm, J. Etnyre, L. Ng, M. Sullivan, Filtrations on the knot contact homology of transverse knots. Math. Ann. **355**(4), 1561–1591 (2013). arXiv:1010.0450

[17] Y. Eliashberg, Invariants in contact topology, in *Proceedings of the International Congress of Mathematicians*, vol. II, Berlin, 1998 (1998), pp. 327–338 (electronic)

[18] Y. Eliashberg, A. Givental, H. Hofer, Introduction to symplectic field theory. Geom. Funct. Anal. (Special Volume, Part II):560–673, 2000. GAFA 2000 (Tel Aviv, 1999). arXiv:math/0010059

[19] J.B. Etnyre, L.L. Ng, J.M. Sabloff, Invariants of Legendrian knots and coherent orientations. J. Symplectic Geom. **1**(2), 321–367 (2002). arXiv:math/0101145

[20] H. Fuji, S. Gukov, P. Sułkowski, Super-A-polynomial for knots and BPS states. Nucl. Phys. B **867**(2), 506–546 (2013). arXiv:1205.1515

[21] S. Garoufalidis, On the characteristic and deformation varieties of a knot, in *Proceedings of the Casson Fest*. Geom. Topol. Monogr., vol. 7 (Geom. Topol. Publ., Coventry, 2004), pp. 291–309 (electronic). arXiv:math/0306230

[22] S. Garoufalidis, The colored HOMFLY polynomial is q-holonomic. arXiv:1211.6388

[23] R. Gopakumar, C. Vafa, On the gauge theory/geometry correspondence, in *Winter School on Mirror Symmetry, Vector Bundles and Lagrangian Submanifolds*, Cambridge, MA, 1999. AMS/IP Stud. Adv. Math., vol. 23 (Amer. Math. Soc., Providence, 2001), pp. 45–63. arXiv:hep-th/9811131

[24] S. Gukov, A. Schwarz, C. Vafa, Khovanov-Rozansky homology and topological strings. Lett. Math. Phys. **74**(1), 53–74 (2005). arXiv:hep-th/0412243

[25] M. Khovanov, L. Rozansky, Matrix factorizations and link homology. II. Geom. Topol. **12**(3), 1387–1425 (2008). arXiv:math/0505056

[26] A. Knapp, Symplectic structures on the cotangent bundles of open 4-manifolds. arXiv: 1209.3045

[27] P.B. Kronheimer, T.S. Mrowka, Dehn surgery, the fundamental group and SU(2). Math. Res. Lett. **11**(5–6), 741–754 (2004). arXiv:math/0312322

[28] P. Lisca, P. Ozsváth, A.I. Stipsicz, Z. Szabó, Heegaard Floer invariants of Legendrian knots in contact three-manifolds. J. Eur. Math. Soc. **11**(6), 1307–1363 (2009). arXiv:0802.0628

[29] W. Magnus, Rings of Fricke characters and automorphism groups of free groups. Math. Z. **170**(1), 91–103 (1980)

[30] L. Ng, Knot and braid invariants from contact homology. I. Geom. Topol. **9**, 247–297 (2005) (electronic). arXiv:math/0302099

[31] L. Ng, Knot and braid invariants from contact homology. II. Geom. Topol. **9**, 1603–1637 (2005) (electronic). With an appendix by the author and Siddhartha Gadgil. arXiv:math/0303343

[32] L. Ng, Conormal bundles, contact homology and knot invariants, in *The Interaction of Finite-Type and Gromov-Witten Invariants (BIRS 2003)*. Geom. Topol. Monogr., vol. 8 (Geom. Topol. Publ., Coventry, 2006), pp. 129–144. arXiv:math/0412330

[33] L. Ng, Framed knot contact homology. Duke Math. J. **141**(2), 365–406 (2008). arXiv: math/0407071

[34] L. Ng, Combinatorial knot contact homology and transverse knots. Adv. Math. **227**(6), 2189–2219 (2011). arXiv:1010.0451

[35] L. Ng, P. Ozsváth, D. Thurston, Transverse knots distinguished by knot Floer homology. J. Symplectic Geom. **6**(4), 461–490 (2008). arXiv:math/0703446

[36] H. Ooguri, C. Vafa, Knot invariants and topological strings. Nucl. Phys. B **577**(3), 419–438 (2000). arXiv:hep-th/9912123

[37] S.Yu. Orevkov, V.V. Shevchishin, Markov theorem for transversal links. J. Knot Theory Ramif. **12**(7), 905–913 (2003). arXiv:math/0112207

[38] P. Ozsváth, Z. Szabó, D. Thurston, Legendrian knots, transverse knots and combinatorial Floer homology. Geom. Topol. **12**(2), 941–980 (2008). arXiv:math/0611841

[39] T. Perutz, The symplectic topology of cotangent bundles. Eur. Math. Soc. Newsl. (75), 30–33 (2010)

[40] D.A. Salamon, J. Weber, Floer homology and the heat flow. Geom. Funct. Anal. **16**(5), 1050–1138 (2006). arXiv:math/0304383

[41] C. Viterbo, Functors and computations in Floer homology with applications. II. Preprint (1996). Available at http://www.math.polytechnique.fr/cmat/viterbo/Prepublications.html

[42] N.C. Wrinkle, The Markov theorem for transverse knots. arXiv:math/0202055

L. Ng (✉)

Mathematics Department
Duke University
Durham, NC 27708
USA

e-mail: ng@math.duke.edu

url: http://www.math.duke.edu/~ng/

MIX
Papier aus verantwortungsvollen Quellen
Paper from responsible sources
FSC® C105338

If you have any concerns about our products,
you can contact us on
ProductSafety@springernature.com

In case Publisher is established outside the EU,
the EU authorized representative is:
Springer Nature Customer Service Center GmbH
Europaplatz 3, 69115 Heidelberg, Germany

Printed by Libri Plureos GmbH
in Hamburg, Germany